Research Notes in Neural Computing

Michael A. Arbib   Jörg-Peter Ewert
Editors

Visual Structures and Integrated Functions

With 174 Illustrations

Springer-Verlag
Berlin Heidelberg New York
London Paris Tokyo
Hong Kong Barcelona
Budapest

Michael A. Arbib
Center for Neural Engineering
University of Southern California
Los Angeles, CA 90089-2520
USA

Jörg-Peter Ewert
Universität Kassel, GhK
FB 19 – Abt. Neurobiologie
Heinrich-Plett-Str. 40
W-3500 Kassel
FRG

Managing Editor
Bart Kosko
Engineering Image Processing Institute
University of Southern California
University Park
Los Angeles, CA 90089-0782
USA

ISBN 3-540-54241-8 Springer-Verlag Berlin Heidelberg New York
ISBN 0-387-54241-8 Springer-Verlag New York Berlin Heidelberg

Library of Congress Cataloging-in-Publication Data
Visual structures and integrated functions / edited by Michael A. Arbib and Jörg-Peter Ewert. -- (Research notes in neural computing; v. 3) "Papers presented at the Workshop on Visual Structures and Integrated Functions held at the University of Southern California in Los Angeles on August 8-10, 1990"--Pref.
ISBN 3-540-54241-8 -- ISBN 0-387-54421-8
1. Visual pathways--Congress. 2. Physiology, Comparative-Congress. I. Arbib, Michael A. II. Ewert, Jörg-Peter, 1938 III. Workshop on Visual Structures and Integrated Functions (1990: University of Southern California) IV. Series. QP475.V6 1991 596'.01823-dc20 91-22312

Typesetting: camera-ready by authors
Printed in the United States of America

33/3140 – 543210 – Printed on acid-free paper

Preface

This volume integrates theory and experiment to place the study of vision within the context of the action systems which use visual information. This theme is developed by stressing:

(a) The importance of situating any one part of the brain in the context of its interactions with other parts of the brain in subserving animal behavior. The title of this volume emphasizes that visual function is to be be viewed in the context of the integrated functions of the organism.

(b) Both the intrinsic interest of frog and toad as animals in which to study the neural mechanisms of visuomotor coordination, and the importance of comparative studies with other organisms so that we may learn from an analysis of both similarities and differences. The present volume thus supplements our studies of frog and toad with papers on salamander, bird and reptile, turtle, rat, gerbil, rabbit, and monkey.

(c) Perhaps most distinctively, the interaction between theory and experiment. Thus we offer a rich array of models in this volume. High-level schema models show the basic functional interactions underlying visuomotor coordination, testable by lesion experiments. Neural network models address data from neurophysiology and neuroanatomy. We also offer fascinating new data with pointers towards the way that models must develop to address them. Although the stress in the modeling is on *Computational Neuroscience* (the use of computational methods to understand neurobiological phenomena), our results have implications for *Neural Engineering* (the use of ideas inspired, but not necessarily constrained, by the study of the brain to design highly parallel, often adaptive, machines).

The present volume comprises the papers presented at the "Workshop on Visual Structures and Integrated Functions" held at the University of Southern California in Los Angeles on August 8-10, 1990. This Workshop is the fourth in a series entitled "Visuomotor Coordination in Frog and Toad: Models and Experiments." These workshops have centered on the evolving body of data about frog and toad, and on the development of a set of models which together constitute *Rana computatrix*, the "frog that computes." The study of *Rana computatrix* has implications that go far beyond the study of frogs and toads, *per se,* and past workshops — and the present volume — have devoted much space/time to these implications. It is not so many years ago that the study of invertebrates was regarded as being, at best, peripheral by the majority of neuroscientists. Yet the increasing attention to mechanisms of neural function has made *Aplysia* and other invertebrates invaluable in the study of basic cellular mechanisms of facilitation, rhythm generation, and habituation. However, drastic differences in organizational principles separate the primate brain from the *Aplysia* nerve net. The study of "creatures" which "evolve in the computer" can provide opportunities for understanding organizational principles which are not to be sought solely in terms of cellular mechanisms but in terms of structural constructs (layers and modules), functional constructs (schemas), and computational strategies (cooperative computation in neural nets, adaptation, etc.).

As organizers of the meeting, we invited a group of scientists who could between them address the issues (a) to (c) above of our continuing scientific enterprise. They responded with lively talks, which generated much discussion to provide "connective tissue" which greatly strengthened the meeting. To help the reader gain an understanding of the connection between the papers and the broader enterprise of which they are a part, we have provided two opening perspectives, with Arbib reviewing the "Neural Mechanisms of Visuomotor Coordination: The Evolution of *Rana computatrix*," while Ewert offers "A Prospectus for the Fruitful Interaction Between Neuroethology and Neural Engineering." Following these, the workshop papers have been grouped into sections, each with a unifying theme as set forth in the following paragraphs:

From the Retina to the Brain: Jeffrey L. Teeters, Frank H. Eeckman, and Frank S. Werblin build on Teeters' work in developing a leaky integrator model of cells in frog retina to show how biophysical modeling can be coupled to experiments to help understand change-sensitive inhibition in ganglion cells of the salamander retina. Frédéric Gaillard and René Garcia discuss properties of retinal cells which suggest that classical ganglion cell types are best seen as representing peaks in a continuum rather than discrete cell types, and extends this scheme to cells of the nucleus isthmi. Thomas J. Anastasio provides further insights into distributed representations in cell populations by showing how the distributed coding in the vestibular nuclei can be understod in terms of an adaptive technique for identifying properties of cells mediating the vestibulo-oculomotor reflex in mammals; while Robert F. Waldeck and Edward Gruberg offer new findings about cells of the nucleus isthmi, studying the effects of optic chasm hemisection on the parsing of visual information.

Approach and Avoidance: Information from the two retinas must be combined and gated in an action-dependent way if it is to serve the needs of the organism. Paul Grobstein has shown a parcellation of tectal output which segregates the heading of a prey object from other data about its position. He further finds that these tectal outputs are coded by overall activity in a pathway, rather than by some variant of retinotopic coding. In their two papers, Michael A. Arbib and Alberto Cobas extend this work by providing models of prey-catching and predator avoidance which stress that, whereas stimulus and response direction are the same for prey, different maps are involved for predator location and escape direction. They then provide models of schema interactions which explain Grobstein's data on the medullary hemifield deficit and suggest new experiments on both approach and avoidance. Jim-Shih Liaw and Arbib then provide neural network models for a number of the schemas involved in predator-avoidance, while David J. Ingle provides exciting new data on "triggering" and "biasing" systems in avoidance behavior. The last two papers in this section introduce a comparative dimension. Paul Dean and Peter Redgrave discuss the involvement of the rat superior colliculus in approach and avoidance behaviors (most mammalian studies note only its role in approach movements, as in visual saccades) while Colin G. Ellard and Melvyn A. Goodale offer data, on the computation of absolute distance in a visuomotor task by the mongolian gerbil, which is most suggestive for future modeling.

Generating Motor Trajectories: The previous sections show how visual data enters the brain, and give a high level view of the overall interactions involved in processing those data to commit the animal to some form of approach or avoidance behavior. This section looks briefly at the actual generation of motor behavior. Simon Giszter, Ferdinando A. Mussa-Ivaldi and Emilio Bizzi report experiments on the coding of motor space in the frog spinal cord which suggest that a wide variety of stimulation sites code for the movement of the frog's leg towards some (site-dependent) equilibrium situation. Reza Shadmehr's theoretical analysis of movement generation with kinematically redundant biological limbs provides a general limb control strategy consistent with this "equilibrium point hypothesis."

Finally, Ananda Weerasuriya focuses on non-limb movements by analyzing data on motor pattern generators in anuran prey capture and providing pointers to neural modeling of their interactions.

From Tectum to Forebrain: In many studies of visumotor coordination in frog and toad, the central role is given to the tectum and its close associates the pretectum and nucleus isthmi. Here we see that forebrain structures must be taken into account as well, and we complement studies of frog and toad with comparative studies of turtles, birds, and monkeys. Ingle explores the role of frog striatum in frog spatial memory, and looks at mammalian homologies which are explored in greater detail in the model by Peter F. Dominey and Arbib of the way in which many brain regions interact in generation of delayed and multiple saccades in primates. Ewert, N. Matsumoto, W.W. Schwippert and T.W. Beneke extend the data base on the interactions of tectum and forebrain in toads by providing intracellular studies on how striato-pretecto-tectal connections provide a substrate for arousing the toad's response to prey. Philip S. Ulinski, Linda J. Larson-Prior and N. Traverse Slater offer insights for comparative modeling with data on cortical circuitry underlying visual motion analysis in turtles, while Toru Shimizu and Harvey Karten's comparative study of evolutionary origins of the representation of visual space show that structures brought together in laminated cortex in mammals may be segregated in other species raises interesting questions about the utility of layered structures in the brain. Finally, Edmund T. Rolls completes our comparative analysis of forebrain mechanisms by with a study of information representation in temporal lobe visual cortical areas of macaques.

Development, Modulation, Learning, and Habituation: The final section of this volume considers how the nervous system changes on a variety of time scales. Sarah Bottjer's study of hormonal regulation of birdsong development provides a comparative dimension for the study by Albert Herrera and Michael Regnier of hormonal regulation of behavior in male frogs, in which they present data on how androgens control the neuromuscular substrate for amplexus (clasping) during the mating season. Gabor Bartha, Richard F. Thompson and Mark A. Gluck provide a study of sensorimotor learning and cerebellum which shows how modeling and data are being integrated in the study of conditioning in rabbits. The volume closes with three papers which exhibit the successful integration of theory and experiment in the study of basic learning behaviors in toads. Francisco Cervantes-Pérez, Angel D. Guevara-Pozas and Alberto A. Herrera-Becerra analyze data and modeling of the modulation of prey-catching behavior. C. Merkel-Harff and Ewert expand the data base with their study of learning-related modulation of toad's responses to prey by neural loops involving the forebrain, while DeLiang Wang, Arbib and Ewert report a dialog between modeling and experimentation in unravelling the dishabituation hierarchy revealed in visual pattern discrimination in toads.

The meeting and this volume were made possible in part by funds from the Industrial Affiliates Program of the Center for Neural Engineering (CNE) and from the Program in Neural, Informational and Behavioral Sciences (NIBS) of the University of Southern California, as well as from NIH grant 1R01-NS24926 to Michael Arbib. As such, they are part of the continuing dialogue between the study of living brains and the study of neural network technology maintained by the faculty and students of CNE and NIBS, as well as part of a continuing pattern of international cooperation in seeking to integrate theory and experiment in our efforts to understand the function of the brain in animal and human behavior, and to probe the implications of that understanding for "perceptual robotics." We would also like to record our warm thanks to Paulina Baligod-Tagle and her assistant Hao Cao for all their aid in the organization and conduct of the Workshop.

Michael A. Arbib
Jörg-Peter Ewert

Table of Contents

Overview

Neural Mechanisms of Visuomotor Coordination: The Evolution of *Rana computatrix*[1]

Michael A. Arbib
Center for Neural Engineering
University of Southern California
Los Angeles, CA 90089-2520, U.S.A.
arbib@rana.usc.edu

Abstract

This paper reviews the "evolution" of Rana computatrix, a set of models of the neural architectures and functions that underlie visually guided behavior in frogs and toads. We introduce arrays of leaky integrator neurons as the style of neural modeling most used in Rana computatrix to date, and briefly discuss the technique of back-propagation. We then present the methodology of schema theory, which integrates perception and action by decomposing an overall behavior into the interaction of functional, neurally explicable, units called schemas. Finally, we present an overview of work on Rana computatrix, making use of data presented in this volume — not only on frogs and toads but comparative studies of other species — to point to future directions for research. We consider models for prey-selection, depth perception, detour behavior, approach and avoidance, tectal columns (for facilitation and pattern recognition), retina, habituation and memory, and generation of motor behavior.

Introduction

The aim of this paper is to offer, in outline, a *computational* theory of the neural mechanisms underlying visuomotor coordination in frog and toad. Since computers may be used to process data of any kind, from chemical readings to anatomical diagrams, it is not the use of the computer that makes this essay in neuroscience computational. Rather, it is the attempt to characterize the brain in terms of precise descriptions of functional and structural subsystems, and to show (whether by mathematical analysis or computer simulation) how the interactions of these subsystems may fully explain some repertoire of behavior. In recent years, models which invoke "parallel distributed processing" in neural networks have attained great visibility, and this article will certainly present a range of models expressed explicitly in terms of the interaction of a few neural networks. But we

1 Preparation of this paper was supported in part by NIH grant 1R01-NS24926.

shall argue that computational neuroscience also needs a less "fine-grain" description to analyze the way in which many neural networks, or even a number of functional patterns that cut across any simple structural decomposition, may cooperate to yield some overall behavior. We offer schema theory (Arbib 1981) as the framework for analysis of perceptual and motor behavior in both hierarchical and distributed terms.

The structure of the paper is as follows: The next section introduces certain key ideas in the modeling of neural networks; in many cases such modeling will be linked to the data of neuroanatomy and neurophysiology. The following section introduces schema theory and relates it to the experimental technique of lesion analysis. The bulk of the paper provides an exposition of the status of *Rana computatrix*, "the frog that computes", presenting a number of models of anuran visuomotor mechanisms at both the schema and neural net levels. This exposition will stress the "evolution" of our models — tracing the way in which new models incorporated or built upon old ones, rather than modeling the evolution of frogs and toads as biological species. This paper is *not* intended as a review of the contents of the present volume, but the survey of the models that constitute *Rana computatrix*, and the discussion of future research, will include many pointers to relevant contributions throughout the book.

Neurons and Neural Networks

There is a whole gamut of models of neurons and neural networks. The simplest is the classic McCulloch-Pitts (1943) neuron, which has just two states, 1 (firing) and 0 (non-firing). It has a discrete time scale, t = 0, 1, 2, 3, ... (with the unit being perhaps a millisecond in a biological model). At each discrete time, a neuron takes its inputs $x_i(t)$, multiplies them by the synaptic weights w_i, adds them up, and tests to see if the sum reaches or passes threshold θ. If yes, the next state $y(t+1)$, which equals the next output, is set to 1; if not, it is set to 0:

$$y(t+1) = 1 \quad \text{if and only if} \quad \Sigma_i w_i x_i(t) \geq \theta \tag{1}$$

A slight generalization is to allow the state/output to vary continuously from 0 to 1, so that the next state is computed by passing the weighted input sum through some suitable (e.g., sigmoid) function f to obtain the next state,

$$y(t+1) = f(\Sigma_i w_i x_i(t)). \tag{2}$$

Rather than offering an exhaustive review of where neural modeling has gone from this basic beginning,[2] this section will have two simple aims. The first subsec-

---

[2] A review of neural modeling anchored by biophysical models is given in the collection edited by Koch and Segev 1989. Arbib 1987,1989 review a number of different neural models. Rumelhart and McClelland 1986 is the standard reference for psychological modeling in terms of neuron-like elements (without claiming that these elements model biological neurons).

tion sets forth the approach to neural networks most used in the neural models of *Rana computatrix* to date. The second reviews the training method that is at present most widely used for *artificial* neural networks, namely back-propagation. Even though this method is not part of the story of *Rana computatrix*, it has become well known to biologists as an integral part of the "neural networks renaissance", and so it seemed worthwhile to clarify the relation of this form of machine learning to the study of biological neural networks.

Arrays of Leaky Integrator Neurons

Our group has used a model of complexity intermediate between the McCulloch-Pitts neuron and the Hodgkin-Huxley equation to help us understand how large populations of neurons cooperate in visually guided behavior. The cells in this "leaky integrator" model interact only via their "firing rates," while the firing rate of a cell depends only on a single membrane potential (each cell is modelled as a single compartment). In computer simulation, we update the state of the network every Δt ms, cycling through two steps of simulation, first combining current values of firing rates to compute new values of membrane potentials, and then forming new values of the firing rates.

Step 1. Updating the Membrane Potentials: The membrane potential of each cell is described by a differential equation of the form

$$\tau_m \, dm(t)/dt = - m(t) + S_m(t).$$

Here m(t) denotes the membrane potential of that cell at time t; τ_m is the time constant for the rate of change of this potential; and $S_m(t)$ is the total input the cell receives from other cells. If this input is 0, m(t) exponentially decays toward 0 with time constant τ_m. If the total input $S_m(t)$ is positive, the equation will drive m(t) toward a higher value; if negative, m(t) will tend toward a lower value. In our computer simulations, we have in the past taken the simplest possible approach to solving the differential equation, using the Euler method. We choose some small time step Δt (say 20 ms of "toad time") and then approximate dm(t)/dt by the ratio $(m(t+\Delta t) - m(t))/\Delta t$. We thus replace our differential equation by the difference equation $\tau_m[(m(t+\Delta t) - m(t))/\Delta t] = - m(t) + S_m(t)$ which, on rearranging terms, yields

$$m(t+ \Delta t) = (1- \Delta t/ \tau_m)m(t) + (\Delta t/ \tau_m)S_m(t). \tag{3}$$

Our latest simulation system, NSL 2.0 (Weitzenfeld 1990), allows one to write the differential equations in a model without reference to any numerical method, allowing the user to choose different numerical methods (trading off, e.g., speed and accuracy) on different occasions without re-specifying the model.

Step 2. Updating the Firing Rates: A more detailed model than ours would explicitly model spike generation in response to changes in membrane potential. However, we use instead a coarse approximation which expresses the intuition that "the higher the membrane potential, the higher the firing rate", using a function f, usually non-decreasing, which converts m(t) into a firing rate M(t) = f(m(t)).

Having specified the two steps in updating neural activity, we now turn to the use of "layers" and "masks" to specify the overall structure of many of our models.

Specifying the inputs from one layer to another: We must also specify how each term $S_m(t)$ incorporates the input from all the other cells to which the given cell is connected. In many models, the connections from one layer to another are uniform in that the synaptic weight connecting from a neuron in layer A to a neuron in layer B depends only on their relative position, so that we can use a "mask" W whose (k,l) entry expresses the strength of the synaptic connection from the neuron at position (i+k,j+l) in layer A to position (i,j) in layer B *independent of* the values of i and j. The notation B = W*A then denotes that the array of activity B is obtained by connecting array A to array B according to the connection weights in W, i.e., the activity of the (i,j) element of the B array satisfies the formula

$$B(i,j) = \Sigma_{k,l} W(k,l) A(i+k,j+l)$$

where the summation runs over the indices (k,l) of the mask W. Thus if layer m receives its input from layers of neurons whose firing rates are given by the arrays A, B, and C, then NSL represents the total input to m by a sum of the form $W_{a.m}*A + W_{b.m}*B + W_{c.m}*C$.[3]

This approach was also used by Teeters (1989) in his model of the frog retina, and sufficed to reproduce available data on the firing of frog ganglion cells. However, in subsequent work on change sensitive responses in salamander retina, Teeters, Eeckman, and Werblin (this volume) include some biophysical detail in the neurons in their model. The point, then, is not that the leaky integrator provides the optimal level of modeling, but rather that it is adequate for a wide range of models of neural mechanisms. A good model omits unnecessary complexities while highlighting those details that are essential to its functioning. There is no virtue in complexity for its own sake. Rather, we see models "evolving" as new features are added only to the extent that they yield insight into new phenomena. In the same spirit, NSL is not a static neural simulation language, but is evolving as our group extends the vocabulary in which our models are expressed.

Back-Propagation

Much work in neural networks uses neurons of the type described in (1) or (2), but embellished with some "learning rule" to automatically adjust the synaptic weights, either "with a teacher" (extending work on the Perceptron [Rosenblatt 1958]) or "without a teacher" (with Hebbian or "anti-Hebbian" synapses based on those of Hebb 1949). Back-propagation (Rumelhart, Hinton, and Williams 1986) takes a

---

[3] NSL can also express more general forms of connectivity, but this uniform connectivity suffices for a wide range of models.

"feedforward" network (the neurons are arranged as a series of layers from input to output, with connections only from one layer to the next, and no loops) in which the neurons are of the form described by (2). If there were no "hidden layers" (i.e., layers intermediate between input and output), the training rule would be like that of the Perceptron, adjusting the weights of the output elements to reduce the "error" between the observed output pattern and the desired output pattern in response to the current input to the network. What back-propagation adds is a method of "credit-assignment" (or "blame-assignment") for the hidden layers. Error assignments are propagated back layer by layer through the hidden layers (which change their weights accordingly) by a rule which, roughly speaking, says that an element is in error to the extent that it has strong connections to neurons which have already been judged to be in error.

Hebbian synapses are employed as plausible models of biological mechanisms of plasticity, but the reverse flow of error signals in back-propagation is not now regarded as biologically plausible. However, back-propagation is used in some biological studies to identify, given the structure and input-output behavior of a network, a set of connection weights that will yield the observed behavior. Here, the *process* of back-propagation is judged biologically implausible as a model of learning or development, but the *result* of the process is used to present connection schemes whose biological plausibility is open to discussion. For example, Anastasio and Robinson (1989) use back-propagation to train a model of the vestibular nucleus to gain insight into distributed parallel processing in the vestibulo-oculomotor system. Anastasio (this volume), noting that the vestibular nuclei transmit pursuit and saccadic, as well as vestibular, signals to extraocular motoneurons, suggests that any possible combination of these signals can be found on vestibular neurons, and shows how this diversity is reproduced in the back-propagation trained model. By contrast, most studies of frog retina have enumerated discrete types of ganglion cell as specific "feature detectors". The truth may lie somewhere in between. Gaillard (this volume) suggests that classical ganglion cell types are best seen as representing peaks in a continuum rather than discrete cell types. In other words, large networks may exhibit *"linear combinations"* of a few cell types, rather than only pure examples of those types, or an unrestricted set of "all possible" features.[4]

Schema Theory and Levels of Modeling

The notion that the brain is to be understood in terms of the interactions of a vast network of neurons is widely accepted. In many neurophysiological studies, recordings will be made in some region of the brain, with the activity of single cells being correlated with features of sensory stimulation or patterns of motor response.

---

4 In the neural models of *Rana computatrix*, all the connections have been developed "by hand", with the modeler trying out a variety of connection schemes in repeated runs on the computer. An interesting question is whether back propagation (or some other identification technique) would "discover" the various cell types seen in tectum and pretectum, or even discover response types not yet identified. Unfortunately, back propagation is a great consumer of computer cycles, and is usually applied to networks far smaller than ours.

However, in brain regions far from the periphery, simple sensory or motor correlates become increasingly illusory. It becomes necessary to relate the activity of a cell to some hypothesis about the role of the region in the brain's mediation of behavior. Moreover, the brain's activity can seldom be interpreted in a simple chain of "boxes" transforming stimuli into responses. Rather, the current states of many brain regions interact with sensory data, motor and internal feedback, and each other to provide the dynamic basis for the way in which these states are updated as the animal's behavior is coordinated with its perception of the world about it. In short, the study of specific neural networks must be complemented by a high-level view of how the functions of these regions contribute to the whole. What further needs stating is that while the brain may be considered as a network of interacting "boxes", namely anatomically distinguishable structures, from an anatomical point of view, there is no reason to expect each box to mediate a single function that is well-defined from a behavioral standpoint. An experimentalist might, for example, approach the cerebellum by postulating that it serves for learning elemental movements, or mediating feedforward, or rendering movement more graceful. It may do just one of these things, but it is more likely that it does none of them by itself, but rather participates in each of them and more besides. We thus need a language which lets us express hypotheses about the various functions that the brain performs that is separated from a commitment to localization of any one function in any one region, but which can nonetheless allow us to express the way in which many regions participate in a given function, while a given region may participate in many functions.

My candidate for this language is called *schema theory* (Arbib 1975, 1981), where the word "schema" is used to denote both functions that subserve perception (*perceptual schemas*) and action (*motor schemas*) and the processes that link them. Since such functions may be studied apart from their neural localization (or, more strictly, neural distribution), schema theory has developed as a methodology for cognitive science and artificial intelligence as well as brain theory — but here we shall emphasize the issues in developing a functional analysis of behavior that could be linked to brain function.[5]

Before characterizing schemas in general, it may help to consider a specific example of how lesion experiments can test a schema model of a simple behavior. Frogs and toads snap at small moving objects and jump away from large moving objects. Thus, a simple schema-model of the frog brain might simply postulate four schemas, two perceptual schemas (processes for recognizing objects or situations) and two motor schemas (controlling some structured behavior). One perceptual

---

[5] Arbib 1989 (especially Section 2.2 and Chapter 5) provides a current review of schema theory which includes, e.g., schema-based approaches to computer vision and human language, as well as schema models linked to neural mechanisms of visuomotor coordination. Schema theory contributes to distributed artificial intelligence (cf. Arbib 1991) even when the individual schemas are *not* implemented by (artificial) neural networks. For example, Minsky 1985 espouses a model of mind in which "members of society" play a role analogous to schemas, while Brooks 1986 builds robot controllers using layers made up of asynchronous modules similar to schemas — yet neither uses neural networks in their designs.

schema recognizes small moving objects, another recognizes large moving objects. The first schema activates a motor schema for approaching the prey; the latter activates a motor schema for avoiding the enemy. And these schemas, in turn, feed the motor apparatus. Lesion experiments can put such a model to the test. It was thought that perhaps the pretectum was the locus for recognizing large moving objects, while the tectum was the locus for recognizing small moving objects. The above model would predict that an animal with lesioned pretectum would be unresponsive to large objects, but would respond normally to small objects. However, the facts are quite different. A pretectum-lesioned toad will approach moving objects large and small, and does not exhibit avoidance behavior. This leads to a new schema model[6] in which a perceptual schema to recognize large moving objects is still localized in the pretectum but the tectum now contains a perceptual schema for *all* moving objects. We then add that activity of the pretectal schema not only triggers the avoid motor schema but also inhibits approach. This new schema model still yields the normal behavior to large and small moving objects, but also fits the lesion data, since removal of the pretectum removes inhibition, and so the animal will now approach any moving object.

This example motivates many (but not all) of the properties of schemas that we now present. Items (a) through (d) list those defining properties shared by schemas whether or not their relation to the brain is under discussion, while item (e) adds the special characteristics that schemas must have when they are used to define a model of the brain — here the brain modeler must not only identify potential schemas for a given behavior but must also show that they pass the extra test of being mappable onto neural circuitry.

a) Where conventional computers store data passively, to be retrieved and processed by some central processing unit, schema theory explains behavior in terms of the interaction of many concurrent activities for recognition of different objects, and the planning and control of different activities. Schemas are modular entities that can become *active* in response to certain patterns of input from sensory stimuli or other schemas that are already active. Thus schema theory views the use, representation, and recall of knowledge about the world as mediated through the activity of a network of interacting schemas which between them provide processes for going from a particular situation and a particular motivational state to a suitable course of action (which may be overt or covert, as when learning occurs without action or the animal changes its state of readiness). This activity may involve passing of messages, changes of state (including activity level), and the activation and deactivation of schemas.

Thus, rather than thinking of computation as serial — here's the next step, here's the next step, do a test to decide what to do next — we are concerned now with a new paradigm in which computation is distributed across the interaction of many systems. In this approach — which I shall call *cooperative computation* —

---

[6] This model is essentially due to Ewert and von Seelen 1974, but theirs is refined to give a more precise account of what stimuli excite the perceptual and motor schemas *and to what extent they excite them.*

the key question is how local interaction of systems can be integrated to yield some overall result *without explicit executive control.* Here, the term "cooperative computation" is really a shorthand for "computation based on the competition and cooperation of concurrently active agents." Cooperation yields a pattern of "strengthened alliances" between mutually consistent hypotheses about aspects of a problem that represent the overall solution to a problem; it is as a result of competition that hypotheses which do not meet the evolving (data-guided) consensus lose activity. Our study of *Rana computatrix* uses two grains of cooperative computation: the fine-grain style in which many neurons work in parallel to process some array of information; and the coarse-grain style, where relatively diverse systems — our schemas — interact.

b) A set of *basic motor schemas* is hypothesized to provide simple, prototypical patterns of movement. These combine with *perceptual schemas* to form *coordinated control programs* which interweave their activations in accordance with the current task and sensory environment. Thus motor schemas in general may be either basic, or built up from other schemas as coordinated control programs. Schema activations in general are largely task-driven, reflecting the the animal's goals and the physical and functional requirements of the task.

c) The activity level of a perceptual schema represents a "confidence level" that the object represented by the schema is indeed present; while that of a motor schema may signal its "degree of readiness" to control some course of action.[7] A schema network does not, in general, need a top-level executor since schema instances can combine their effects by distributed processes of competition and cooperation (i.e. interactions which, respectively, decrease and increase the activity levels of these instances). Such cooperative computation enables schemas to modulate one another to compensate for the incompleteness of information available to each one. Motor schemas may cooperate by summing their effects on the motor apparatus, while vision involves both competition between schemas offering discordant interpretations of part of an image, as well as cooperation between schema instances providing contextual cues for one another.

d) In a general setting, there is no fixed repertoire of basic schemas. Rather, new schemas may be formed as assemblages of old schemas; but once formed a schema may be tuned by some adaptive mechanism. This tunability of schema-assemblages allows them to start as composite but emerge as primitive, much as a skill is honed into a unified whole from constituent pieces. For this reason, a model expressed in a schema-level formalism may only approximate the behavior of a model expressed in a neural net formalism (cf. the Section of Arbib 1981 on "Program Synthesis and Visuomotor Coordination").

---

[7] The activity level of a schema may be but one of many parameters that characterize it. Thus a schema for "ball" might include parameters for its size, color, and velocity. If a schema is implemented as a neural network then *all* the schema parameters could be implemented via patterns of neural activity. It is thus important to distinguish "activity level" as a particular parameter of a schema from the "neural activity" which will vary with different neural implementations of the schema.

Note that, in (a) through (d), as distinct from their footnotes, the words "brain" and "neural" do not appear. We may regard neural schema theory as a specialized branch of schema theory, just as neuropsychology is a specialized branch of psychology. In (e) below, we spell out just what makes a schema-theoretic model into a model of brain mechanisms. But first it is worth stressing that a schema is, as the name suggests, schematic. Initially, it may be very schematic indeed, but, as we come to know more, we can refine its description. One may go from simply specifying how the schema responds to a single sensory stimulus to specifying how the schema responds to a spatial pattern on a sensory array. It is then the disposition of activity in an array that constitutes the inputs or outputs of the various schemas which can be related by some overall specification — which will be further refined as we "fill in" the neural networks in the fashion that is now described:

e) A given schema defined functionally may involve the cooperation of multiple brain regions, while, conversely, a given region may participate in a variety of schemas. A top-down analysis may advance specific hypotheses about the localization of (sub)schemas in the brain, and these may be tested by lesion experiments, with possible modification of the model (e.g., replacing one schema by several interacting schemas with different localizations) and further testing. Once robust hypotheses are obtained about the localization of schemas, we may then model a brain region by seeing if its known neural circuitry can implement the posited schemas. In some cases the model will involve properties of the circuitry that have not yet been tested, thus laying the ground for new experiments. A neuron may participate in the implementation of multiple schemas. For example, certain neurons whose activity correlates with that of the perceptual schema for predators will also, via an inhibitory pathway, contribute to the perceptual schema for prey.

In this paper, and in this volume, we may think of each basic schema as having its own dedicated neural circuitry, with more complex schemas being realized by patterns of activity across the circuits which realize the schemas of the coordinated control program which defines it. In particular, if the *same* basic schema occurs more than once in some coordinated control program, then we imply that the program will require the activity of only one of these *instances* at any one time.[8] For example, in the paper by Arbib and Cobas (this volume), the overall schema for "prey-capture and predator avoidance" contains a subschema for "prey-capture" and another for "predator avoidance", and — greatly refining the simple model discussed above — each of these contains the "Orient" motor schema. But the overall schema is so structured that at most one of the subschemas for "prey-

---

[8] This is not true in general. Given a schema that represents information about some object (e.g., how to recognize it and how to act accordingly), we may need several *different* instances, each suitably tuned, to subserve our perception of several instances of that object (cf. Arbib 1989, Section 5.2). Similarly, we cannot think of the linkage of schema instances in an assemblage as always corresponding to fixed anatomical connections between the circuitry implementing the given schemas. This latter point is related to Lashley's 1951 discussion of the problem of repetition of action in a sequence of behaviors, and is taken up by Arbib 1990 in the context of "motor set".

capture" and "predator avoidance" is active at any one time, and so the circuitry for "Orient" schema will be activated either with parameters for orienting toward the prey, or with parameters for orienting away from the predator. Moreover, should the competition between the "prey-capture" and "predator avoidance" subschemas be unsuccessful in such a way that both activate the Orient schema, it will simply mean that the same circuitry receives simultaneous, conflicting, commands — in which case it might, e.g., orient the animal to the "average" direction.

We again see that a schema may be incomplete. Here, we might have specified how the Orient schema will generate a motor command when given a single heading parameter as input, yet have said nothing about how it will react to multiple inputs. However, if we had given a more detailed specification in terms of, say, a neural network receiving an array of headings as input, we might well find that, having tuned the network to respond correctly to "correct" inputs, it might then predict responses in "aberrant" situations, predictions that could then be put to experimental test. Note that the result of this work may or may not require refinement of the neural network model, but will certainly allow the extension of the original schema-level specification to encompass a wider range of situations.

To close this section, it must be stressed that schema theory is *not* a model of the brain. It is a language for *expressing* models of the brain. In this sense, schema theory is more like group theory than relativity theory. Relativity theory is a model of the physical world — it can be falsified or revised on the basis of physical experiments. However, group theory stands or falls for the scientist seeking to explain the world (as distinct from the mathematician proving theorems) not by any criterion of whether it is true or false, but rather on whether its terminology and theorems aid the expression of successful models. Schema theory does not yet have the rigor or stock of theorems of group theory, but it will be my aim to show in what follows that the success of brain models using the language of instantiation, modulation, activity levels, etc., of (a) to (e) above strengthens the argument that schema theory is a valuable complement to neural network theory in the development of computational neuroscience.

Rana computatrix

Following these short introductions to neural modeling and schema theory, we now review many of the models which constitute *Rana computatrix*, and use data presented in this volume to point to directions for further research.

A Methodological Perspective

The review starts with two early models which provide some insight into the processes of competition and cooperation at the level of neural nets. The Didday model of prey-selection embodies pure competition (the neuron that "wins" the competition determines which prey the frog will snap at). The Dev model of stereopsis embodies both competition (between neurons encoding different depths in a given direction) and cooperation (so that neurons encoding similar depths in nearby directions will excite each other, thus favoring stable states that encode surfaces rather than rapid fluctuations in depth with changing visual direction).[9] Each of these models presents a single neural network. They thus define schemas which are used in later studies, but are not themselves constituted by an assemblage of schemas. Their neurons are leaky integrator neurons, but are not linked to identified neuron types in actual nervous systems.

The next section, on "Depth and Detours", first presents two models of depth perception which use cooperation between schemas to yield a form of "sensor fusion", showing how the binocular cue of disparity and the monocular cue of accommodation may be used to complement each other to yield a system that achieves more accurate depth estimates than could a system relying on a single cue. The Cue Interaction Model couples two copies of the Dev "schema", one driven by disparity and the other by accommodation; while the Prey Localization Model couples two copies of the Didday schema, one driven by the left eye and the other by the right eye, with a controller for the accommodation of the two eyes, to interweave the selection of a single prey with its localization in space. (Note how this use of the Dev and Didday models exemplifies what we called the "evolution" of our models — as new models incorporate or build upon old ones.) We then turn to the Orientation Model and the Path-Planning Model, which provide alternative hypotheses for how a toad, confronted by a worm behind a semi-transparent barrier, can "choose" whether to approach the worm directly or to make a detour. The important methodological point here is that these two models demand separate schemas for mapping the location of barriers and for locating a prey, but do not depend on the use of a neural network to implement them. This corresponds to the way in which a neurophysiologist may have information about the type of signals encoded by certain afferents to the region under current study, yet know nothing of the circuitry which produces them. A schema may summarize such data, and thus act as a placeholder for later neural analysis of the region under current study.[10] The following section on "Approach and Avoidance" introduces modeling that is purely at the schema level, and shows how such models may be tuned in the light of data on lesion experiments, and make new predictions. A crucial point is that schemas

---

9 Amari and Arbib 1977 provided a stability analysis unifying the Didday and Dev models. This general mathematical analysis has provided a reference point for much of our later work, especially that on "Depth and Detours". In fact, most of the models described below which use the Didday schema are implemented with a variant network due to Amari and Arbib. This reinforces the point that a model of interacting schemas remains robust if a schema is implemented with different (neural) networks so long as its interactions with other schemas and the environment remain unchanged. However, differing implementations may show differing responses to input patterns that were not included in the original specification of the schema, and thus may suggest new experiments to extend our understanding of the brain.

10 All the models introduced so far, as well as the facilitation and prey-recognition models described below, are presented at greater length (and with figures) in Arbib 1989a, my introductory contribution to the volume based on the previous Workshop in our series. Other models presented in this paper are based on work, most of which is reported in this volume, conducted since that Workshop was held in 1987.

which, with hypotheses about their neural localization,[11] survive testing of the behavior of normal and lesioned animals provide constraints on the function of a brain region which offer valuable cues for the neuroanatomist or neurophysiologist seeking to probe the region's circuitry.

In the previous paragraph, I have followed a line of research which builds explicitly on the early work on prey-selection and stereopsis, using formal neural models and interacting schemas. The next section describes a parallel line of research centered around neural networks based on detailed physiological studies of different cell types. Our model of facilitation uses a single "tectal column", a small circuit combining a vertical sample of tectal cell types, while our model of prey recognition exemplifies our "evolutionary" approach to the development of models by interconnecting an array of such tectal columns and then subjecting them to inhibitory modulation from a model of pretectum. However, this model of tectal-pretectal interactions represented the retina by a simple schema rather than a neural network, a schema which simply tabulated the average firing rates of a variety of tectal ganglion cells to a limited set of stimuli passing through their receptive fields. This sufficed to tune the original model of pattern recognition and so made the use of this simpler schema worthwhile. However, it does not let us predict responses to novel stimuli, nor does it let us model phenomena where the time course of retinal response makes an important contribution to the animal's behavior, and we have thus developed the neural model of the retina to be described below.

The final two sections of our overview may be described briefly. "Habituation, Memory, and Modulation" takes us beyond the study of non-adaptive networks of schemas and/or neurons to offer a detailed neural model of habituation in the toad, and show how its further development involves the integration of theory and experiment, and looks at new data on spatial memory in frogs and on hormonal modulation of the behavior of male frogs during the mating season to point the way to further modeling. Then, noting that all the above models all describe the processing of visual input or the generation of commands for motor schemas, but do not describe the neural analysis of these motor schemas, we close with a section on "Generation of Motor Behavior" which presents a variety of challenging data as well as a generic model of the control of limb movements that may serve as a building block for a variety of motor schemas for *Rana computatrix*.

---

[11] Recall that our form of localization is not of the crude "one function, one box, one brain region" variety, but rather seeks to understand how a function is realized in the cooperative computation of multiple brain regions/schemas. We also understand that, as the cycle of modeling and experimentation proceeds, our definition of a function may change to better accommodate the body of relevant data — just as, in our motivating example, our original tectal schema for recognizing *small* moving objects was replaced by one for all moving objects.

Competition and Cooperation in Neural Nets

Prey-Selection

When confronted with two "flies" within its snapping zone, the frog may snap at one of the flies, not snap at all, or snap in between at the "average fly" (Ingle 1968). Didday (1976) offered a simple *distributed* model of this "choice" behavior. The input to the tectum is a position-tagged "foodness array", while the layer of cells that yields the input to the motor circuitry is said to code "relative foodness". The transformation from foodness to relative foodness employs an array of S-cells,[12] in topographic correspondence with the other layers. These cells are introduced to allow different regions of the tectum to compete so that normally only the most active region provides an above-threshold input to the motor circuitry. Each S-cell inhibits the activity of cells in its region of the relative-foodness layer by an amount that increases with greater activity *outside* its particular region. This ensures that high relative-foodness activity persists only if surrounding areas do not contain sufficiently high activity to block it. Plausible interconnection weights ensure that if input activity in one region far exceeds activity in other regions then this region eventually overwhelms the others, directing the frog to snap at the corresponding target. If two regions have sufficiently similar activity levels then they may both (providing they are very active) overwhelm all others and simultaneously take command, or the two most active regions may simply turn down each other's activity to the point that neither is sufficient to take command. Buildup of inhibition on the S-cells precludes the system's quick response to new stimuli — there is hysteresis. Didday thus introduced an "N-cell" for each S-cell to monitor temporal changes in the activity of its region. Should it detect a large increase in this activity, it then overrides the inhibition on the S-cell and permits this new level of activity to enter the relative foodness layer.

Stereopsis

Although it is based on human psychophysics and mammalian data, Dev's (1975) model of stereopsis played an important role in the development of *Rana computatrix*. The problem for many models of binocular perception is to suppress the "ghost" targets that arise from matching the image of one stimulus on one retina with the image of a different stimulus on the other retina. There is an array of neurons in which each neuron receives maximal stimulation from the two eyes for a stimulus located at a particular depth and in a particular visual direction. The

---

[12] Actually, Didday called them sameness cells because their inclusion in the model was motivated by the data of Lettvin et al. 1959 on sameness cells. However, I call them S-cells to emphasize that they have a precise computational role in the model irrespective of their relation to neurophysiological data (and similarly for N-cells and newness cells). Unfortunately, no neurophysiological tests have been made of network properties observed in the model.

essence of the Dev scheme is to have those neurons which represent similar features at nearby visual directions and approximately equal depths excite each other (they cooperate), whereas those neurons which correspond to the same visual direction but different depths are (via interneurons) mutually inhibitory (they compete). The Dev model may thus be viewed as comprising an array of Didday's maximum selectors, one for each visual direction, with excitatory cross-coupling biasing nearby selectors to choose a similar maximum. In this way, neurons which could represent elements of a surface in space will cooperate, whereas those which would represent paradoxical surfaces at the same depth will compete, thus ensuring that the world is, where appropriate, "seen" in terms of continuous surfaces rather than scattered points at random depths. The result is that, in many cases, the system will converge to an adequate depth segmentation of the image. However, such a system may need extra cues. In animals with frontal facing eyes ambiguity can be reduced by the use of vergence information to drive the system with an initial depth estimate. Another method is to use accommodation information to provide the initial bias for a depth perception system; this is more appropriate to the amphibian, with its lateral-facing eyes.

Depth and Detours

Collett (1982) has shown that a toad, confronted with a barrier beyond which a worm can be seen, may sidestep the barrier and then approach the prey. Even in a setup where, soon after its initial movement, the toad can no longer see the worm, it proceeds along a trajectory whose final stage clearly indicates that the animal has retained an accurate representation of their position — though the final approach is aborted by the lack of adequate stimuli. Epstein (1979) adapted the Didday model by positing that each prey provides excitatory tectal input with a sharp peak while each barrier provides a trough of inhibition whose tectal extent is slightly greater, retino-topically, than the extent of the barrier in the visual field. When the model tectum acts upon the resultant input pattern, it will choose either a prey-locus or the edge of a barrier as the maximum. However, given that the toad's behavior depends upon the distance of the worms relative to the barrier, a full model must incorporate depth perception. Arbib and House (1987) gave two models for detour behavior which make use of separate "depth schemas" for prey and barriers. Before presenting these models of detour behavior, we will describe specific models of depth perception in frog and toad — even though the detour models do not depend on specific choices of neural mechanism for depth perception.

A monocular frog can snap fairly accurately at prey presented within its monocular field (Ingle 1976), and Collett (1977) showed, with prisms and lenses placed in front of the eyes of the toad, that the toad relies mainly on stereopsis in its binocular field, while the monocular toad makes depth judgments based on accommodation. We describe two such models (House 1989, summarizing work conducted in 1982-84) which function on accommodation cues in the monocular animal but are otherwise most dependent upon binocular cues.

The Cue Interaction Model

The Cue Interaction Model uses cooperative computation between two schemas, each based on Dev's stereopsis model, to build a depth map of the visual field, and so is appropriate for representing barriers. An accommodation-driven field, M, receives information about accommodation (the sharper the image at a particular depth in a given direction, the greater the activity of the neuron corresponding to that spatial position), while a system, S uses disparity information as input. The initial state of the accommodation field is blurred, representing the lack of fine tuning offered by accommodation. Targets are better tuned in the stereopsis field, but they offer ghost images in addition to the correct images. However, the systems are so intercoupled that a point in the M field will excite the corresponding point in S, and vice versa. As a result, ghost targets are suppressed while accommodation information is sharpened. Localization is now precise and unambiguous, and can be used to guide the behavior of the animal.

The Prey Localization Model

Collett and Udin (1983) postulated that the toad may use triangulation to locate its prey, rather than a process of disparity matching, suggesting a depth resolving system (for prey, not for barriers) that does not produce a complete depth map but instead selects a single prey and locates it in space. In the Prey Localization Model (House 1989; for a mathematical analysis, see Chipalkatti & Arbib 1987), each side of the brain selects a prey target based on output of the contralateral retina, and computes a depth estimate by triangulation to adjust lens focus (thus coupling two copies of the Didday schema with a motor schema for control of lens accommodation). If the selected retinal points correspond to the same prey-object, then the depth estimate will be accurate and the object will be brought into clearer focus, "locking on" to the target. If the points do not correspond, lens adjustment will tend to bring one of the external objects into clearer focus, and the two halves of the brain will tend to choose that object over the other.

While conceding that each of the above models needs further development (and further data to constrain it), we argue that, instead of there being a single general depth-perception mechanism, there are various neural strategies to cope with the vast array of visuo-motor tasks required of the animal. The present volume has relatively little material which advances our understanding of depth computation in anurans. Gaillard extends the analysis of cell types in n. isthmi, but does not address their role in depth perception. However, Ellard and Goodale provide data which is most suggestive for future modeling. Where the *Rana* models are based on disparity and accommodation, Ellard and Goodale show that mongolian gerbils, trained to estimate the distance of a visual target in order to make a ballistic jump to it, often execute a series of "head bobs" that increase in number, size, and velocity as the distance to be jumped increases. This suggests that gerbils produce retinal motion in order to compute the distance to the target, and points up the importance of optic flow in depth estimation (cf. Arbib 1989, Section 7.2 for a review).

Moreover, Ellard and Goodale find that retinal image size is also used as a cue to distance. They thus argue, in the spirit of cooperative computation at the level of schemas, that information from disparate sources is blended to arrive at the final distance estimate and suggest that the gerbil may have specialized subcortical circuitry for distance computation that is independent of the cortical mechanisms responsible for recognizing the properties of objects.

With this, we now turn to models of detour behavior which use "depth schemas" as vital subsystems.

The Orientation Model

In the Orientation Model, the retinal output of both eyes is processed for "barrier" and "worm" recognition to provide separate depth mappings. The barrier map B (the output of the barrier-depth schema) is convolved with a mask I which provides a (position-dependent) inhibitory effect for each fencepost; while the worm depth map W is convolved with a mask E which provides an excitatory effect for each worm.[13] The total excitation $T = B*I + W*E$ is summed in each direction, and then the Didday schema chooses that direction with maximal activity. If this corresponds to the prey, the animal will approach and snap, otherwise, further processing is required.

Detour behavior is an example of the coordination of motor schemas, where the sidestepping schema acts to modulate the orienting schema. Ingle observed that a lesion of the crossed-tectofugal pathway will remove orienting; lesion of the crossed-pretectofugal pathway will block sidestepping; while lesion of the uncrossed-tectofugal pathway will block snapping. Ingle (1982) suggests a model for detour behavior based upon principles similar to those employed in our simple orientation model. Wide-field tectal neurons driven by retinal R2 ganglion cells provide a spread effect. He proposes that if inhibition from pretectal cells driven by barrier detectors is sufficient to suppress excitation in narrow-field tectal neurons, the effect of the wide-field tectal neurons will be to provide a lateral shift of the locus of tectal excitation which would translate into a corresponding shift in turning angle.

The Path-Planning Model

The Path-Planning Model associates with each point of the two depth maps (for prey and barriers) a two-dimensional vector to indicate the preferred direction of the animal were it to follow a path through that corresponding position. Each prey sets up an attractant field, while each fencepost sets up a field for a predominantly lateral movement around the post. The summed field is then used to determine targets for forward and lateral motion. This model is computationally adequate in that it provides a parallel computation scheme for converting the perception of prey and barriers into the parameters that characterize an appropriate trajectory. However, it

---

[13] If a prey-localization schema is used to generate information about worm position, then W will contain a peak for at most one worm.

is not structured to conform to, e.g., Grobstein's lesion data showing that heading information is carried by a separate pathway from information about elevation and distance. Future work will thus not only explore the extension to side-stepping, turning and snapping but also seek new data to show how the components of the various vectors are distributed across different regions of the brain.

Lara et al. (1984) have provided schemas for a wider range of toad behavior which shows how the toad's response to prey is modified by intervening chasms as well as barriers. The model postulates perceptual schemas for gaps which provide a target for detour behavior, rather than having this behavior driven by inhibition/repulsion derived from perceptual schemas for barriers. Ballistic movement is obtained in their model by a sequence of several schema activations. We now turn to a new schema-level model which can activate several motor schemas concurrently, and with variable intensity, to yield more flexible and varied behavioral patterns than the purely sequential approach.

Schemas for Approach and Avoidance

Grobstein has repeatedly stressed that information from the two retinas must be combined and gated in an action-dependent way to serve the needs of the organism, and has shown a parcellation of tectal output segregating prey heading from other data about the prey. He also finds (this volume) that these tectal outputs are coded by overall activity in a pathway, rather than by some variant of retinotopic coding. He thus sees the central representation as involving neither pure place-coding nor pure distributed activity coding (cf. our earlier discussion of the work of Anastasio), but rather as containing elements of both. He suggests that the representation is specifically three dimensional, but this is somewhat misleading. True, he has separated the coding of heading from that for depth and elevation for a single prey object, but far greater dimensionality is required in representing a complex environment containing prey, predators and barriers. Arbib and Cobas (this volume) thus extend this work by providing models of prey-catching and predator avoidance at the level of maps and schemas, which is the right level to capture and extend data from lesion and behavioral experiments of the kind conducted by Grobstein (they do *not* offer neural network models). The motor schemas are driven by specific internal maps which between them constitute a distributed internal representation of the world. These maps collectively provide the transition from topographically-coded sensory information to population-coded inputs to the diverse motor schemas that drive muscle activity. In particular, following Grobstein, one pathway encodes the heading of the prey. However, whereas stimulus and response direction are the same for prey, different maps are involved for predator location and escape direction. They thus distinguish between the *Positional Heading* hypothesis (heading codes the position of the object) and the *Motor Heading* hypothesis (each system has a separate projection pathway converging in a different way onto a heading map coding the required motor response). Cobas and Arbib (this volume) follow the latter hypothesis, with motor actions constructed through the interaction of different motor schemas via competition and cooperation. Two or more schemas may cooperate to yield the final motor pattern. The model generates different motor zones for prey-catching behavior which match those observed in normal conditions and in the medullary hemifield deficit, and offers predictions for experiments on both approach and avoidance behaviors.

Ingle ("Control of Frog Evasive Direction", this volume) shows that midbrain systems for triggering either feeding or evasive behaviors interact with modulating or "biasing" systems in determining the final choice of direction. Some biasing effects seem to act directly upon the tectum (e.g., those from pretectum and from n. isthmi) while others (including pretectum and striatum) may converge in the brainstem with tectal efferents to guide orientation. New data suggests that both cerebellum and striatum bias avoidance directions via a common center in tegmentum.

Dean and Redgrave discuss the involvement of rat superior colliculus (SC) in approach and avoidance (most mammalian studies note only its role in approach movements, as in visual saccades), with at least two classes of response to novel sensory stimuli. One class contains the orienting response, together with movements resembling tracking or pursuit; the second contains defensive movements such as avoidance or escape. The two response systems appear to depend on separate output projections, and are probably subject to different sensory and forebrain influences. These findings clearly emphasize the similarities between the SC and the optic tectum in frog and toad, but further modeling must address the different efferents which distinguish rat SC from anuran tectum even as we use the rat data to extend the methodology of Cobas and Arbib to mammalian forms.

Liaw and Arbib (this volume) have developed a neural net model which refines certain of the schemas modeled by Cobas and Arbib. They focus on neural mechanisms for the detection of a looming stimulus, and the determination of escape direction. (Ulinski, Larson-Prior and Slater [this volume] offer insights for comparative modeling with data on cortical circuitry underlying visual motion analysis in turtles.) The model incorporates known data on neurons in toad's Retina, Tectum and THalamus-pretectum. The visual stimulus in the model is transmitted to the avoidance circuitry via R3 and R4 ganglion cells. T3 neurons detect movement toward the eye and T6 neurons monitor stimulus activity in the upper visual field. These tectal signals, along with depth information, converge onto the TH6 neurons which determines whether the visual stimulus is a looming threat. The topography of the T3 neurons provides the spatial map of the stimulus location. This signal is projected onto a motor heading map which specifies the direction of the avoidance movement.

Neural Models based on Anatomy and Physiology

Having thus made the transition to neural models which incorporate data on tectal and pretectal circuitry, we now present the "tectal column models" which first introduced anatomically motivated circuitry into the design of *Rana computatrix*, and a neural network model of the retina.

Facilitation and the Tectal Column

The first tectal column model was motivated by a facilitation effect. Presenting a worm to a frog for 0.3 s may yield no response, whereas orientation is likely to result from a 0.6 s presentation. However, if a worm is presented initially for 0.3 s, then removed, and then restored for only 0.3 s, the second presentation suffices to elicit a response if the intervening delay is at most a few seconds. Ingle (1975) observed tectal cells whose time course of firing accorded well with this facilitation effect, leading Lara et al. (1982) to a model in which the "short-term memory" is encoded as reverberatory activity in a neural circuit. Their *tectal column model* is abstracted from the anatomy of Székely and Lázár (1976) — and thus is currently being updated in the light of new anatomical data (e.g., as reviewed by Lázár 1989).

Each column comprises one pyramidal cell (PY) as sole output cell, one large pear-shaped cell (LP), one small pear-shaped cell (SP), and one stellate interneuron (SN). All cells are modeled as excitatory, save for the stellates. The retinal input to the model is a lumped "foodness" measure, and activates the column through glomeruli with the dendrites of the LP cell. LP axons return to the glomerulus, providing a positive feedback loop. A branch of LP axons also goes to the inhibitory SN cell. There is thus competition between "runaway positive feedback" and the stellate inhibition. PY is excited by both SP and LP, and requires sufficient activation from both of them to trigger a motor response.

The overall dynamics of the tectal column (each neuron is represented by the leaky integrator model) will depend upon the actual choice of synaptic weights, as well as the choice of membrane time constants. Appropriate choices ensure that excitation of the input does not lead to runaway reverberation between the LP and its glomerulus. There is, as in the experimental data, a critical presentation length below which there is no pyramidal response. Input activity activates LP, which re-excites the glomerulus but also excites the SN, which reduces LP activity. But if input continues, it builds on a larger base of glomerular activity, and so over time there is a build-up of LP-SN alternating firing. If the input is removed too soon, the reverberation will die out without activating SP enough for its activity to combine with LP activity and trigger the pyramidal output. If input is maintained long enough, the reverberation may continue, though not at a high enough level to trigger output. However, re-introducing input shortly after this "subthreshold" input can indeed "ride upon" the residual activity to build up to pyramidal output after a presentation time too short to yield output with an initial presentation.

In a related vein, Cervantes-Pérez et al. (this volume) study the temporal dynamics of the processing of visual information to explain processes related to learning and short- and long-term memory.

Prey Recognition, Arrays of Tectal Columns and Pretectal Modulation

Ewert studied the behavior of a toad in a perspex cylinder from which it could see a stimulus object being rotated around it, measuring how often the animal would respond with an orienting movement (the more frequent the turn, the more "prey-like" the object) for different stimuli. A worm-like stimulus (rectangle moved in the direction of its long axis) is increasingly effective with increasing length; whereas for 8° or more extension on its long axis, the antiworm stimulus (rectangle moved in the orthogonal direction) proved ineffective in releasing orienting behavior. The square showed an intermediate behavior, the response it elicits rising to a maximum at 8°, but being extinguished by 32°. Ewert also recorded the activity of different types of neurons in toads. TH3 neurons have a response that is uniform for worms, increases with increasing length for antiworms, but is greatest for squares, increasing with their size. The activity of T5.1 seems to correlate fairly well with increases in the length of the stimulus in the direction of motion, but it is the T5.2 neurons whose overall rate of *neural* response seems to best match the overall frequency of the *behavioral* response. T5.2 pattern discrimination is essentially abolished by pretectal lesion.

Ewert and von Seelen (1974) define a "prey schema" by using a linear filter tuned to respond maximally to worms and a linear filter tuned to respond to increasing extent of a stimulus in the antiworm direction, with the "worm schema" exciting, and the "antiworm schema" inhibiting, the output of the overall schema. Where Lara et al. model the temporal dynamics of a single tectal column, Cervantes-Pérez et al. (1985) go a step further in the evolution of *Rana computatrix*, extending the Ewert and von Seelen model by using an 8x8 array of tectal columns (which actually serves more like an "all moving objects" detector than a "worm filter"), receiving inhibitory modulation from a pretectum modeled by an array of TH3 cells receiving R3 and R4 input with synaptic connections tuned to make it act like an "antiworm filter". The retinal input in *this* model is based on ganglion cell response curves (Ewert 1972), where only the average rate of firing of a cell is given for each stimulus, rather than the temporal pattern of that response.

With appropriate setting of the various masks and time constants, the model does indeed exhibit in computer simulation responses to moving stimuli of different types that provide a good match to the neural data. However, the model is only approximate at a quantitative level, so that — if our goal is prediction of detailed neural firing rather than just a general understanding of pattern recognition networks — further work must be done on tuning the model parameters. Cervantes-Pérez al. (1985) provide results of further simulation, a demonstration of directional invariance of response, and a discussion of motivation.

Retina

Since the response of tectum and pretectum depends on the spatiotemporal pattern of retinal ganglion cell firing, and not just a schematic presentation of the average rate of firing of a cell is given for each stimulus, Teeters (1989; see also Teeters and Arbib 1990) extended the work of Lee (1986) and an der Heiden and Roth (1987) to develop a detailed model of frog retina which accounts for many characteristic response properties used to classify anuran ganglion cell types in a fashion consistent with data concerning interneurons. Receptors are modeled as responding with a step change to changes in intensity, while horizontal cells respond only to global changes in intensity brought about by full field illumination changes — yet accurate ganglion cell responses are obtained even with these simplifications. Hyperpolarizing and depolarizing bipolar cells are generated by subtracting local receptor and horizontal potentials. Two classes of transient amacrine cells (On and Off) are generated using a high-pass filter with a thresholded output which responds to positive going changes in the corresponding bipolar cell potentials. The model shows how a selective combination of bipolar and amacrine channels can account for the response properties used to classify the anuran ganglion cell types and makes several experimental predictions. The importance of this model is that it now allows us to model the response of *Rana computatrix* to a wide variety of visual stimuli, not just those for which experimental data are already well tabulated.

Habituation, Memory, and Modulation

Habituation

Ewert and Kehl (1978) discovered a "dishabituation hierarchy" which motivated Lara and Arbib (1985) to "evolve" the tectal array with plasticity and forebrain modulation. Habituation is an elementary form of learning in which response to a stimulus will diminish with repeated presentation of the stimulus if there is no punishment or reward associated with the presentations. Habituation has been much studied in Aplysia for insight into molecular mechanisms of synaptic plasticity (see Bailey and Kandel, 1985, for a review), and seems to be independent of the specific patterning of stimuli. Habituation in mammals is stimulus specific: given two different patterns A and B , the animal can exhibit this specificity by the phenomenon of dishabituation (i.e., stimulus B can release behavior despite habituation to A). Moreover, in mammals, if stimulus A can dishabituate stimulus B, then stimulus B can dishabituate stimulus A (Sokolov 1975). However, Ewert and Kehl (1978) found a dishabituation hierarchy in toads: a worm-like pattern higher in the hierarchy can dishabituate another pattern lower in the hierarchy, whereas a stimulus lower in the hierarchy cannot dishabituate habituation to a "higher" stimulus.

Sokolov (1975) suggested that stimulus-specific habituation was based on the formation of a neural model of the habituated stimulus, with behavior being released to the extent that the current stimulus differs from the stored model. However, such a model yields symmetric dishabituation, and so Lara and Arbib (1985) suggested that the habituated stimulus was stored in the toad as a model of *intensity* of some brain variable, with a stimulus yielding dishabituation only to the extent that its encoding is *stronger* than the stored model. Unfortunately, their model of the neural basis for this ordering was flawed, in part because it used an *ad hoc* model of the toad retina. The subsequent availability of a detailed retinal model (Teeters 1989) provides the basis for a new model to be described below.

Merkel-Harff and Ewert (this volume) present a number of extensions to the data base for our modeling of habituation with studies of how, in an individual toad, responses to visual objects can be influenced by various modes of learning or by changes in motivation or attention. These modulatory processes are provided by neural loops involving structures of the telencephalon (e.g., posterior ventral medial pallium, also called the "primordium hippocampi", and ventral striatum) that communicate with the optic tectum via diencephalic nuclei (e.g., anterior thalamus). Such results, especially those on the role of anterior thalamus and medial pallium, motivated a new approach to modeling habituation by Wang and Arbib (1990) which provides the basis for the report by Wang, Arbib and Ewert (this volume) of a dialog between modeling and experimentation. Here habituation is posited to depend on the average firing rates of cells of the anterior thalamus. Simulation demonstrates that the retinal response to the trailing edge of a stimulus is as crucial for pattern discrimination as the response to the leading edge, and predicts new dishabituation hierarchies based on reversing stimulus-background contrast and shrinking stimulus size. After the predictions were made, Wang and Ewert designed a group of behavioral experiments to test the predictions. In particular, they selected a pair of stimuli from the predicted hierarchy with contrast reversal which exhibits the opposite ordering to that in the original hierarchy. The experimental result was as predicted. Other results suggest extensions to the model, such as having habituation exhibit a measure of size invariance.

Cervantes-Pérez et al. (this volume) study how the intensity and duration of a toads' response is modulated by learning, when the animal is repeatedly stimulated with a visual worm-like dummy. They find that the toad's motor response to a moving worm-like stimulus gradually decreases until complete inhibition, that the intensity of the current prey-catching response depends on the consequences of previous motor actions intended to catch the prey, as well as a number of other results which point to further modeling.

Spatial Memory

David Ingle ("The Striatum and Short-Term Spatial Memory", this volume) finds that frogs show a robust memory for the position of recently-seen obstacles which not only may last for at least 60 seconds after their removal but can also compensate for passive rotation of the frog. He offers a speculative "sliding map" network model, and provides evidence that this short-term memory depends upon the frog's striatum. Behavioral tests also reveal a second mode of spatial memory based upon landmarks. These results are the basis for analogies between these two species of frog memory and contrasting classes of mammalian memory linked to caudate and hippocampus, respectively. Such mammalian homologies provide a natural link to the model by Dominey and Arbib (this volume) of the way in which many brain regions interact in generation of delayed and multiple saccades in primates. Of particular relevance here is their model of a striato-thalamo-cortical system for saccade target memory, and an application of ideas of Berthoz and Droulez (1989) to suggest how a retinotopic map of visual targets in parietal cortex may be updated to take account of eye movements. In carrying these ideas back to the development of *Rana computatrix*, an important role will be played by the studies of Ewert, Matsumoto, Schwippert and Beneke (this volume) on striatal/pretectal/tectal connectivity — with body movements replacing eye movements.

Hormonal Modulation

To close this prospectus for modulation of circuitry, we note another time scale of change in the nervous system, that occasioned by seasonal changes in hormonal modulation, such as occur in males during the mating season. In some species of anurans, the so-called "explosive breeders", prey catching does not occur during the breeding period. Amplexus, the stereotyped behavior in which males clasp onto the dorsal side of females for prolonged periods during mating, is controlled by seasonal fluctuations in androgen levels.

Among explosive breeders, males will approach and attempt to clasp practically any moving object. They apparently do not discriminate visually between males and females (see Kondrashev 1987). When they clasp another male, it gives a stereotyped release call. Thus, mate recognition among explosive breeders is non-specific whereas prey recognition is highly specific (although subject to conditioning: toads allowed to eat mealworms out of an experimenter's hand can be conditioned to respond to the moving hand alone). Betts (1989) has thus suggested that prey and mate are part of the same stimulus continuum for the toad, with recognition of these two types of stimuli implemented by the same tectal circuit. Specifically, T5 circuitry is modulated by pretectal circuitry. The less the pretectal inhibition, the larger the stimulus that will be "recognized" as prey. Cervantes-Pérez et al. (1985) view pretectal inhibition as a specific local measure of "antiworm-ness", but they also use a global shift in the level of this habituation to represent motivational state, e.g., hunger. The hungrier the animal, the lower the inhibition, and thus the larger the stimulus at which it will snap. Betts adds the notion that pretectal inhibition (or T5 threshold) is lowered by hormonal influences during the mating season to allow indiscriminate responses. However, Betts' discussion lacks a crucial element. If his model is correct, then the same population of cells which normally drives the motor pattern generators (MPGs) for prey-catching must, in the male frog during the mating season, drive the MPG for clasping. The simplest explanation would be that the T5.2 population has efferents to both sets of MPGs, but that it requires high androgen levels to simultaneously disable the MPG for prey-catching and enable the MPG for clasping. At present, no data are available on any of these posited hormonal effects. However, Herrera and Regnier (this volume) have studied the cellular mechanisms by which androgens change the effector part of the system, the motor unit. (Bottjer [this volume], working with zebra finches, shows that both song learning and its neural substrate are sensitive to gonadal hormones such as testosterone and estrogen.) They show that amplexus represents the output of an androgen-sensitive system that extends from the hypothalamus, down the neuraxis to the motoneurons, out to the periphery via spinal nerves, and ends in the flexor muscles of the male chest and forearm. Androgen yields a hypertrophy of these muscles. With this, we now turn to issues in motor control.

Generation of Motor Behavior

Previous sections show how visual data enter the brain, and model a variety of interactions involved in processing those data to commit the animal to some form of approach or avoidance behavior. An important target for future work on *Rana computatrix* is to model the neural implementation of motor schemas. Here, we briefly review how papers in this volume suggest a number of promising directions for future research. (As the one study of motor learning in this book, Bartha, Thompson, and Gluck study classical conditioning of rabbit eyelid closure as an example of sensorimotor learning involving the cerebellum.)

Weerasuriya focuses on motor pattern generators in anuran prey capture. The outputs of the relevant sensory analyzers share common access to the various motor pattern generators. Each component of the motor output has its own characteristic degree of variation in its duration and execution. The coordinated spatio-temporal sequence of the skeletal muscle activity underlying approaching and orienting is generated by adjustable, loosely coupled, motor pattern generators which reflect the perceived relative location of the prey. The motor components for approach and orientation, while being similar in that they have variable outputs, are dissimilar with respect to the degree of modifiability during execution. Weerasuriya offers a number of data suggestive of extensions to the work of Cobas and Arbib and which may aid in the neural analysis of the various motor schemas.

The papers by Giszter et al. and by Shadmehr address the specific issue of controlling the movement of a single limb. Giszter et al. report experiments on the coding of motor space in the frog spinal cord which suggest that a wide variety of sites for stimulation of spinal premotor interneurons code for the movement of the frog's leg towards some (site-dependent) equilibrium situation. In contrast, stimulation of motoneuron nuclei or ventral nerve trunks gave parallel or divergent fields without stable equilibria. However, the performance of the limb may differ when it is returned to its start position after many manipulations. Given this hysteresis, further study is needed on the role of proprioceptive information. Shadmehr's theoretical analysis of movement generation with kinematically redundant biological limbs provides a general limb control strategy consistent with the "equilibrium point hypothesis." Since there is so much redundancy, the problem of controlling limb movement is extremely "ill posed." Specifying stiffness is, Shadmehr shows, one way to define a unique solution to these problems. However, unique solutions to "overcomplete" problems are not necessary, and future work may use an adaptive mechanism to manage the feed-forward, static transformation, as well as tuning via dynamic feedback. Most neuronal models (our own included) contain the implicit assumption that this objective is always achieved by the same movement. However, the objective can be achieved by any of an array of different movements (the phenomenon of "motor equivalence") and future neural models must address the issue of this variability.

A current target for modelling the role of limb movements in adaptive behavior is the wiping reflex in the frog. In our group, we have a preliminary model

(Shadmehr 1989) of a one-dimensional restriction of the scratch reflex in the cat, in which a simple circuit provides flexor drive during the positioning phase and then an alternating flexor and extensor drive during the rhythmic phase of scratching. In this model, it is the transformation of position on the body into time constants of the network that yields appropriate positioning. Fukson et al. (1980) find that the wiping reflex in the spinal frog depends on the body schema, with the trajectory used by the tip of the hindlimb to reach for the forelimb shoulder and move along the limb to remove an irritant changing appropriately when the position of the frog's body and limbs changes. Thus, we must address the subtle question of how somatotopy is modulated by representation of the current state of the body.

Conclusions

Our studies of *Rana computatrix* show the excitement of "evolving" an integrated account of a single animal, meeting the challenges of coordinating different aspects of vision with mechanisms for control of an expanding repertoire of behavior. They also pose many challenges for future modeling. To single out just five for mention: How are the models of depth and detours to be integrated with tectal column models? How will our increasing understanding of motor pattern generators constrain our models of how the visual system provides their control signals? To what extent does our model of habituation extend to models of, for example, conditioning, and to what extent will other types of learning demand the analysis of radically different mechanisms? How must the leaky integrator model be expanded to address the data on hormonal (and other) modulation? And when will our models begin to make sense of neural form-function relations? All these modeling questions in turn raise new questions for experimental analysis.

But *Rana computatrix* is not just an evolving model of a single animal. Rather, through its integrated use of schema theory and neural modeling, it provides a paradigm for computational neuroscience in general, wherever it is desired to link complex behavior (i.e., behavior that cannot be reduced to the function of a single circuit) to neural circuitry. To see this, one need only read Dean and Redgrave's pointers to *Rattus computator,* or read Ingle's "The Striatum and Short-Term Spatial Memory" conjointly with Dominey and Arbib's account of monkey superior colliculus as pointing the way to *Macaca computatrix,* to see the basis for a comparative computational neuroscience of visuomotor coordination. Elsewhere (Arbib 1989, Sections 5.4 and 8.3), I offer the even bolder claim that the roots of our intelligence in visuomotor coordination point the way to a theory of higher mental functions based on this new paradigm. *Ex Rana computatrix ad omnia.*

References

Amari, S. and Arbib, M.A., 1977, Competition and cooperation in neural nets, in *Systems Neuroscience* (Metzler, J., ed.), pp. 119-165, Academic Press, New York

an der Heiden, U., and Roth, G., 1987, Mathematical model and simulation of retina and tectum opticum of lower vertebrates. *Acta Biotheoretica,* 36:179-212.

Anastasio, T.J., and Robinson, D.A., 1989, Distributed Parallel Processing in the Vestibulo-Oculomotor System, *Neural Computation,* 1:230-241.

Arbib, M. A., 1975, Artificial intelligence and brain theory: unities and diversities, *Ann. Biomed. Eng.* 3: 238-274.

Arbib, M.A., 1981, Perceptual structures and distributed motor control, in *Handbook of Physiology - The Nervous System II. Motor Control* (V.B. Brooks, Ed.), Amer. Physiol. Soc. pp. 1449-1480.

Arbib, M.A., 1987, *Brains, Machines, and Mathematics,* 2nd Edition, Springer-Verlag.

Arbib, M.A., 1989, *The Metaphorical Brain 2: Neural Networks and Beyond,* Wiley-Interscience.

Arbib, M.A., 1989a, Visuomotor Coordination: Neural Models and Perceptual Robotics, in *Visuomotor Coordination: Amphibians, Comparisons, Models, and Robots,* (J.-P.Ewert and M.A.Arbib, Eds.), Plenum Press, pp.121-171.

Arbib, M.A., 1990, Programs, Schemas, and Neural Networks for Control of Hand Movements: Beyond the RS Framework, in *Attention and Performance XIII. Motor Representation and Control* (M. Jeannerod, Ed.), Lawrence Erlbaum Associates, pp.111-138.

Arbib, M.A., 1991, Schema Theory, *Encyclopedia of Artificial Intelligence,* Second Edition, Wiley-Interscience (in press).

Arbib, M.A., and House, D.H., 1987, Depth and Detours: An Essay on Visually Guided Behavior, in *Vision, Brain and Cooperative Computation,* (M.A. Arbib and A.R. Hanson, Eds.), A Bradford Book/The MIT Press, pp. 129-163.

Bailey, C.H., and Kandel, E.R., 1985, Molecular approaches to the study of short-term and long-term memory, in *Functions of the Brain* (C.W. Coen, Ed.) Oxford University Press, pp.98-129.

Berthoz, A. and Droulez, J., 1989, The concept of Dynamic Memory in Sensorimotor Control, *Dahlem Conf. on Principles and concepts in motor control,* Berlin.

Betts, B., 1989, The T5 base modulator hypothesis: A dynamic model of T5 neuron function in toads, in *Visuomotor Coordination: Amphibians, Comparisons, Models, and Robots,* (J.-P.Ewert and M.A.Arbib, Eds.), Plenum Press., pp.269-307.

Brooks, R.A., 1986, A robust layered control system for a mobile robot, *IEEE Journal of Robotics and Automation,* RA-2:14-23.

Cervantes-Pérez, F., Lara, R., and Arbib, M.A., 1985, A Neural Model of Interactions Subserving Prey-Predator Discrimination and Size Preference in Anuran Amphibia, *J. Theor. Biol.,* 113:117-152.

Chipalkatti, R., and Arbib, M.A., 1987, The Prey Localization Model: A Stability Analysis, *Biological Cybernetics,* 57:287-300.

Collett, T., 1977, Stereopsis in toads, *Nature,* 267: 349-351

Collett, T., 1982, Do toads plan routes? A study of the detour behavior of *Bufo Viridis, J. Comp. Physiol.* 146: 261-271

Collett, T., and Udin, S., 1983,The role of the toad's nucleus isthmi in prey-catching behavior. *Proceedings of second workshop on visuomotor coordination in frog and toad: Models and Experiments* (R. Lara and M.A. Arbib, Eds), COINS-Technical Report 83-19, University of Massachusetts, Amherst, Massachusetts

Dev, P. , 1975, Perception of depth surfaces in random-dot stereograms: A neural model, *Int. J. Man-Machine Studies,* 7: 511-528

Didday, R., 1976, A Model of visuomotor mechanisms in the frog optic tectum, *Mathematical Biosciences,* 30:169-180.

Epstein, S., 1979, Vermin users manual. Unpublished project report, Computer and Information Science, University of Massachusetts at Amherst.

Ewert, J.-P., 1972, Zentralnervöse Analyse und Verarbeitung visueller Sinnesreize, *Naturwiss. Rundsch.*, 25:1-11.

Ewert, J.-P., 1989, The release of visual behavior in toads: Stages of parallel/ hierarchical information processing, in *Visuomotor Coordination: Amphibians, Comparisons, Models, and Robots* (J.-P. Ewert and M.A. Arbib, Eds.), Plenum Press, pp.39-120.

Ewert, J.-P., and Kehl, W., 1978, Configurational prey-selection by individual experience in the toad *Bufo bufo*. *J Comp Physiol* 126: 105-114.

Ewert, J.-P. and von Seelen, W., 1974, Neurobiologie und System-Theorie eines Visuellen Muster-Erkennungsmechanismus bei Kröte, *Kybernetik*, 14:167-183

Fukson, O.I., Berkinblit, M.B., and Feldman, A.G., 1980, The spinal frog takes into account the scheme of its body during the wiping reflex, *Science*, 209:1261-1263.

Hebb, D.O., 1949, *The Organization of Behavior*, John Wiley & Sons.

House, D., 1989, *Depth Perception in Frogs and Toads: A Study in Neural Computing, Lecture Notes in Biomathematics* 80, Springer-Verlag, Berlin.

Ingle, D., 1968, Visual releasers of prey catching behavior in frogs and toads, *Brain, Behav., Evol.*, 1: 500-518

Ingle, D., 1975, Focal attention in the frog: behavioral and physiological correlates, *Science*, 188: 1033-1035

Ingle, D., 1976, Spatial visions in anurans, in *The Amphibian Visual System* (K. Fite, Ed.), Academic Press: New York, pp. 119-140

Ingle, D. J., 1982, Organization of visuomotor behaviors in vertebrates, in *Analysis of Visual Behavior* (D. J. Ingle, M.A. Goodale, and R. J. W. Mansfield, Eds.), pp. 67-109, MIT Press.

Koch, C., and Segev, I., Eds., 1989, *Methods in Neuronal Modeling: From Synapses to Networks*, The MIT Press.

Kondrashev, S.L., 1987, Neuroethology and color vision in amphibians. A commentary, *Behav. Brain Sci.*, 10: 385

Lara, R. and Arbib, M.A., 1985, A model of the neural mechanisms responsible for pattern recognition and stimulus specific habituation in toads, *Biological Cybernetics* , 51:223-237.

Lara, R., Arbib, M.A., and Cromarty, A.S., 1982, The role of the tectal column in facilitation of amphibian prey-catching behavior: a neural model, *J. Neurosci.*, 2: 521-530.

Lara, R., Carmona, M., Daza, F., and Cruz, A., 1984, A global model of the neural mechanisms responsible for visuomotor coordination in toads, *J. Th. Biol.* 110:587-618.

Lashley, K.S., 1951, The problem of serial order in behavior, in *Cerebral Mechanisms in Behavior: The Hixon Symposium* (L. Jeffress, Ed.), Wiley, pp. 112-136.

Lázár, Gy., 1989, Cellular architecture and connectivity of the frog's optic tectum and pretectum, in *Visuomotor Coordination: Amphibians, Comparisons, Models, and Robots*, (J.-P.Ewert and M.A.Arbib, Eds.), Plenum Press., pp.175-199.

Lee, Y., 1986, *A neural network model of frog retina: a discrete time-space approach*, Technical Report TR-86-219, Computer Science Department, University of Southern California.

Lettvin, J. Y., Maturana, H., McCulloch, W. S. and Pitts, W. H., 1959, What the frog's eye tells the frog brain, *Proc. IRE.* 47: 1940-1951.

McCulloch, W.S., and Pitts, W.H., 1943, A logical calculus of the ideas immanent in nervous activity. *Bull. Math. Biophys.* 5: 115-133.

Minsky, M.L., 1985, *The Society of Mind*, Simon and Schuster.

Rosenblatt, F., 1958, The perceptron: A probabilistic model for information storage and organization in the brain, *Psychol. Rev.*, 65:386-408.

Rumelhart, D.E., Hinton, G.E., and Williams, R.J., 1986, Learning Internal Representations by Error Propagation, in *Parallel Distributed Processing: Explorations in the Microstructure of Cognition* (Rumelhart, D., and McClelland, J., Eds.), The MIT Press, Vol.1:318-362.

Rumelhart, D.E., and McClelland, J.L., Eds., 1986, *Parallel Distributed Processing: Explorations in the Microstructure of Cognition,* The MIT Press.

Shadmehr, R., 1989, A neural model for generation of some behaviors in the fictive scratch reflex, *Neural Computation*, 1:242-252.

Sokolov, E., 1975, Neuronal mechanisms of the orienting reflex, in *Neuronal mechanisms of the orienting reflex*, (E. Sokolov and O.Vinogradova, Eds.) Erlbaum, pp.217-235.

Székely, G. and Lázár, G., 1976, Cellular and synaptic architecture of the optic tectum, in *Frog Neurobiology* (Llinás, R. and Precht, W., Eds.). Springer Verlag, pp. 407-434

Teeters, J.L., 1989, *A simulation system for neural networks and model for the anuran retina*,Technical Report 89-01, Center for Neural Engineering, University of Southern California, Los Angeles.

Teeters, J.L., and Arbib, M.A., 1990, A model of anuran retina relating interneurons to ganglion cell responses, *Biol.Cybern.* (in press).

Wang, D.L., and Arbib, M.A., 1990, How does the toad's visual system discriminate different worm-like stimuli? *Biol.Cybern.* (in press).

Weitzenfeld, A., 1990, *NSL, Neural Simulation Language, Version 2.0*, TR 90-01, Center for Neural Engineering, University of Southern California.

A Prospectus for the Fruitful Interaction Between Neuroethology and Neural Engineering

J.-P. Ewert

Neurobiology, FB 19, University of Kassel
D-3500 Kassel, FR of Germany

Abstract. *There are two important goals in neuroscience. One concerns the understanding of functions related to animal and human behavior - a research topic of neuroethology; the other, based on results of the former, seeks to develop strategies for the construction of so-called intelligent machines - the research field of neural engineering. The idea to date is not to build human's cognition or toad's pattern recognition into a technical device, rather to look for and to take advantage of certain task-oriented operational principles which, implemented by neuronal networks, are more economical and presumably even simpler than those an engineer faced with a comparable technical problem might have thought of. Selecting visually guided behaviors of toads, in this chapter we emphasize the need for models of brain functions, explain the advantage of the dialogue between experimentalists (neuroethologists) and modelers (computational neuroscientists), and point to the scope of questions and joint projects for the near future. The topics are: 1) Parallel distributed processing of visual signals; 2) implication of retinal on- and off-channels; 3) centrifugal control of retinal function; 4) forebrain-involved modulation of stimulus-response mediating tectal networks; 5) combinatorial control of motor pattern generating systems. Each topic addresses (a) neuroethological data, (b) current neuroethological projects, and (c) questions for neural engineering.*

1. The Metaphorical Brain

Michael Arbib emphasized in his book "The Metaphorical Brain 2" that in trying to understand principles of brain function we rely on two metaphors (Arbib 1989): (1) The metaphor "humans are animals" points to the evolutionary aspect of brain and behavior, suggesting that comparative studies of behaviorally relevant neuro-principles in homologue brain structures of differently organized vertebrates contribute to the understanding of comparable functions of the human's brain; this is one of the goals of *neuroethology* (e.g., see Ewert 1980; Guthrie 1987). (2) The metaphor "humans are [bio-]machines" regards brains as highly complex (adaptive, integrative, interactive, and parallel distributed processing) computers, suggesting that the understanding of neuro-principles underlying goal oriented behaviors in animals and humans may give us ideas for the construction of so-called intelligent machines, such as perceptual robots; this is the domain of *neural engineering* (e.g., see Arbib 1987, 1989, 1990; Eckmiller et al. 1990).

Controversy surrounds this concept. For example, drawing on the technical advances based on the developments of mathematical neuronal ("neuroid") networks, critics predict that engineers by their *intuition* will be able to construct cognitively operating robots long before neuroscientists begin to understand the neurophysiological basis of "thought". From my point of view, this statement is vague and misleading. Putting aside the question about the various definitions of "thought" (e.g., Walker 1983), let me emphasize on two aspects. First, thinking in terms of engineer*logic* follows rules and algorithms that are fundamentally different from the bio*logic* of the neuronal information processing that underlies engineer's thinking (for a discussion of this topic see Creutzfeldt 1983). Second, although an ultimate goal in neuroscience is to tackle the global question of how the brain works, our objective here is to learn how brains solve particular problems in their specific interactions with their environments and how they deal with changing conditions, that is to analyze task-related neuroarchitectures and to try to understand the algorithms of the related information processing principles. These may then inspire computer scientists and engineers to implement the "philosophy" of such a principle - which by intuition they would not have thought of - into a robot for the solution of a comparable technical problem.

2. Goals in Neural Engineering

In neural engineering, the development of perceptual robotics plays a central role, studying robots which explore, navigate, and (re)act autonomously in natural or artificial structured environments, thereby *avoiding* obstacles, *turning toward* a target *approaching* the target, and *grasping* it. Neurobiological principles already considered for their technical application deal with three-dimensional evaluations of visual space and obstacle avoidance by means of *inverse perspective mapping* (Mallot & v.Seelen 1989) or navigation with *perceptual and motor schemas* (Arbib 1981, 1990; Arkin 1989), to mention two examples.

The technical developments of the next decades will increasingly count on *adaptive* robotic systems (e.g., see Schmidhuber 1990). The required presently available adaptive neuronal networks (ANNs) depend on an external "teacher" who knows and supervises how each input pattern has to be responded to by the ANN. For a given task this may be sufficient. For sequential tasks, where the output changes the input in an unpredictable manner, the problem is different. A solution may take advantage of dynamic adaptive neuronal networks (DANNs) that apply "reinforcement"(R)-learning whereby an R-algorithm, for example, evaluates success and failure. Such RANNs are comparable with biological learning systems. An alternative solution with a biological background is provided by Arbib's (1981) schema-theoretical approach, e.g. in a manner that a sensory schema A triggers a corresponding motor schema A* whereby completion of A* leads to another sensory schema B that triggers motor schema B* etc., or a given schema assemblage may integrate motor schemas with the perceptual schemas that supply the necessary parameters for the task.

3. Challenges from Neuroethology

Not much to our surprise organisms are faced with and have to solve in a certain sense comparable problems to the ones mentioned for robots: sensory pattern recognition, its modification by learning, and sensorimotor coordination. What are the pecularities of biological neuroarchitectures?

One is *self-organizing control*. Let me explain this with an example. If an engineer is asked to construct a system that in response to a signal executes a motor-driven action, he would connect an adequate sensor with an appropriate motor. The biological system is different: it too has filter and motor systems, but in between and in parallel other systems are integrated whose aim - most of the time - is to "interrupt" sensorimotor action. That means, there is an excitatory stimulus-response pathway (e.g., eyes feeding information to muscles) and in connection to it there are brain structures whose main functions are *inhibition*-assisted sensory filtering, storage, and response control. The neurophysiologist Richard Jung described this illustratively:

> "Without inhibition, any excitation would propagate in the nervous system like an avalanche, thus driving the synaptic powder keg toward explosion already in response to small stimuli, comparable to an epileptic seizure. For the neuronal functional order of the central nervous system, therefore, inhibitory processes at the synapses must be the rule, whereas excitatory transmission is the exception, as Tönnies has formulated it first [Tönnies 1949]" (Jung 1953, p.11/12; Engl. transl.).

During evolution, a general philosophy of the brain probably was to inhibit safely. Inhibition of inhibitory processes - "disinhibition" - thus may be an important prerequisite for the readiness to (re)act.

Another pecularity, refers to the fact that sensory information entering the brain is processed in a *parallel distributed* manner in structures at various stages of integration and loop-operated interaction under different aspects of significance, such as: relational assignments of object features, distinction between signal induced and self-induced stimulation, signal localization, stimulus specific habituation to repetitive occurrence of the same signal, conditioning to the combined occurrence of different signals, and the readiness to respond. The parallel/integrative paradigm allows the brain to run certain kinds of information processing simultaneously and to optimize fast decisions. For example, a toad faced with a stripe moving prey-like in the direction of the longer axis responds with prey-catching, while faced with the same stripe when its longer axis is oriented perpendicular to the direction of movement, the behavior immediately "freezes"; however, it "thaws" within less than a second when the same stripe again is moving prey-like (Ewert 1968; Ewert & IWF 1982). Similarly fast in toads is the retrieval of information related to an associatively stored property, e.g., bee-avoidance due to certain negative preying experience (Cott 1936). In classical computers, recalling information when the address is not known takes a relatively long time since all addresses must be examined, a process whose duration is positively correlated with the number of addresses. If we understand the neurobiological operational principles, at least in terms of their "philosophy", this can be useful for adaptive perceptual robotic systems of the sixth generation (e.g., see Arbib 1990).

Trying to understand neuroprinciples of behavior, our view is often restricted. Mao Tse Tung (1935) illustrating the limitations of our view of the world used the analogy of a frog sitting in a well: the frog says that the sky does not extend beyond the brim of the well. Experimenter's view can be compared with "key hole peeping", whereby the peephole of the "black box" is defined by the experimental method, in particular if only one technique is applied to understand brain functions. But we can reduce this problem by "multiple hole peeping": the application of different experimental techniques then allows the experimenter a description from various points of view, e.g. morphological, distributed functional, integrative, interactive, connectional, synaptic, dynamic, pharmacological, metabolic, etc. In our laboratory, the following experimental methods are currently in use: (1) Quantitative ethological studies with moving two- or three-dimensional objects or computer generated moving video patterns; (2) regional mapping of energy metabolism with the [^{14}C]-2-deoxyglucose(2DG) method in the behaving animal; (3) extracellular recordings from immobilized animals in response to perimetrically moved two-dimensional objects or computer generated video patterns; (4) intracellular recording and iontophoretic labeling with Co^{3+}-lysine in immobilized animals; (5) ortho- and antidromic electro-stimulation/recording; (6) antero- and retrograde labeling of pathways with horseradish peroxidase, HRP; (7) extracellular recordings in behaving animals; (8) behavioral and neurophysiological studies in brain lesioned animals; (9) pharmacological studies; (10) combination of (1)-(9) whereby the dialogue with modeling computer neuroscientists yields so-to-speak a "cognitive umbrella".

A promising example of model-assisted experimentation is provided by the contribution of Wang (this volume). He instructively shows modeling - supported by computer simulation - as a process of transforming theory derived from data into testable structures. Where empirical data are lacking, or are difficult to obtain because of structural and/or technical constraints, the modeler makes assumptions and approximations that, by themselves, are a source of hypotheses. If the neural model is tied to empirical data, it is used to predict results and hence again becomes subject to experimental tests whose resulting data in turn lead to further improvements of the model. In this process modeling also takes advantage of predictions which subsequently are not confirmed by experimental data. The collaboration between our neurobiological group at the University of Kassel and the neurocomputer science group at the University of Southern California is beginning to show success in various domains of this dialogue (see also Lara & Arbib 1985; Cervantes-Pérez 1989; Lara 1989).

4. Neuroethological Data, Projects, and Questions

Investigating visually guided prey-catching behavior of toads *Bufo bufo*, in the following I shall select some aspects of visuo-motor functions: 1) parallel distributed processing of object features; 2) implication of on- and off-channels; 3) centrifugal control of retinal function; 4) forebrain-loop involved modulation of stimulus-response mediating tectal networks; 5) combinatorial control of bulbar/spinal motor pattern generating systems. Each topic addresses (a) neuroethological data, (b) current neuroethological projects, and (c) questions for neural engineering.

4.1 Parallel Distributed Processing of Object Features

In anuran amphibians (Ranidae, Bufonidae) millions of years ago, eyes had evolved with at least four kinds of filters according to the properties of four classes of retinal ganglion cells R1 to R4 on the input side (Lettvin et al. 1959), as had bulbar/reticulo-spinal systems responsible for the generation of various motor patterns at the output side (Edinger 1908; Rubinson 1968). With respect to Barlow's (1953) "fly" [R3] and Lettvin's "bug" [R2] and "enemy" [R4] detector hypotheses, such sensory systems, in principle, were sufficient to select two kinds of essential behavior: prey-catching to *small* objects and predator avoidance to *big* objects. This concept is brilliant, and one would not be surprised when direct retino-bulbar/spinal connections exist, if there so-to-speak were not brain in between, particularly the diencephalon in German

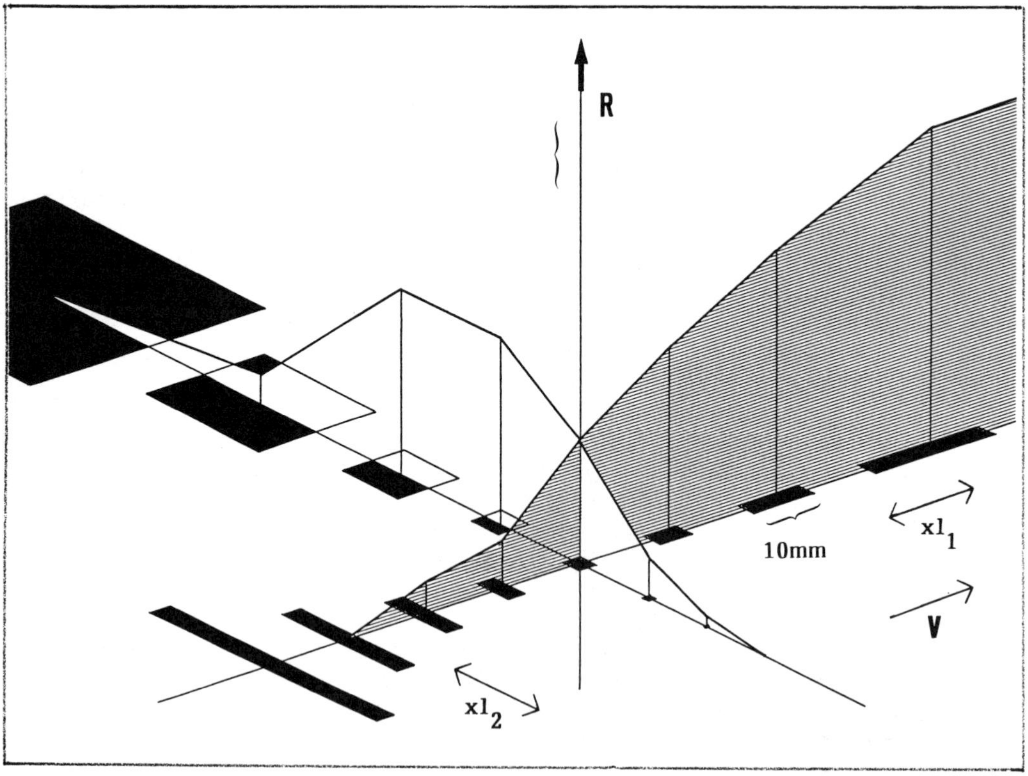

Fig. 1. The species-universal configural prey-schema of common toads *Bufo bufo bufo*. In this diagram, a black rectangular object moving at velocity (v) can be evaluated for its release of prey-catching (R) in comparison with the set of objects it is assigned to, namely by variation of its configural features xl_1 and xl_2 along a continuum. The prey-catching activity R is a measure of resemblance of an object with prey (scale: 5 orienting responses per min). Of particular importance is the algorithm that determines R in response to *variation* of xl_1; xl_2. (After Ewert 1968, 1969).

called *Zwischenhirn* (= "between brain") with its connections to the telencephalic forebrain and to the midbrain tectum/tegmentum, forming a macro-network that analyzes visual features and secures the organism's (re)actions (see also Ewert, Matsumoto, Schwippert & Beneke, this volume).

a) Neuroethological Data

(1) The Feature Assignment Principle

Our quantitative investigations on visually guided prey-catching in toads (Ewert 1968, 1969) have shown that their visual system distinguishes prey, nonprey, and predator not just in terms of "small" or "big", but by an evaluation of moving configural stimuli, i.e. features in relation to each other. The critical features are object expansion parallel to the direction of movement (xl_1), perpendicular to the direction of movement (xl_2), the relation ($xl_1 : xl_2$), and the stimulus area ($xl_1 . xl_2$). Within behaviorally relevant limits of size, $xl_1 > xl_2$ signals *prey*, $xl_1 < xl_2$ *nonprey*, whereas xl_1 ($=xl_2$) $>$ 50 mm *predator* (Fig. 1). If a small stripe is moved in direction of its longer axis ($xl_1 > xl_2$), we refer in laboratory jargon to "worm-configuration", W, and if the longer axis of the same stripe is oriented perpendicular to the direction of movement ($xl_2 > xl_1$) to "antiworm-configuration", A. The preference of W *vs.* A in prey-catching is maintained for variable movement vector, so far the speed is detectable (Ewert et al. 1979; Burghagen & Ewert 1983). This relational principle is also valid for moving textured patterns (Ewert et al. 1970, ref. in Ewert 1984). In addition to this principle, object size also plays a role, whose estimation in terms of "size constancy" (Ingle 1976; Ewert & Gebauer 1973, ref. in Ewert 1984) requires depth measurements for which binocular vision is not essential (see also Collett 1977).

Since adult toads *Bufo bufo*, just captured in the field, all show the same type of pattern discrimination as depicted in Fig. 1, it can be concluded that this ability is *species universal*, by which we denote "universal" to all members of the same species *bufo*. The ability is based either on genetic dispositions whereby all individuals have the same experiences, or on an innate property as developmental studies suggest (Traud 1983). Generally, we can say that increase of xl_2 in particular has a strong inhibitory effect on prey-catching behavior in amphibians, probably meaning to the animal "be cautious" (Ewert & Traud 1979, ref. in Ewert 1984).

(2) Object Movement vs. Self-Induced Movement

The fact that the object features xl_1, xl_2 are related to the image shift across the retina, implies a further pecularity: in a homogeneous untextured environment toads do not distinguish between the retinal image produced by a moving object and the same image induced by the moving toad in front of the stationary object. The origin of the movement is not evaluated in this case (Burghagen & Ewert 1983). Consequently, an actively moving or a passively moved (Fig. 2Ab) toad distinguishes different stationary objects in the same manner as if these objects were moving (Fig. 2Aa). At a first glance this seems to be a drawback, but the conditions for such "confusions" are rare - for example, if a toad would live in a homogeneous untextured environment. The toad's biotope, however, provides a structured stationary surrounding, which leads to different retinal images in cases that an object moves or the toad moves in front of the stationary object. In the former case, a prey-like object is extracted from its background and adequately responded to; in the latter case, the object is masked by the moving background structures due to surround inhibition, whereby the parameter xl_2 plays a dominant role (Ewert et al. 1970, ref. in Ewert 1984). A comparable ethological phenomenon is inhibition of prey-catching if a toad is faced with several simultaneously moving prey objects, the "swarm effect". We conclude that object- and self-induced moving retinal images, here, are not distinguished by a computation based on a "reafference principle" (Holst & Mittelstädt 1950) using proprioceptive information from the animal's movement, as has been suggested by Manteuffel (1989) for salamanders.

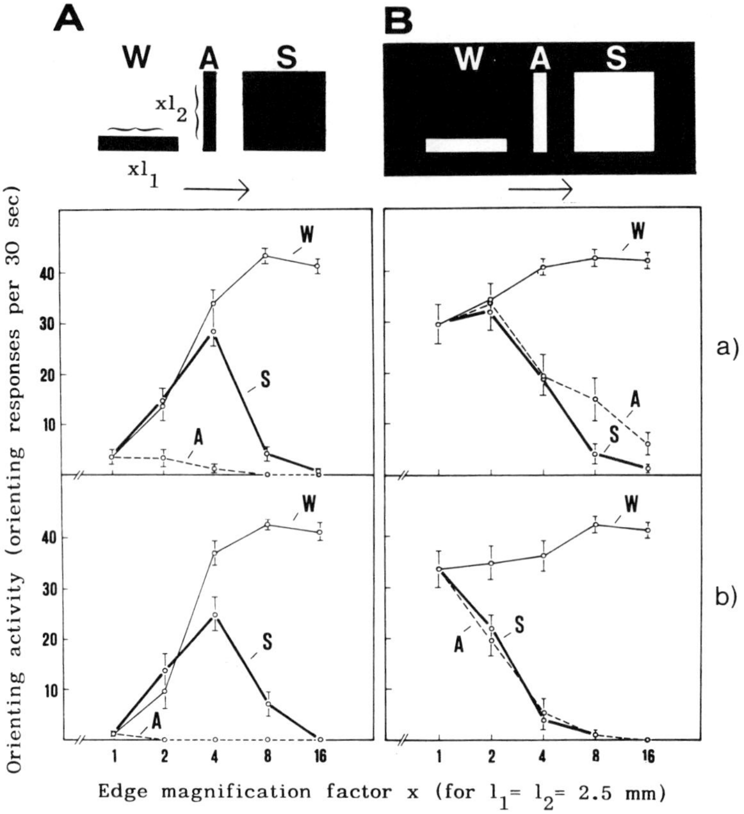

Fig. 2. Discrimination of moving (a) and self-induced moving (b) black objects against a white background (A) or white against black (B). In (a) the object was moved around the toad (sitting in a cylindrical glass vessel) at constant angular velocity of 10 deg/sec. In (b) the object was stationary and the toad itself was moved by rotating (at 10 deg/sec) the platform on which the toad was sitting. (Note that the same results were obtained if the platform rotated stepwise or at variable acceleration to increase vestibular activity). Ordinate: average prey-catching activity R [orienting responses per 30 sec] of 20 toads. Abscissa: edge length of stripes *W* of 2.5 mm constant width xl_2 and different length xl_1 in horizontal direction; stripes *A* of 2.5 mm constant width xl_1 and different length xl_2 in vertical direction; squares *S*, xl_1=xl_2 [*W* refers to worm-configuration, *A* to antiworm-configuration]. Arrow: direction of moving retinal image. (From Burghagen & Ewert 1983).

(3) The Neuronal Network

Retina-filtered visual information provided by retinal ganglion cell classes alone is not sufficient to explain the above described behavioral stimulus-response relationships. The decisive evidence is provided by recordings in freely moving toads. Schürg-Pfeiffer (1989) has shown that the retinal so-called "bug detectors" [class R2] are optimally activated if a small black stripe is moved with its longer axis perpendicular to the direction of movement (providing this axis fits the diameter of the neuron's excitatory receptive field, ERF); but the probability of prey-catching is very low. If this stripe traverses the ERF center in the direction of its longer axis, the R2 neuron's discharge frequency is somewhat less, while the toad readily responds with prey-catching. Furthermore, the inhibition provided by R2 neuron's inhibitory receptive field, IRF, that surrounds the ERF, is not sufficient to explain the algorithm that underlies toad's prey selection. We have found that object-defining features are processed in a parallel/distributed manner in mesencephalic tectal and diencephalic pretectal retinal projection fields (for review see Ewert 1987): Configurally, the feature xl_1 is evaluated in tectal structures and expressed by tectal class T5.1 neurons; xl_2 is evaluated in pretectal thalamic structures and expressed by class TH3 neurons (Fig. 3A) which are also

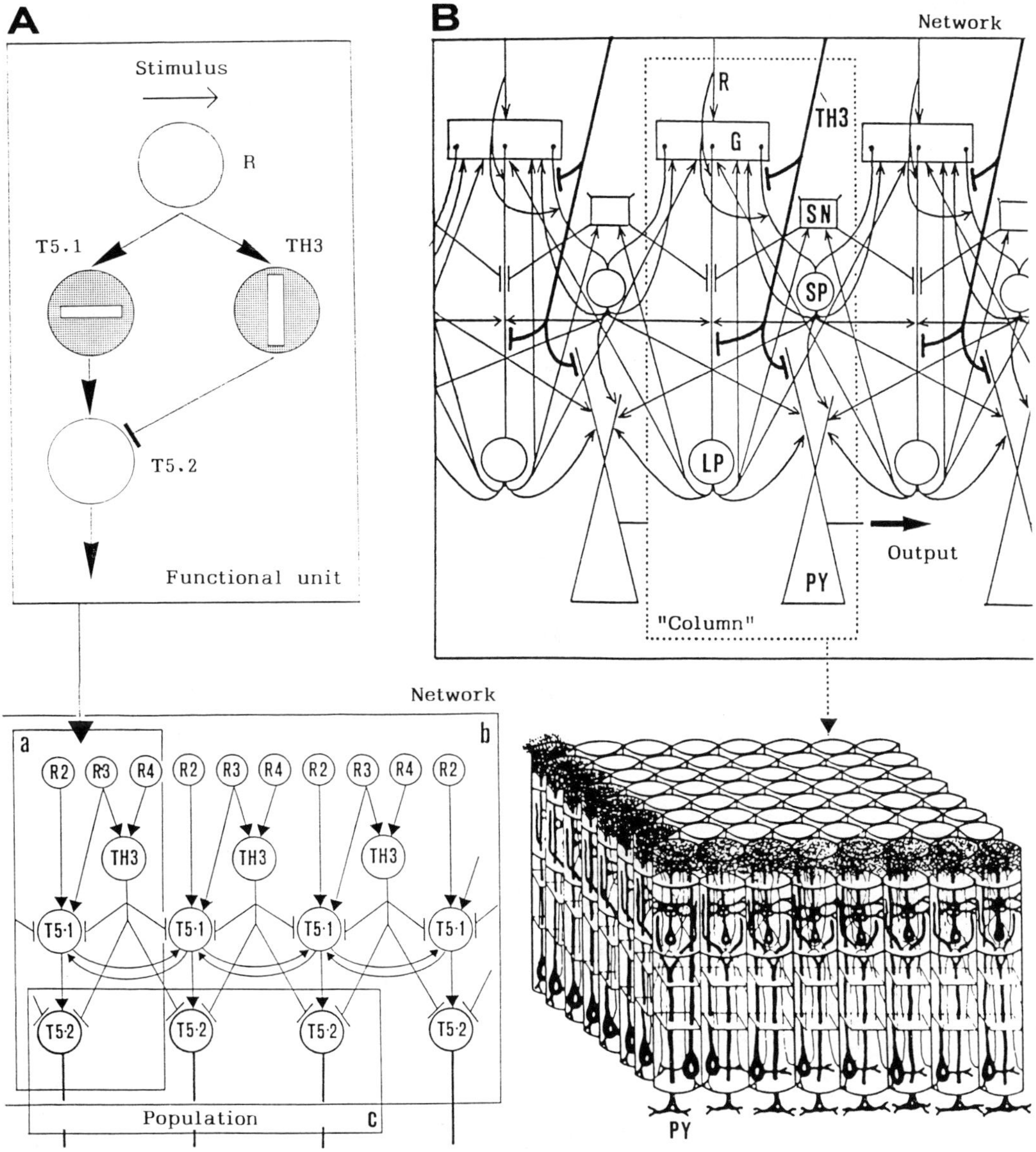

Fig. 3. Models for the discrimination of moving configural objects through analysis in a pretectal/tectal network in toads. A) In a functional unit receiving retinal (R) input, the different kinds of sensitivity of pretectal thalamic (TH3) and tectal (T5.1) neurons to the configural features xl_1 and xl_2 of a stimulus object moving in horizontal direction are illustrated by 'window discriminatiors'; the feature relationships are expressed by the response of T5.2 neurons; arrows indicate excitatory and lines with bars inhibitory connections (from Ewert 1974). Bottom: *Functional unit* (a) integrated in a pretectal/tectal *network* (b) whose output is mediated by a *population* (c) of T5.2 neurons. The lateral excitatory and inhibitory connections are not restricted to immediately adjacent neurons (for a systems theoretical analysis see Ewert & v. Seelen 1974). Inhibitory tectal interneurons T5.3 which in response to TH3 input may inhibit T5.1 and T5.2 (Ewert 1987) are not shown. B) Advanced computer model of the feature-analyzing network: a functional unit (column) - consisting of glomerulus (G), tectal small pear-shaped neurons (SP), large pear-shaped neurons (LP), stellate neurons (SN), and pyramidal neurons (PY) - receives retinal (R) and pretectal (TH3) input at different levels (retinal input to PY not shown here). LP/SP correspond to T5.1-types and PY to T5.2. Below: Array of 8 x 8 interacting columns. (After Lara et al. 1982.)

strongly sensitive to moving big objects and textured surfaces (Tsai & Ewert 1988). The TH3 neurons are located in the lateral posterodorsal (Lpd) and the lateral (but not central) part of the posterior (P) pretectal thalamic nuclei; we call this the Lpd/P region (for anatomy see Neary & Northcutt 1983). These T- and TH-types show no functional anisotropies regarding shape of ERFs and direction of ERF traverses.

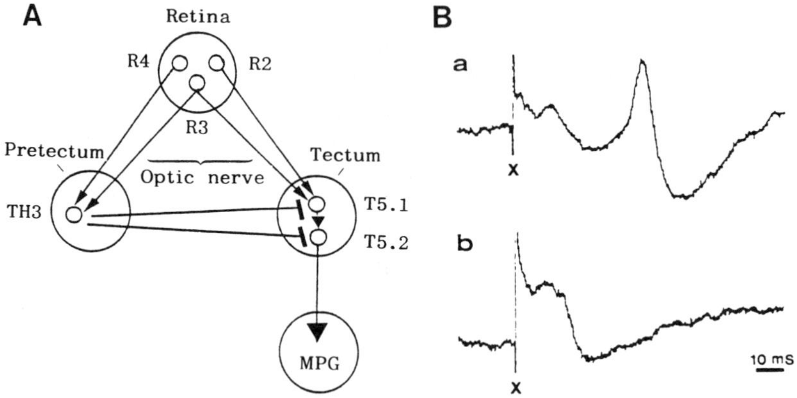

Fig. 4. Evidence of pretecto-tectal postsynaptic inhibitory input. A) Parallel innervations of toad's pretectal Lpd/P region (involving TH3 cells) and optic tectum (involving T5.1 and T5.2 cells) via the optic nerve mediated by R2, R3, and R4 retinal ganglion cell axons, and pretecto-tectal connections; *arrows* denote excitatory and *lines with bars* inhibitory influences. B) Sequential EPSP/IPSP activity of a tectal cell intracellularly recorded to electrical stimulation "x" of the entire contralateral optic nerve (a); the mainly IPSP activity of the same tectal cell to electrical stimulation in the ipsilateral pretectum (b). EPSP, excitatory, IPSP, inhibitory postsynaptic potential; MPG, motor pattern generator. (From Ewert et al. 1990a).

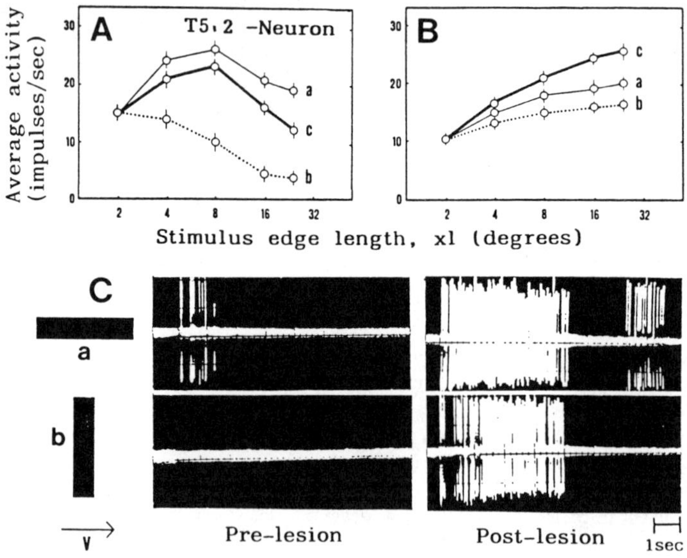

Fig. 5. A) The activity of common toad's prey-selective tectal T5.2 neurons (n=20) reflects the probability that the configuration of a moving visual object fits prey (cf. Fig. 1); *a*, black stripes of variable legth xl_1 (worm-configuration); *b*, stripes of variable length xl_2 (antiworm-configuration); *c*, squares of different size. B) Impairment of toad's T5-neurons' object discrimination after ipsilateral direct-current induced pretectal lesions; averages of n=20 neurons. (From Ewert & v. Wietersheim 1974). C) Records from a grass frog's T5.2 cell in response to stripes moving (*v*) in worm-configuration (*a*) or in antiworm-configuration (*b*) before and after an ipsilateral pretectal lesion applied by injection of kainic acid.

The above described visual perceptual properties for prey-catching can be explained basically by tectal intrinsic mutual excitation and subtractive network interactions in terms of pretecto-tectal inhibition; this computation is expressed in particular by class T5.2 neurons that are integrated in the retino-pretectal/tectal network (Fig. 3). Experimentally, we have provided evidence of pretecto-tectal inhibitory connections with the aid of various stimulation, recording, and lesion methods, most recently by intracellular recording inhibitory postsynaptic potentials, IPSPs, of tectal cells in response to pretectal stimulation (Fig. 4b). The activity of T5.2 cells in response to configural moving objects reflects the probability that the object fits the prey-schema (cf. Figs. 1 and 5A); after pretectal lesions, tectal neurons lose their configural object discrimination (Fig. 5B,C), and the animal's prey recognition becomes agnostic, i.e. prey, predator or moving background structures are all responded to with prey-catching (Ewert 1968, reviews see 1984, 1987; Schürg-Pfeiffer 1989). As with Jung's (1953) analogy, after release of pretecto-tectal inhibition each visual moving retinal image drives the tectal mutually-excitatory "powder keg" to explosion (see Fig. 5C, post-lesion), expressed by disinhibited prey-catching behavior (see also Ewert, Matsumoto, Schwippert & Beneke, this volume).

In freely moving toads (in contrast to pharmacologically immobilized ones) T5.2 neurons display a sensitivity to the real size of objects moving in an investigated range of 3.5 to 25 cm from the animal (Schürg-Pfeiffer, Spreckelsen & Ewert 1990). It is suggestive, but not yet proved, that "size-constancy" in prey-catching behavior results from object distance dependent network interactions in terms of pretecto-tectal inhibition. We hasten to admit that besides the classes TH3, T5.1, and T5.2, other feature sensitive neurons exist (Ewert 1984, 1987) whose properties result from various kinds of computation in the pretectal/tectal network system. Furthermore, feature sensitive/selective cells may be modulated by pretectal or hypothalamic influences involving telencephalic structures (see also contributions by Ewert, Matsumoto, Schwippert & Beneke and Merkel-Harff & Ewert, this volume).

b) Current Neuroethological Projects

In the following, we list questions for experiments which will shed further light on the functional neuro-architecture responsible for object discrimination:

1) Simultaneous intracellular records with two microelectrodes from a pretectal thalamic TH3 and a tectal T5.2 neuron in response to stripes moving in prey and nonprey configuration, respectively: question of temporal correlation between pretectal and tectal responses in terms of excitatory (EPSPs and spikes) and inhibitory (IPSPs) activities.

2) Intracellular recording TH3 putative pretecto-tectal projection neurons in their antidromic responses to tectal stimulation and subsequent iontophoretical labeling the neuron with Co^{3+}-lysine: question of pretectal axonal projection patterns.

3) Simultaneous recordings from TH3 and T5.3 neurons to prey- or nonprey-like moving stripes: question of direct or T5.3-relayed pretectotectal inhibition.

4) Intracellular records of tectal neurons to moving visual objects after pretectal ablation: question of accumulating EPSPs in favor of IPSPs.

5) Neuropharmacological studies of pretecto-tectal projection neurons: e.g., question of GABA-mediated neurotransmission.

6) Drawing on the phenomenon that lesions to the anterior dorsal thalamus (including the nucleus of Bellonci) have a similar prey agnostic effect like pretectal thalamic lesions - which, however, is restricted to the frontal visual field - connectional studies by means of electro-stimulation/recording and labeling methods are required: question of excitatory anteriorthalamo-pretectal and/or inhibitory anteriorthalamo-tectal pathways.

7) Measuring activities of pretectal neurons in response to objects at various distances in behaving toads: question of depth-dependent properties.

8) Investigation of time courses of functional recovery of object discrimination after anteriorthalamic and/or pretectal lesions of different sizes: question and evaluation of self-organizing processes.

9) Recordings in young postmetamorphic toads: question of T- and TH-neuronal properties.

c) Questions for Neural Engineering

1) With respect to neuro-computing, the question arises whether the described behavioral data - including "size constancy" - can be all simulated by one extended model (Arbib 1982; Cervantes-Pérez 1989), simultaneously asking for predictions in terms of feedback for new experiments.
2) An engineer would solve the problem of direction-invariant discrimination of a stripe whose longer axis is oriented either parallel or perpendicular to the direction of movement *explicitly* with the aid of a huge number of serially connected asymmetric processors working like orientation ("line") detectors. The toad's brain performs such a task economically by *implicit* computing (Stevens 1987) using symmetric processors (TH3, T5.1, and T5.2) characterized by approximately circular ERFs without IRFs and lacking directional selectivities. The required anisotropy draws on the relational property of the stimulus features xl_1 and xl_2 and their parallel processing in connection with subtractive interaction. We ask whether the "philosophy" of *implicit computing* of object features may be useful for certain tasks in perceptual robotics.
3) With respect to (2), we have learned from ethology that objects are not necessarily recognized *in toto*, rather by certain features in their spatial or spatio-temporal relations, and this is - within limits - size-invariant. Regarding machine vision, we ask whether the description of an object in terms of feature relationships may have advantages compared to the current evaluation of an object in terms of pixels. Approaches using vector computation may be one promising approach.
4) Pattern recognition in machine vision based on specific detectors requires an enormous expense in configuring computing, which calls for unique functional architectures that are flexible for changes in goal oriented tasks. In this context, we ask how models deal with object recognition by *cell population analyses* or with dynamic principles such as provided by the *modulator hypothesis* (Betts 1989).

Fig. 6. Rubin's "Face/Vase" psychophysical phenomenon.

4.2 Implication of On- and Off-Channels

The classical example of Rubin's (1915) "Face/Vase" psychophysical phenomenon shows us that our visual system at black/white contrast borders gets into a conflict regarding the interpretation of object and background. The picture in Fig. 6 can be interpreted either as *face silhouettes* facing each other against a white background or as a *vase* against a black background, whereby the former is often preferred. In some respect, a comparable figure/ground confusion can be obtained in prey-catching toads in their responses to a black prey-like stripe moving against a white background or a white one moving against black.

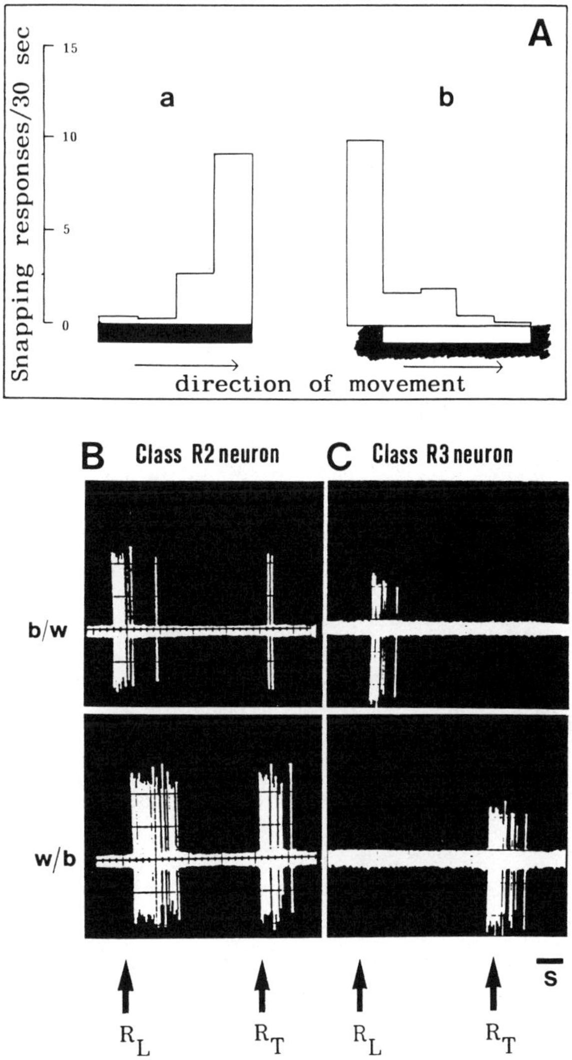

Abb. 7. Contrast-direction dependent edge preferences. A) Head preference (a) in the common toad's snapping response to a black stripe moving prey-like against a white background; the "head" is determined by the stripe's leading edge; arrow: direction of movement. (b) Tail preference in response to a white stripe of the same size moving against a black background; the tail is determined by the stripes's trailing edge. This phenomenon is independent of the direction of movement. (From Burghagen & Ewert 1982). B) Traversing the center of a retinal class R2 neuron's excitatory receptive field, ERF, a black stripe against a white background (b/w) or a white stripe against black (w/b) elicits stronger spike activity at its leading edge (R_L) *vs.* the trailing edge (R_T). [In order to check - here not shown - whether the relatively weak spike activity, R_T, is due to neuronal adaptation after the response to the leading edge, the stripe was stopped for 60 sec before the trailing edge traversed the ERF; then it was found for b/w and w/b that R_L is not statistical significantly different from R_T]. C) In the non-adaptating class R3 neurons, it was found that $R_L > R_T$ for b/w and $R_T > R_L$ for w/b. (From Tsai & Ewert 1987).

a) Neuroethological Data

(1) Contrast-Direction Dependent Edge Preference

Ingle & McKinley (1978) have shown that marine toads preferentially orient, fixate, and snap toward the leading edge of a black prey-like stripe moving against a white background. Whereas Ingle referred to an invariant "head" preference phenomenon, we found in common toads that if the contrast direction between object and background is reversed, the toad predominantly or even exclusively snaps toward the trailing edge of the stripe: the animal in particular fixates the black background at the moving contrast border as if this were the leading edge of the stimulus object (Fig. 7Ab). The significant features of the targets in both cases (Fig. 7Aa,b) are off-effects at a moving contrast border. (For details see Burghagen & Ewert 1982).

(2) Information Provided by Retinal Ganglion Cells

The retinal ganglion cell classes R2 and R3 provide the main visual inputs to the pretectal/tectal network responsible for configural object discrimination. In the ERFs of these cells, on- and off-patches in different proportion are irregularly distributed (Grüsser & Grüsser-Cornehls 1976). Brisk onset or offset of diffuse retinal illumination leads in class R2 neurons to a short on-response, whereas class R3 neurons show either a short off-burst or a strong off- and a weak on-burst. Traversing the ERFs of these neurons with a black stripe (length about twice the ERF diam.) against a white background or with a white stripe against a black background (Tsai & Ewert 1987), the class R2 neurons respond more strongly to the leading edge, due to neuronal adaptation (Fig. 7B b/w and w/b). The non-adaptating off-dominated class R3 neurons, in contrast, respond in both cases predominantly (or exclusively) to the edge producing an off-effect (Fig. 7C), quite comparable to tectal T5.2 neurons and toad's prey-catching. We suggest that R2 neurons contribute information to tectal networks mainly about the length of relatively small objects, while R3 neurons contribute information about off-positions at contrast borders of small or large objects (see also Barlow 1953). Comparing these neurophysiological results with prey-catching behavior, it is evident that not only R2 but also R3 neurons feed significant excitatory input to the tectal prey-catching releasing systems.

(3) Implications for Configural Object Discrimination

The off-preference at a moving object's contrast border has an impact for the discrimination of moving configural objects, such as squares or stripes of different orientation to the direction of movement: white objects against black background are less acurately discriminated in prey-catching than black objects against white. More specifically, because in response to white objects toads are mainly judging the black area behind the object: (i) white stripes oriented perpendicular to the direction of movement ($xl_2 > xl_1$) are largely confused with white squares ($xl_1 = xl_2$) of a comparable extension xl_2 (Fig. 2Ba; *A*, *S*); (ii) extending a small white area in the direction of movement does not remarkably increase the prey-catching activity, so that prey-like stripes of different length are not well distinguished (Fig. 2Ba; *W*); in addition, this may be influenced by a ceiling effect due to the already relatively strong releasing value of small white objects, compared to black ones of a comparable size and the same amount of the object/background contrast. Interestingly, for humans a white object against a black background looks bigger than the same object with reversed contrast direction, a phenomenon which is explained by "optical irradiation".

b) Current Neuroethological Projects

The fact that toads, during prey-selection, take significant advantage of visual information provided by "off-channels" raises a variety of questions:

1) Previous studies on prey-catching have shown that the preference of small white moving objects against a black background *vs.* black against white is pronounced in the toad's main hunting season, however, reversed toward the hibernation season. A comparable phenomenon has been obtained in class R2 neuron's responses (Ewert & Siefert 1974, ref. in Ewert 1984); furthermore, studies in R2 neurons have shown a gradient of w/b sensitivity in the toad's ventrodorsal visual field. This raises, for example, questions regarding seasonal dependent preferences of moving contrast edges.

2) Since this perceptual phenomenon (1) is coincident with the occurrence of male's pigmentation of the thumbs of their forelimbs, introducing breeding motivation, studies on contrast-direction dependent phenomena after application of sex-hormones will be conducted. Furthermore, comparative studies in juvenile toads are required.

3) Regarding a plasticity of retinal functional properties, both the origin and the meaning of efferent or systemic control systems must be investigated.

c) Questions for Neural Engineering

The phenomenon of contrast direction influenced object discrimination - which seems to be common in visual systems of vertebrates including humans - raises a variety of questions if machine vision is to take advantage of principles from vertebrate vision:

1) Do current models of the anuran's object discrimination (Arbib and co-workers) consider the significance of off-channels and, if so, do the computer simulations accord with the neuro-behavioral data? Which predictions does the model provide? (For an example see the contribution by Wang, Arbib & Ewert, this volume).

2) Is contrast-direction sensitive object discrimination neurophysiologically an "artifact" or is it meaningful? In the former case, models should be developed that avoid this problem for machine vision. In the latter case, models should explain the advantage.

3) The flexible contrast-direction dependent discrimination of moving objects for a dorsal-ventral gradient in the toad's visual field calls for an explanation in terms of a principle for task-specific adaptation useful in perceptual robots.

4.3 Centrifugal Control of the Retina

The neural network of the retina is, from a developmental point of view, a protuberance of the diencephalon. Hence, it is not surprising that on the one hand pre-processed visual information in terms of retinofugal afferents is fed to various diencephalic and mesencephalic structures (e.g., Lázár 1971, 1989; Fite & Scalia 1976), and that on the other hand the retinal network obtains retinopetal efferents (Maturana 1958).

a) Neurobiological Data

(1) Origins of Retinopetal Projections

In the vertebrate literature there are several reports pointing to projections to the retina from different brain areas (for a review see Uchiyama 1989). Among lower vertebrates, in cyclostomes, teleosts, and sauropsids there are projections from the isthmo-optic nucleus - the ION-type retinopetal system - that itself obtains tectal input. In salamanders, Fritzsch & Himstedt (1981) have described retinopetal neurons in a pretectal area. In frogs retinopetal neurons using a substance P related neuropeptide were found in a septal area, whereas retinopetal neurons probably using a gonadotropin-releasing hormone seem to have their origin in the preoptic area (Uchiyama et al. 1981, 1988).

(2) Retinal Neuromodulation

The fact that dopamine plays a role in the retinal network as the neurotransmitter of certain interplexiform and amacrine cells or as a modulator of spatial summation in horizontal cells (Bodis-Wollner 1990; Yasui, pers. comm.) suggests flexibility in retinal information processing. This is consistent with the idea that retinal ganglion cells in their various properties display to some extent a continuum, so far R1 and R2 neurons and the various subtypes of R3 and R4 neurons are concerned (for a discussion see Gaillard, this volume). The suggestion of flexibility in retinal networks is also consistent with recent findings in our laboratoy (Glagow 1990) showing that systemic application of the dopamine agonist apomorphine dosage-dependent selectively influences properties of R2 neurons of the retino-tectal projection: the discharge frequency in response to a moving visual stimulus generally increases, the ERF size increases, and correspondingly these neurons are optimally tuned to bigger objects. (For a review on dopamine neurons in retinal function see Bodis-Wollner 1990).

b) Current Neuroethological Projects

The finding that neurons from different brain structures project to the retina - involving different neurotransmitters - points to various types of modulatory processes whose behavioral significance is largely unknown. Let me select three topics of our main interest:
1) The retinopetal projection from the preoptic area should be investigated with respect to seasonal changes of contrast-direction dependent preferences for moving objects.
2) To investigate retinopetal influences, retinal ganglion cell properties will be studied in toads after lesions to septal, preoptic, pretectal, or isthmic nuclear areas, and in forebrain-less preparations.
3) The fact that among retino-tectal projection neurons, class R2 can be "pushed" pharmacologically in their visual activity is a useful tool to evaluate the function of these neurons for visuomotor functions. Studies in our laboratory show that toads after systemic application of apomorphine exhibit dosage-dependent an increase of snapping- *vs.* orienting-activity, a decrease in the snapping latency, and a retarded habituation.

c) Questions for Neural Engineering

Since modulations of retinal properties are as yet not considered in models of visual object discrimination in anurans, questions of their significance for network models arise:
1) Implementing retinopetal modulatory functions into the current model, which predictions can be made for visual information processing in the pretectal/tectal network?
2) What may be the advantage of modulating network properties at "low" (retinal) *and* at "high" (pretectal/tectal) levels of integration for visual pattern discrimination and the release of prey-catching?
3) Regarding (1) and (2), can we derive from multilevel modulation of neural networks a sort of "philosophy" in terms of *modulation patterns* suitable for implementation in perceptual robots to adapt their pattern recognition for task-oriented operations?

4.4 Forebrain-Mediated Modulation of the Pretectal/Tectal Network

From a historical point of view, there are interesting notions about the relevance of the telencephalic forebrain of "lower vertebrates" claiming that it can be entirely removed without obvious behavioral changes: in 1888, the anatomist Steiner remarked that the telencephalon of bony fish is a "Form ohne Inhalt" (volume without contents); the physiologist Schrader (1887), investigating visually guided behavior in frogs, came to a comparable conclusion. With the improvement of surgical techniques, Blankennagel (1931) then clearly demonstrated that frog's prey-catching behavior failed to occur after bilateral telencephalic ablation. In toads, we have shown that prey-catching activity progressively decreased with bilateral telencephalic ablations in a rostro-caudal direction (Ewert 1965, ref. in 1984). More specifically, after unilateral telencephalic ablation, (i) toads ignored prey only toward stimuli presented to the eye contralateral to the ablated side, suggesting interruption of ipsilateral information flow between forebrain and tectum; (ii) prey stimuli presented to the ipsilateral eye in the same preparation elicited prey-catching but at lower response frequency, suggesting that information transfer between both brain hemispheres (e.g., mediated via the anterior commissure) is required for normal prey-catching activity. Since we have shown that visual prey-catching in telencephalic bilaterally ablated *and* pretectally lesioned toads is hyperexcited (Ewert 1967), it is obvious that telencephalic structures are no serial components of the visuomotor pathway.

a) Neuroethological Results

Sophisticated neuroanatomical studies in anurans by means of modern techniques involving degeneration, horseradishperoxidase HRP, Co^{3+}-lysine tracing, and immunocytochemical studies (e.g., see

Table 1. Qualitative survey of experimental paradigm-related changes of glucose utilization across various brain structures vMP to OT in naive [Nav], prey-habituated [Hab], and prey/predator conditioned [Cond] toads *Bufo bufo*. Glucose utilization was determined in relation to a "neutral" reference structure. In [Hab] and [Cond], the *change* of [^{14}C]-2DG uptake was calculated in brain-to-brain comparison with [Nav]. In [Nav], the [^{14}C]-2DG uptake was measured in brain-to-brain comparison with visually unstimulated toads. An *increase* "+" of [^{14}C]-2DG uptake (in comparison to controls) is interpreted as an increase of neuronal activity; *no change* "0" means no evaluable change of activity; a *decrease* "-" suggests inhibition. Furthermore, we must consider that the [^{14}C]-2DG uptake represents the entire sum of the metabolic activity in neuron populations of (or within) a brain structure. Abbreviations: Telencephalon: vMP, posterior ventral medial pallium; AL, nucleus amygdalae pars lateralis; vSTR, posterior ventral striatum. Diencephalon: dHYP, dorsal hypothalamic nucleus; La, lateral anterior thalamic nucleus; Lpd, pretectal lateral posterodorsal thalamic nucleus; Lpv, pretectal lateral posteroventral thalamic nucleus; P, pretectal posterocentral thalamic nucleus. Mensencephalon: OT, optic tectum, ventral-cental layers. For quantitative results and methodological details see Finkenstädt & Ewert 1988a,b; Finkenstädt 1989a,b; Finkenstädt et al. 1985, 1986).

			[^{14}C]-2DG uptake in:							
Animal	Stimulus	Behavior	vMP	AL	vSTR	dHYP	La	Lpd	Lpv	OT
Nav	W	prey-catching	-	0	++	(+)	++	(+)	(+)	++
Nav	A	no response	0	0	+	(+)	+	+	(+)	+
Nav	S	avoidance	-	0	++	(+)	+	++	(+)	+
Hab	W	no response	++	0	-	++	+	0	0	-
Cond	S	prey-catching	++	++	0	++	+	-	++	++

Halpern 1972; Kicliter & Ebbesson 1976; Northcutt & Kicliter 1980; Wilczynski & Northcutt 1977, 1983a,b; Lázár 1989) revealed a wealth of connections between telencephalon and optic tectum, mediated via diencephalic structures, which suggests various loop operations (Ewert 1987). Their investigation requires a method that allows one to test the activity patterns in these brain structures during the release of behavioral responses. The [^{14}C]-2DG-technique is an adequate tool to monitor the pattern of energy metabolism throughout the brain during a test period with a visual moving object in naive toads (monitoring species-universal properties) and in conditioned toads (monitoring experience-related changes). Table 1 shows examples of changes of glucose utilization in some brain structures mapped in response to *prey* (black stripe moving prey-like in worm-configuration, W), *nonprey* (same black stripe moving so-to-speak in antiworm-configuration, A), and *predator* (moving large black square, S) in naive [Nav] toads, prey-habituated [Hab] toads (non-associative learning), and prey/predator conditioned [Cond] toads (associative learning to treat a predator stimulus as prey). Based on the data partly obtained from behavioral investigations, [^{14}C]-2DG studies, pathway tracings, single neuron recordings, and lesion experiments, we hypothesize loops with different properties (arrows indicate excitatory influences and lines with cross bars inhibitory influences):

(i) In *naive animals*, for the visual release of prey-catching the following loop operation may control the retino(R)-pretectal(Lpd)/tectal(OT) network in a manner of priming and arousing: in response to retinal (R) input, the tectal (OT) prey-catching releasing system is inhibited by Lpd. The system can be primed/aroused by the telencephalic caudal ventral striatum vSTR which inhibits Lpd in response to tectal information mediated via diencephalic structures, such as the lateral anterior (La) thalamic nucleus:

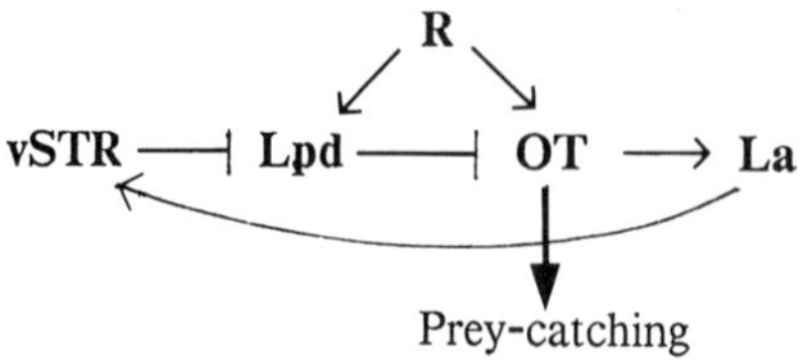

Lesion test. Bilateral OT ablation: visual prey neglect; bilateral vSTR-lesions: visual prey neglect; bilateral Lpd-lesions: binocular disinhibition of visual prey-catching; unilateral Lpd lesion: disinhibition of prey-catching toward any objects moving in the visual field of the eye contralateral to the lesion; combined vSTR/Lpd-lesion: disinhibition of prey-catching comparable to the behavior after Lpd-lesion; vMP-lesions: normal prey recognition (for details see Ewert, Matsumoto, Schwippert & Beneke, this volume).

(ii) For *prey-habituation* (Ewert & Kehl 1978), the following loop is suggestive: with the repetitive presentation of the same prey stimulus, the telencephalic posterior ventral medial pallium vMP (in response to visual information mediated via anterior thalamus AT) decides about the novelty character of the stimulus by comparing its cues with stored ones. In case of persisting sameness, vMP via dorsal hypothalamus (dHYP) inhibits the tectal prey-catching releasing system:

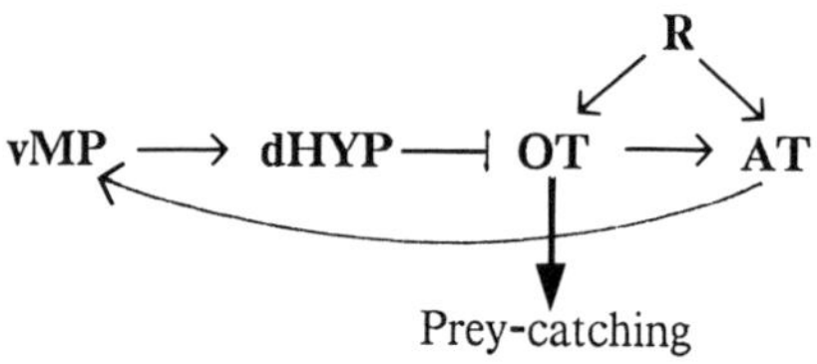

In case of newness - defined by certain "novel" cues of a prey stimulus - vMP is not activated (cf. Table 1) and prey-catching proceeds (for details see Wang, Arbib & Ewert, this volume).
Lesion test. Bilateral vMP-lesions: weak prey habituation in naive toads and dishabituation in pre-lesion habituated ones; bilateral AT-lesions: weak prey habituation in naive animals (Ewert 1984; Finkenstädt & Ewert 1983, 1988a).

(iii) For *learning to associate* a visual predator stimulus with prey (Brzoska & Schneider 1978; Burghagen 1979), the following loop is suggestive: in the course of repetitive feeding a toad with a mealworm (prey) presented with the experimenter's moving hand (predator), the vMP - informed by OT via AT - takes part in prey/predator association. As a consequence, vMP becomes sensitive to the hand or to other large moving objects and facilitates prey-catching by an inhibitory influence on Lpd. The telencephalic pars lateralis of the nucleus amygdalae (AL) - connected to Lpd - is not yet considered:

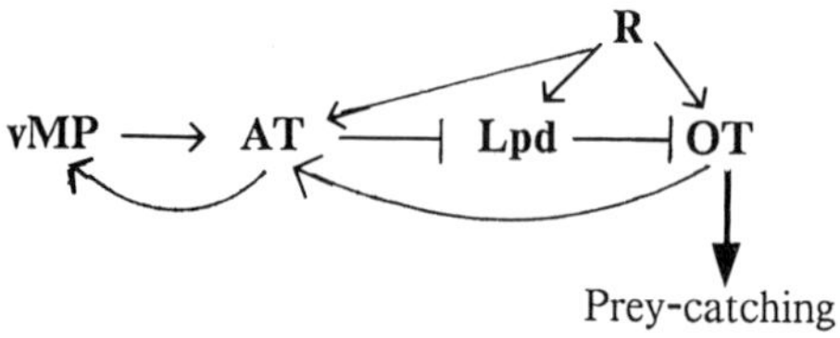

Lesion test. Bilateral vMP-lesions: no associative learning in naive toads and naive responsiveness in trained toads (for details see Finkenstädt & Ewert 1988b; Finkenstädt 1989b).

Summarizing the results, we envision the integrative properties of a "minimal network" for visual information processing in prey-catching as follows:

1) The optic tectum disconnected from the telencephalon and the diencephalic (extrahypothalamic) forebrain elicits prey-catching in response to retinal input from any moving stimulus, due to tectal intrinsic mutual excitatory processes, reminding Jung's "powder keg" analogy.
2) The inhibitory input to the tectum from the pretectal (Lpd) thalamic nucleus in response to retinal information limits the spread of tectal self-excitation and allows the animal visually to distinguish between moving configural stimuli and to select prey from nonprey.
3) Controlling Lpd by inhibitory influences from the ventral striatum primes and arouses visual prey-catching (see Ewert, Matsumoto, Schwippert & Beneke, this volume).
4) During prey/predator conditioning, vMP-mediated "associative inhibition" of Lpd modifies the prey-recognition system - as after pretectal lesions - in a manner of prey "generalization".
5) In the course of prey habituation, vMP-evalutated sameness of a repetitively presented stimulus leads via dHYP to a stimulus-specific inhibition of the tectal prey-catching releasing system in a manner of prey "specification" (see Wang, Arbib & Ewert, this volume).
6) The three discussed loops cannot be adequately evaluated in isolation as scetched here; for example, we have to consider that the loop-operation involving STR is integrated in loops that take advantage of the functions of vMP.
7) Various forebrain structures (e.g., vMP, vSTR, dHYP, P, AT) are partly involved in other sensory detection and learning processes (see also Ingle, this volume).
8) Other types of learning (e.g., associating visual and olfactory cues in prey-catching) involve further structures and loop operations (see Merkel-Harff & Ewert, this volume).

b) Current Neuroethological Projects

The loop operations suggested here require further justification by the application of specific experimental techniques:

1) Regarding a priming/arousing striato-pretecto-tectal disinhibitory pathway, intracellular recording/labeling studies are suitable to examine whether disinhibition is brought about by inhibitory modulation of tonic pretecto-tectal inhibition (see Ewert, Matsumoto, Schwippert & Beneke, this volume).
2) In connection with (1), intracellular recording/labeling a pretectal thalamic TH3 neuron in response to striatal stimulation (orthodromic) and subsequently to tectal stimulation (antidromic) and tracing the recorded neuron's fibers will contribute information about its integration in the global neural network. Pharmacological studies on striato-pretectal and pretecto-tectal neurotransmitters are the next step. Furthermore, the function of the entopeduncular nucleus relaying between vSTR- and Lpd-neurons must be carefully examined.
3) The question of "interocular" transfer after monocular prey/predator association learning has to be investigated from different points of view: a) comparison between object discrimination with the "trained eye" and the "naive eye", respectively; b) study of interhemispherical asymmetries in [^{14}C]-2DG uptake; c) investigation of effects of transections of commissures (e.g., anterior, posterior, postoptic, and tectal commissures) pre- and post-learning, respectively, in connection with [^{14}C]-2DG studies.
4) Using brain "landmarks" from [^{14}C]-2DG studies, recordings in freely moving toads in response to a conditioned stimulus during short-term learning will contribute to the question of associative (e.g., Hebbian) synapses. In this context, we ask to which extent structures of the primary visual pathway *and* structures extrinsic to it are involved in associative learning processes.
5) Investigations of other learning paradigms (e.g., visual/olfactory association; see Merkel-Harff & Ewert, this volume) will ask for nuclei that are generally involved in learning and others that may be specific to certain learning paradigms, whereby (3) and (4), too, are of special importance.
6) Regarding stimulus-specific prey habituation, we will record from relevant nuclei (e.g., vMP, AT, OT) to "trace" the phenomenon of stimulus-specificity and to elucidate networks involved in the storage and in the comparison of information related to visual cues. Regarding AT, in this context it is important

physiologically to distinguish between the dorsal anterior thalamic nucleus and the nucleus of Bellonci. For example, our current hypothesis suggests that in the naive toad structures of AT exert excitatory influences on Lpd, but (other structures of AT) under certain conditions of associative learning may inhibit Lpd. Furthermore, questions of interocular transfer in habituation will be studied.
7) Habituation of the "release call", displayed by males mainly in the breeding season in response to repetitive clasping of their lateral body flanks, will be investigated comparably.
8) The time course of extinction following associative and non-associative learning, as well as the question of recovery of learning following brain lesions (e.g., vMP) are necessary to examine processes of functional recovery and self-organization.

c) Questions for Neural Engineering

The neuroethological data and suggestions concerned with arousing visuomotor activity and with associative or non-associative learning processes in object discrimination, raise questions both for modeling and neural engineering:
1) Do the present models consider parallel-distributed and loop-operated information processing regarding stimulus features and their evaluation under the various aspects of associative and non-associative learning? In this context, the advantages of three possible principles should be evaluated: (i) in connection to the primary sensory pathway, learning proceeds *extrinsically* and modulates/modifies the mode of sensory information processing; (ii) learning proceeds *intrinsically* in the sensory information processing network; (iii) combination of (i) and (ii). Which predictions can be made for decisive experiments? Which kinds of network properties allow the storage of visual information and the comparison between incoming and stored signals?
2) Are principles derived from (1) useful for perceptual robotics in terms of adaptive flexibility of object discrimination in association with *different coincident* sensory stimuli or in specific adaptation to the features of *same sequential* sensory stimuli?
3) How can the operational principle of a striato-pretecto-tectal priming/arousal mechanism be integrated in the current model of object discrimination and its modification through learning under the aspect of sharing certain neural structures for different task-related computations? Can a "philosophy" derived from such a principle be suitable for perceptual robots?

4.5 Combinatorial Control of Motor Pattern Generating Systems

The classical idea of command neurons (Wiersma & Ikeda 1964) suggests that behaviorally relevant sensory information is filtered and fed to specific command neurons that trigger a corresponding motor pattern generator. The important idea behind this concept is still valid if the following additional criteria are considered: (i) *command systems* (Kupfermann & Weiss 1978) may consist of collectively operating neurons (command elements) of the same or similar filter type, whereby these neurons are integrated in a network and thus express in their responses certain network computations; (ii) *coded commands* in the manner of "command releasing systems" (Ewert 1980, 1987) take advantage of different types of command elements that evaluate different aspects of an object, such as configural features and position in space; (iii) *cooperation of local distributed* different command elements may be required for certain tasks (DiDomenico & Eaton 1987); (iv) *state-dependent* inputs related to arousal, attention, and motivation are integrated; (v) *feedback* to command systems from the motor systems they excite may be essential in certain cases (Davis & Kovac 1981); (vi) *economy* in terms of "neural parsimony" probably plays a role (Comer 1987) whereby certain command elements are shared by different command releasing systems; (vii) *safety* is taken into account in such a manner that the same motor pattern can be triggered by different command releasing systems involving different sensory modalities such as vision or touch (Grobstein et al. 1983); (viii) *flexibility* in terms of adaptation of a motor pattern to changing external conditions can be achieved, for example, by loop-operated associative changes in the feature processing network structure that determines the properties of a feature-analyzing command element (Ewert 1987), or by level-settings of command elements, or by tuning of modulatory command elements (Kupfermann & Weiss 1978).

Regarding such an "evolving command concept" (Eaton 1983), the operations between sensory analysis and motor responses represented by *sensorimotor codes* (Ewert 1987) are in a broad sense related to what we call cognition. Paillard (1987) points out that the tuning of "sensorimotor instruments" can be regarded as a primordial form of an organism's adaptation to environmental constraints which had prepared the way for the emergence of cognitive instruments dealing with the control of perception and action.

a) Neuroethological Results

(1) Command Signal Carrying Tecto-Bulbar/Spinal Cells

In toads, the command concept is supported by experiments in which electro-stimulation of tectal or pretectal thalamic areas elicits prey-catching (orienting, snapping) or avoidance behaviors (turning away, ducking, jumping), respectively (Ewert 1967c, ref. in 1984; Ewert et al. 1983). Tecto- and pretecto-bulbar/spinal projections have been shown by anatomical tracing studies using HRP (Weerasuriya & Ewert 1981; Ingle 1983) and Co^{3+}-lysine (Tóth et al. 1985). The descending character of certain physiologically classified tectal cells (such as T5.2) could be determined by antidromic activation of these cells in response to electro-stimulation of tecto-bulbar/spinal tracts in the medulla oblongata (Satou & Ewert 1985). A contribution of as yet hypothetical tegmental neurons, assumed to precisely describe the object position in space, is discussed by Grobstein (this volume).

(2) Premotor/Motor Network

Extracellular recordings (Ewert et al. 1990b) in comparison with results from intracellular recording/labeling studies (Schwippert et al. 1990) of the medullary medial reticular formation, MRF, have shown cells that display properties of tectal projection neurons in various combination together with medullary (bulbar) characteristics (Fig. 8). This is consistent with the concept of "sensorimotor codes" (Ewert 1987). Some of these cells, especially adjacent to or within the branchiomotor column, display a special property in that visual input (e.g., from a prey-like object) elicits periodic bursts as a post-stimulus event consisting of groups of spikes interrupted by short intervals (Fig. 9). Histologically, most MRF neurons are characterized by (differently extending) richly arborized dendritic trees which indicate highly integrative functions. Interestingly, the medial reticular premotor/motor network does not show any ordering in terms of an obvious sensory map or of clusters of neuron types that share a response property; exceptions are certain directionally selective monocular widefield neurons (probably responsible for optokinetic nystagmus) located in the median MRF, and the "bursters" located mainly adjacent to or within the cranial motonerve nuclei. All this literally justifies the term "reticular" formation which is involved in the generation of visually released motor patterns such as head/neck movements and snapping (e.g., Edinger 1908; Ebbesson 1976; Weerasuriya 1989 and this volume; Matsushima et al. 1989; Schwippert et al. 1990; Ewert et al. 1990b).

b) Current Neuroethological Projects

The fact that aperiodic sensory input to MRF may trigger periodic patterns of activity, raises two main questions:
1) Is cyclic bursting a general property of bulbar/spinal motor networks? More specifically, does the same network, depending on certain peripheral and/or internal inputs, trigger cyclic bursting for periodic motor patterns (e.g., breathing, calling, gulping, hopping, walking) and depending on other input conditions elicit

one premotor burst for the initiation of an action pattern (e.g., for a ballistic head movement, or a tongue flip)? This requires, for example, multichannel recordings in succinylcholine immobilized animals (equipped for recording in the freely moving mode) during the fading paralyzing pharmacological influence. For example, the *repetitive* bursting after presentation of a prey stimulus (Fig. 9) may be due to proprioceptive feedback conditions that are linked with the toad's immobilization.
2) Are bursting neurons coupled and, if so, to which extent and in which manner? Parallel recordings from different reticular neurons, including bursters, will provide clues on parallel/convergent information processing, as well as motor coordination.

c) Questions for Neural Engineering

Especially in this domain of sensorimotor integration, setting-up decisive experiments is very difficult, since all information relevant for motor responses passes the narrow space of the medulla to select appropriate motor coordinations. Looking at the fine histology in this region - after reconstruction of the dendritic trees of adjacent neurons - one is inclined to compare this with a section through a clew of wool. Classical investigation techniques (electro-stimulating, lesioning) are therefore less useful tools for a functional approach. The most significant questions are:
1) How does a comprehensive model deal with "command releasing systems" in terms of *sensorimotor codes* that (i) specify objects in a task- and goal-directed manner [comparable to perceptual schemas] and (ii)

	a_0	a_1	a_2	a_3	a_4	b_1	b_2	b_3	c	d	e	f	g	h	i	j	k	
a_0	3																	Monocular small visual ERF: S=W=A
a_1	×	7																Monocular small visual ERF: S>W>A
a_2	×	×	6															Monocular small visual ERF: W>S>A
a_3	×	×	×	4														Monocular small visual ERF: S>A>W
a_4	×	×	×	×	7													Monocular small visual ERF: S>A=W=no efficacy
b_1		1				4												Binocular small visual ERF: OR-gate (object far)
b_2						×	3											Binocular small visual ERF: OR-gate (object close)
b_3			1			×	×	4										Binocular small visual ERF: AND-gate (object close)
c	2	1	1		1				7									Frontal wide visual ERF
d	3	2	2	1	3					6								Horizonal wide visual ERF
e					7						5							Dorsal wide visual ERF
f		**17**	5	5	8						1	11						Total visual ERF
g	2	**25**	**12**	9	8	1			4	8	8	**16**	6					Cutaneous tactual ERF
h	6	**24**	**11**	7	**13**	2			2	8	**10**	**32**	**57**	30				Tonic discharging
i	5	**12**	3	5	8				2	5	3	**25**	**13**	**37**	4			Warming-up firing
j		6	5	3	7					3	6	**13**	**32**	**58**	**12**	5		Cyclic bursting
k					4	1						8	3	1		4	24	Sluggish spiking

Fig. 8. Combinations of properties in medullary neurons obtained from extracellular recordings in immobilized common toads. Matrix 'neurogram' of properties (a)-(k) of 341 medullary neurons. White patches refer to counts of combination pairs; for example, a neuron showing [f, h, j] appears three times in the neurogram, as f and h, f and j, h and j; the most frequent combinations are in bold face. Crossed patches represent mutually exclusive categories, i.e., assignments to the same properties. Black patches of the matrix refer to counts of neurons showing only one of the properties (note that these neurons could be recorded long enough for adequate study). W, A, and S denote the efficacy of worm, antiworm, and square objects for the activation of neurons; ERF, excitatory receptive field. (From Ewert et al. 1990b).

select the appropriate response [comparable to motor schemas in Arbib's terminology]? For example, do different (types of) command elements converge on one type of cell or on different cells of the motor pattern generating circuit. Is coincidence of activity in all command elements - like an *AND* gate - a necessary requirement for motor pattern activiation?

2) It should be tested by means of models, whether in certain cases different motor patterns may take advantage of a global motor pattern generating structure (e.g., in correspondence with the breathing pattern generator). We then may ask how this global motor network depending on "sensorimotor codes" activates and selects [via coordinated control programs] the corresponding motor patterns for the action patterns. Alternatively, the model should make predictions for parcellated interacting motor pattern generators.

3) The results of (1) and (2) could be of importance for artificial adapting neuronal networks (ANNs) or dynamic ones (DANNs) and even for reinforcement guided RANNs. Specifically, we may ask whether target-oriented turning, approaching, grasping, or avoiding in perceptual robots can take advantage of coded commands, whereby in response to a target a *sensorimotor code* commands the release of a specific action; after its execution, the new sensorial conditions correspond with another code for a different action. For example, a sensorimotor code {u,w,x,y,z} commands turning; after the turn, the peripheral conditions are adequate for {u,v,w,z} commanding approach; thereafter, the conditions are adequate for {u,v,w} that

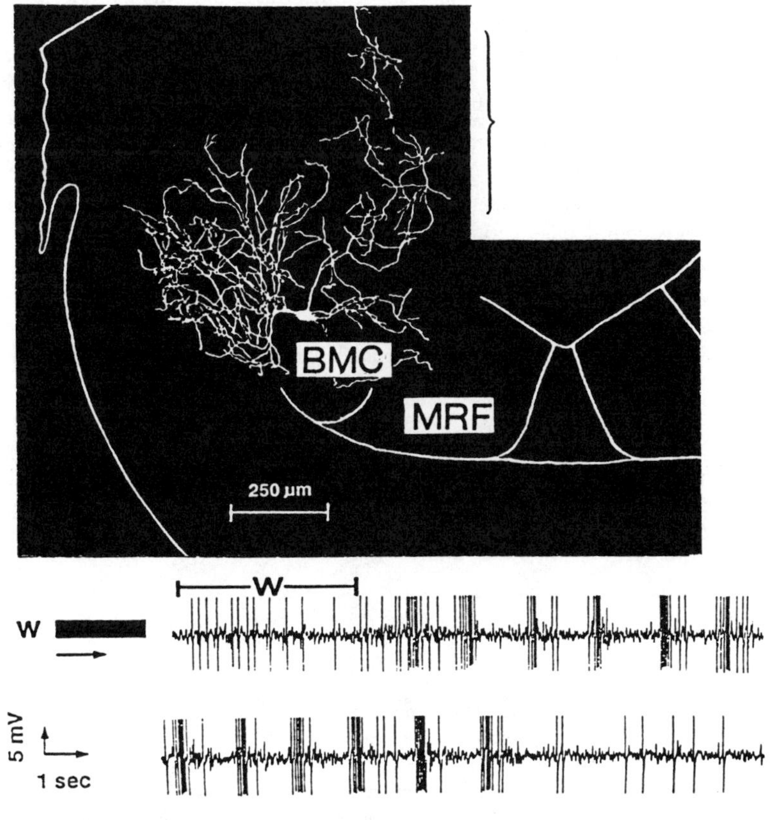

Fig. 9. Reconstruction of a T5.2 property displaying cell of the branchiomotor column in the medulla oblongata (histological semitransverse section) after intracellular recording and Co^{3+}-lysine iontophoretic labeling in an immobilized toad. The soma is located in the Vth cranial motor nucleus. Note the huge dendritic arborizations of which one tree (see bracket) could be followed rostrally via several transverse sections. In response to a worm-like (W) moving stripe, the neuron discharged and suddenly responded with cyclic bursts of spikes as a post-stimulus event. (From Schwippert et al. 1990).

commands grasping. The tasks and the sequence of tasks, here, are flexible and adaptive to changing targets. Since the ongoing sensorial conditions determine the kind of action, the sequence of actions can be different depending on the target's "behavior". If the target moves around the robot at a constant distance, then {u,w,x,y,z} triggers permanent turning; the robot would stop turning, taking advantage of "software" implementing feature specific habituation to "u". A different goal directed task requires modification of the operation related to "u", which may proceed in connection with associative processes.

5. Concluding Remarks

Having pointed out some aspects of the challenges of vertebrate (amphibian) neuroethology for information science in the sense of a fruitful dialogue between experimentalists (*neurobiologists*) and computational neuroscientists (*modelers*) on the one hand, and the perspectives of implementing useful principles of task-solving in technical devices (*neural engineering*) on the other hand, many questions are still open. In the following, I shall focus on a general one, namely the "potential capacity" in neuronal networks.

The objectives of modeling behavior are not only simulation and prediction, but also looking for so-called "hidden functions" inherent in the network for which we have not asked. More specifically, we are just beginning to learn from artificial neuronal ("neuroid") networks that with the implementation of a new algorithm, these networks not only execute the desired function, but also may display other properties not considered for the network construction. In the future, therefore, I see a great advantage in experiments with biological neuronal network models, in order to test their *potential capacity*, asking for "new functions" in terms of operational principles which the neurobiologist - from his biological point of view - would not have thought of. This objective has three implications. One concerns new ideas for technological developments so-to-speak "created by neural engineering". The other objective points to the brain. For example, amphibian brains may harbor functional principles or primordial ones (Ploog & Gottwald 1974) of which amphibians take *little* advantage, but phylogenetically advanced classes like reptiles or mammals may take *significant* advantage to deal with their special problems. So, if we are going in the future to "construct brains" in terms of functional macro-networks, I see a challenge to discuss questions of "minimal", "optimal", and "supra-optimal" systems.

ACKNOWLEDGEMENTS. I thank Mrs. Gerda Kruk and Mrs. Gisela Kaschlaw for technical assistance and animal care, and Mrs. Ursula Reichert and Mrs. Gudrun Frühauf for text processing. The work was supported by the DFG and by the Foundation's Fund for Research in Psychiatry.

References

Arbib MA (1981) Perceptual structures and distributed motor control. In: Brooks VB (ed) Handbook of physiology. The nervous system II. Motor control. American Physiological Society Bethesda MD, pp 1449-1480

Arbib MA (1982) Modelling neural mechanisms of visuomotor coordination in frog and toad. In: Amari S, Arbib MA (eds) Competition and cooperation in neural nets. Springer, Berlin Heidelberg New York, pp 342-370

Arbib MA (1987) Brains, machines, and mathematics. Springer, New York Berlin Heidelberg London Paris Tokyo

Arbib MA (1989) The metaphorical brain 2: neural networks and beyond. J Wiley & Sons, New York Chichester Brisbane Toronto Singapore, 458 pp

Arbib MA (1990) Neural computing: the challenge of the sixth generation. EDUCOM Bulletin 23: 2-12

Arkin R (1989) Neuroscience in motion: the application of schema theory to mobile robotics. In: Ewert J-P, Arbib MA (eds) Visuomotor coordination: amphibians, comparisons, models, and robots. Plenum Press, New York London, pp 649-671

Barlow HB (1953) Summation and inhibition in the frog's retina. J Physiol (Lond) 173: 377-407

Betts B (1989) The T5 base modulator hypothesis: a dynamic model of T5 neuron function in toads. In: Ewert J-P, Arbib MA (eds) Visuomotor coordination: amphibians, comparisons, models, and robots. Plenum Press, New York, pp 269-307

Blankenagel F (1931) Untersuchungen über die Großhirnfunktion von *R temporaria*. Zool Jahrb 49: 271-322

Bodis-Wollner I (1990) Visual deficits related to dopamine deficiency in experimental animals and Parkinson's disease patients. TINS 13: 296-302

Brzoska J, Schneider H (1978) Modification of prey-catching behavior by learning in the common toad (*Bufo b bufo* L, Anura, Amphibia): changes in response to visual objects and effects of auditory stimuli. Behav Processes 3: 125-136

Burghagen H (1979) Der Einfluß von figuralen, visuellen Mustern auf das Beutefangverhalten verschiedener Anuren. PhD Thesis, Univ Kassel

Burghagen H, Ewert J-P (1982) Question of "head preference" in response to worm-like dummies during prey capture of toads *Bufo bufo*. Behav Processes 7: 295-306

Burghagen H, Ewert J-P (1983) Influence of the background for discriminating object motion from self-induced motion in toads *Bufo bufo* (L). J Comp Physiol 152: 241-249

Cervantes-Pérez F (1989) Schema theory as a common language to study sensori-motor coordination. In: Ewert J-P, Arbib MA (eds) Visuomotor coordination: amphibians, comparisons, models, and robots. Plenum Press, New York London, pp 421-450

Collett TS (1977) Stereopsis in toads. Nature 267: 349-351

Comer CM (1987) Sensorimotor functions: what is a command, that a code may yield it? A commentary. Behav Brain Sci 10: 372

Cott HB (1936) The effectiveness of protective adaptations in the hive-bee, illustrated by experiments on the feeding reactions, habit formation and memory of the common toad (*Bufo bufo bufo*). Proc Zool Soc (London) 1: 113-133

Creutzfeldt OD (1983) Cortex cerebri. Springer-Verlag, Berlin Heidelberg New York, 484 pp.

Davis WJ, Kovac MP (1981) The command neuron and the organization of movement. TINS 4: 73-76

DiDomenico R, Eaton RC (1987) Toward a reformulation of the command concept. A commentary. Behav Brain Sci 10: 374-375

Eaton RC (1983) Is the Mauthner cell a vertebrate command neuron? A neuroethological perspective on an evolving concept. In: Ewert J-P, Capranica RR, Ingle DJ (eds) Advances in vertebrate neuroethology. Plenum Press, New York, pp 629-636

Ebbesson SOE (1976) Morphology of the spinal cord. In: Llinás L, Precht W (eds) Frog neurobiology. Springer, Berlin Heidelberg New York, pp 679-706

Eckmiller R, Hartmann G, Hauske G (1990) (eds) Parallel processing in neural systems and computers. North-Holland, Amsterdam New York Oxford Tokyo

Edinger L (1908) Vorlesungen über den Bau der nervösen Zentralorgane des Menschen und der Tiere, Vol.2. Vogel, Leipzig

Ewert J-P (1967) Untersuchungen über die Anteile zentralnervöser Aktionen an der taxisspezifischen Ermüdung der Erdkröte (*Bufo bufo* L). Z Vergl Physiol 57: 263-298

Ewert J-P (1968) Der Einfluß von Zwischenhirndefekten auf die Visuomotorik im Beute- und Fluchtverhalten der Erdkröte (*Bufo bufo* L). Z Vergl Physiol 61: 41-70

Ewert J-P (1969) Quantitative Analyse von Reiz-Reaktions-Beziehungen bei visuellem Auslösen der Beutefang-Wendereaktion der Erdkröte (*Bufo bufo* L). Pflügers Arch 308: 225-243

Ewert J-P (1974) The neural basis of visually guided behavior. Sci Amer 230: 34-42

Ewert J-P (1980) Neuroethology: an introduction to the neurophysiological fundamentals of behavior. Springer, Berlin Heidelberg New York

Ewert J-P (1984) Tectal mechanisms that underlie prey-catching and avoidance behaviors in toads. In: Vanegas H (ed) Comparative neurology of the optic tectum. Plenum Press, New York London, pp 247-416

Ewert J-P (1987) Neuroethology of releasing mechanisms: prey-catching in toads. Behav Brain Sci 10: 337-405

Ewert J-P, IWF (1982) Gestalt perception in the common toad. I) Innate prey recognition. Institut für den Wissenschaftlichen Film IWF, Göttingen, Film No C 1430

Ewert J-P, Kehl W (1978) Configurational prey selection by individual experience in the toad *Bufo bufo*. J Comp Physiol 126: 105-114

Ewert J-P, Seelen W v (1974) Neurobiologie und System-Theorie eines visuellen Muster-Erkennungsmechanismus bei Kröten. Kybernetik 14: 167-183

Ewert J-P, Wietersheim A v (1974) Der Einfluß von Thalamus/Praetectum-Defekten auf die Antwort von Tectum-Neuronen gegenüber bewegten visuellen Mustern bei der Kröte (*Bufo bufo* L). J Comp Physiol 92: 149-160

Ewert J-P, Arend B, Becker V, Borchers H-W (1979) Invariants in configurational prey selection by *Bufo bufo* (L). Brain Behav Evol 16: 38-51

Ewert J-P, Burghagen H, Schürg-Pfeiffer E (1983) Neuroethological analysis of the innate releasing mechanism for prey-catching behavior in toads. In: Ewert J-P, Capranica RR, Ingle DJ (eds) Advances in vertebrate neuroethology. Plenum Press, New York London, pp 413-475

Ewert J-P, Schwippert WW, Beneke TW (1990a) Parallel distributed processing of configural moving objects in the toad's visual system. In: Eckmiller R, Hartmann G, Hauske G (eds) Parallel processing in neural systems and computers. North-Holland, Amsterdam New York Oxford Tokyo, pp 109-112

Ewert J-P, Framing EM, Schürg-Pfeiffer E, Weerasuriya A (1990b) Responses of medullary neurons to moving visual stimuli in the common toad. I) Characterization of medial reticular neurons by extracellular recording. J Comp Physiol, in press

Finkenstädt T (1989a) Stimulus-specific habituation in toads: 2DG studies and lesion experiments. In: Ewert J-P, Arbib MA (eds) Visuomotor coordination: amphibians, comparisons, models, and robots. Plenum Press, New York London, pp 767-797

Finkenstädt T (1989b) Visual associative learning: searching for behaviorally relevant brain structures in toads. In: Ewert J-P, Arbib MA (eds) Visuomotor coordination: amphibians, comparisons, models, and robots. Plenum Press, New York London, pp 799-832

Finkenstädt T, Ewert J-P (1983) Visual pattern discrimination through interactions of neural networks: a combined electrical brain stimulation, brain lesion, and extracellular recording study in *Salamandra salamandra.* J Comp Physiol 153: 99-110

Finkenstädt T, Ewert J-P (1988a) Stimulus-specific long-term habituation of visually guided orienting behavior toward prey in toads: a ^{14}C-2DG study. J Comp Physiol 163: 1-11

Finkenstädt T, Ewert J-P (1988b) Effects of visual associative conditioning on behavior and cerebral metabolic activity in toads. Naturwissenschaften 75: 95-97

Finkenstädt T, Adler NT, Allen TO, Ebbesson SOE, Ewert J-P (1985) Mapping of brain activity in mesencephalic and diencephalic structures of toads during presentation of visual key stimuli: a computer assisted analysis of ^{14}C-2DG autoradiographs. J Comp Physiol 156: 433-445

Finkenstädt T, Adler NT, Allen TO, Ewert J-P (1986) Regional distribution of glucose utilization in the telencephalon of toads in response to configurational visual stimuli: a ^{14}C-2DG study. J Comp Physiol 158: 457-467

Fite KV, Scalia F (1976) Central visual pathways in the frog. In: Fite KV (ed) The amphibian visual system: a multidisciplinary approach. Academic Press, New York San Francisco London, pp 87-118

Fritzsch B, Himstedt W (1981) Pretectal neurons project to the salamander retina. Neurosci Lett 24: 13-17

Glagow M (1990) Der Einfluß von Apomorphin auf die visuelle Erregbarkeit im rezeptiven Feld retinaler Klasse-2-Neuronen der Erdkröte *Bufo bufo spinosus* L. Staats-Exam Thesis, Univ Kassel

Grobstein P, Comer C, Kostyk SK (1983) Frog prey capture behavior: between sensory maps and directed motor output. In: Ewert J-P, Capranica RR, Ingle DJ (eds) Advances in vertebrate neuroethology. Plenum Press, New York, pp 331-347

Grüsser O-J, Grüsser-Cornehls U (1976) Neurophysiology of the anuran visual system. In: Llinás L, Precht W (eds) Frog neurobiology. Springer, Berlin Heidelberg New York, pp 297-385

Guthrie DM (1987) (ed) Aims and methods in neuroethology. Manchester Univ Press, Manchester

Halpern M (1972) Some connections of the telencephalon of the frog *R pipiens.* Brain Behav Evol 6: 42-68

Holst E v, Mittelstaedt H (1950) Das Reafferenzprinzip. Naturwissenschaften 37: 464-476

Ingle DJ (1976) Spatial vision in anurans. In: Fite KV (ed) The amphibian visual system: a multidisciplinary approach. Academic Press, New York San Francisco London

Ingle DJ (1983) Brain mechanisms of visual localization by frogs and toads. In: Ewert J-P, Capranica RR, Ingle DJ (eds) Advances in vertebrate neuroethology. Plenum Press, New York London, pp 177-226

Ingle DJ, McKinley D (1978) Effects of stimulus configuration on elicited prey catching by the marine toad (*Bufo marinus*). Anim Behav 26: 885-891

Jung R (1953) Einführung in die allgemeine Neurophysiologie. In: Gauer OH, Kramer K, Jung R (eds) Physiologie des Menschen 10. Allgemeine Neurophysiologie. Urban & Schwarzenberg, München Berlin Wien, pp 1-31

Kicliter E, Ebbesson SOE (1976) Organization of the "nonolfactory" telencephalon. In: Llinás L, Precht W (eds) Frog neurobiology. Springer, Berlin Heidelberg New York, pp 946-972

Kupfermann I, Weiss KR (1978) The command neuron concept. Behav Brain Sci 1: 3-39

Lara R (1989) Learning and memory in the toad's prey/predator recognition system: a neural model. In: Ewert J-P, Arbib MA (eds) Visuomotor coordination: amphibians, comparisons, models, and robots. Plenum Press, New York London, pp 833-855

Lara R, Arbib MA (1985) A model of the neural mechanisms responsible for pattern recognition and stimulus specific habituation in toads. Biol Cybern 51: 223-237

Lara R, Cervantes F, Arbib MA (1982) Two-dimensional model of retinal-tectal-pretectal interactions for the control of prey-predator recognition and size preference in amphibia. In: Amari S, Arbib MA (eds) Competition and cooperation in neural nets. Springer, Berlin Heidelberg New York, pp 371-393

Lázár G (1971) The projection of the retinal quadrants on the optic centers in the frog: a terminal degeneration study. Acta Morph Acad Sci Hung 19: 325-334

Lázár G (1989) Cellular architecture and connectivity of the frog's optic tectum and pretectum. In: Ewert J-P, Arbib MA (eds) Visuomotor coordination: amphibians, comparisons, models, and robots. Plenum Press, New York London, pp 175-199

Lettvin JY, Maturana HR, McCulloch WS, Pitts WH (1959) What the frog's eye tells the frog's brain. Proc Inst Radio Engin 47: 1940-1951

Mallot HA, Seelen W v (1989) Why cortices? Neural networks for visual information processing. In: Ewert J-P, Arbib MA (eds) Visuomotor coordination: amphibians, comparisons, models, and robots. Plenum Press, New York London, pp 357-382

Manteuffel G (1989) Compensation of visual background motion in salamanders. In: Ewert J-P, Arbib MA (eds) Visuomotor coordination: amphibians, comparisons, models, and robots. Plenum Press, New York London, pp 311-340

Mao Tse Tung (1935) Worte des Vorsitzenden. Verlag für fremdsprachige Literatur (German transl. Peking 1971)

Matsushima T, Satou M, Ueda K (1989) Medullary reticular neurons in the Japanese toad: morphology and excitatory inputs from the optic tectum. J Comp Physiol A 166: 7-22

Maturana HR (1958) Efferent fibers in the optic nerve of the toad (*Bufo bufo* L). J Anat 92: 21-26

Neary T, Northcutt RG (1983) Nuclear organization of the bullfrog diencephalon. J Comp Neurol 213: 262-278

Northcutt RG, Kicliter E (1980) Organization of the amphibian telencephalon. In: Ebbesson SOE (ed) Comparative neurology of the telencenphalon. Plenum Press, New York London, pp 203-255

Paillard J (1987) Cognitive versus sensorimotor encoding of spatial information. In: Ellen P, Thinus-Blanc C (eds) Cognitive processes and spatial orientation in animal and man Vol II. Martinus Nijhoff Publ, Dordrecht

Ploog D, Gottwald P (1974) Verhaltensforschung: Instinkt, Lernen, Hirnfunktion. Urban & Schwarzenberg, München

Rubin E (1915) Synsoplevede Figurer. Glyden dalske, Copenhagen

Rubinson E (1968) Projections of the tectum opticum of the frog. Brain Behav Evol 1: 529-561

Satou M, Ewert J-P (1985) The antidromic activation of tectal neurons by electrical stimuli applied to the caudal medulla oblongata in the toad *Bufo bufo* L. J Comp Physiol A 157: 739-748

Schmidhuber J (1990) "Reinforcement" - Lernen und adaptive Steuerung: Ein Überblick. Nachrichten Neuronale Netze 2: 2-4

Schrader MEG (1887) Zur Physiologie des Froschgehirns. Pflügers Arch 41: 75-90

Schürg-Pfeiffer E (1989) Behavior-correlated properties of tectal neurons in freely moving toads. In: Ewert J-P, Arbib MA (eds) Visuomotor coordination: amphibians, comparisons, models, and robots. Plenum Press, New York London, pp 451-480

Schürg-Pfeiffer E, Spreckelsen C, Ewert J-P (1990) Tectal small-field neurons recorded in prey-catching toads are sensitive to the real object size. Ann ENA meeting, Oxford Press

Schwippert WW, Beneke TW, Ewert J-P (1990) Responses of medullary neurons to moving visual stimuli in the common toad. II) An intracellular recording and cobalt-lysine labeling study. J Comp Physiol, in press

Stevens KH (1987) Implicit versus explicit computation. A commentary. Behav Brain Sci 10: 387-388

Tönnies JF (1949) Die Erregungssteuerung im Zentralnervensystem. Erregungsfokus der Synapse und Rückmeldung als Funktionsprinzipien. Arch Psychiatr 182: 478-535

Tóth P, Csank G, Lázár G (1985) Morphology of the cells of origin of descending pathways to the spinal cord in *Rana esculenta*. A tracing study using cobalt-lysine complex. J Hirnforsch 26: 365-383

Traud R (1983) Einfluß von visuellen Reizmustern auf die juvenile Erdkröte (*Bufo bufo* L). PhD Thesis, Univ of Kassel

Tsai H-J, Ewert J-P (1987) Edge preference of retinal and tectal neurons in common toads (*Bufo bufo*) in response to worm-like moving stripes: the question of behaviorally relevant "position indicators." J Comp Physiol 161: 295-304

Tsai H-J, Ewert J-P (1988) Influence of stationary and moving textured backgrounds on the response of visual neurons in toads (*Bufo bufo* L). Brain Behav Evol 32: 27-38

Uchiyama H (1989) Centrifugal pathways to the retina: influence of the optic tectum. Vis Neurosci (submitted)

Uchiyama H, Sakamoto N, Ito H (1981) A retinopetal nucleus in the preoptic area in a teleost *Navodon modestus*. Brain Res 222: 119-124

Uchiyama H, Reh TA, Stell WK (1988) Immunocytochemical and morphological evidence for a retinopetal projection in anuran amphibians. J Comp Neurol 274: 48-59

Walker S (1983) Animal thought. Routledge & Kegan Paul, London Boston Melbourne Henley

Weerasuriya A (1989) In search of the motor pattern generator for snapping in toads. In: Ewert J-P, Arbib MA (eds) Visuomotor coordination: amphibians, comparisons, models, and robots. Plenum Press, New York, pp 589-614

Weerasuriya A, Ewert J-P (1981) Prey-selective neurons in the toad's optic tectum and sensorimotor interfacing: HRP studies and recording experiments. J Comp Physiol 144: 429-434

Wiersma CAG, Ikeda K (1964) Interneurons commanding swimmeret movements in the crayfish, *Procambarus clarkii* (Girard). Comp Biochem Physiol 12: 509-525

Wilczynski W, Northcutt RG (1977) Afferents to the optic tectum of the leopard frog: an HRP study. J Comp Neurol 173: 219-229

Wilczynski W, Northcutt RG (1983a) Connections of the bullfrog striatum: afferent organization. J Comp Neurol 214: 321-332

Wilczynski W, Northcutt RG (1983b) Connections of the bullfrog striatum: efferent projections. J Comp Neurol 214: 333-343

From the Retina to the Brain

A computer model to visualize change sensitive responses in the salamander retina

Jeffrey L. Teeters[*+]
Frank H. Eeckman[*]
Frank S. Werblin[+]

[*]*Lawrence Livermore National Laboratory*
[+]*University of California at Berkeley*

Abstract. Visualization of the two dimensional dynamic pattern of activity in change sensitive amacrine and ganglion cells of the salamander retina is achieved through simulation of a computational model. The model is constructed from neurophysiological data suggesting that a four neuron circuit is involved in the detection of change and the feedforward of change sensitive inhibition to ganglion cells. Two versions of the model are implemented: one using a Cray computer which allows inclusion of much biological detail, the other using a PIPE image processor, which allows real time simulation with dynamic input images. The models show that the four neuron circuit can account for previously observed movement sensitive reductions in ganglion cell sensitivity and allow visualization and prediction of the spatio-temporal pattern of activity in change sensitive retinal cells.

Introduction

In ganglion cells of the salamander retina (Werblin, 1972; Werblin & Copenhagen, 1974) and cat (Jakiela, 1978), movement in the receptive field surround decreases the response to stimuli presented at the receptive field center as illustrated in Fig 1. Wunk & Werblin (1979) showed that all ganglion cells receive change-sensitive inhibition, and Barnes & Werblin (1987) implicated a change-sensitive amacrine cell with widely distributed processes. The change-sensitivity of these amacrine cells has been traced in part to a truncation of synaptic release from the bipolar terminals that presumably drive them (Maguire *et al.*, 1989). The transient response of these amacrine cells, mediated by voltage gated currents (Barnes & Werblin, 1986; Eliasof *et al.*, 1987) also contributes to this change sensitivity.

What is the two-dimensional pattern of dynamic activity generated by the interactions of the bipolar, amacrine and ganglion cells implicated in the change-sensitive system? To address this question we have constructed a computational

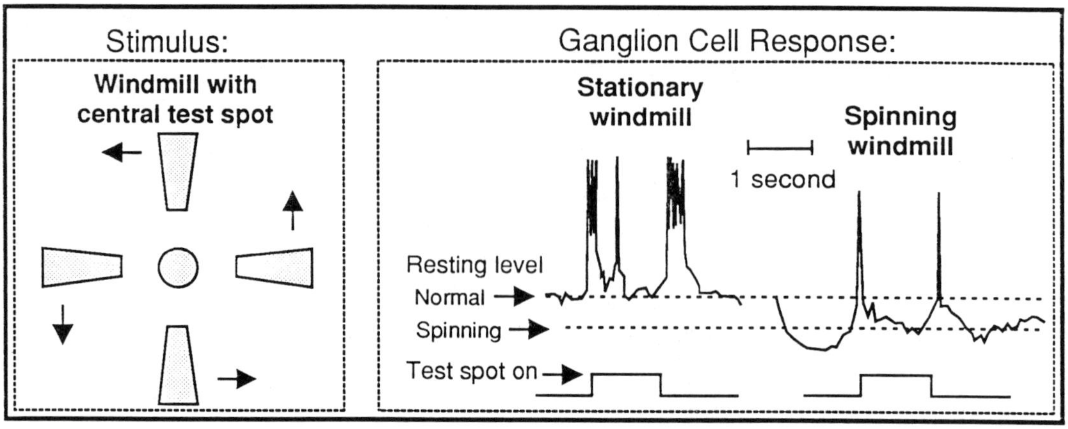

Fig. 1 - Change-sensitive inhibition reduces ganglion cell response. Data is from Werblin (1972).

model that incorporates the anatomical and physiological characteristics described previously for these cells.

Two versions of the computational model were built. The first version is simulated on a Cray computer, incorporates much biological data, but is unable to process dynamic images in real time. The second version of the model is implemented on a PIPE image processor. It is less biologically faithful than the first model, but allows real time processing of CCD input images enabling better visualization of retina processing in natural environments.

Cray Model

Implementing the experimental data and hypothetical circuit

To implement the Cray version of the model, a change sensitive inhibitory module was generated based upon earlier physiological studies, and its characteristics were tuned to match experimental evidence. This module was then inserted into a more complete model of the retina where it acted to inhibit ganglion cells.

The change-sensitive circuit proposed earlier (Werblin *et al.*, 1988; Maguire *et al.*, 1989) is reproduced in Figure 2. This is meant to describe a very local region of the retina where the receptive fields of the two bipolar cells are spatially overlapping. When a visual target enters this receptive field, the bipolar cells are both depolarized. The sustained bipolar cell activates the narrow field amacrine cell that, in turn feeds back to the synaptic terminal of the transient bipolar cell to truncate transmitter release after a brief (ca. 100 msec) delay. Because the signal reaching the wide field amacrine cell is truncated after about 100 msec, the wide

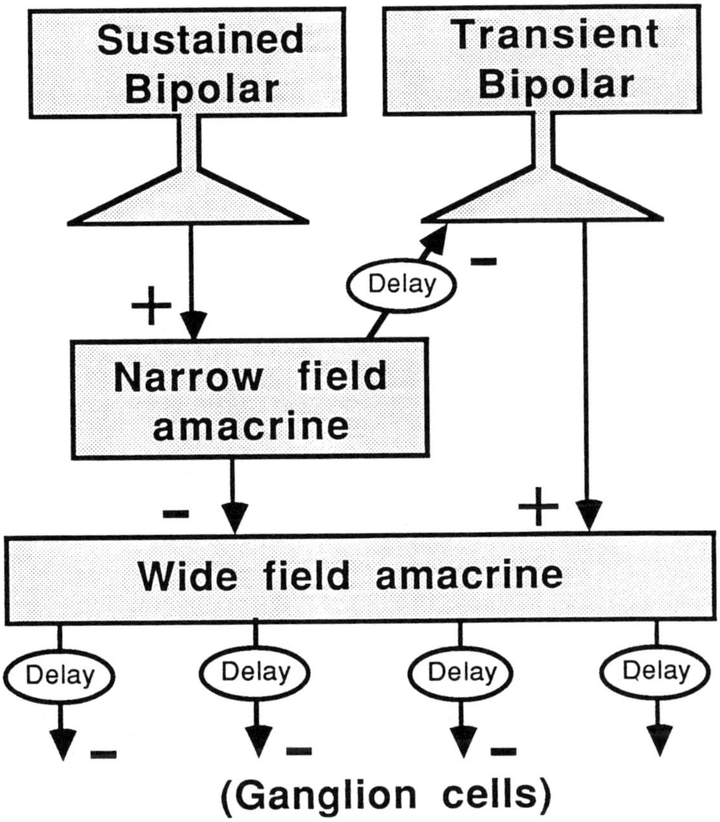

Fig. 2 - Circuitry to be analyzed.

field amacrine cell will receive excitation when the target enters the receptive field, but will not continue to respond in the presence of the target.

The spatial profiles of synaptic input and output for the cell types involved in the model are summarized in Figure 3. The bipolar and narrow field amacrine cell sensitivities extend over a region corresponding roughly to their dendritic spread. The wide field amacrine cell appears to receive input over a local region near the cell body, but delivers its inhibitory output over a much wider region corresponding the the full extent (ca. 500 mm) of its processes.

Fig 4 shows the electrical circuit model for each cell type, and also illustrates the pathways for interactions between these cell types that are implemented in the model. In Figure 4, the boxes contain the circuit for each cell and the arrows between boxes represent synaptic interactions thought to occur between cells as determined through experiments in which a neurotransmitter is puffed onto the bipolar dendrites. Bipolar cells are modeled using two compartments, corresponding to the cell body and axon terminal as suggested in Maguire *et al.* (1989). Amacrine cells are modeled using only one compartment as was done in Eliasof *et. al.* (1987).

Several, probably unphysiological, simplifications and hypotheses were made to facilitate implementing the model illustrated in Figure 4. Only three currents are used to approximate transient spikes in the wide field amacrine cell rather than the five currents which Eliasof *et. al.* (1987) demonstrated are involved. It is assumed that conductances in the bipolar terminals do not appreciably influence the potential in the cell body and that Calcium channels in the transient bipolar terminal are presumed to quickly depolarize the terminal when a threshold is reached. For many aspects of the model, lack of data necessitated the use of "reasonable guesses" which will surely change as more is learned. Despite these difficulties, the general framework and much of the known data concerning the circuit shown in Figure 4 are incorporated. Details of the model are given in the appendix.

Testing the circuit elements in the computational model

Computer simulation was used to test whether the single cell properties and proposed interactions between cells shown in Figure 4 are consistent with the responses recorded from the neurons during applications of a neurotransmitter puff.

The results of these simulations are shown in Figure 5. The left column of Figure 5 describes an experiment performed, the middle column shows the experimentally obtained responses, and the right column shows the simulated responses. For simplicity, scales giving the magnitudes of the responses are not shown here, but in all traces the magnitude of the simulated response is very close to that of the observed response.

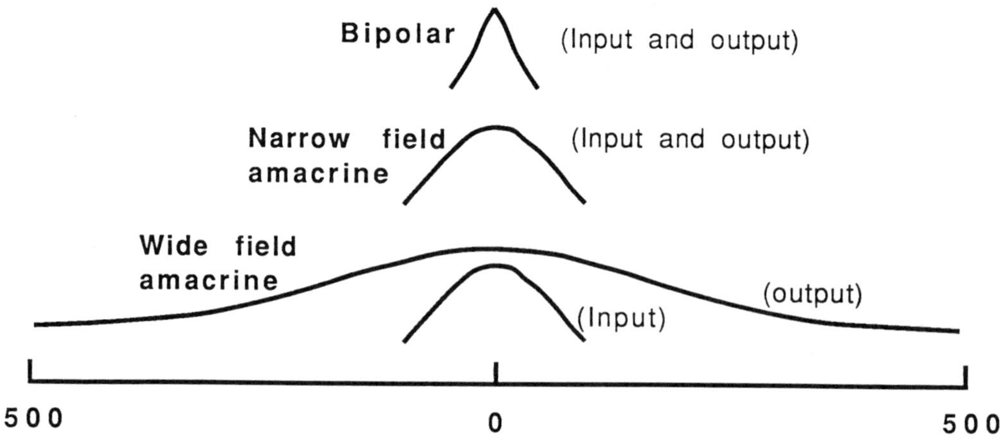

Fig. 3 - Distance from cell center (μm)

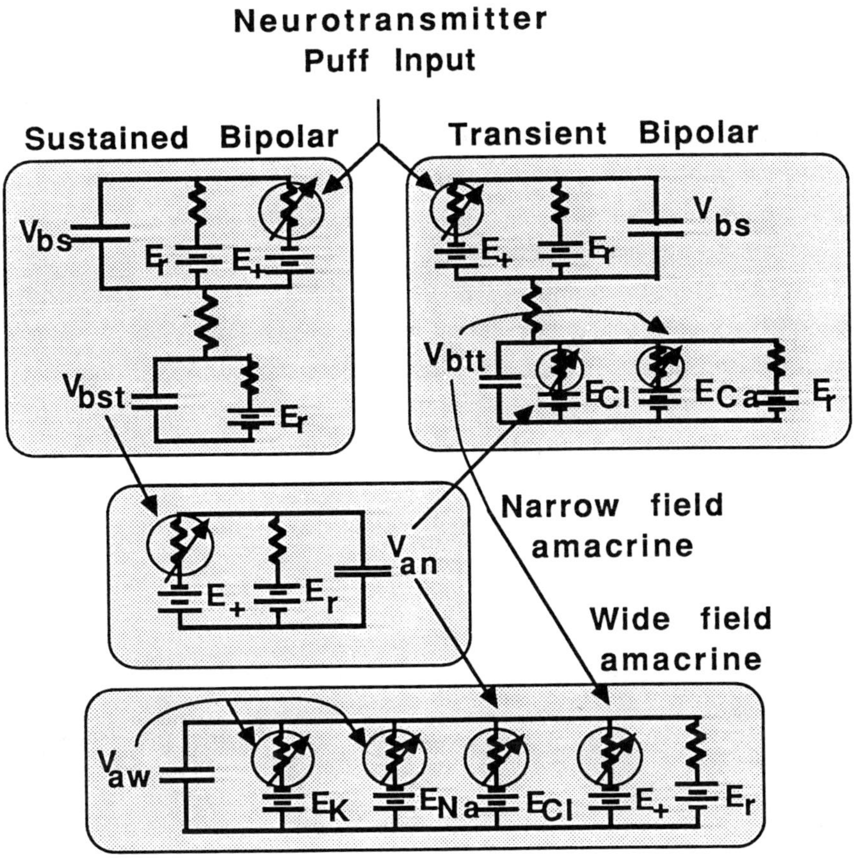

Fig. 4 - Details of Circuitry

The match between the simulated and observed responses voltage clamps of the wide field amacrine shown in the fourth row is not so close because there is a sustained outward current which is observed experimentally that is not apparent in the simulations. This shows that the model is not perfect and is something that needs to be investigated further.

Fortunately, this difference between the model and observed response does not prevent the hypothesized function of the circuit from being simulated. This is shown on the bottom row where both the observed and simulated voltage responses from the wide field amacrine are transient.

Integrating the individual cells into a change sensitive circuit

Figure 5 illustrates that we have to a large degree succeeded in combining the characteristics of single cells into a model which can explain many of the observed properties thought to be due to the interaction between these cells.

The next step in our analysis is to investigate how this circuit of cells influences the response of ganglion cells in a functionally significant way. To do this requires simulating the input to the bipolar dendrites and simulating the ganglion cells which receive the transient inhibition generated by the wide field amacrine. This amounts to a integrated model of an entire patch of retina, including receptors, horizontal cells, the four neuron circuit discussed earlier, and ganglion cells. The manner in which we accomplish this is illustrated in Figure 6.

The left side of figure 6 shows the elements in the model. The receptors and horizontal cells were modeled very simply as low pass filters with different time constants and different areas of spatial input. The ganglion cell was modeled as receiving a transient excitatory input which is generated phenomenologically by a thresholded high pass filter from the transient bipolar cell. The inhibitory input to the ganglion cell is implemented as coming from the transient wide field amacrine cells described previously. For simplicity, voltage gated currents and spiking are not implemented in the ganglion cell model, and only the off bipolar pathways are simulated.

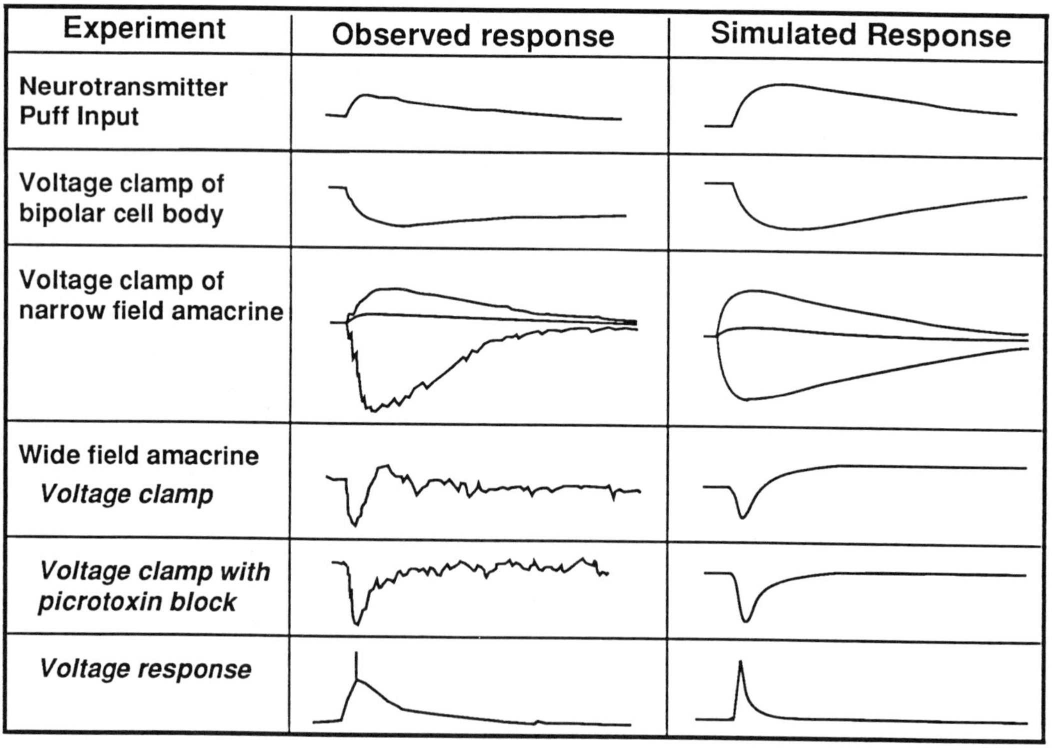

Fig. 5 - Example puff simulations.

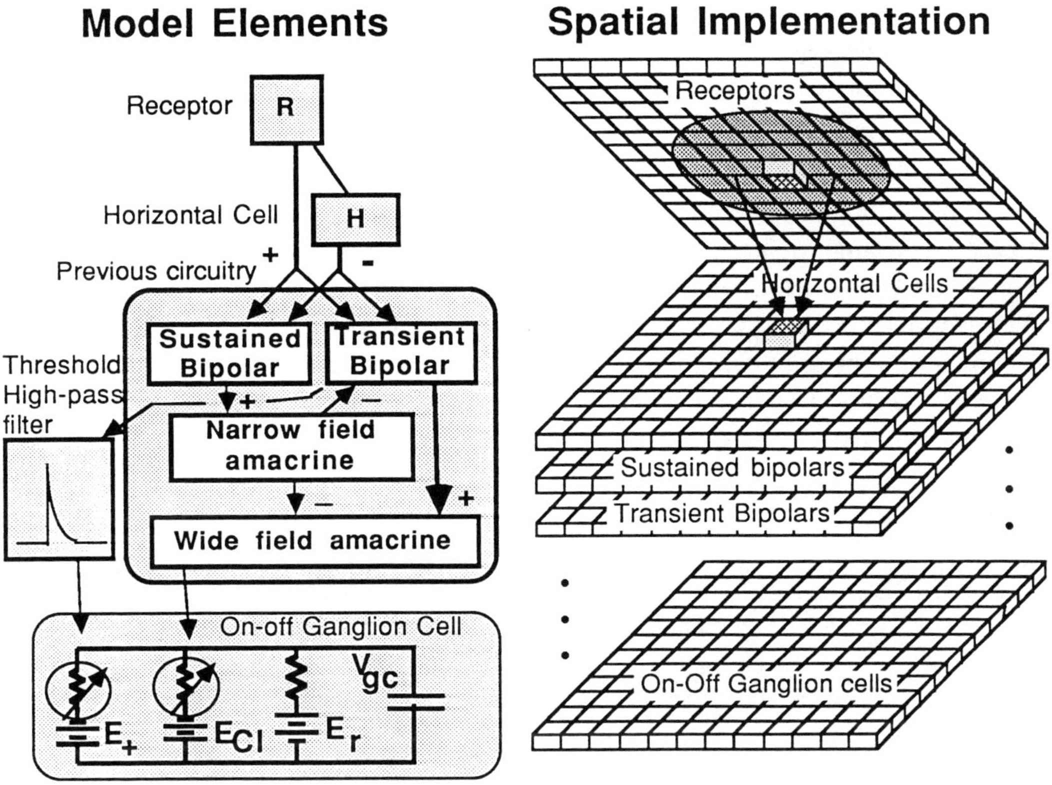

Fig. 6 - Integrated retinal model.

The right hand of Figure 6 illustrates how the model retina is implemented spatially. The circuit for each cell type is duplicated across the entire retina patch in a matrix format. The known spatial properties of each cell, such as the spatial range of transmitter sensitivity and release are incorporated into the model. Details of the model are given in the appendix.

Testing the integrated model

The spatial properties of the integrated model provides us with a new method of visualization the responses of retinal neurons. Rather than have a single curve representing the response of a single unit over time we can make snapshots of the simultaneous pattern of activity in many neurons spatially distributed across the retina patch.

Figure 7 shows such a spatial snapshot taken of three different neuron responses just after a central light spot is turned on, during the presence of a stationary and spinning surround windmill. The neuron responses are the transient bipolar terminal, the wide field amacrine neurotransmitter release, and the ganglion cell voltage response. On the left column is shown the response to a flashing spot when the windmill is stationary. On the right is shown the response to the same flashing spot but under conditions in which the windmill is spinning.

When the windmill is stationary, the transient bipolar terminal responds only to the center flash. Responses to the windmill vanes are suppressed by the narrow field amacrine cell causing the appearance of four regions of hyperpolarizing responses around the center. The wide field amacrine responds to the central test flash and releases transmitter as shown in the second row. The array of ganglion cells responds to both the excitatory input generated by the spot at the bipolar terminals and the inhibitory input generated by the wide field amacrines. Because the wide field inhibition has not yet taken effect at this point in time, the ganglion cells respond well to the flashing spot.

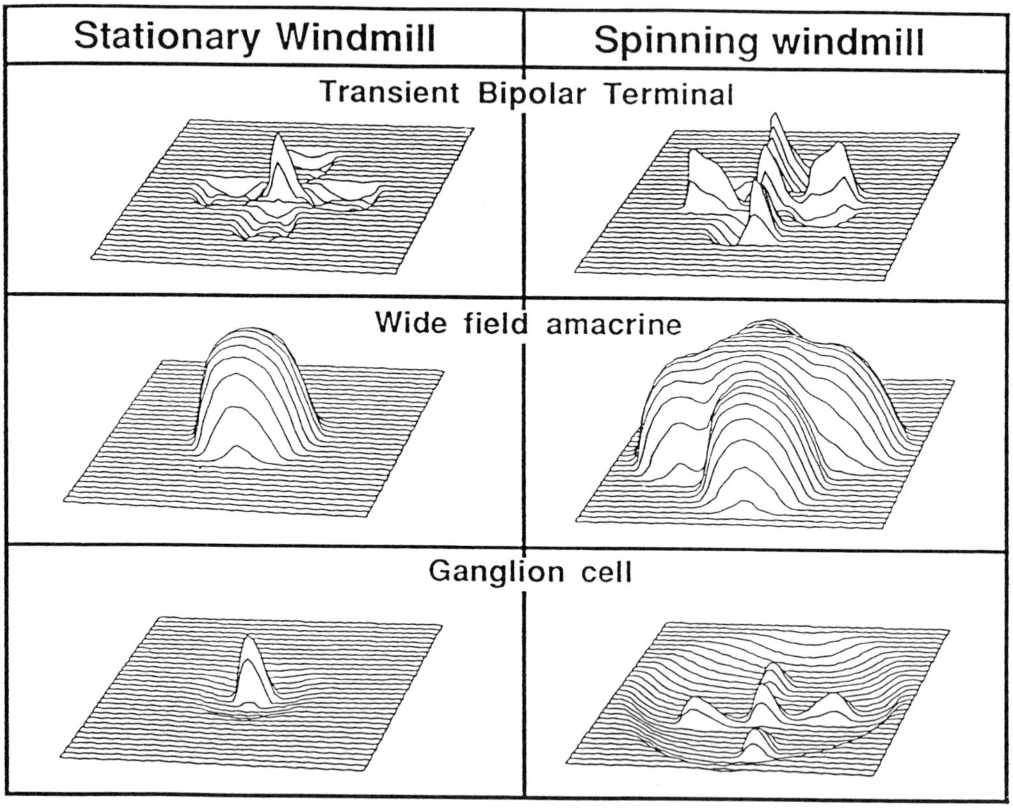

Fig. 7 - Control of sensitivity, spatial response

When the windmill is spinning, as is shown on the right hand column, the transient bipolar terminals generate a response to the leading edge of the windmill vanes. The wide field amacrine cells receive excitatory input from the transient bipolar terminal responses to the vane, and consequently release inhibitory neurotransmitter over a wide area as shown in in the right column. Because inhibition is being continuously generated by the spinning windmill, the response of the ganglion cells across the retinal patch has a large bowl shaped area of hyperpolarization which reduces the ganglion cell response of the cells to the central test flash. This is seen by the fact that the height of depolarization in the centrally located ganglion cells is much smaller under conditions of a spinning windmill than if the windmill is stationary. This is consistent with the results found experimentally which are illustrated in Figure 1. Experimental data not yet attained, but which are predicted by the model simulations illustrated in Figure 7, are the spatial patterns of activity generated in the bipolar, amacrine, and ganglion cells in response to the different stimuli.

While spatial snapshots such as those given in Figure 7 are more informative than traces from single units in that they allow visualization of the activity in many cells at once, they also have the drawback of showing activity at only one particular instant in time. To view both the spatial and temporal responses of neurons requires a movie consisting of a series of spatial snapshots taken at successive instants of time. Movies allow visualization in a much better way than is possible with either single unit recordings or static spatial snapshots. For the spinning windmill and other simple stimuli, we have generated such movies by recording successive frames of simulation results onto video tape.

PIPE Model

While the Cray model gives valuable insight into the system response to simple stimuli, it currently does not show the model response to complex stimuli found in the real world. While it is possible to test the Cray model using a pre-recorded digitized sequence of real world images, this is laborious and does not allow one to interactively change the stimulus based on the the model response.

To overcome this limitation, we have implemented a second, less biologically detailed model on a PIPE image processing computer. The PIPE, (manufactured by ASPEX Inc. of New York, NY), is designed for the real-time processing of image data. The version of the PIPE we used has the following features: (1) A TV camera provides input at the rate of sixty 256x256 pixel frames each second with each pixel having 8 bits of grey level information. (2) There are four internal processing units, each of which can execute operations on entire frames of data in 1/60 of a second.

Operations include arithmetic (+, -, *, /), special functions (sin, cos, exp, thresholding, and others), and convolution with an arbitrary 3x3 mask. (3) There is memory storage for 64 frames of data. This memory is used to store the input and output images and intermediate results used in future computations. (4) Output from any computation may be routed to a TV monitor for display.

The algorithm for the retina model and change sensitive circuit was modeled after the Cray version, but made less detailed because of the limitations of the computations that one can perform on the PIPE as compared to the general purpose computations allowed on the Cray. Specifically, all spatial properties implemented on the PIPE are generated through one or more convolutions with 3x3 masks rather than explicit interactions that can be specified on the Cray. Reversal potentials, conductances, cell capacitances, and synaptic functions were all lumped into simple difference equations. While these simplifications prevent the PIPE model from being used for quantitative simulation of neurophysiological transmitter puff experiments similar to those performed using the Cray version, they do allow testing the model with more complex stimuli and the formation of visualizations and insights that were not possible with the Cray implementation.

This is shown in Figures 8a-d which are the output of some layers in the pipe model. The input image in all cases is a human face, in various positions and states of movement. Figure 8a shows an unprocessed input image. Figure 8b shows the result of photoreceptor discretization. The image is made much coarser because of the large field of view (0.5 visual degrees) of each tiger salamander receptor. Figure 8c shows the output of the bipolar cell layer. The different time constants of receptor and horizontal cells cause the enhancement of moving edges in the bipolar cell layer. Unfortunately, this is difficult to see in static pictures shown here.

The dynamic aspect of the model is better illustrated in Figure 8d which shows the widespread ganglion cell output (white areas) to a sudden movement of the head. If there is no movement, no response occurs and the image is all black. In this picture, the inhibitory influence of the wide field amacrine cells has not yet arisen. The effect of this inhibition is to turn off ganglion cells, thus greatly increasing the transient nature of the ganglion cell response.

One discovery that made with the PIPE model which was not obvious from the Cray simulations, is that once the wide field amacrines fire, the response of transient ganglions is substantially inhibited for the duration of the movement. The ganglion cells will not fire strongly again until the stimulus stops and allows the movement detecting wide field amacrines to recover. This in essence makes these ganglion cells respond stronger to an initial movement rather than to continuous movement.

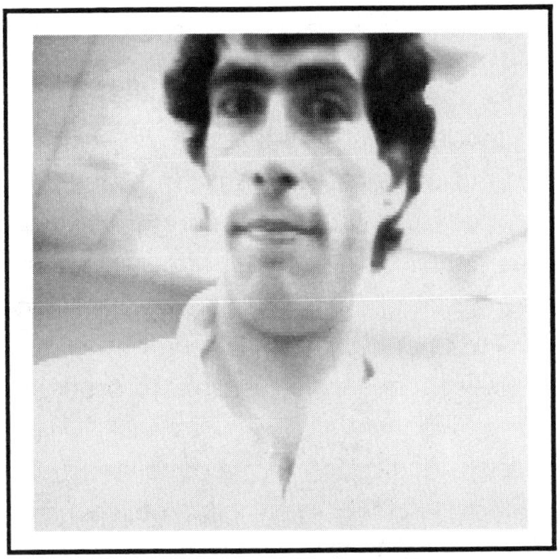

(a) Original input

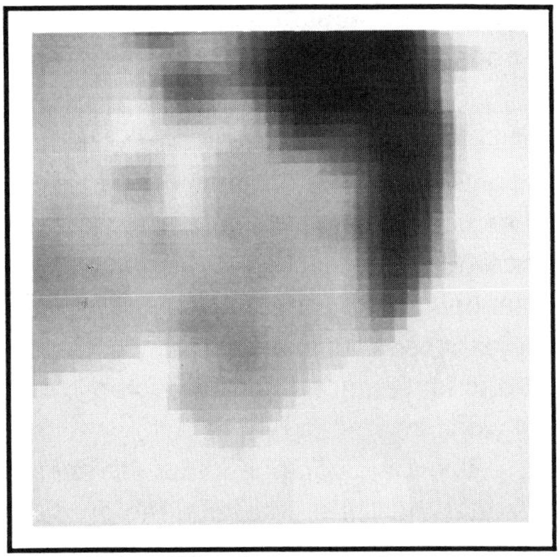

(b) Photoreceptors

(c) Bipolar Cells

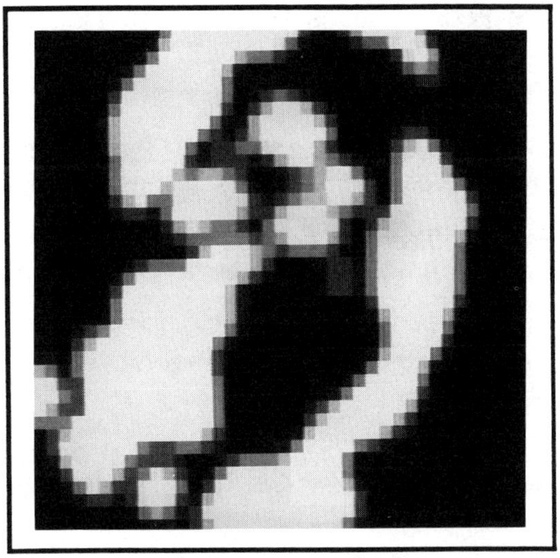

(d) Ganglion Cells

Fig. 8 - Simulations from PIPE retinal model

Summary

Using computer simulation of a neurophysiologically based model, we are able to visualize the two-dimensional spatio-temporal pattern of activity in movement sensitive neurons in the salamander retina. The model accounts for observed control of retinal sensitivity in ganglion cells in response to a spinning windmill stimulus and demonstrates that the experimental data describing properties of four neurons in the inner retina are compatible with the hypothesis that these neurons are involved in the detection of change and the feedforward of change-sensitive inhibition to ganglion cells. Two versions of the model were presented. First, a Cray computer was used to model the four neurons and demonstrate that the proposed interactions between them are sufficient to reproduce many of the observed network properties in response to a puff of neurotransmitter. The Cray model was also expanded to integrated model of the retina allows testing their influence on ganglion cell response during a light stimuli. Second, a PIPE image processing computer was used to implement a simplified version of the Cray model which is able to process real world images in real time.

The main contribution of the models are their verification of the consistency of presently available data, and the prediction of neural activity in ganglion cells which is subject to refutation or verification by new experiments. We are currently recording the spatio-temporal response of ganglion cells to moving stimuli so that direct comparisons to these model predictions can be made.

References

Barnes, S. and Werblin, F.S. (1986). Gated Currents Generate Single Spike Activity in Amacrine cells of the Tiger Salamander. *Proc. Natl. Acad. Sci. USA* **83**: 1509 - 1512.

Barnes, S. and Werblin, F.S. (1987). Direct Excitatory and Lateral Inhibitory Synaptic Inputs to Amacrine Cells in the Tiger Salamander Retina. *Brain Res.* **406**: 233 - 237.

Eliasof S., Barnes S. and Werblin, F.S. (1987). The Interaction of Ionic Currents Mediating Single Spike Activity in Retinal Amacrine Cells of the Tiger Salamander. *J. Neurosci.* **7**: 3512 - 3524.

Jakiela, H.G. (1978). The Effect of Retinal Image Motion on the Responsiveness of Retinal Ganglion Cells in the Cat. Ph.D. Thesis Northwestern Univ., Evanston, Ill.

Maguire, G. , Lukasiewicz, P. and Werblin F.S. (1989). Amacrine Cell Interactions Underlying the Response to Change in the Tiger Salamander Retina. *J. Neurosci.* **9**: 726 - 735.

Werblin, F.S. (1972). Lateral Interactions at Inner Plexiform Layer of a Vertebrate Retina: Antagonistic Response to Change. *Science.* **175**: 1008 - 1010.

Werblin, F.S. and Copenhagen, D.R. (1974). Control of Retinal Sensitivity. III. Lateral Interactions at the Inner Plexiform Layer. *J. Gen. Physiol.* **63**: 88 - 110.

Werblin, F.S., Maguire, G., Lukasiewicz, P., Eliasof, S., and Wu, S. (1988). Neural Interactions Mediating the Detection of Motion in the Retina of the Tiger Salamander. *Visual Neurosci.* **1**: 317 - 329.

Wunk, D.F. and Werblin, F.S. (1979). Synaptic Inputs to Ganglion Cells in the Tiger Salamander Retina. *J. Gen. Physiol.* **73**: 265 - 286.

Appendix: Details of the Models

The algorithms for the Cray and PIPE versions of the model are given here. Variables in the equations stand for 2-D arrays which correspond to the 2-D spatial arrangement of each cell type. In both models, "i" is input array of light stimulus. All other variables, and parameters are defined in comments (!) interspersed with the equations. To simulate the models, the input (i) and all equations are updated each simulation time step. Simulation time step for the Cray and PIPE models were respectively 20 and 66 ms. Mask(d,s) in the Cray model refers to a 2-D Gaussian convolution mask of diameter d and standard deviation s. Formally, the magnitude at distance x from the mask center is: $\exp(-1/2*(x/s)^2)$ if x <= d, 0 otherwise. * stands for 2-D convolution, • for multiplication. "norm" is a function normalizes x to the range 0-1. Formally norm(x,lo,hi) = 0, if x <= lo; (x-lo)/(hi-lo), if lo < x < hi; and hi, if x >= hi. Other functions are: max(a,b) = a, if a > b; b otherwise. pos(a) = 1, if a > 0, 0 otherwise, and the if-then-else construct.

Cray Model:

Receptor (r).

$$\frac{1}{T} \cdot \frac{dr}{dt} = (1 - i) - r \qquad ! \; T = 30 \text{ ms}$$

Horizontal cell (h).

$$\frac{1}{T} \cdot \frac{dh}{dt} = \text{mask}(1600,2500) * r - h \qquad ! \; T = 180 \text{ ms}$$

Bipolar cell body (sustained and transient, b).

pbin = mask(110,50) * max(r-h, 0)

$$\frac{C}{(Wr + We \cdot pbin)} \cdot \frac{db}{dt} = \frac{Vr \cdot Wr + Ve \cdot We \cdot pbin}{Wr + We \cdot pbin} - b$$

With: Vr = -60mv, Wr = 0.5ps, Ve = 0mv, We=0.9ps, C = 5pf (cell capacitance)

Sustained bipolar cell terminal (bst).

$$\frac{C}{(Wr + Wb)} \cdot \frac{d\,bst}{dt} = \text{mask}(110,55) * \frac{Vr \cdot Wr + b \cdot Wb}{Wr + Wb} - bst$$

With: C = 0.1pf, Vr = -60mv, Wr = 0.1ps, Wb = 3ps (Cell body => Axon Terminal Cond).

Synapse from bipolar terminal to narrow field amacrine.

$$\frac{1}{T} \frac{d\,fbs}{dt} = bst - fbs \qquad ! \; T = 20\text{ms}$$

obs = norm(fbs,Vlo,Vhi) ! Vlo = -60mv, Vhi = 0mv

Narrow field amacrine.

nobs = mask(200,138) * obs

$$\frac{C}{(Wr + nobs \cdot We)} \cdot \frac{d\,an}{dt} = \text{mask}(200,138) * \frac{Vr \cdot Wr + Ve \cdot We \cdot nobs}{Wr + nobs \cdot We} - an$$

With: Vr = -65mv, Wr = 0.5ps, Ve = -10mv, We=21ps, C = 5pf

Synapse from narrow field amacrine (oan).

$$\frac{1}{T} \frac{d\,fan}{dt} = \text{mask}(200,138) * an - fan \qquad ! \; T = 80\text{ms}$$

oan = norm(fan,Vlo,Vhi) ! Vlo=-30mv, Vhi=0mv

Transient bipolar cell terminal (btt).

soan = mask(110,55) * oan ! input from narrow field amacrine

$$\frac{C}{(Wr+Wb+Wi \cdot soan)} \cdot \frac{d\,btt}{dt} = mask(110,55) * \frac{Vr \cdot Wr+b \cdot Wb+Vi \cdot Wi \cdot soan}{Wr+Wb+Wi \cdot soan} - btt$$

btt = if btt <= Thr then btt else btt/Div ! Ca channels activate above Threshold Thr

With: C=.1pf, Vr=-60mv, Wr=.1ps, Wb=3ps, Vi=-75mv, Wi=22ps, Thr=-45mv, Div=5

Synapse from transient bipolar terminal (obt).

fbt = max(btt,Thr) ! Thr = -50
obt = norm(fbt,Vlo,Vhi) ! Vlo=-50, Vhi=10

Thresholded High pass filter input to ganglion cell (hbt).

$$\frac{1}{T} \cdot \frac{d\,hbx}{dt} = obt - hbx$$! T = 400ms

hbt = max(obt - hbx, 0)

Wide field amacrine input and pre-spike potential (aw).

woan = mask(200,138)*oan ! spatial input from narrow field amacrine
wobt = mask(200,138)*obt ! " " transient bipolar terminal

$$\frac{C}{(Wr+wobt \cdot We+woan \cdot Wi)} \cdot \frac{d\,aw}{dt} = \frac{Vr \cdot Wr+wobt \cdot Ve \cdot We+woan \cdot Vi \cdot Wi}{Wr+wobt \cdot We+woan \cdot Wi} - aw$$

With: C=20pf, Vr=-70mv, Wr=1.052ps, Vi=-70mv, Wi=17ps, Ve=-10mv, We=21ps

Wide field amacrine spiking and lateral transmission (aws).

```
if aw-Vlo-awk > Thr then      ! Thr = 30mv, Vlo = -70mv
        aws = 0
        awk = Thr
else
        aws = mask(1000, 330) * aw
```

$$\frac{1}{T} \cdot \frac{d\,awk}{dt} = -\,awk$$! T = 10ms

endif

Synapse from wide field amacrine (oaw).

oawb = norm(aws_{t-1},Vlo,Vhi) ! Vlo=-70 Vhi=10
oaw = oawb / (oawb + Vmid) ! Vmid = 0.05 (Nonlinear transmitter release).

Ganglion cell (g).

ghbt = mask(266,95) * hbt ! Spatial input from bipolar

$$\frac{1}{T}\frac{d\,goaw}{dt} = mask(266,95) * oaw$$! Spatial input from amacrine. T=200ms

$$g = \frac{Vr \cdot Wr+Ve \cdot We \cdot ghbt+Vi \cdot Wi \cdot goaw}{Wr+We \cdot ghbt+Wi \cdot goaw}$$

With: Vr=-32mv, Wr=2.225ps, Ve=75mv, We=25ps, Vi=-60mv, Wi=25ps

PIPE Model:

Receptors.

$L = squeeze(A*i)$

$S_{t+1} = w \bullet (A*S_t) + (1-w) \bullet L_{t+1}$! w=.1

$R_{t+1} = A*S_t$

Where squeeze(x) selects every other element in both horizontal and vertical directions (similar to all black squares on a chess board). A is the 3x3 averaging convolution mask with all elements equal to 1/9, * denotes convolution, • denotes multiplication, and w is a constant between zero and one.

Horizontal Cells.

$P_{t+1} = w \bullet (A*P_t) + (1-w) \bullet R_{t+1}$! w=.6

$Q_{t+1} = w \bullet (A*Q_t) + (1-w) \bullet A*P_{t+1}$! w=.7

$H_{t+1} = A*Q_t$

Bipolar Cells.

$BH_t = R_t - H_t$

$BD_t = H_t - R_t$

Bipolar Cell terminals.

$PH_{t+1} = w \bullet PH_t + (1-w) \bullet Thresh(BH_{t+1}, 20)$! w=.6

$TH_{t+1} = Thresh(BH_{t+1}, 20) - PH_t$

$PD_{t+1} = w \bullet PD_t + (1-w) \bullet Thresh(BD_{t+1}, 20)$! w=.6

$TD_{t+1} = Thresh(BD_{t+1}, 20) - PD_t$

With Thresh(x,y) = x if x >= y, zero otherwise

Wide field amacrine cell.

$W1_{t+1} = spread(GE_t \text{ or } W3_t)$

$W2_{t+1} = spread(W1_t)$

$W3_{t+1} = W2_t / 2$

$W4_{t+1} = w \bullet (A*W4_t) + (1-w) \bullet W3_{t+1}$! w=.1

$GI_{t+1} = Tri(W4_t)$! Inhibitory input to On-Off Ganglion cell

Where spread(x) = sets each bit in xi,j to 1 if it or the corresponding bit in any neighbor element is equal to 1. Formally, xi,j = ($x_{i-1,j-1}$ or $x_{i-1,j}$ or $x_{i-1,j+1}$ or $x_{i,j-1}$ or $x_{i,j}$ or $x_{i,j+1}$ or $x_{i+1,j-1}$ or $x_{i+1,j}$ or $x_{i+1,j+1}$). "or" is the logical bitwise or operation. Tri(x) = 0 if x < 3, 140 otherwise).

On-Off Ganglion cell.

$GE_{t+1} = Spike(TH_t) \text{ or } Spike(TD_t)$! Excitatory input to On-Off ganglion cell.

$GC_{t+1} = Max(GE_t - GI_t, 0)$

where: spike(x) = FF_{16} if x > 2, 0 otherwise
"or" is logical bitwise or operation.
Max(a,b) = a if a >= b, otherwise b.

Properties of Retinal and Retino-tecto-isthmo-tectal Units in Frogs

Frédéric Gaillard & René Garcia

*Laboratory of Neurophysiology, UA CNRS 290, Faculty of Sciences,
40 avenue du Recteur Pineau ; F-86022 Poitiers, France.*

Abstract: *In the frog Rana esculenta, only two main populations of units can be defined without ambiguity through the parameters of their velocity function, ERF sizes, responses to diffuse light illumination, responses to stationary targets, and optimal stimulus diameter. Parameters of the velocity function cannot be used alone to determine with confidence the precise functional class (R1-R3) of a given retinal cell. Units found in the ipsilateral visuotectal projection respond in a manner strongly similar to contralateral units. Present data suggest that the major retinal informations are preserved through their crossed isthmotectal transfert.*

1. Introduction

Retinal ganglion cells in frogs that terminate in the superficial neuropile of the optic tectum are classically divided into four main groups (Hartline 1938; Lettvin et al. 1959; Maturana et al. 1960; Grüsser & Grüsser-Cornehls 1976). Class R1 neurons display a sustained discharge when a moving target is suddenly stopped within the receptive field (ERF). The response ceases when the ambiant light is turned off but reappears at light on. Class R2 neurons behave like R1 neurons but their tonic firing to stationary targets is erased after a transient step in darkness. Class R3 neurons show a phasic response to punctate or diffuse ON and OFF light stimulations but do not discharge to stationary targets. Finally, class R4 neurons exhibit a long lasting firing when lights are turned off. All these retinal ganglion cells respond vigorously to targets moved in any direction throughout their ERF. This commonly used classification has been challenged by many authors (even by the pionneers themselves) who either found retinal neurons with intermediate functional properties (Maturana et al. 1960; Pomeranz 1972; Reuter & Virtanen 1972; Backström & Reuter 1975) or rejected the existence of a clear separation between some classes (Keating & Gaze 1970a; Witpaard & Ter Keurs 1975; Hodos et al. 1982).

In parallel, ipsilaterally-driven visual units can also be recorded throughout the optic tectum except caudolaterally. The ipsilateral visuotectal linkage (IRTT pathway) is an indirect route which needs first the integrity of the contralateral retinotectal system; second, tectal cells able to transfert the retinal signals toward the nucleus isthmi; and third, isthmic cell axons to drive informations from there to the opposite tectum (ipsilateral to the stimulated eye in fact). This neuronal circuitry, first specified by Keating & Gaze (1970b) through electrophysiological mapping studies, is valid for both optic lobes so that any point in the fronto-superior visual field is binocularly seen at single tectal loci and bitectally represented (Gaze & Jacobson 1962). Two types of visually driven ipsilateral (IRTT) units have been described: a superficial "sustained, tonic" I1 type located in layer 9A and a deep "event, transient " I2 type located in layers 9F and 8 (Gaze & Keating 1967, 1970; Grüsser & Grüsser-Cornehls 1968, 1969; Gruberg & Lettvin 1980).

In the seventies, a decisive contribution to our knowledge of the functional properties of single visual units in frogs was also brought by Grüsser et al. (1965 -76). They showed that under constant experimental conditions, (1) the relationship between the mean firing frequency (R) of any visual neuron and the angular velocity of the target (V) fits well a power function:

$$R = k(v) . V^{\alpha} \text{ (impulses / s) (Eq. 1)}$$

where the exponent α appears highly selective for each class of cell and is independent of the experimental conditions whereas the constant k(v) also but less specific (it represents nevertheless the neuronal discharge for a stimulus velocity equal to unity) varies with physiological and technical parameters;

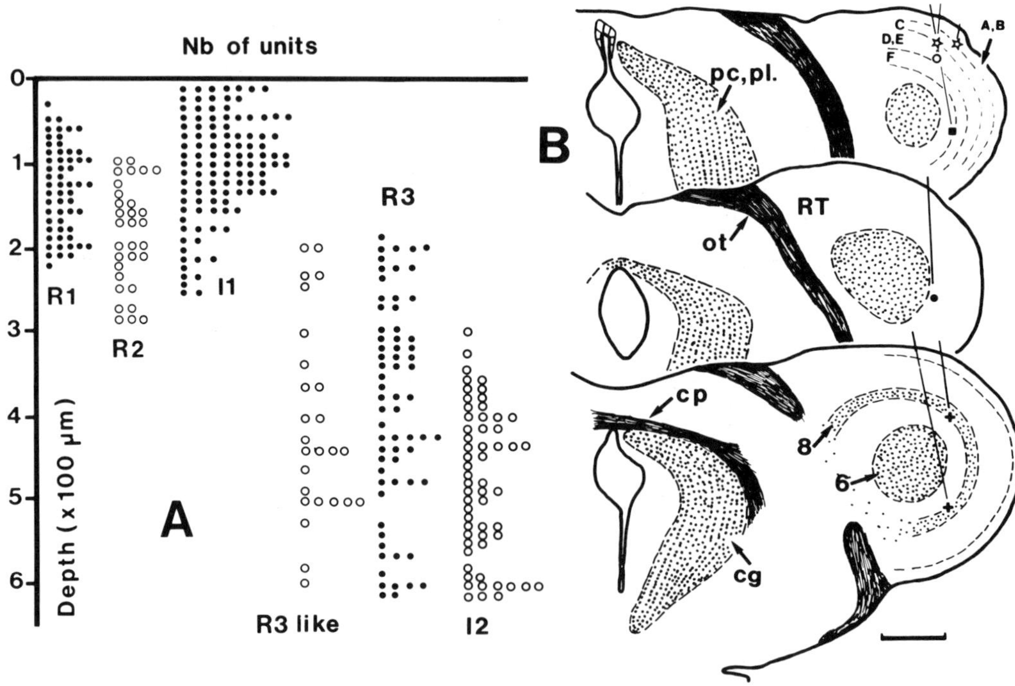

Figure 1. Recording sites. (A) Location in depth of the units studied as deduced from micromanipulator readings. (B) Camera lucida drawings of some unit positions in the right tectum (RT) as given from Alcian Blue deposits. Stars, filled square, open circle and black dot represent respectively I1, I2, R1, R2, and R3 units. Crosses in layers 6 and 8 designe tectal neurons. Scale bar: 350 µm (Garcia & Gaillard 1989a).

and (2) that the relationship between the mean firing frequency (R) and the size of a moving stimulus (D) can be expressed at best by a logarithmic function

$$R = k(d) - (a)\ \mathrm{Log}\,|\,(D / ERF)\,| \quad \text{(impulses/s)} \quad \text{(Eq. 2)}$$

where ERF is the area of the excitatory receptive field, k(d) and (a) are constants. The optimal response is obtained for a characteristic Dopt value.

As a consequence, quantitative studies on velocity and area functions might appear as a powerful tool for classification as well as a useful argument to prove that visual afferents to a "binocular tectal column" respond in the same way to moving objects (Gruberg & Lettvin 1980). This paper reports data obtained the last five years in the common water frog, *Rana esculenta* .

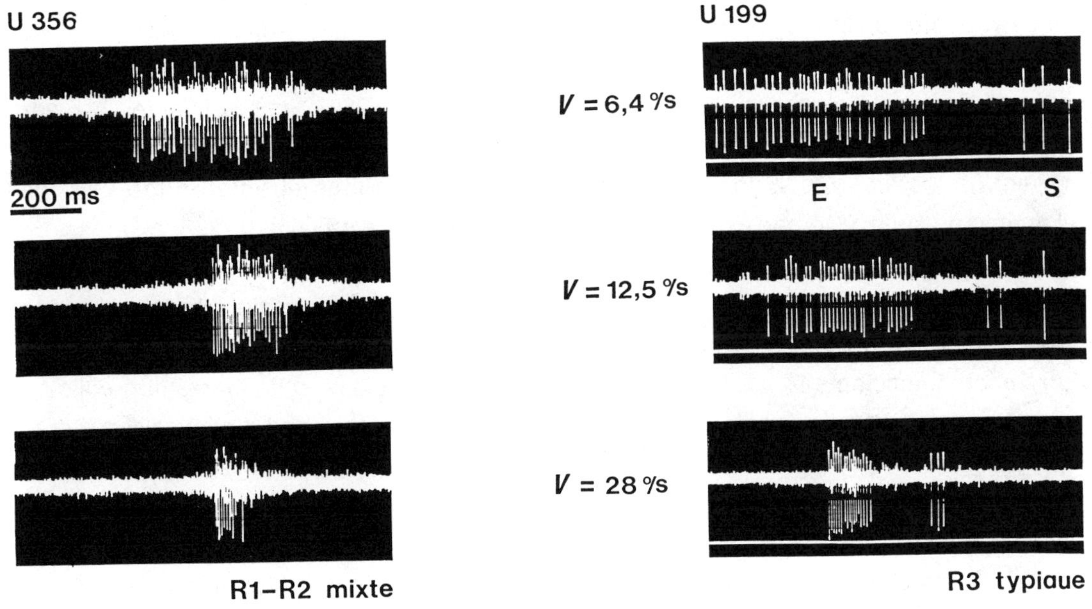

Figure 2. Typical recordings from R2 (D = 0.64°) and R3 (D = 4.6°) units. Note frequent responses to the trailing edge (S) of the stimulus in R3 cells (Garcia 1988).

2. Experimental procedure

Experiments were performed on about two hundred adult frogs (50g b.w.) kept in a vivarium under standard laboratory conditions (water temperature 16°C ; normal light-dark rhythm). After surgery, frogs were placed on a micromanipulator platform facing 25 cm away an X-Y plotter used as stimulating screen (homogeneous white background; luminance Lb = 7.4 cd.m^{-2}). Visual responses were recorded with glass micropipettes filled with 2M NaCl (tip diameter 2.5-7.0 μm; impedance ≈ 2.5 MΩ at 1000Hz) and fed onto a standard electrophysiological channel (cut-off frequency 150-4000 Hz). Recordings were made from the superficial neuropile of the rostrolateral part of the optic tectum (Fig. 1). Visual units were first classified using qualitative criteria (Maturana et al. 1960) and then stimulated with black disks (0.6° - 15° in diameter; luminance Ls = 0.9 $cd.m^{-2}$; contrast | C | = (Lb - Ls) / (Lb + Ls) = 0.8) automatically moved in the vertically upwards through the ERF at various linear velocities (Vl = 0.07 -37 $cm.s^{-1}$; Fig. 2). Test stimuli were each presented 1-3 times per session and in a pseudo-random order. Successive trials were separated by a 2 min interval to prevent neuronal adaptation..

The main stimulus parameters were calculated as described earlier (Garcia 1988; Garcia et al. 1988 ; Garcia & Gaillard, 1989a, b). The value of the stimulus visual angle (D) was given knowing its absolute diameter and the distance between the frog's eye and the ERF center. The angular velocity (V) of the stimulus was derived from the linear amplitude of the stimulus velocity with the appropriate tangent correction depending on the position of the ERF center on the experimental screen. The activity of any single ganglion cell was expressed usually as the mean impulses per second measured from the first to the last spike of the whole discharge (Method A; Grüsser et al. 1967; Schürg-Pfeiffer & Ewert, 1981) excepted when discharges at both the leading and trailing edges of the target were present. In that case, only the spike train elicited when the target entered the ERF was taken into account (Method B'; ≈ 30-40% of the units; Garcia & Gaillard, 1989a).

One hundred and ninety three retinal ganglion cells were recorded long enough to be quantitatively studied. Data were computed as follows. First, linear stimulus-response relationships were determined for each single unit by regression analysis. Results with correlation coefficients $r^2 < 0.95$ were rejected. For each class, data were then collected according to appropriate target diameters (Table 1) and proceeded statistically either (1) from the formerly calculated {k(v); α} values so that an "average" velocity function for that group was given by:

$$R = [1/n. \sum k(v)] . V^{(1/n. \sum \alpha)} \quad \text{(Eq. 3)}$$

or (2) by a direct fit of all the experimental { R; V }-paired values, { k(v); α } values being in that case least squares estimates of the parameters of the regression line. Both methods gave nearly identical results (Table 1). Statistics and regression analysis were performed using Statworks™ and Anova™ programs adapted for a MacIntosh microcomputer (Apple).

3. Existence of two broad populations of retinal ganglion cells

When pooled together irrespective of the functional properties of each retinal ganglion cell under investigation, all the values of the exponent α were found to be between 0.30 and 1.13. These values (N = 193) were however not homogenously distributed around the mean (p = 0.03; Chi-square test) but the observed distribution (Fig. 3A) failed to show distinct neuronal classes. As for the values of the constant k(v) , a bimodal repartition of the retinal ganglion cells could only be suspected (Fig. 3D). Values of the exponent α of the velocity function of any retinal neuron appeared thus not selective enough to define a valid classification when used singly. Most of the ERF sizes ranged between 2° and 8° (Fig.3G).

Introducing now for classification either the "sustained"' or the "event" qualitative criterion (Keating & Gaze, 1970a), two main populations of visual units emerge clearly: (1) units producing a tonic, sustained response to movement-gated stationary targets but no reaction to diffuse light ON-OFF stimulations. They show α values between 0.30 and 0.75 (mean = 0.52; S.D. = 0.10; N = 98; Fig. 3B); and (2) units responding to diffuse light ON-OFF stimulations but without sustained reaction to movement-gated stationary targets. They show α values between 0.56 and 1.13 (mean = 0.83; S.D. = 0.13; N = 95; Fig. 3C). The difference between these two mean values of the exponent α is highly significant ($p = 0.001$; df = 191; t-test). The addition of a rather simple qualitative criterion to each value of the exponent α of the velocity function appeared therefore of a critical importance to reveal at least the existence of two main populations of retinal ganglion cells in frogs. But no more because (i) the distribution of the α values is normal in the population of the sustained units ($p = 0.20$; Chi-square test); and (ii) the existence of two peaks in the distribution of the α values of the event units is barely significant ($p = 0.05$; Chi-square test).

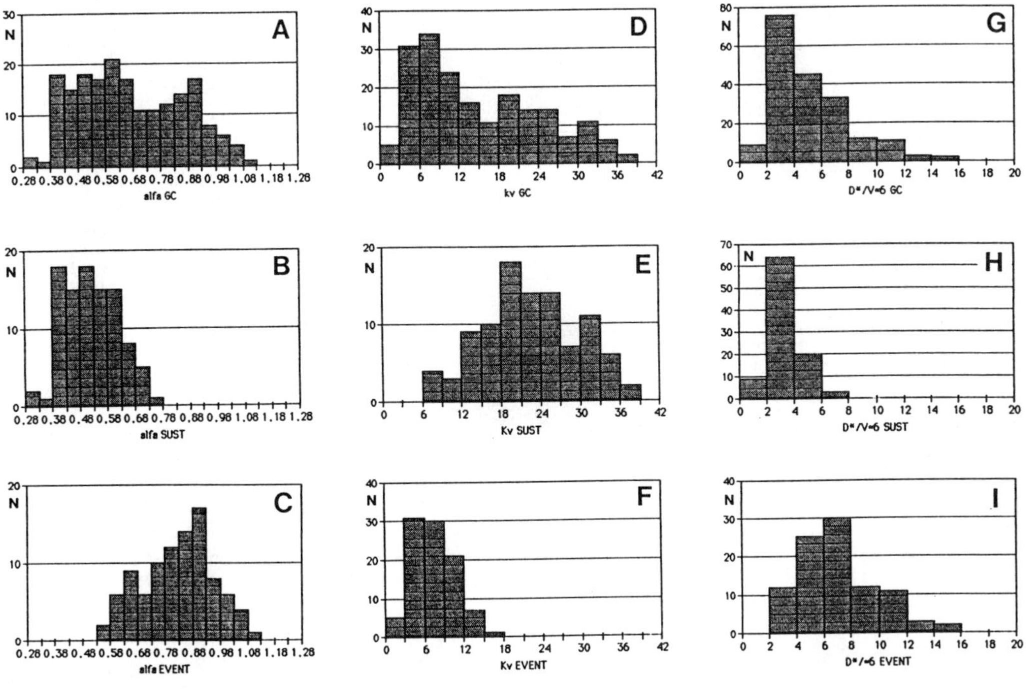

Figure 3. Distributions of α, k(v) and D* values as collected from each experiment. First row: whole population of the ganglion cells (GC). Second row: sustained units (SUST). Third row: event units (EVENT). D* values are those obtained with small targets (D = 2°-4°) moved at V ≈ 6°/s

In parallel, three other criteria studied quantitatively support unequivocally the existence of two functional populations of retinal cells in the frog. First, independent of the experimental conditions, values of the constant k(v) in the sustained units were found to be between 8.2 and 37.7 (mean = 22.6; S.D. = 7.4; N = 98; Fig. 3E) while event units have k(v) values between 1.46 and 17.3 (mean = 7.4; S.D. = 3.2; N = 95; Fig. 3F). Both distributions appear unimodal. The difference between mean values is statistically significant (p = 0.001; df = 191; t-test). Second, there is also a strong dichotomy between the ERF sizes (D*) when measured with targets moved at a low velocity (V ≈ 6°/s; Schürg-Pfeiffer & Ewert 1981; Grüsser-Cornehls, 1988). Sustained units have rather small ERF (D* = 3.3°; S.D. = 1.1° ; N = 98), most of the collected values (90 %; Fig. 3H) being included between 2.0° and 6.0°. In the opposite, event units have larger ERF (D* = 7.1°; S.D. = 2.6° ; N = 95) distributed over a wide range (3.0°-15.0°), the most frequent values (80 %; Fig. 3I) being found between 4.0° and 12.0°. Distributions of the ERF sizes between both groups are also significantly different (p = 0.001; df = 189; t-test). Third, optimal target diameters able to induce a maximal reponse in each group of cells are clearly distinct. Sustained and event units are maximally activated by 2.0°-3.0° and 4.5°-7.5° targets, respectively (see below).

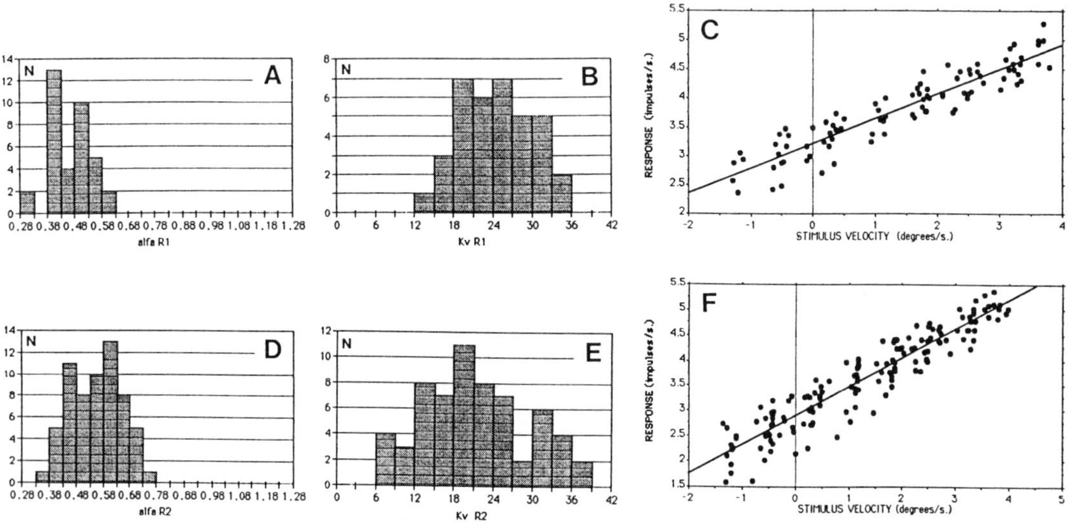

Figure 4. Velocity function studies on sustained retinal ganglion cells. Upper row: class R1 units. Lower row: Class R2 units; (A, D) Distributions of the values of the exponent α. (B, E) Distributions of the constant k(v). (C, F) Velocity functions as obtained by regression anlysis of all the individual {R,V}-paired values. Units stimulated with 0.6°<D<1.6°. Constants {k(v), α} are {24.9,0.43} and {18.2,0.57} respectively. Abscissa: log. of the stimulus velocity. Ordinate: log. of the response.

4. Velocity function in "sustained" ganglion cells

Sustained units could be divided into two major classes when using the erasable character of their tonic firing to targets stationary held in the ERF as the main discriminating criterion. Class R1 neurons were found to be between 50 and 200 μm in depth from the pia. ERF sizes were about 1.7°-5.8° (D* = 3.2° ± 1.1°; N = 36). Responses to movement-gated standing objects lasted approximately 30-50 s. and were not erasable by a transient step to darkness. No response to diffuse light ON-OFF stimulations was observed. Such units have α values between 0.30 and 0.60 (mean = 0.46; S.D. = 0.07; N = 36; Fig. 4A). These values seem to increase slightly with the target diameter from 0.43 (0.6° < D < 1.6°; D/D* ≈ 0.3; Fig. 4C) to 0.50 (5.6° < D < 7.6°; D/D* ≈ 2.0), both values being statistically different ($p = 0.04$; df = 19; t-test; Table 1). Values of the constant k(v) are distributed between 14.9 and 35.5 (mean = 24.7 ; S.D. = 5.7 ; N = 36; Fig. 4B) with a maximum for medium-sized targets (1.6° < D < 3.6°; D /D* ≈ 0.73).

Class R2 neurons were found slightly deeper in the superficial neuropile (100-300 µm). ERF were about 3.4° in diameter (S.D. =1.1°; N = 62). They habituated rapidly, ceasing to respond after the fourth or fifth stimulus presentation at a rate of 0.5/s and the sustained, movement-gated reaction to stationary targets (usually 20-50 s. in duration) always disappeared with a transient step to darkness. Here again, no response to diffuse light ON-OFF stimulation was observed. These units have α values between 0.36 and 0.75 (mean = 0.56 ; S.D. = 0.10; N = 62) i.e. lower than those reported currently (range = 0.62-0.72 ; Grüsser et al., 1967; 1968a, b). The distribution of the values of the exponent α is normal ($p = 0.5$; Chi-square test) suggesting the existence of a unique type of erasable units (56 % of them having however "R1-like" α and k(v) values; Fig. 4D-E; Garcia & Gaillard, 1989b). Values of the exponent α are rather constant independent of the target diameter, at least whithin the range of the stimulus sizes presently used. Values of the constant k(v) in the sustained erasable units appear widely distributed between 8.2 and 37.7 (mean = 21.3; S.D. = 7.9; N = 62; Fig. 4E). They show a maximum for medium-sized targets (1.6° < D < 3.6°; D/D* ≈ 0.73) and this value is statistically higher than those found either for small (0.6° < D < 1.6°; D/D* ≈ 0.30) or for large (5.6° < D < 7.6°; D/D* ≈ 2.0) targets ($p = 0.04$ in both cases; t-test). Velocity curves of both R1 and R2 ganglion cells (Fig. 4C, F) are statistically different when tested with small stimuli (D ≤ 3.5°; i.e. D/D* ≈ 0.73) but not when large targets are used (D ≥ 5.6°).

5. Velocity function in "event" ganglion cells

As suggested by the bimodally peaked distribution of the values of the exponent α of the velocity function (Fig. 3C), there would exist two functionally different groups of event units. These two groups may be better defined considering their reaction to various stimulus conditions. Event units in the major group (69.5 %) were recorded at depths between 200 and 600 µm. ERF were about 6.6° in diameter (S.D. = 2.6°; N = 66; Fig. 5C). A tonic response was well obtained to moving but not to stationary targets whatever was their diameter (D = 0.6°-30°). Adaptation to repetitive stimulations at a rate of 0.5/s was absent. They showed always a strong but brief phasic response at both ON and OFF light stimulations (diffuse or point source). These units have α values ranged between 0.73 and 1.13 (mean = 0.89; S.D. = 0.08 ; N = 66; Fig. 5A).The exponent α is constant, independent of the target diameter, the observed variations being not statistically significant (an unexpected low value of the exponent α was however obtained for stimuli approximatively equal to the ERF size). These units also have a constant k(v) of the velocity function between 1.5 and 11.2 (mean = 6.1; S.D. = 2.3; N = 66; Fig. 5B). This constant k(v) is dependent of the stimulus diameter being maximum with targets equal to the ERF size (Table 1). These units are "typical" class R3 units.

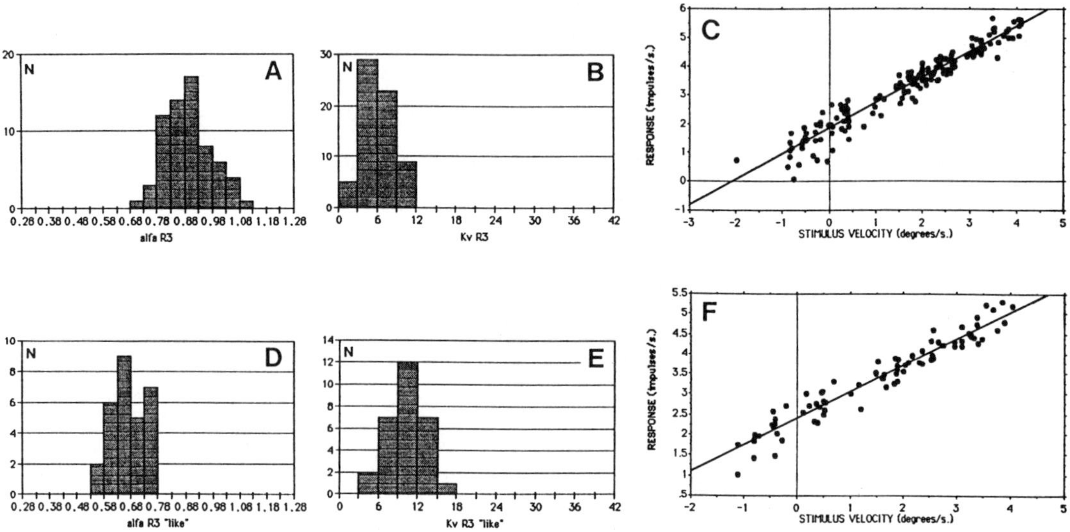

Figure 5. Velocity function studies on event retinal ganglion cells.Upper row: class R3 units. Lower row: class R3-like units. (A, D) Distributions of the exponent α . (B, E) Distributions of the constant k(v). (C, F) Velocity functions as determined in Figure 4. Event units stimulated with 3.6°<D<5.6°. Constants {k(v), α} are {6.4, 0.89} and {11.1, 0.66} respectively.

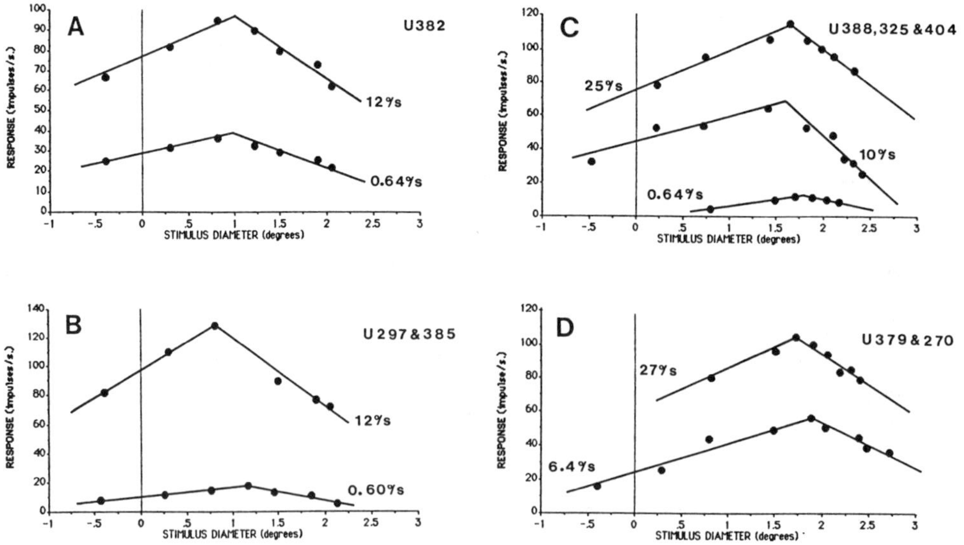

Figure 6. Area functions studies on retinal ganglion cells. Curves obtained for various stimulus velocities. (A) R1 units. (B) R2 units. (C) R3 units. (D) R3-like units. Note specific Dopt values in sustained and event units. Abscissa: logarithm of the stimulus diameter

Event units in the minor group (30.5 %) were recorded among the preceeding R3 ganglion cells and displayed the same functional properties except for (1) a tendency to possess larger ERF sizes ($D^* \approx 8.1° \pm 2.6°$; N = 29); and (2) a more discrete reaction to diffuse light stimulations: some of them responded at both ON and OFF, one component of the response showing a strong adaptation (55%); while others responded either at ON (28%) or OFF (17%) only. Slowly moved stimuli were always preferred (Garcia and Gaillard, 1989a). Values of the exponent α and of the constant k(v) are distributed between 0.56 and 0.77 (mean = 0.67; S.D. = 0.06; N = 29; Fig. 5D) and between 3.7 and 17.3 (mean =10.4; S.D. = 2.8; N = 29; Fig. 5E), respectively. Average values of the exponent α as well as of the constant k(v) remain stable with the target diameter (statistically non significant) although maximal k(v) values are always observed in single unit studies with targets 5°-7° in diameter. Finally, α and k(v) values obtained in this "R3-like" group of event units are systematically different from those obtained in "typical" R3 units (Garcia and Gaillard, 1989a) independent of the stimulus diameter (Fig. 5F; Table 1).

6. Area function in "sustained" and "event" ganglion cells

The area function of retinal units was further investigated on 16 class R1, 11 class R2, 4 "R3-like" and 22 "typical" R3 ganglion cells by moving stimuli of increasing diameters (D) at a constant angular velocity (choosen between 0.6 and 30°/s; Fig. 6).

As currently reported, the mean firing frequency of all these retinal cells increases with D up to a maximal value (Rmax) obtained with an optimal stimulus size (Dopt) and then decreases for larger stimuli. Dopt values are about 2.5° (range = 2.0°-3.2°; Fig. 6A-B) in R1 and R2 "sustained" units and about 6.0° (range = 4.5°-7.5°;Fig. 6C-D) in both groups of "event" units. In single units, increasing and decreasing R values fit the formula

$$R = k[d\ (1,2)] \pm [a\ (1,2)]\ .\ \log D \text{ (impulses / s)} \quad \text{(Eq. 4)}$$

Rising and descending slopes [a (1,2)] of the area function as well as Rmax are velocity dependent, but not Dopt. Searching for an adequate solution to express at best our experimental data lead us to the following observations: (1) for a given stimulus velocity, Rmax corresponds to the unit response obtained when the stimulus conditions are optimal, i.e. mainly when D = Dopt; (2) Rmax is velocity dependent so that it can be expressed as a simple power function with k(v) equal to k(opt); (3) there is a strong relationship between Dopt and the ERF size of any visual neuron so that Dopt = r . D*; (4) finally, it was assumed (Grüsser et al. 1968a; Ewert et al. 1979b) that D* values give accurate estimations of the "functional" receptive field sizes when small ($D \approx 3$-6°) and slowly ($V <$ 10°/s) moved targets are used as stimuli. As a consequence, we suggest

$$R = k(opt)\ .\ V^{\alpha} - |\ a\ (1,2)\ .\ \log\ (D\ /\ r\ .\ D^*)\ | \quad \text{(Eq. 5)}$$

as a possible mathematical expression of quantitative stimulus-response relationships for ganglion cells in the frog retina. Under our experimental conditions, values for {k(opt); α ; r} would be { 27; 0.46; 0.80}, { 24; 0.56; 0.75}, {11; 0.67; 0.75} and {7; 0.89; 0.90} in R1, R2, R3-like and R3 types, respectively.

7. Velocity function in ipsilaterally-driven units

Preliminary studies having shown that velocity and area functions of the ipsilaterally driven units projecting at tectal level through the IRTT pathway could be expressed at best by Eq. (1, 2), responses to these stimulus parameters were quantified and compared to the above data using identical experimental protocols (Fig. 7). When pooled together, one can observe that the values of the exponent α of the velocity of all the investigated IRTT units are largely distributed between 0.46 and 1.28 (Fig. 8A). However, this distribution is clearly bimodally peaked for α values equal about to 0.60 and 0.90 respectively ($p = 0.005$; Chi-square test). Similar observations can be drawn for the two other criteria represented in Figure 8D, G: (1) values of the constant k(v) (range = 0.46-20.5) although less clearly are also bimodally peaked; and (2) there is a strong dichotomy between the ERF sizes ($V \approx 6.0°/s$) some of them being around $D^* \approx 2°$-$6°$, and the other around $D^* \approx 10°$-$14°$. As reported above for the retinal ganglion cells, data derived from our velocity function studies would indicate that there would exist at least two populations of IRTT units at tectal level. Fortunately, only two families of ipsilateral units with more or less distinct properties have been reported up to now in the literature.

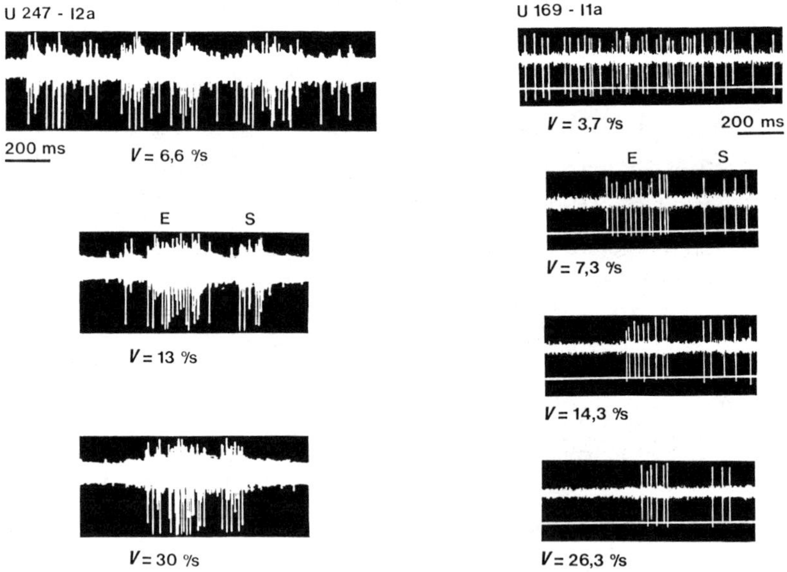

Figure 7. Typical recordings from type I2 units (D = 4.6°) and type I1 units (D = 2.7°) at different stimulus velocities.

Type I1 units (N = 58) were located immediately below the tectal surface, up to 250 μm in depth, 86% of the recorded sample being found between 50-150 μm (Fig. 1). ERF sizes were about 1.5°-10.5° (mean value = 4.30°; S.D. = 2.1°; Fig. 8H), 73% of them having $2° < D^* < 6°$. These units responded to moving as well as to movement-gated stationary targets 2°-4° in diameter. They could be thus considered as "sustained" ipsilateral units. However, the response duration to stationary objects was about 10- 15 s only, i.e much shorter than was observed initially by Gaze & Keating (1967). Complete adaptation to repetitive stimulations (f = 0.5 Hz) occured rapidly. Type I1 units failed to respond to changes in diffuse illumination but small spots (2° in dia.) evoked weak ON but strong OFF responses.

Most of the recorded type I1 units (N = 41; 71%) have α values ranged between 0.46 and 0.75 (mean = 0.60; S.D. = 0.07; N = 41; Fig. 8B) and k(v) values between 6.4 and 20.5 (mean = 12.6; S.D. =3.4; N = 41; Fig. 8E). The exponent α appears relatively constant as the stimulus size is increased although low values were obtained with medium targets (1.6° < D < 3.6°; D/D* ≈ 0.55). In the opposite, k(v) values show a maximum for 1.6 < D < 5.6° but this average value is not satistically different from those obtained either with small (D < 1.6°) or large (D > 5.6°) targets. However, a net increase for k(v) but a weak decrease of α was observed in all the single units when tested first with a 0.6° target and then with a 2.2° target. Power functions (Fig. 9A) are usually valid between 0.3-30°/s.

The remaining recorded units (N = 17; 29% of our sample) did not fit this general scheme: in 8 units, high α values (0.8-0.9) and low k(v) values (7-11) were found; in 3 more units, discharges were not affected by the stimulus velocity; in 3 other units, responses showed considerable variations; and in the last 3 units, results could not be expressed by a power function .At present, we have too few data to decide if these units are representative subgroups of ipsilateral type I1 units.

Type I2 ipsilateral units (N = 29) were found at a deeper level in the optic tectum (between 300 and 600 μm below the pia when measured from micromanipulator readings; Fig. 1), admixed with "event" ganglion cells. ERF sizes ranged between 5.0°-18° (mean value D* = 11.7°; S.D. = 3.5°; Fig. 8I), the majority of them (70%) being included between 8 and 14°. Type I2 units responded well to black disks of any size (2°< D < 30°) moved through their ERF, with a net preference for the leading edge. Neither contrast nor directional selectivity was observed. Complete adaptation to repetitive stimulations (f = 0.5 Hz) occured rapidly. They were also activated by brisk changes in the background illumination as well as by light spot stimulations. Finally, these units were not sensitive to movement-gated stationary targets, so that they could be considered as "event" units.

All but two type I2 units showed values of the exponent α of the velocity function distributed between 0.74 and 1.28 (mean = 0.93; S.D. = 0.12; N = 27; Fig. 8C, 9B). A net peak can be observed for $0.8 < \alpha < 1.0$ (70% of the values). Values of the exponent α are constant, independent of the stimulus diameter. At the same time, values of the constant k(v) were found to be between 0.5 and 8.7 (mean = 3.7; S.D. = 2.0; N = 27; Fig. 8F), 89% of them being inferior to 6. These values also are independent of the stimulus size (statistically non significant) although largest values were obtained with the most efficient stimuli (D ≈ 6°). The two remaining units exhibited $\alpha \approx 0.6$.

From these velocity function studies, we can thus say that type I1 ipsilateral units seem to respond to moving visual targets as "sustained" R2 ganglion cells do (p= 0.09; t-test) especially when targets 1.6°-3.6° in dia. are used. Conversely, type I2 ipsilateral units show values of the exponent α of the velocity function strongly similar to the values found in "event" R3 ganglion cells (p = 0.1; t-test). In both cases however, k(v) values are lower than those observed for the corresponding retinal neurons, meaning that IRTT units really have a weaker and a more sluggish firing pattern than retinal ganglion cells (Gruberg & Lettvin 1980).

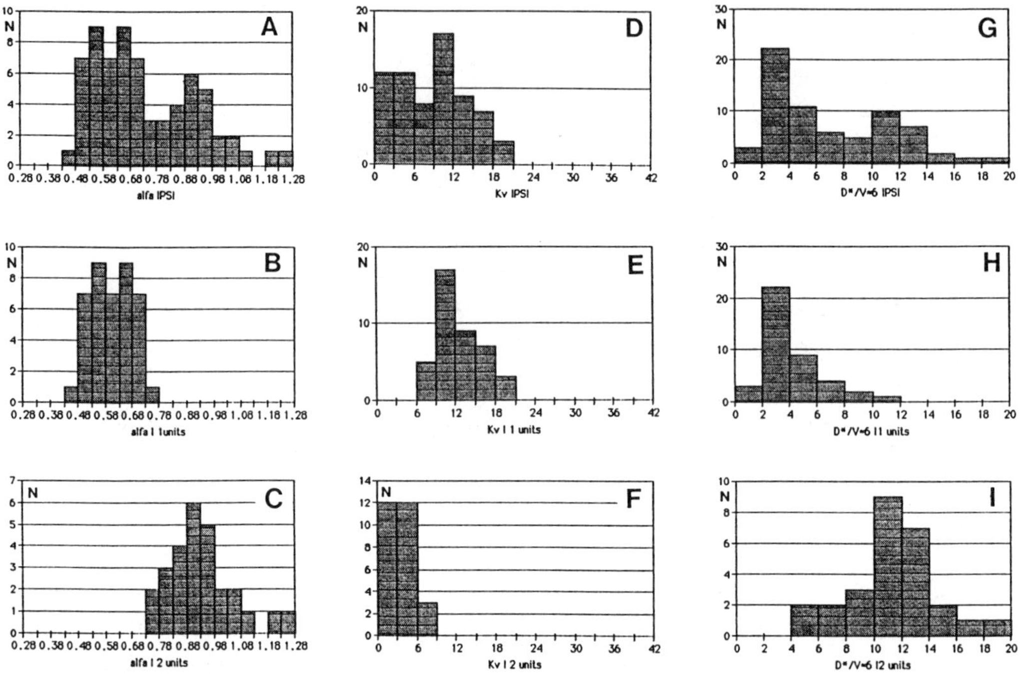

Figure 8. Distributions of α, k(v) and D* values as collected from each experiment in the ipsilateral projection. First row: in the whole population of the ipsilateral units (IPSI). Second row: in the sustained I1 units. Third row: in the event I2 units. D* values are those obtained with small targets (D = 2°- 4°) moved at V ≈ 6°/s.

8. Area function in ipsilaterally driven units

The area function of the IRTT units was further studied on 28 type I1 units and 10 type I2 units by moving stimuli of increasing diameters (D) at constant angular velocities. The mean firing frequency of these ipsilateral units also increased with the size of the test stimulus up to a maximal value (Rmax) obtained for a specific target diameter (Dopt) and then decreased rapidly. Stimulus-response relationships in these units fit well Eq.4 valid for retinal ganglion cells. Dopt values were found to be about 2°-4° (mean ≈ 3.0°) and to be about 4.5°-7.5° (mean ≈ 6.5°) for type I1 units and I2 units, respectively (Fig. 9C-D). Rising and descending slopes [a (1,2)] of the area function as well as Rmax but not Dopt values are velocity dependent. Under our experimental conditions, a good fit between the unit responses and the two main stimulus parameters studied above was obtained by Eq. 5 previously established for retinal cells with {13.5; 0.60; 0.70} and {4.0; 0.93; 0.55} for the constants {k(opt); a ; r} in I1 and I2 units, respectively.

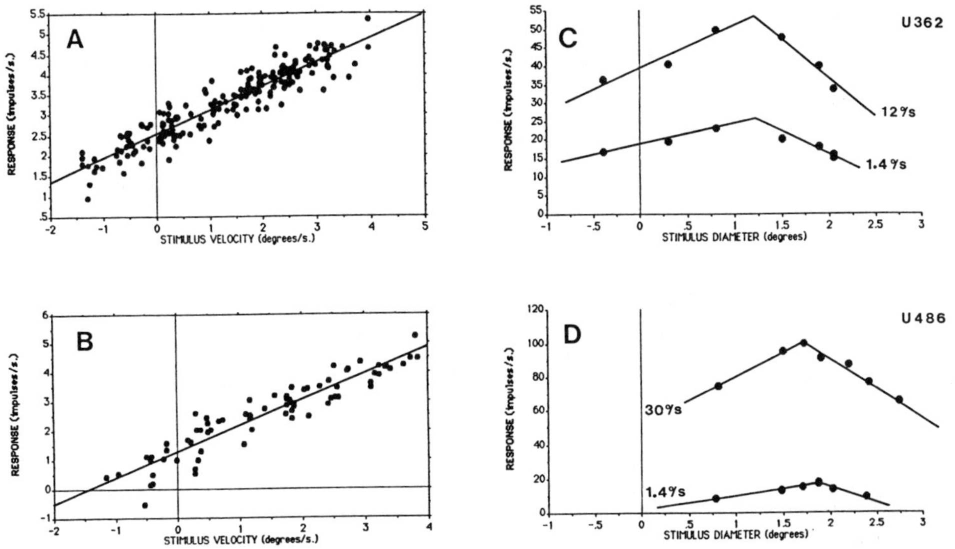

Figure 9. Velocity and area functions in ipsilateral I1 units (upper row) and in ipsilateral I2 units (lower row). (A, B) Velocity functions calculated as explained in Figure 4. Values of the constants {k(v); α} are {12.9; 0.59} and {3.6; 0.89} for I1 and I2 types, respectively. Units stimulated with $1.6° < D < 3.6°$. Abscissa: logarithm of the stimulus velocity. Ordinate: logarithm of the response. (C, D) Area functions expressed for different stimulus velocities. Note similarities in Dopt values with sustained and event ganglion cells, respectively. Abcissa: logarithm of the stimulus diameter.

9. "Sustained" R1 and R2 cells: comparison with previous data

The most helpful criterion to discriminate between class R1 and class R2 ganglion cells seems thus to test the erasibility power of a transient step in darkness on the neuronal tonic discharge elicited in response to targets stationary held in the field. As a matter of fact, this criterion appeared more satisfactory for us than for previous authors (Keating & Gaze 1970a; Pomeranz 1972; Reuter & Virtanen 1972; Witpaard & Ter Keurs, 1975) since 90% out of the sustained units have been classified undoubtedly either as R1 or R2 units.

Parameters of the velocity function of class R1 (non-erasable) ganglion cells reported in this work are roughly of the same order of magnitude than those currently assigned to this neuronal class. However, it is difficult to discuss the origin of the minor variations observed (75 % of the units in our sample have α values between 0.34 and 0.52) because the unique reference value reported at present for *Rana esculenta* (α = 0.5) has been deduced from studies on five single units only (Grüsser et al. 1968a). Further experiments are consequently needed to solve this problem.

Class R2 (erasable) ganglion cells represent the prominent part (63%) of the sustained ganglion cells population. They are also at the origin of a potent but unexpected divergence with currently known data. Finkelstein and Grüsser (1965) as well as Grüsser et al. (1967, 1968a, b) reported (1) that all the α values were betwen 0.62 and 0.78 (α mean = 0.7) and k(v) $\approx$ 20; and (2) that changing the contrast, the stimulus size or shape as well as the path through the ERF affects only k(v).

Our experiments give us a different picture since a vast majority of the recorded R2 units (56%) show both k(v) and α values similar to those found in class R1 units (Fig. 3A). These units, called "mixed R1-R2" units have been described in a recent paper (Garcia & Gaillard 1989b) despite there was really no experimental evidence to scatter the population of the erasable units in two distinct groups. In fact, the question to be raised here is why such units have never been listed before? A possible explanation is that group of "mixed R1-R2" units corresponds to sustained units showing "ambiguous" qualitative properties (Maturana et al. 1960; Pomeranz 1972; Reuter & Virtanen 1972; Backström & Reuter 1975; Gordon & Hood 1976). We cannot reject this hypothesis since six out of the recorded "mixed" erasable units ($\approx$ 10%) clearly showed such "ambiguous" properties. It is quite possible that some minor characteristics, potentially representative of "ambiguity" might have been underestimated during preliminary testing. However, too much low α values have been obtained on R2 (erasable) cells studied carefully to take this hypothesis seriously into consideration. We believe rather that the range of the exponent α of the velocity function of the erasable R2 units extends undoubtedly far below 0.62, overlapping thus widely the distribution of the α values found in class R1. This result, together with the fact that none of the other parameters presently studied show significant differences with those obtained in class R1, lead us to consider R1 and R2 units as two prominent types of a single functional family of "sustained" retinal ganglion cells as suggested originally by Maturana et al. (1960) as well as later by Keating & Gaze (1970a).

10. Validity of the classification of the "event" retinal cells

When qualitatively studied, event units resembled strongly each other but could be nevertheless differentiated into two groups through their reaction to diffuse light stimulation. Typical R3 ganglion cells reacted briskly but strongly to such a stimulation whereas the response of R3-like units was more sluggish and capricious. They also appeared different through their velocity function. Typical R3 units show $\alpha = 0.89$ whereas R3-like units have $\alpha = 0.67$ i.e. quite similar to that found in the uppermost R2 units of our scheme (Fig. 3B,C). Moreover, the distribution of all the α values collected for the event units appears slightly bimodally peaked. As a result, event units might be the unique ganglion cells family which could be divided in two groups on the basis of velocity function studies.

As a matter of fact, there are electrophysiological (Barlow 1953; Maturana et al. 1960; Ewert et al. 1979a; Witpaard & ter Keurs 1975; Stiles et al. 1985; Grüsser & Grüsser-Cornehls, 1976; Watanabe & Murakani, 1984; Grüsser-Cornehls & Saunders 1981; Grüsser-Cornehls & Langeveld 1985) as well as anatomical (Kahle & Grüsser-Cornehls 1988) data supporting the idea that the population of event units in *Rana* is heterogeneous. Most of the "deviating" R3 neurons recorded up to now resemble R2 units in their response to moving stimuli. There would be thus two groups of "event, changing contrast" detectors in frogs, one of them responding in a manner similar to "sustained, erasable" units. However we think there are actually not enough close correlations between all these experimental results to assume there exist two independent, clearly separated functional groups of class R3 units. Our impression is that quantitative stimulus-response relationships are a bit more powerful than qualitative (subjective) criteria to show subclasses in "event" ganglion cells, the reverse argument being valid for "sustained" ganglion cells.

Stim. dia.		R1	R2	R3L	R3	I1	I2
	N	13	20	1	2	6	
(A)	Kv	25.3 ± 6.2	19.2 ± 6.5	9.16	3.2 ± 1.6	10.7 ± 3.4	
	α	0.43 ± 0.08	0.56 ± 0.08	0.74	0.85 ± 0.1	0.64 ± 0.05	
		(α = 0.43)	(α = 0.57)				
	N	15	34	12	30	26	10
(B)	Kv	26.5 ± 4.9	23.8 ± 8.3	9.4 ± 2.8	5.9 ± 2.5	13.4 ±3.5	3.8 ± 2.4
	α	0.46 ± 0.07	0.55 ± 0.10	0.68 ± 0.05	0.90 ± 0.06	0.59 ± 0.08	0.92 ± 0.14
		(α = 0.46)	(α = 0.54)	(α = 0.67)	(α = 0.86)	(α = 0.59)	(α = 0.89)
	N		1	13	21	4	7
(C)	Kv		12.25	11.5 ± 2.9	6.6 ± 2.3	13.1 ± 2.2	3.5 ± 1.5
	α		0.67	0.67 ± 0.07	0.90 ± 0.09	0.59 ± 0.02	0.95 ± 0.11
			(α = 0.66)	(α = 0.89)			
	N	8	7	2	7	5	8
(D)	Kv	20.6 ± 4.7	16.6 ± 6.6	11.1 ± 2.3	7.34 ± 1.4	10.7 ± 2.8	4.2 ± 2.0
	α	0.50 ± 0.06	0.57 ± 0.10	0.69 ± 0.06	0.83 ± 0.06	0.63 ±0.05	0.89 ± 0.08
		(α = 0.50)	(α = 0.57)	(α = 0.68)	(α =0.80)		
	N			1	3		2
(E)	Kv			9.0	5.8 ± 0.8		1.5 ± 0.7
	α			0.71	0.94 ± 0.06		1.12 ± 0.22
					(α = 0.92)		
	N				3		
(F)	Kv				4.50 ± 1.4		
	α				0.85 ± 0.20		
ΣN		36	62	29	66	41	27
Kv mean		24.7 ± 5.7	21.3 ± 7.9	10.4 ± 2.8	6.1 ± 2.3	12.6 ± 3.4	3.7 ± 2.0
α mean		0.46 ± 0.07	0.56 ± 0.10	0.67 ± 0.06	0.89 ± 0.08	0.60 ± 0.07	0.93 ± 0.12

Table 1. Values of the constant k(v) (± S.D.) and of the exponent α (± S.D.) of the velocity function as found in the different unit classes according to the stimulus diameter. In parenthesis, values calculated by linear regression of individual {R;V}-paired values. (A) 0.6°<D<1.6° (B) 1.6°<D<3.6° (C) 3.6°<D<5.6° (D) 5.6°<D<7.6° (E) 7.6°<D<9.6° (F) D≥ 9.6°.

11. Implications for modelling binocular tectal columns

With the exception of the dimming units not studied here (only four R4 units were recorded), this work shows that retinal ganglion cells as well as IRTT units which project both at a single binocular tectal locus can be classified into two main populations having highly specific but antagonistic features when at least five out of the most frequently used criteria are quantitatively computed.

In each afferent system, sustained and event types of visual units can be characterized without ambiguity. Sustained ganglion cells and IRTT units show in common: (1) no response to diffuse light stimulations; (2) a tonic discharge to movement-gated stationary targets; (3) ERF sizes distributed between 2° and 6°; (4) an optimal stimulus diameter of about 2°-3°; and (5) low α values (mean = 0.52 and 0.60, respectively) but rather high k(v) values (mean = 22.5 and 12.5, respectively). In opposition to this, ganglion cells and IRTT units of the "event" type exhibit (1) brief responses to changes in the background illumination; (2) no reaction at all to movement-gated stationary objects; (3) larger ERF sizes from 4° to 12°; (4) a maximum activity for targets 6° in dia.; and (5) high α values (mean = 0.83 and 0.93, respectively) but rather low k(v) values (mean = 7.4 and 3.7, respectively).

In each of these four populations of visual units, (1) ERF sizes increase linearly with the stimulus diameter; (2) Dopt values are independent of the stimulus velocity; (3) values of the exponent α are roughly constant, independent of the stimulus diameter; whereas (iv) the constant k(v) of the velocity function shows a maximum when the target size is optimal. Figures 3 and 8 as well as Table 1 show clearly that the distributions of the k(v) values of the sustained and event types both in the contralateral as well as in the ipsilateral pathway are higly different.

As a result, it appears thus tempting to us to agree a current opinion developped earlier by Keating & Gaze (1970a, c) and by Hodos et al. (1982) to consider only sustained, event and dimming types as the three major functional components of the retinotectal projection in frogs. As Grüsser & Himstedt (1973) and Grüsser-Cornehls (1988) stated, these three basic components may constitute a general "funktioneller Bauplan", each being represented at best by one or more units subtypes (classes) depending on the ecological and behavioral adaptation of the animal under study.

In frogs, sustained and event ganglion cell types (at least) have functional homologues in the IRTT projection. Ipsilateral and contralateral projecting fibers are thus not only in-register in the horizontal plane but are also congruent in their functional properties according to their layering in depth. It may be asked whether these functional similarities may help to the formation of superimposed tectal maps during development if, as assumed by Udin (1985, 1990) and Udin et al. (1990), anatomical and spatial congruence is based on correlated activity patterns between fibers synapsing at identical dendritic positions.

A further and yet actual question concerns the neuronal types involved in the tecto-isthmal transfert. Gruberg & Lettvin (1980) reported that the sole input to the isthmic nucleus comes mainly from pyramidal cells located in layer 6 and below although ganglionic and bipolar cells are also candidates. In parallel, Matsumoto et al. (1986) found that pyramidal neurons exhibited always T5(2) functional properties, whereas the other anatomical types showed T5(1,3,4) properties. In the future, it would be of interest to test IRTT units with configurational stimuli to show (1) if these specific properties are preserved or lost after isthmal tranfert; and (2) if sustained and event ipsilateral unit types originate from different types of intratectal neurons.

The last but not least unsolved problem deals with the functional properties of the tectal binocular cells as well as the origin of their inputs. These neurons have received up to now a rather limited attention and stimulus-response relationships are still lacking.

Preliminary velocity function studies performed on a few T1B neurons have revealed (unpublished studies) that they can be divided into "sustained" and "event" types as far as the exponent α is an adequate criterion: one half of these neurons showed $0.35 < \alpha < 0.65$ whereas the other half exhibited $0.70 < \alpha < 1.0$. Ipsilateral and contralateral responses had always similar α values. The response obtained both eyes open was equivalent to the monocular responses, independent of the receptive field disparity. Would it thus possible that binocular neurons receive primarily direct informations from "sustained" and "event" afferent channels? Direct informations from these channels are probably best candidates but inputs will be delayed by about 30 ms at their arrival on a cell dendrite. No information is actually available on the physiological significance of this delay. Are there any intermediate neurons between visual inputs and binocularly activated neurons? To our knowledge none of the tectal neurons recorded up to now are excited ipsilaterally only. However, addition of a tectal neuron in the contralateral pathway may suppress this delay and is frequently used in some computed schemas (House 1988). Nevertheless such schemas appear in contradiction with theories on the development of isthmotectal fibers. Future experiments have to be done to solve this problem of binocular connectivity.

Acknowledgements. Results exposed in this work were obtained in fulfilment of the requirements for the Doctoral Dissertation of R. Garcia. The authors would like to thank two reviewers who recently scrutinized part of this paper and made helpful criticisms.

13. References

Backström A-C, Reuter T (1975) Receptive field organization of ganglion cells in the frog retina: contributions from cones, green rods and red rods. *J Physiol (Lond.)* 246: 79 -107

Barlow HB (1953) Action potentials from the frog's retina. *J Physiol (Lond.)* 119: 58-68

Butenandt E, Grüsser O-J (1968)The effect of stimulus area on the response of movement detecting neurons in the frog retina. *Pflügers Arch* 298: 283-293

Ewert J-P, Borchers HW, von Witersheim A (1979a) Directional sensitivity, invariance and variability of tectal T5 neurons in response to moving configurational stimuli in the toad *Bufo bufo (L.)*. *J. Comp Physiol* 132: 191-201

Ewert J.-P, Krug H, Schönitz G (1979b) Activity of retinal class R3 ganglion cells in the toad *Bufo bufo (L)* *J. Comp Physiol* 129: 211-215

Finkelstein D, Grüsser O-J (1965) Frog retina: Detection of movement. *Science* 150: 1050-1051

GarciaR. (1988) *Etude quantitative du codage de l'information visuelle dans le système retinotecto tectal ipsilatéral de Rana esculenta (Amphibien anoure): comparaison avec la projection contralatérale directe* . Doctoral dissertation, Univ of Poitiers

Garcia R, Gaillard F (1989a) Quantitative studies on ipsilateral type 2 retino-tecto-tectal units in the frog. Homologies with R3 ganglion cells. *J Comp Physiol* 64: 377-389

Garcia R, Gaillard F (1989b) Quantitative stimulus-response studies on sustained ganglion cells in the frog retina. *Visual Neurosci* 2: 455-463

Garcia R, Gaillard F, Jacquenod J-C (1988) A quantitative analysis of the ipsilateral tectal unit responses to moving stimuli in frogs. *J Comp Physiol* 162: 443-451

Gaze RM, Jacobson M (1962) The projection of the binocular visual field on the optic tecta of the frog. *Q J Exp Physiol* 47: 273-280

Gaze RM, Keating MJ (1967) Visual responses from ipsilateral tectal units in the frog *J Physiol (Lond.)* 192: 52-53P

Gaze RM, Keating MJ (1970) Receptive field properties of single units from the visual projection to the ipsilateral tectum in the frog. *Q J Exp Physiol* 55: 143-152

Gordon J, Hood DC (1976) Anatomy and physiology of the frog retina. In: Fite KV (ed.) *The amphibian visual system. A multidisciplinary approach*. Academic Press, New-York, pp. 29 - 86

Gruberg ER, Lettvin JY (1980) Anatomy and physiology of a binocular visual system in the frog *Rana pipiens* . *Brain Res* 192: 313-325

Grüsser O-J, Grüsser-Cornehls U (1968) Die Informationsverarbeitung im visuellen Sys -tem des Frosches. In: Marko H, Färber G (eds) *Kybernetic* . Oldenburg, München, pp 331-360

Grüsser O-J, Grüsser-Cornehls U (1969) Neurophysiologie des Bewegungssehens. *Erg Physiol* 61: 178-265

Grüsser O-J, Grüsser-Cornehls (1973) Neuronal mechanisms of visual movement percep -tion and some psychophysical and behavioral correlations. In: Jung R (ed) *Handbook of Sensory Physiology. Vol. VII/3.* Springer, Berlin, Heidelberg, New-York, pp 333-349

Grüsser O-J, Grüsser-Cornehls U (1976) Neurophysiology of the anuran visual system. In : Llinas R, Precht W (eds) *Frog Neurobiology.* Springer, Berlin Heidelberg, New-York, pp 297-385

Grüsser O-J, Grüsser-Cornehls U, Finkelstein D, Henn V, Patutschnik M, Butenandt E. (1967) A quantitative analysis of movement detecting neurons in the frog's retina. *Pflügers Archiv* 293: 100-106

Grüsser O-J, Finkelstein D, Grüsser-Cornehls U (1968a) The effect of the stimulus velocity on the response of movement sensitive neurons of the frog's retina. *Pflügers Archiv* 300: 49-66

Grüsser O-J, Grüsser-Cornehls U, Licker MD (1968b) Further studies on the velocity function of. movement detecting class-2 neurons in the frog retina. *Vision Res* 8: 1173-1185

Grüsser-Cornehls U (1988) Neurophysiological properties of the retinal ganglion cell clas -ses of the Cuban treefrog, *Hyla septentrionalis. Exp Brain Res* 73: 39-52

Grüsser-Cornehls U, Himstedt W (1973) Responses of retinal and tectal neurons of the salamander *(Salamandra salamandra L.)* to moving visual stimuli. *Brain Behav*

Evol 7: 145-168
Grüsser-Cornehls U, Langeveld S (1985) Velocity sensitivity and directional selectivity of frog retinal ganglion cells depend on chromaticity of moving stimuli. *Brain Behav. Evol* 27: 165-185
Grüsser-Cornehls U, Saunders RMcD (1981) Chromatic subclasses of frog retinal ganglion cells: studies using black stimuli moving on a monochromatic background. *Vision Res* 21: 469 - 478
Hartline HK (1938) The response of single optic nerve fibers of the vertebrate eye to illumi -nation of the retina. *Am J Physiol* 121: 400-415
Hodos W, Dawes EA, Keating MJ (1982) Properties of the receptive fields of the frog retinal ganglion cells as revealed by their response to moving stimuli. *Neuroscience* 7: 1533 - 1544
House DH (1988) A model of the visual localization of prey by frog and toad. *Biol Cybern* 58: 173 - 192
Kahle G, Grüsser-Cornehls U (1988) Electrophysiological and morphological correlation of frog retinal ganglion cell classes. *Naturwissenschaften* 75: 465-468
Keating MJ, Gaze RM (1970a) Observations on the 'surround' properties of the receptive fields of frog retinal ganglion cells. *Quart J Exp Physiol* 55: 129-142
Keating MJ, Gaze RM (1970b) The ipsilateral pathway in the frog. *Q J Exp Physiol* 55: 284 -292
Keating MJ, Gaze RM (1970c) The depth distribution of visual units in the contralateral optic tectum following regeneration of the optic nerve in the frog. *Brain Research* 21: 197-206
Lettvin J., Maturana HR, McCulloch WS, Pitts WH (1959)What the frog's eye tells the frog's brain. *Proc. IRE* 47: 1940-1951
Matsumoto N, Schwippert WW, Ewert J-P (1986) Intracellular activity of morphologically identified neurons of the grass frog's optic tectum in response to moving configura tional visual stimuli. *J Comp Physiol A* 159: 721-739
Maturana HR, Lettvin JY, Mc Culloch WS, Pitts WH (1960) Anatomy and physiology of vision in the frog (*Rana pipiens*). *J. Gen. Physiol* 43: 129-175
Pomeranz B (1972) Metamorphosis of frog vision: changes in ganglion cell physiology and anatomy. *Exp Neurol* 34: 187-199
Reuter T, Virtanen K (1972) Border and colour coding in the retina of frog. *Nature* 239: 260 - 263.
Schürg-Pfeiffer E, Ewert J-P (1981) Investigation of neurons involved in the analysis of ges talt prey-features in the frog, *Rana temporaria* . *J Comp Physiol* 141:139-152
Stiles M, Tzanakou E, Michalak R, Unnikrishnan K, Goyal P, Harth E (1985) Periodic and non-periodic burst responses of frog (*Rana pipiens*) retinal ganglion cells. *Exp Neurol* 88: 176 - 197
Udin SB (1985) The role of visual experience in the formation of binocular projections in

frogs. *Cell Mol Neurobiol* 5: 85-102

Udin SB (1990) Plasticity in the ipsilateral visuotectal projection persists after lesions of one nucleus isthmi in *Xenopus. Exp Brain Res* 79: 338-344

Udin SB, Fisher MD, Norden JJ (1990) Ultrastructure of the crossed isthmotectal projection in *Xenopus* frogs. *J Comp Neurol* 292: 246-254

Watanabe S-I, Murakami M (1984) Synaptic mechanisms of directional selectivity in gan -glion cells of frog retina as revealed by intracellular recordings. *Jpn J Physiol* 34: 497-511

Witpaard J, ter Keurs HEDJ (1975) A reclassification of retinal ganglion cells in the frog, based upon tectal endings and response properties. *Vision Res* 15: 1333-1338

Distributed Processing in Vestibulo-Ocular and Other Oculomotor Subsystems in Monkeys and Cats

THOMAS J. ANASTASIO

Departments of Otolaryngology and Head and Neck Surgery,
and Physiology and Biophysics
University of Southern California Medical School
Los Angeles, CA 90033, USA

Abstract. *Neurons that subserve oculomotor functions present a complex array of behavior patterns. Neural network models of the vestibulo-ocular reflex, programmed from an initially semi-random state with the back-propagation learning algorithm, reproduce the complexity of vestibular nucleus neurons and suggest that it may be a consequence of distributed processing. Because vestibulo-ocular and other oculomotor subsystems are similarly complex, it is probable that the latter also process information in a distributed manner. This short survey briefly reviews the results of neural network modeling studies of the vestibulo-ocular reflex, and presents examples of distributed processing in other oculomotor subsystems throughout the central nervous systems of monkeys and cats.*

The Oculomotor System

The control of conjugate eye movements, which has been studied most thoroughly in monkeys and cats, involves a small set of different functions (Carpenter 1978). The vestibulo-ocular reflex (VOR) maintains visual image stabilization during head movements. The optokinetic response (OKR) complements VOR by inducing following of the moving visual surround. Fixation denotes the maintenance of gaze at a stationary point in space, and the saccadic system rapidly changes fixation from one point to another. In some foveate animals, such as primates, the pursuit system allows following of a target that is moving against a stationary background.

Many regions throughout the central nervous system contribute to eye movement control. These include the brainstem, tectum, cerebellum, basal ganglia and cortex. These regions differ in the level of control they exert over eye movements, and in the context under which they operate. For example, burst neurons in the reticular formation project directly to extraocular muscle motoneurons, and their activity is mechanistically related to all saccadic eye movements (Hepp et al. 1989). In contrast, neurons in frontal (Goldberg and

Segraves 1989) and parietal (Andersen and Gnadt 1989) cortices have enhanced activity associated with goal directed saccades, and their discharge is related in a more general way to eye movements.

Brain regions that control eye movements are similar in that they all contain neurons with complex and highly variable behavior patterns. Rather than falling into well defined functional groups, neurons in the same region appear to mix and match attributes idiosyncratically, each making their own unique contribution to eye movement control. To gain insight into the complex representation of eye movement commands by vestibular nucleus neurons (VNNs), the horizontal and vertical VORs were modeled as neural networks programmed from an initially semi-random state with the back-propagation learning algorithm (Anastasio and Robinson 1989ab, 1990). The resulting networks contained model VNNs that varied in their behavior patterns, just as their real counterparts do. Eye movement commands in the simulated networks had been distributed nonuniformly among the multitude of model neurons, so that their coordinated activity produced orderly eye movements, even though their organization seemed erratic. The models suggest that VNN behavior patterns vary because these neurons process eye movement commands in a distributed manner. Variability in the properties of neurons in other oculomotor subsystems is comparable to that observed in the vestibular nuclei. This may be an indication that they are also distributed processors.

Distributed Processing in the Vertical Vestibulo-Ocular Reflex

The VOR is capable of producing compensatory eye movements in three dimensions. Two sets of three quasi-orthogonal semicircular canal sensors provide information about head rotational velocity in space. This information is used to rotate the eyes at an equal and opposite velocity, via controlled contractions of the three pairs of quasi-orthogonal, agonist/antagonist extraocular muscles of each eye. The head rotational velocity signal is transmitted from the canals to the extraocular muscles along a three neuron chain composed of semicircular canal afferents, VNNs and extraocular muscle motoneurons. The VOR is schematized in Fig. 1; canal and muscle geometry (left eye only) is for the cat (Ezure and Graf 1984).

The actions of the canals and muscles are described by sensitivity and action vectors perpendicular to their geometric planes. The canals are maximally activated by head rotations about their sensitivity-vectors. The muscle action-vectors are the axes of rotation of the eye when each muscle acts alone. These vectors define a coordinate frame for the canals in which head rotations will be represented, and another for the muscles to represent eye movement commands. Because canal and muscle frames are nonorthogonal and differ from one another, a sensorimotor transformation must occur along the VOR pathway.

This sensorimotor transformation can be brought about by appropriately combining canal signals at the VNNs, the motoneurons, or both. The sensitivity-vectors of the canal afferents and the action-vectors of the motoneurons are determined by canal and muscle

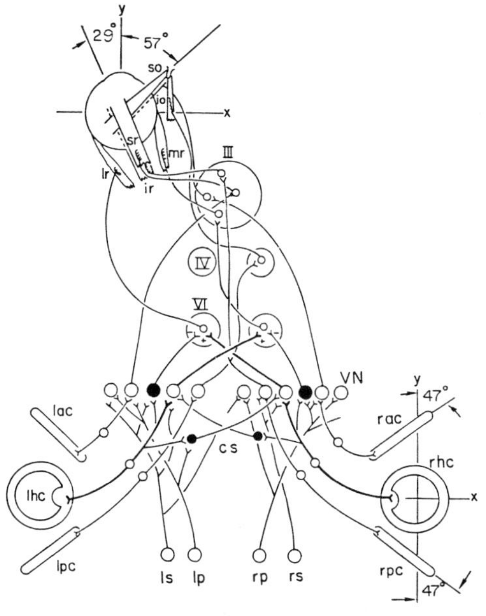

Fig. 1. Schematic diagram of the vestibulo-oculomotor system (Anastasio and Robinson 1989a). Inputs to the vestibular nuclei (VN) include afferents from the six semicircular canals and projections from the pursuit and saccadic oculomotor subsystems. Outputs of the VN project to the extraocular muscle motoneurons of which are shown the six to the left eye. The major connections of the horizontal VOR are shown by heavy lines. Filled cells are inhibitory. lac, lpc, lhc, rac, rpc, rhc, left or right anterior, posterior and horizontal canals. sr, ir, lr, mr, so, io, left superior, inferior, lateral and medial recti, and superior and inferior oblique muscles. III, IV, VI, oculomotor, trochlear and abducens nuclei. ls, lp, rs, rp, left or right pursuit and saccadic inputs. cs, feedforward inhibitory commissural fibers.

geometry. The sensitivity-vectors of the VNNs might be expected to align with the canal or muscle vectors, or to assume some intermediate position. Although the sensitivity-vectors of VNNs in cat have a weak tendency to cluster around those of the canals, they are dispersed in many directions, as shown in Fig. 2 (Baker et al. 1984).

To gain insight into this complex encoding of spatial information, the sensorimotor transformation from the vertical canals to the vertical muscles was modeled as a three-layered neural network (Anastasio and Robinson 1990). The input units represented afferents from the left and right anterior and posterior canals, and the output units represented motoneurons of the superior and inferior rectus and oblique muscles of the left eye (Fig. 1). Units in the hidden (middle) layer represented VNNs. To represent the abundance of VNNs, the hidden layer contained 40 units.

The canal afferent responses and motoneuron activations that would correspond to head rotations and their appropriate compensatory eye movements, were calculated for head rotations about various vectors in the horizontal plane. The synaptic connections from the hidden to the output units were arranged so that half of the hidden units reciprocally innervated the vertical rectus motoneurons while the other half reciprocally innervated the oblique motoneurons. Prior to training, the input-to-hidden connections were randomized. The network was then trained to produce the correct motoneuron activations for each pattern of canal afferent responses. The backpropagation algorithm (Rumelhart et al. 1986) was used to iteratively adjust the values of the input-to-hidden weights so that the errors between the actual and calculated output activations were minimized.

Fig. 2. Normalized sensitivity-vectors of vestibular nucleus neurons in the cat (Baker et al. 1984). lac, lpc, lhc, sensitivity vectors of the left anterior, posterior and horizontal canals.

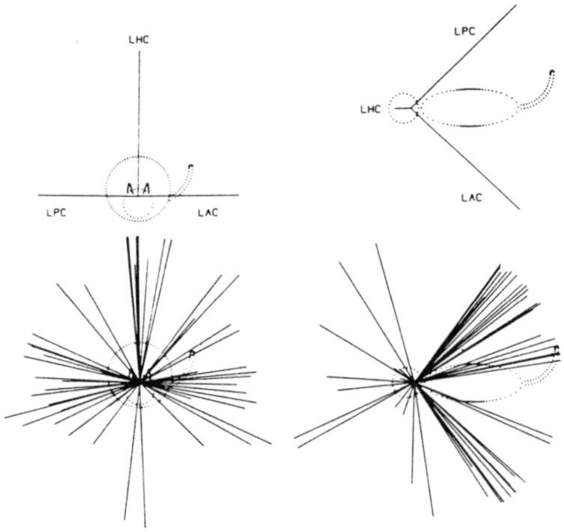

With the hidden-to-output connections segregated, canal convergence was limited to the hidden layer. With this constraint, the sensorimotor transformation had to occur at the hidden unit level. It might have been expected, therefore, that the hidden unit vectors would align with the action-vectors of the outputs. Although model VNN sensitivity-vectors did have a weak tendency to cluster around those of the canals and muscles, they diverged considerably (Fig. 3). The distribution of model VNN sensitivity-vectors is qualitatively similar to that of real VNNs in the cat (Fig. 2).

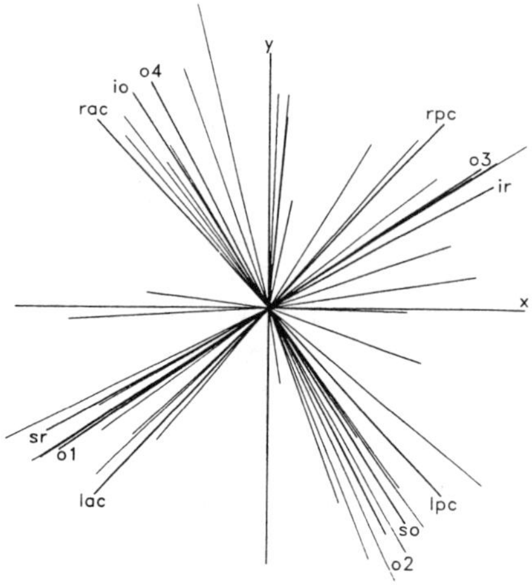

Fig.3. Scaled sensitivity-vectors of vestibular nucleus neurons in a neural network model of the cat vertical VOR (Anastasio and Robinson 1990). rac, rpc, sensitivity-vectors of the right anterior and posterior canals. sr, ir, so, io, action-vectors of the superior and inferior rectus and oblique muscles. o1, o2, o3, o4, action-vectors of the moto-neurons. The motoneuron and muscle action-vectors do not coincide because the muscles form a nonorthogonal coordinate system. x, right. y, anterior. Other abbreviations as in previous figure.

This variability is a consequence of distributing the sensorimotor transformation over many (in this case 40) separate elements. If there had been only two hidden units, one reciprocally innervating the vertical recti and the other the obliques, then each motoneuron pair would receive its input from only one hidden unit. In this case, the sensitivity-vectors of the hidden units would have to be aligned with the motoneuron action-vectors in order to produce the correct output activations. With 40 hidden units, however, each motoneuron receives inputs from 20 model VNNs. There is now no requirement that the hidden unit sensitivity-vectors be aligned with motoneuron action-vectors, as long as their combined inputs produce the correct motoneuron activations. Of course, the hidden unit sensitivity-vectors could be aligned with the action-vectors of the motoneurons they innervate, but this is only one of an infinite number of model VNN sensitivity-vector distributions that would work. Because it is unnecessary to require that VNNs have tight distributions of sensitivity-vectors, the nonuniform, divergent configuration is preferred by the learning algorithm and by nature as well.

Distributed Spatial Representations in Other Oculomotor Areas

As is the case for the VOR, other oculomotor subsystems also represent motor and sensory frames in a distributed way. Neurons in the accessory optic and pretectal nuclei process moving, full-field visual stimuli and subserve the OKR (Simpson 1984). The accessory optic system (AOS) is important for the control of the vertical OKR (ibid.). It might be expected that AOS neurons would be responsive to visual surround rotations in the planes of the vertical canals or extraocular muscles. It has been suggested that the AOS is organized in the canal coordinate frame (ibid.). Although the preferred stimulus motion directions of cat AOS neurons do tend to cluster around vectors perpendicular to canal sensitivity-vectors, the distribution is divergent (Fig. 4) (Grasse and Cynader 1982). Interestingly, the excitatory and inhibitory preferred-directions of each neuron are separated by more than 180 degrees, and the excitatory directions of the population (Fig. 4, A) diverge more widely than the inhibitory directions (Fig. 4, B). Neurons in the nucleus of the optic tract (NOT) are important for the control of the horizontal OKR (Simpson 1984). The preferred-directions of NOT neurons in cat are also divergent (Hoffmann and Schoppmann 1981).

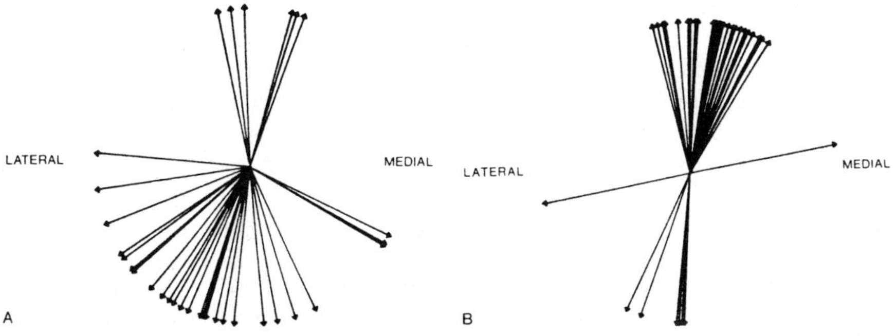

Fig. 4. Normalized excitatory (A) and inhibitory (B) preferred-directions of accessory optic neurons in the cat midbrain (Grasse and Cynader, 1982).

Burst neurons in the reticular formation encode saccadic eye movement commands, and relay them directly to extraocular motoneurons (Hepp et al. 1989). On-direction is defined as the direction of eye movement for which the discharge of a premotor or motoneuron is maximal. Vertical saccades are controlled primarily by burst neurons in

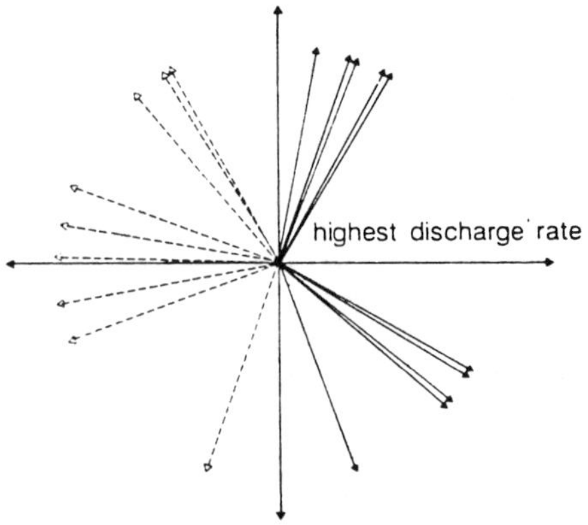

Fig. 5. Normalized on-directions of burst neurons in the right (solid lines) and left (dashed lines) rostral interstitial nuclei of the medial longitudinal fasciculus in the monkey (Hepp et al. 1988).

the rostral interstitial nucleus of the medial longitudinal fasciculus (riMLF) (ibid.). Rather than aligning with the on-directions of vertical rectus and oblique motoneurons, the on-directions of riMLF burst neurons in monkeys are widely divergent (Fig. 5) (Hepp et al. 1988). Neurons in the paramedian pontine reticular formation (PPRF) control horizontal saccadic eye movements (Hepp et al. 1989). The on-directions of PPRF neurons are also divergent in monkeys (Strassman et al. 1986ab). Like VNNs, pretectal and midbrain neurons contributing to the OKR, and burst neurons in the reticular formation controlling saccades, all have activity-vector distributions that are widely divergent. This indicates that these regions may all encode information about the direction of eye movements in a distributed way.

Complex Oculomotor Signal Combinations in the Vestibular Nuclei

The vestibular nuclei transmit pursuit and saccadic eye velocity commands, as well as vestibular (head rotational velocity) information, to the extraocular motoneurons (Fig. 1). Many VNNs also carry eye position signals. Motoneurons, the final common pathway to the extraocular muscles, carry all three velocity commands as well as the eye position command (Robinson 1970). It might be expected that the VNNs, like the motoneurons, would carry all four commands in equal proportions. Alternatively, VNNs could comprise four separate classes, each specialized to carry a different command: vestibular, pursuit, saccadic or eye position. In fact, primate VNNs vary considerably in their oculomotor behavior patterns (Fuchs and Kimm 1975, Lisberger and Miles 1980). Some VNNs do carry all four commands, and have discharge patterns almost indistinguishable from motoneurons. Other VNNs appear to be specialized for one particular command, such as pure vestibular or pure pursuit cells. Still others carry various combinations of commands in varying proportions. Almost any combination of the four eye movement signals can be found among VNNs.

Motoneurons always increase their firing rate for eye movements in the same direction, whether these are vestibular, pursuit or saccadic (Robinson 1970). For example, a left abducens motoneuron that increases its firing rate for rightward head rotations (and produces a leftward compensatory eye movement) will also carry a leftwardly directed pursuit command. Thus motoneurons have vestibular (head rotation) sensitivity and pursuit (eye rotation) activity in opposite directions. This is generally true for VNNs also. A small percentage of VNNs, however, actually have vestibular sensitivity and pursuit signals in the same direction, apparently encoding a compensatory eye movement in one direction and a pursuit eye movement in the opposite direction (Lisberger and Miles 1980).

Motoneurons fire a burst for saccades in one direction and pause for saccades in the other (Robinson 1970). Some VNNs also have this type of saccade related behavior, but for others the situation is more complicated. Some VNNs burst in one direction but don't pause in the other and vice-versa. Others burst or pause for saccades in both directions, and still others lack saccade related activity altogether (Fuchs and Kimm 1975, Lisberger and Miles 1980).

To gain insight into this complex organization, the horizontal vestibulo-oculomotor system of the monkey was modeled as a three-layered neural network (Anastasio and Robinson 1989b). The input units represented afferents from the left and right horizontal canals, and neurons carrying pursuit and saccadic eye velocity commands to the vestibular nuclei (Fig. 1). The eye position signal is believed to be derived from the three eye velocity commands by a process of neural integration (in the mathematical sense), involving feedback interconnections among VNNs and other brainstem neurons (Robinson 1989). The neural network models reviewed here did not incorporate feedback interconnections, so the eye position commands were left out. Only the three velocity commands (vestibular, pursuit and saccadic) were considered. The output units represented motoneurons of the lateral and medial rectus muscles of the left eye (Fig. 1). Again, 40 hidden units represented the population of VNNs. As before, the connections from the hidden to the output units were arranged in a reciprocal pattern. The saccadic inputs also projected directly to the outputs to reflect the direct projection of burst neurons onto motoneurons (Hepp et al. 1989). The input-to-hidden connections were randomized and the network was trained using back-propagation to produce vestibular (compensatory), pursuit and saccadic eye movements following activation of the appropriate inputs.

The behavior patterns of the model VNNs were also complex (Fig. 6, A and B). The vestibular and pursuit activity of the hidden and output units is shown in Fig. 6, A. The model motoneurons (indicated with a + sign) have vestibular and pursuit components (gains) that are equal and opposite, as demanded by training. In contrast, the VNNs vary widely with regard to these gains. VNNs falling near the diagonal line connecting the outputs have approximately equal and opposite vestibular and pursuit components. Most hidden units have unequal vestibular and pursuit gains. Units near the vertical axis would appear as pure vestibular cells; units near the horizontal axis as pure pursuit. Some units even fall into quadrants one and three. These anomalous units would encode compensatory and pursuit eye movements in opposite directions. Thus the complex patterns of vestibular and pursuit behaviors of real VNNs are reproduced in the neural network simulation. Rather than falling into discrete subpopulations, the various cell "types" appear as points along a continuum of possible VNN vestibular and pursuit configurations.

The saccade related activity of the hidden and output units is shown in Fig. 6, B. In this diagram the vestibular (V) and pursuit (P) gains of each unit are compressed by the ratio: $|P|-|V|/|P|+|V|$, where $||$ denotes absolute value. This ratio will be negative one for pure vestibular cells (P=0), positive one for pure pursuit cells (V=0), and zero for cells having equal vestibular and pursuit gains (V=P). The saccadic activity of each unit is represented by a vertical line connecting its activity for saccades in either direction; a positive value is a burst, negative a pause. These vertical lines are located along the horizontal axis according to the values of their vestibular/pursuit ratios.

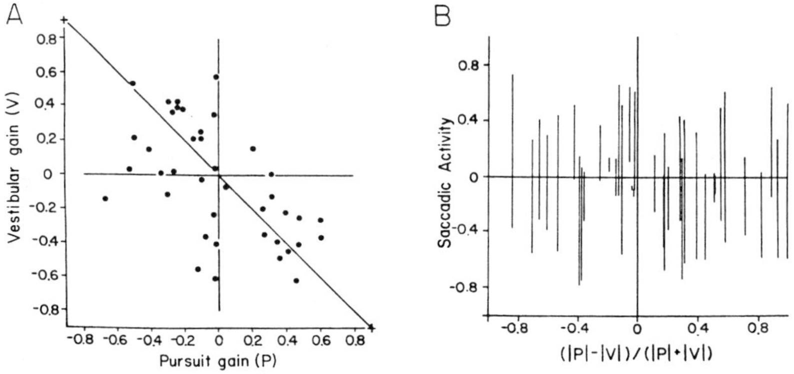

Fig. 6. Scatter plots of the vestibular and pursuit components (A) and saccadic activity (B) of vestibular nucleus neurons in a neural network model of the horizontal vestibulo-oculomotor system of the monkey (Anastasio and Robinson 1989b).

The long line down the center of Fig. 6, B represents the outputs, which have equal vestibular and pursuit gains, and burst and pause for saccades in opposite directions. The hidden unit saccadic activity lines are scattered fairly evenly along the horizontal axis, reflecting the continuum of possible vestibular/pursuit configurations. Fewer than half of these lines extend well above and below zero. These units burst for saccades in one direction and pause in the other, as do the motoneurons. Other cells burst but don't pause, or pause but don't burst. Still others burst or pause in both directions, and a few have practically no saccade related activity at all. Thus the full range of saccade, as well as vestibular and pursuit, related activities of VNNs is reproduced in this neural network model. The three eye movement duties are distributed nonuniformly over the population of VNNs, both in the model and in the actual vestibular nuclei. Such a nonuniform sharing of functions is also characteristic of neurons in other eye movement regions.

Complex Oculomotor Signal Combinations in the Flocculus

The floccular lobes of the cerebellum are involved in the generation of pursuit eye movements and in VOR modulation (Zee 1982). As in the vestibular nuclei, floccular Purkinje cells in monkeys vary widely in their oculomotor behavior patterns (Lisberger and Fuchs 1978, Miles et al. 1980). The main cell "type" in the flocculus is the gaze-velocity Purkinje cell, which has head velocity sensitivity and pursuit activity in the same direction. Thus gaze-velocity Purkinje cells encode the velocity of the eye in space, as the sum of head-in-space and eye-in-head velocities. During the VOR, when head-in-space and eye-in-head velocities are equal and opposite, true gaze-velocity Purkinje cells should exhibit no discharge modulation.

Most floccular Purkinje cells, however, have vestibular and pursuit components that are not equal (ibid.). Consequently, most of them do respond to some degree during the VOR (Lisberger and Fuchs 1978). Further, many floccular Purkinje cells that are not classified as gaze-velocity cells, have pure vestibular or pure pursuit behaviors. Some even have head velocity sensitivity and pursuit commands in opposite directions (Miles et al. 1980), as do motoneurons and most VNNs. It may be that floccular Purkinje cells fall along a vestibular\pursuit continuum, similar to that of VNNs (Fig. 6, A), except that most cells would fall in quadrants one and three.

Other floccular Purkinje cells that do not have vestibular or pursuit components, do have eye position or saccade related activity, or both. Saccade related activity in the flocculus is complex (ibid.) and resembles that in the vestibular nuclei. Further, many gaze-velocity Purkinje cells have eye position and/or saccade related activity in addition to vestibular and pursuit components (Lisberger and Fuchs 1978, Miles et al. 1980). Thus gaze-velocity Purkinje cells may not be a distinct "type", but may instead predominate a behavior pattern continuum like that observed in the vestibular nuclei (Fig. 6, A and B).

A Common Distributed Processing Mode in the Saccadic System

Recent studies, primarily in monkeys, reveal that the control of saccades involves multiple regions throughout the central nervous system (Wurtz and Goldberg 1989). Some of these regions contain neurons with a similar spectrum of response properties. These regions include (but are not limited to) the frontal eye fields (Goldberg and Segraves 1989), posterior parietal cortex (Andersen and Gnadt 1989) the deep layers of the superior colliculus (Sparks and Hartwich-Young 1989) and the substantia nigra pars reticulata (Hikosaka and Wurtz 1989).

The frontal eye fields (FEF) contain a diverse assortment of cells. FEF neurons can have presaccadic, postsaccadic, pursuit or eye position related activity, or can respond to foveal visual or auditory stimuli, and various combinations of these attributes are also observed (Bruce and Goldberg 1985). The presaccadic neurons appear to be those which are involved in the generation of goal directed saccade commands. Presaccadic FEF neurons fall into three categories: visual, movement and visuomovement. Visual cells respond to visual stimuli, movement cells burst prior to saccadic eye movements, and visuomovement cells exhibit both behaviors to varying degrees. A scatter plot of the visual and movement components of a population of FEF cells is shown in Fig. 7 (ibid.). This plot indicates that visual and movement functions are distributed nonuniformly over the population of FEF cells. Rather than representing distinct cell "types", visual, movement and visuomovement behavior patterns represent points along a continuum (ibid.). In addition to visual and motor activities, some presaccadic FEF cells in each category show anticipatory activity associated with an intended saccade to a predicted or remembered target (ibid.). These neurons have a sustained discharge increase during the interval preceding an intended but delayed saccade. Thus sensory, motor and cognitive functions are distributed nonuniformly over the population of FEF neurons.

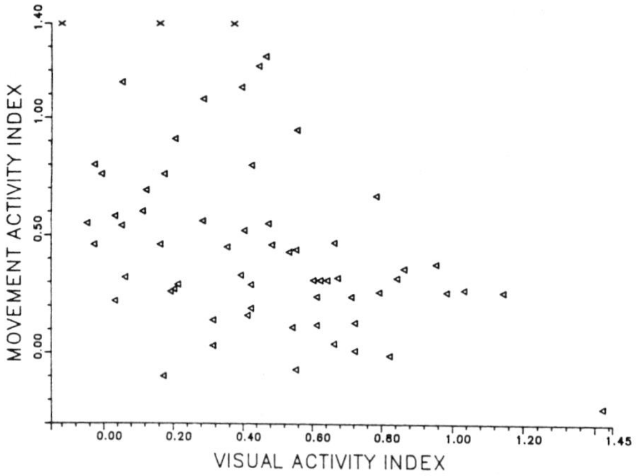

Fig. 7. Scatter plot of the visual and motor components of presaccadic neurons in the frontal eye fields of the monkey (Bruce and Goldberg 1985).

Subregions of the posterior parietal cortex that are involved in saccadic programming include area 7a and the lateral bank of the intraparietal cortex (LIP) (Andersen et al. 1987, Gnadt and Andersen 1988, Andersen and Gnadt 1989). Neurons in these areas display a continuum of behavior patterns that is strikingly similar to that observed in FEF. Posterior parietal neurons that combine visual responses with eye position related activity may be computing the position of targets in head centered coordinates (rather than in retinal coordinates) (Andersen and Gnadt 1989). A back-propagation programmed neural network model demonstrated how visual and eye position information may be distributed over neurons in area 7a (Zipser and Andersen 1988). Presaccadic motor and intended movement activities are also distributed nonuniformly over the cells in area 7a and LIP (Andersen and Gnadt 1989).

Neurons in the deep layers of the superior colliculus can be classified as visual, saccade or visual-motor (Mays and Sparks 1980). Some saccade cells, however, also have visual responses, and it appears that these cell "types" may in fact represent points along a visuomotor continuum, as in FEF and posterior parietal cortex. Collicular neurons that have been called quasi-visual exhibit sustained activity associated with intended saccadic eye movements (ibid.). However, some visual cells have a tonic, sustained discharge for intended targets, and some visual-motor and saccade cells have low level, preburst activity for intended saccades. As in FEF and posterior parietal cortex, sensory, motor and cognitive functions are distributed nonuniformly over the population of collicular neurons.

Many neurons in the deep layers of the superior colliculus are multimodal, responding to auditory and somatosensory as well as visual stimuli (Meredith and Stein 1986). All combinations of these responses can be found among deep collicular neurons in cats (Fig. 8) (ibid.). The deep colliculus is topographically organized in motor coordinates (Sparks and Hartwich-Young 1989). The site of premotor activity within the colliculus specifies saccade direction and amplitude. Sensory receptive fields and intended saccade activity are continuously updated for changes in eye position to conform to the motor map. Although the receptive fields of collicular neurons are topographically organized, the proportions of response modality combinations are similar throughout the deep layers of the colliculus (Meredith and Stein 1986). Also, collicular location does not appear to determine the particular combination of visual, motor and cognitive functions that a particular cell may exhibit, although it does determine the cell's receptive and/or movement field in motor coordinates. It can be concluded, therefore, that a nonuniform distribution of sensory (various modalities), motor and cognitive attributes is superimposed upon the topographic order of the colliculus.

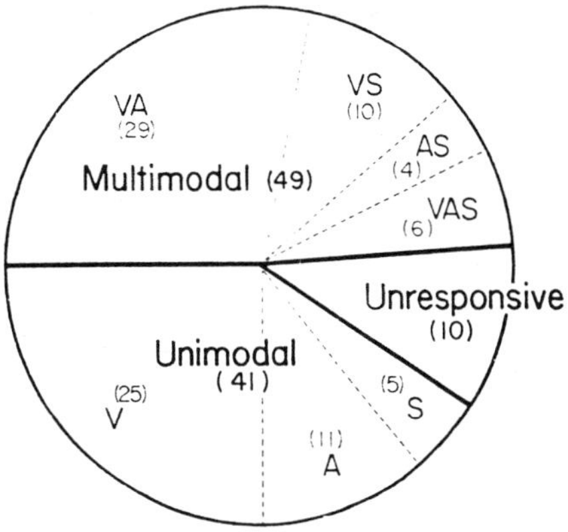

Fig. 8. Pie chart of percentages (in parenthesis) of unimodal and multimodal cells in the deep superior colliculus of the cat (Meredith and Stein 1986). V, visual, A, auditory, S, somatosensory.

The substantia nigra pars reticulata (SNr) influences the generation of saccadic eye movement commands by the superior colliculus (Hikosaka and Wurtz 1989). It appears that SNr cells tonically inhibit collicular neurons in the deeper layers (Hikosaka and Wurtz 1983d). The response of SNr neurons, which is a decrease in firing rate, releases collicular neurons from tonic inhibition (ibid.). As in the FEF, posterior parietal cortex and deep colliculus, sensory (mostly visual), presaccadic motor and intended movement activities are distributed nonuniformly among SNr neurons (Hikosaka and Wurtz 1983abc). In addition, SNr neurons can have responses contingent upon active fixation, or the presence of a visual or a remembered target, and these response types are also distributed over the pop-

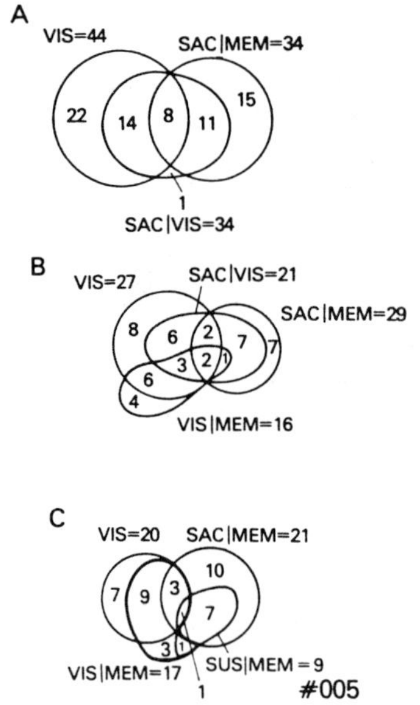

Fig. 9. Venn diagrams showing the combinations of response types of cells in the substantia nigra pars reticulata of monkey #005 (Hikosaka and Wurtz 1983c). The number within each area indicates the number of cells showing a particular response combination. VIS, simple visual response, SAC/VIS, visually-contingent saccade response, SAC/MEM, memory-contingent saccade response, VIS/MEM, memory-contingent visual response, SUS/MEM, memory-contingent sustained (intended movement) response.

ulation (ibid.). Fig. 9 illustrates how visual, saccadic and memory related activities are distributed over a population of SNr neurons in one monkey (Hikosaka and Wurtz 1983c).

In this very short review, some of the unique features of each area participating in saccade command generation have been suppressed, in order to emphasize their similarities. All of these regions appear to process sensory and cognitive information to produce saccadic eye movement commands, and in each, sensory, motor and cognitive functions are distributed nonuniformly over the neural population. As distributed processors, these regions do more than encode or transmit motor commands. They appear to subserve a selection function, a higher-order process that culminates in a "decision" to produce one saccade as opposed to another. As such, these regions may be thought of as nonuniformly distributed winner-take-all networks.

Implications of Distributed Processing in Theory and Experiment

An essential feature of all of the oculomotor neurons described above, whether they serve a primarily sensory or motor role or both, is that they are broadly tuned. Although a particular neuron will have a "best" direction, it will also exhibit activity for nonoptimal stimuli and/or eye movements. Thus any given stimulus or eye movement is encoded not by one, but by a population of neurons. This aspect of distributed coding has been described for the superior colliculus (McIlwain 1982, Sparks and Hartwich-Young 1989), and is generally true for other oculomotor regions as well.

The purpose of this paper is to consider the following question: Given that oculomotor (and many other neural) functions are subserved by populations of neurons operating in parallel, what is the principle underlying their organization? From the examples given above, it appears that in each oculomotor region, the duties to be performed are distributed nonuniformly over the population of neurons. In a distributed network, such a nonuniform representation will work as well as a uniform or segregated one. This insight may mitigate the tendency of experimentalists to place individual neurons into separate functional categories, and of theorists to lump neural populations into discrete "black boxes".

The central problem in any distributed representation is the coordination of the population of elements. The main assumption underlying the theoretical approach reviewed and proposed here, is that this coordination is achieved by error-driven learning. This assumption is valid considering the strong evidence for plasticity throughout the nervous system (Byrne and Berry 1989). The technique of modeling an oculomotor (or other) subsystem as a neural network, programmed from an initially semi-random state with an appropriate learning procedure, can be used to understand the complex organization of a population of neurons operating in parallel. The coordination of distributed, nonuniform elements by adaptive learning emerges as an organizational principal that can help interpret data and improve the veracity of models of oculomotor and other neural systems.

Acknowledgments. This work was supported by grants from the National Eye Institute of the National Institutes of Health, and the Whitaker Foundation. Special thanks to authors and editors for permission to reproduce their figures.

References

Anastasio TJ, Robinson DA (1989a) Distributed parallel processing in the vestibulo-oculomotor system. *Neural Comp* 1: 230-241

Anastasio TJ, Robinson DA (1989b) The distributed representation of vestibulo-oculomotor signals by brain-stem neurons. *Biol Cybern* 61: 79-88

Anastasio TJ, Robinson DA (1990) Distributed parallel processing in the vertical vestibulo-ocular reflex: learning networks compared to tensor theory. *Biol Cybern* 63: 161-167

Andersen RA, Essick GK, Siegel RM (1987) Neurons of area 7 activated by both visual stimuli and oculomotor behavior. *Exp Brain Res* 67: 316-322

Andersen RA, Gnadt JW (1989) Posterior parietal cortex. In: Wurtz RH, Goldberg ME (eds) *The neurobiology of saccadic eye movements*. Elsevier, Amsterdam, pp 315-335

Baker J, Goldberg J, Hermann G, Peterson B (1984) Optimal response planes and canal convergence in secondary neurons in vestibular nuclei of alert cats. *Brain Res* 294: 133-137

Bruce CJ, Goldberg ME (1985) Primate frontal eye fields. I. single neurons discharging before saccades. *J Neurophysiol* 53: 603-635

Byrne JH, Berry WO (eds) (1989) *Neural models of plasticity*. Academic Press, San Diego

Carpenter RHS (1978) *Movements of the eyes*. Pion Press, London

Ezure K, Graf W (1984) A quantitative analysis of the spatial organization of the vestibulo-ocular reflexes in lateral and frontal-eyed animals. I. orientation of semicircular canals and extraocular muscles. *Neuroscience* 12: 85-93

Fuchs AF, Kimm J (1975) Unit activity in the vestibular nucleus of the alert monkey during horizontal angular acceleration and eye movement. *J Neurophysiol* 38: 1140-1161

Gnadt JW, Andersen RA (1988) Memory related motor planning activity in posterior parietal cortex of macaque. *Exp Brain Res* 70: 216-220

Goldberg ME, Segraves MA (1989) The visual and frontal cortices. In: Wurtz RH, Goldberg ME (eds) *The neurobiology of saccadic eye movements*. Elsevier, Amsterdam, pp 283-313

Grasse KL, Cynader MS (1982) Electrophysiology of medial terminal nucleus of accessory optic system in the cat. *J Neurophysiol* 48: 490-504

Hepp K, Vilis T, Henn V (1988) On the generation of rapid eye movements in three dimensions. *Ann N Y Acad Sci* 545: 140-153

Hepp K, Henn V, Vilis T, Cohen B (1989) Brainstem regions related to saccade generation. In: Wurtz RH, Goldberg ME (eds) *The neurobiology of saccadic eye movements*. Elsevier, Amsterdam, pp 105-212

Hikosaka O, Wurtz RH (1983a) Visual and oculomotor functions of monkey substantia nigra pars reticulata. I. relation of visual and auditory responses to saccades. *J Neurophysiol* 49: 1230-1253

Hikosaka O, Wurtz RH (1983b) Visual and oculomotor functions of monkey substantia nigra pars reticulata. II. visual responses related to fixation of gaze. *J Neurophysiol* 49: 1254-1267

Hikosaka O, Wurtz RH (1983c) Visual and oculomotor functions of monkey substantia nigra pars reticulata. III. memory-contingent visual and saccade responses. *J Neurophysiol* 49: 1268-1284

Hikosaka O, Wurtz RH (1983d) Visual and oculomotor functions of monkey substantia nigra pars reticulata. IV. relation of substantia nigra to superior colliculus. *J Neurophysiol* 49: 1285-1300

Hikosaka O, Wurtz RH (1989) The basal ganglia. In: Wurtz RH, Goldberg ME (eds) *The neurobiology of saccadic eye movements*. Elsevier, Amsterdam, pp 257-281

Hoffmann K-P, Schoppmann A (1981) A quantitative analysis of the direction-specific response of neurons in the cat's nucleus of the optic tract. *Exp Brain Res* 42: 146-157

Lisberger SG, Fuchs AF (1978) Role of primate flocculus during rapid behavioral modification of vestibuloocular reflex. I. Purkinje cell activity during visually guided horizontal smooth-pursuit eye movements and passive head rotation. *J Neurophysiol* 41: 733-763

Lisberger SG, Miles FA (1980) Role of primate medial vestibular nucleus in long-term adaptive plasticity of vestibuloocular reflex. *J Neurophysiol* 43: 1725-1745

Mays LE, Sparks DL (1980) Dissociation of visual and saccade-related responses in superior colliculus neurons. *J Neurophysiol* 43: 207-232

McIlwain JT (1982) Lateral spread of neural excitation during microstimulation in intermediate gray layer of cat's superior colliculus. *J Neurophysiol* 47: 167-178

Meredith MA, Stein BE (1986) Visual, auditory, and somatosensory convergence on cells in superior colliculus results in multisensory integration. *J Neurophysiol* 56: 640-662

Miles FA, Fuller JH, Braitman DJ, Dow BM (1980) Long-term adaptive changes in primate vestibuloocular reflex. III. electrophysiological observations in flocculus of normal monkeys. *J Neurophysiol* 43: 1437-1476

Robinson DA (1970) Oculomotor unit behavior in the monkey. *J Neurophysiol* 33: 393-404

Robinson DA (1989) Integrating with neurons. *Ann Rev Neurosci* 12: 33-45

Rumelhart DE, Hinton GE, Williams RJ (1986) Learning internal representations by error propagation. In: Rumelhart DE, McClelland JL, PDP Research Group (eds) *Parallel distributed processing: explorations in the microstructure of cognition, vol 1: foundations*. MIT Press, Cambridge, pp 318-362

Simpson JI (1984) The accessory optic system. *Ann Rev Neurosci* 7: 13-41

Sparks DL, Hartwich-Young R (1989) The deep layers of the superior colliculus. In: Wurtz RH, Goldberg ME (eds) *The neurobiology of saccadic eye movements*. Elsevier, Amsterdam, pp 213-255

Strassman A, Highstein SM, McCrea RA (1986a) Anatomy and physiology of saccadic burst neurons in the alert squirrel monkey. I. excitatory burst neurons. *J Comp Neurol* 249: 337-357

Strassman A, Highstein SM, McCrea RA (1986b) Anatomy and physiology of saccadic burst neurons in the alert squirrel monkey. I. inhibitory burst neurons. *J Comp Neurol* 249: 358-380

Wurtz RH, Goldberg ME (eds) (1989) *The neurobiology of saccadic eye movements*. Elsevier, Amsterdam

Zee DS (1982) Ocular motor abnormalities related to lesions in the vestibulocerebellum in primate. In: Lennerstrand G, Zee DS, Keller EL (eds) *Functional basis of ocular motility disorders*. Pergamon Press, Oxford, pp 423-430

Zipser D, Andersen RA (1988) A back propagation network that simulates response properties of a subset of posterior parietal neurons. *Nature* 33: 679-684

Optic Chiasm Hemisection and the Parsing of Visual Information in Frogs

Robert F. Waldeck and Edward R. Gruberg
Biology Department, Temple University

Abstract. *We made cuts of the frog optic chiasm and optic tract and studied subsequent visual behavior. After transecting either the posterior or anterior half of the optic chiasm, animals responded to visually presented prey in the entire ground level visual field. However, most animals were unresponsive to visually presented threat stimuli anywhere in the visual field. Animals with complete transection of the optic chiasm were unresponsive to both prey and threat stimuli. Animals with unilateral transection of the optic tract were unresponsive to prey and threat stimuli in the contralateral hemifield. After hemisection of the optic chiasm we used anterograde methods to label optic nerve fibers. Posterior hemisection resulted in labeled tissue in the dorsomedial and ventrolateral regions of the optic tectum. Anterior hemisection resulted in labeled tissue in the dorsolateral region of the tectum. These results imply that there is a retinotopic distribution of retinotectal fibers in the optic chiasm. However, using a retrograde method, we found that retinotectal fibers that project to a circumscribed region of the tectum are spread out over a considerable extent of the optic chiasm.*

1. Introduction

In this paper we will describe changes in the visual behavior of leopard frogs after partial cuts of their optic chiasms. The results were unexpected and challenged some of our assumptions about the organization of visual inputs to the brain. We came to do such experiments by an indirect pathway.

At the Kassel meeting in 1987 we had discussed the role of nucleus isthmi on tectal visual function (Gruberg, 1989). Nucleus isthmi is a tegmental nucleus which receives a topographically organized ipsilateral input from the optic tectum and projects bilaterally and topographically to the optic tectum. Fibers from n.

isthmi terminate in superficial tectal layers. Unilateral ablation of n. isthmi causes a scotoma to visually presented prey and threat stimuli in the contralateral monocular field. The changes are similar to the changes that occur after unilateral tectal lobe ablation where there is also a scotoma to prey and threat in the contralateral monocular field. Outside the scotoma visual function is normal. It appears that ipsilaterally projecting isthmotectal fibers modulate tectal function and in their absence a scotoma develops.

What is the role of contralaterally projecting isthmotectal fibers in visually guided behavior? These fibers course forward from n. isthmi and run along the ventral-lateral surface of the midbrain and diencephalon before crossing in the ventral anterior diencephalon. We hoped to selectively cut these fibers at their zone of decussation in the ventral diencephalon and study post-lesion behavior. However, we discovered that contralaterally projecting isthmotectal fibers cross in the posterior part of the optic chiasm admixed with optic nerve fibers (Gruberg et al., 1989). As is well known, retinofugal fibers reach the brain via the optic nerve, and most cross in the optic chiasm and then course in the optic tract to various retino-recipient structures in the diencephalon and midbrain (including the tectum). At each intermediate position (optic nerve, chiasm, optic tract) between the eye and retinorecipient targets there is a discernible, sometimes crude retinotopic order to the distribution of the optic fibers (Reh et al., 1983; Scalia and Arango, 1983). For instance, Montgomery and Fite (1989) have observed that in the optic chiasm "the fibers from each of the four retinal quadrants appear as bands with the nasal quadrant entering the chiasmal anterior pole followed by ventral, temporal and dorsal quadrants". Despite the mix of isthmotectal axons and crossing optic fibers we cut the posterior half of the chiasm to see if there were changes in visually guided behavior particularly those behaviors that are considered mediated by the tectum: prey catching and threat avoidance. We also carried out correlated anatomy in order to histologically assess the tectal distribution of spared retinotectal fibers. We will describe these experiments and complementary follow-up experiments. We will also discuss puzzles that emerged from this work.

2. Results

Behavioral testing. Normal leopard frogs (*Rana pipiens*) were screened for responses to visually presented prey and threat stimuli. Animals were placed in a wooden rectangular arena 70 cm x 1 m with 55 cm high walls. For prey stimulus we used live crickets tethered on a long string which was connected to a rod. The crickets could be placed anywhere around the animal and frogs would turn, approach and/or snap at crickets depending on their horizontal eccentricity and distance. The frogs did not respond to the rod and string without the cricket. For threat stimulus we used a solid black object 10 cm x 8 cm x 3 cm attached to the end of a meter stick. The object could be moved directly toward the animal from any direction. The animal would respond to the looming object jumping away (or turning and jumping away if the object approached from a frontal direction). If the animal was threatened frequently it would puff up or shift its weight in response. We selected normal animals that responded consistently to prey and threat stimuli at all horizontal eccentricities.

Behavioral effects of cutting the posterior half of the optic chiasm To make chiasm cuts each animal was anesthetized and a midline slit was made in the skin of the upper mouth exposing the bone adjacent to the optic chiasm. A patch of bone was removed exposing the chiasm. The chiasm stands out as a flattened x-shaped structure against a darker ventral brain surface. The dura over the the chiasm was cut along the midline and retracted laterally. We inserted two sharpened stainless steel pins into the sagittal plane of the chiasm. One pin was placed at the posterior edge of the chiasm, the other pin midway anterior/posterior. The two pins were brought together severing the intervening fibers. The bone patch was replaced, the skin was sutured and the animal was left to recover overnight at 4°C. Testing of hemisected animals began the following day. We studied 22 such animals. In 19 of these animals post-lesion prey response was normal throughout the ground level visual field (see figure 1 for a typical example). Animals responded accurately at all distances tested (1 cm to 25 cm) and at all horizontal

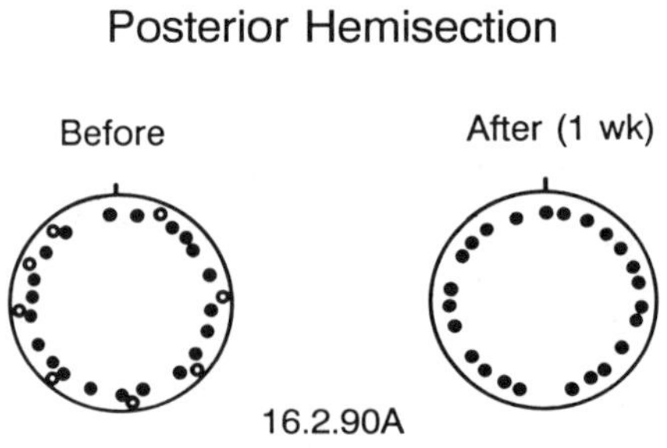

Figure 1. Angular position of ground level prey stimuli (filled circles) and looming threat stimuli (open circles) from which a response could be elicited before and after section of the posterior half of the optic chiasm. A point midway between the eyes is the center point. Tick mark on circle is directly in front of the animal. After hemisection the frog could still respond to prey stimuli but no longer responded to threat stimuli.

eccentricities. When simultaneously presented with 2 crickets in the frontal visual field, the animals attended to, or snapped at, one of the two crickets. The three remaining animals of the 22 inconsistently responded to prey in various parts of the visual field. There were too few responses to determine whether these animals could respond at all horizontal eccentricities and they will not be discussed further.

While visually guided prey catching was normal, avoidance of visually presented threat stimuli was absent throughout the visual field in 14 of the 19 animals (figs 1 and 2). The frogs did not jump away from approaching threat stimuli. When touched, the animals appropriately responded by jumping away, indicating that the motor system was intact. In normal animals prey responses are diminished after threat avoidance testing. In those animals tested, threat stimuli did not appear to inhibit prey catching responses. Four of the 14 frogs occasionally snapped at oncoming threat stimuli approaching from a restricted angle (less than 45°) of the frontal field during the first 3 to 4 presentations during a test session. They would subsequently not respond to threat stimuli

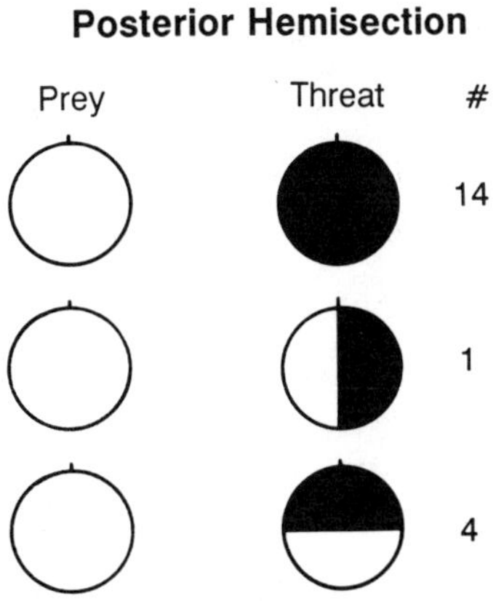

Figure 2. Summary of 19 animals with transection of posterior half of the optic chiasm. Dark areas represent scotomas. Tick marks on circles are directly in front of animals. # is number of animals. Animals respond to prey at all horizontal eccentricities tested. Fourteen of the 19 animals did not avoid threat stimuli from any direction tested.

from any direction. Another animal of the group of 19 retained a normal threat response in the left hemifield. The four remaining animals retained threat responses in all or part of the caudal hemifield.

Of the 14 animals with no obvious change in prey response but complete loss of threat avoidance behavior, 9 were used for neuroanatomical study and had survival periods from 3 to 28 days. Four of these 14 animals were tested until they responded to threat from any direction tested. In these animals recovery was respectively: 3 weeks, 8 weeks, 8 weeks and 20 weeks post-lesion. One animal showed no sign of threat recovery for 3 months. However, 18 days post-lesion it became unresponsive to prey also.

These results show that severing of the posterior part of the chiasm results in continued responses to prey stimuli and loss of responses to threat stimuli. Since prey catching is spared everywhere after chiasm hemisection the fibers of the chiasm are not organized into a single, simple retinotopic map. One way of explaining why, after posterior optic chiasm transection, visually guided prey catching is spared and visually guided threat avoidance is lost is to suppose that optic fibers in the chiasm are functionally segregated. That is, axons in the posterior half of the optic chiasm originating in the retina and/or n. isthmi might mediate information about threat stimuli while the anterior fibers might carry prey stimuli information. The obvious follow up experiment was to sever the front half of the optic chiasm and test visually-guided behavior.

Effects of transection of the anterior half of the optic chiasm We cut the anterior half of the chiasm in an additional set of 7 animals. Post-lesion behavior with anterior chiasm lesions was similar to that with posterior chiasm lesions (fig. 3). In all seven animals there was retention of prey responses

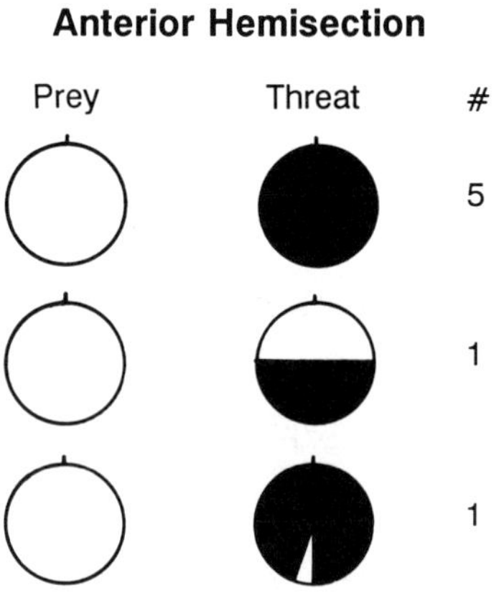

Figure 3. Summary of 7 animals with transection of anterior half of the optic chiasm.

throughout the visual field. In 5 of these animals there was an absence of normal threat avoidance responses throughout the visual field. However, 3 of these 5 animals occasionally snapped at threat stimuli from restricted regions of the frontal field. In the remaining two animals of the 7, one avoided threat stimuli in the frontal hemifield and the other avoided threat stimuli only in a 15° region in the extreme left caudal field.

These results suggest that there is no obvious functional segregation of optic fibers into "threat" and "prey" groups. A possible way of explaining why visually guided prey catching is spared by hemisection of the optic chiasm is that prey catching is mediated by optic fibers that do not cross in the optic chiasm. However, Ingle (1983) has shown that after complete transection of the optic chiasm, visually guided prey catching, as well as threat avoidance, and optokinetic nystagmus are abolished. We have confirmed this result by transecting the chiasm in three animals. There were no responses to prey (and no responses to threat stimuli) at any horizontal eccentricity.

Unilateral section of the optic tract We have seen no significant alteration in the accuracy and latency of responses in animals with chiasm hemisection. In three animals, instead of using sharpened pins, we severed the anterior half of the optic chiasm by making a set of electrolytic lesions down the midline. These lesions caused significant damage laterally. This resulted in loss of prey responses in regions of the visual field. Such lesions interrupt non-crossing fibers as well as crossing fibers. As an extreme case, in three animals we tested the effect of unilaterally cutting the optic tract as it emerges from the chiasm area in three animals. Short term behavioral testing showed that these animals had a scotoma to both prey and threat (fig. 4). But unlike optic nerve section, where the resulting scotoma is restricted to the ipsilateral monocular field, optic tract cut results in a scotoma in the entire contralateral hemifield. This is an unexpected result since it is well known that each tectal lobe receives retinotopic input from the entire contralateral retina.

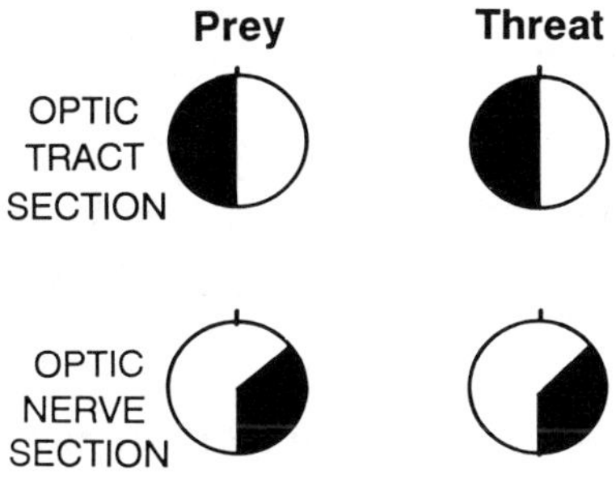

Figure 4. Scotomas (dark areas) to prey and threat stimuli after unilateral optic tract section and after unilateral optic nerve section.

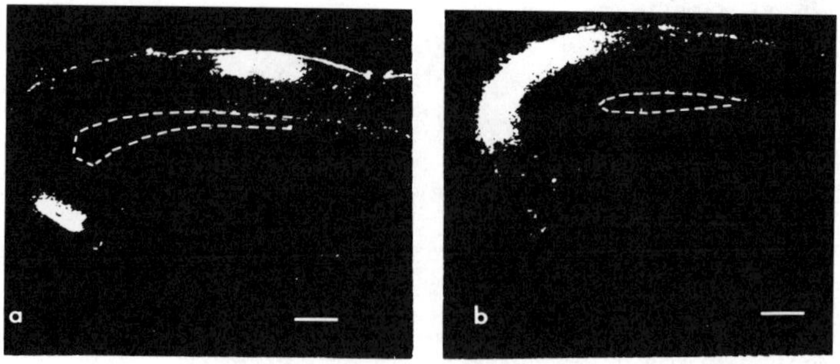

Figure 5. Transverse mid-tectal darkfield autoradiographic sections showing distribution of spared fibers after optic chiasm hemisection and injection of ^{3}H-proline into each eye. a. from animal with the posterior half of the optic chiasm transected. Light area shows the autoradiographic counts. Dashed lines outline tectal ventricles. b. Similar midbrain section in an animal with an anterior chiasm transection. Scale 225 μm.

Anatomy Both prey and threat visual information appear to be mediated by crossing optic fibers but our results suggest that such information is not segregated between anterior and posterior parts of the chiasm. If we assume that retinotectal fibers mediate these responses, our behavioral results imply that retinotectal fibers are not organized in a simple retinotopic map as they cross in the chiasm. Otherwise, frogs with partial section of the optic chiasm should have a scotoma to prey in some part of the visual field. We anatomically determined the tectal distribution of retinotectal fibers in animals with hemisection of the chiasm using ocular injections of ^{3}H-proline and an autoradiographic method (Cowan et al, 1972). In normal animals after intraocular injection there is a characteristic distribution of label in the superficial tectal layers that encompasses the entire contralateral tectal lobe. We studied 6 animals with cut posterior half of the optic chiasm and two animals with cut anterior half. In each animal both eyes were injected and maintained for three to eight days survival at 22°C. Surprisingly, there was a clear difference in the distribution of retinotectal fibers in the two groups. Frogs with posterior hemisection had the dorsomedial and ventrolateral tectum labelled. Three of these animals had no label above background level in the dorsolateral tectum while the remaining animals had sharply reduced labeling in the dorsolateral tectum. This area of the tectum is responsive to visual stimuli in most of the ground level visual field. Animals with anterior hemisection had complementary labeling: the dorsolateral area of the tectum was labeled, the dorsomedial and ventrolateral regions were not. Figure 5 shows the distribution of label. In other animals with hemisected optic chiasms we inserted horseradish peroxidase (HRP) into the optic nerve and reacted the whole brain for HRP activity. A pattern of complementary labeling was seen with anterior and posterior hemisected chiasms which was in agreement with the autoradiographic method. Our anatomical results suggest that retinotectal fibers crossing in the anterior half of the optic chiasm and the posterior half of the optic chiasm terminate in mutually exclusive regions of the retinotopic tectum. It appears that animals with optic chiasm hemisection can respond to prey at all ground level positions, even though the retinotopically organized tectum has a large area that is apparently deafferented.

We were uncomfortable with such a conclusion and used another anatomical approach. For retinal ganglion cells projecting to a circumscribed region of the tectum we determined the distribution of their axons in the optic chiasm and their cell bodies in the retina. To do this, we inserted an HRP crystal at one tectal location per animal. After uptake into terminals HRP is retrogradely transported in axons back to cell bodies. The HRP was visualized using a standard staining method (Adams, 1977). We found that an HRP crystal placed at a particular location of the tectum retrogradely labels cells in a small area of the retina. However, across the sagittal face of the optic chiasm there is proportionally a much larger area containing labeled fibers. Shown in figure 6 is an example. An HRP crystal was placed in dorsolateral midtectum. After survival time of 7 days, labeled ganglion cells were found in central retina. These cells occupied 1.8% of the retina. However, the retrogradely labeled fibers occupy 18% of the area of the optic chiasm as viewed in the sagittal plane. Thus optic fibers projecting to a circumscribed area of the tectum are quite spread out in the chiasm. This result helps explain why prey catching is spared after hemisection of the chiasm. It does not explain why there is such a difference between anatomical methods that rely on orthograde and retrograde transport.

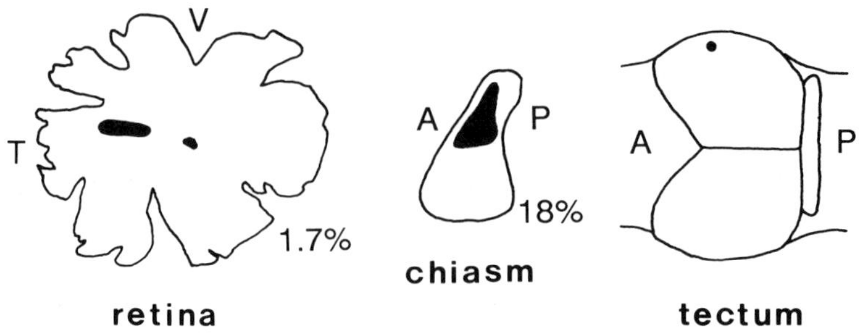

Figure 6. Distribution of labeled tissue (dark areas) in retina and sagittal plane of optic chiasm after insertion of HRP crystal into dorsolateral tectum (filled circle). Each number is the percentage of area occupied by labeled material. A, anterior; P, posterior; T, temporal; V, ventral.

3. DISCUSSION

Our results have uncovered several puzzles regarding the nature of visual inputs to the frog brain. 1) Why do optic chiasm hemisections spare prey catching and lose threat avoidance? 2) Why is there a hemifield scotoma when the optic tract is severed but a monocular field scotoma when the optic nerve is severed? 3) What is the role of contralaterally projecting n. isthmi fibers?

Prey vs Threat We can not explain why prey catching is spared and threat avoidance is lost after optic chiasm hemisection. Perhaps more transtectal integration is needed for processing threat stimuli than for prey stimuli and this is a population dependent process: if there is less than a certain number of retinotectal fibers nothing happens. However, that is not a very satisfying explanation. In population dependent processes as one loses more of the population one expects a degradation of the process not simply a loss of a response (see for instance,Kostyk and Grobstein, 1987; Masino and Grobstein, 1989; Rumelhart and McCleland, 1986).

One might suppose that information about prey stimuli is distributed over a smaller area of the tectum than threat stimuli. However, a cricket placed close to the frog (where the frog is most likely to snap) subtends as large an angle on the retina as a looming threat stimulus when it evokes an escape response in the frog.

Optic chiasm hemisection reveals a dichotomy between prey and threat stimuli. In a majority of our animals there are no responses to threat stimuli while prey responses look quite normal. However, in 7 animals there is transitory snapping at threat stimuli somewhat reminiscent of the behavior of anurans after lesions of the posterior dorsal thalamus. Such lesions "disinhibit" animals and they respond to threat as if it were prey (Ewert 1970; Ingle, 1973). The "antiworm" elicits a response characteristic of "worm". Perhaps those animals with hemisected chiasms that snapped at threat stimuli had lost a significant proportion of the optic fibers that project to dorsal posterior thalamus.

Scotoma After Optic Nerve Section vs Optic Tract Section It is well known that eye enucleation or optic nerve cut leads to a scotoma in the monocular field of that eye. We expected a similar result from cutting the optic tract. After optic nerve section or optic tract section the nondeafferented tectal lobe should be receiving its normal input from the contralateral retina. Our finding that such animals only respond in the hemifield suggests that visual inputs to other retinoreceptive targets or the crossed isthmotectal projection must facilitate the response. Such a hemifield scotoma is commonly found in mammals where corresponding parts of each retina project to one side of the brain. Functionally, the frog seems to be a closet mammal.

Contralaterally Projecting Isthmotectal Fibers Our initial motivation for carrying out the current experiments was to investigate the role of contralaterally projecting isthmotectal fibers. Since we see no difference in behavior when we cut the posterior half of the chiasm (which contains the crossing isthmotectal fibers) and when we cut the anterior half of the chiasm (which does not contain the crossing isthmotectal fibers) we still can not attribute any function to these fibers. Collett et al (1987) found that bilateral n. isthmi lesions in anurans did not disrupt an animal's ability to use binocular cues to judge distance.

ACKNOWLEDGEMENT This work was supported by NIH grant EY04366

BIBLIOGRAPHY

Adams, J.C. (1977) Technical considerations on the use of horseradish peroxidase as a neuronal marker. Neurosci. 2: 141-145.

Collett, T.S., S.B. Udin and D.J. Finch (1987) A possible mechanism for binocular depth judgements in anurans. Exp. Brain Res. 66: 35-40.

Cowan, W.M., D.I. Gottlieb, A.E. Hendrickson, J.L. Price and T.A. Woolsey (1972) The autoradiographic demonstration of axonal connections in the central nervous system. Brain Res. 37: 21-51.

Ewert, J.-P. (1970) Neural mechanisms of prey catching and avoidance behavior in the toad (Bufo bufo, L.) Brain Behav. Evol. 3: 36-56.

Gruberg, E.R. (1989) Nucleus isthmi and optic tectum in frogs. In, Visuomotor Coordination J.-P. Ewert and M.A. Arbib eds. Plenum, NY pp 341-356.

Gruberg, E.R., M.T. Wallace and R.F. Waldeck (1989) Relationship between isthmotectal fibers and other tectopetal systems in the leopard frog. J. Comp. Neurol. 288: 39-50.

Ingle, D. (1973) Disinhibition of tectal neurons by pretectal lesions in the frog. Science 180: 422-424.

Ingle, D. (1983) Brain mechanisms of visual localization by frogs and toads. In, Advances in Vertebrate Neuroethology, J.-P. Ewert, R.R. Capranica and D.J. Ingle, eds Plenum, NY pp 177-226.

Kostyk, S.K. and P. Grobstein (1987) Neuronal organization underlying visually elicited prey orienting in the frog: II Anatomical studies on the laterality of central projections. Neurosci. 21: 57-82

Masino, T. and P. Grobstein (1989) The organization of descending tectofugal pathways underlying orienting in the frog, Rana pipiens: I Lateralization, parcellation, and an intermediate spatial representation. Exp. Brain Res. 75: 227-244.

Montgomery N. and K.V. Fite (1989) Retinotopic organization of central optic projections in Rana pipiens J. Comp. Neurol. 283: 526-540.

Reh, T.A., E. Pitts and M. Constantine-Paton (1983) The organization of the fibers in the optic nerve of normal and tectum-less Rana pipiens J. Comp. Neurol. 218: 282-296.

Rumelhart, D.E. and J.L. McCleland (1986) Parallel distributed processing: explorations in the microstructure of cognition. M.I.T. Press, Cambridge.

Scalia, F. and V. Arango (1983) The anti-retinotopic organization of the frog's optic nerve. Brain Res. 266: 121-126.

Approach and Avoidance

Directed Movement in the Frog: A Closer Look at a Central Representation of Spatial Location

Paul Grobstein
Department of Biology
Bryn Mawr College
Bryn Mawr, Pennslyvania 19010

Abstract. *The neuronal circuitry underlying directed, ballistic movements in the frog includes a stage in which information about target location is represented in a form which is both experimentally distinguishable from spatial representations closer to the sensory and motor sides of the nervous system, and distinctive in its organization. Three dimensional location is represented in a distributed fashion, in terms of independent orthogonal components Each component appears to be population coded, apparently as the total activity in a particular neuronal structure. These findings are discussed in relation to related findings in other systems, with the objectives of identifying possible generalizations about spatial representations involved in sensorimotor processing and of defining directions for future research based on these.*

INTRODUCTION

The nature of the neuronal circuitry underlying directed, ballistic movements is currently under active investigation in a variety of organisms and situations (cf. Georgopoulos, 1986; Berkinblitt et al., 1986; Ewert, 1987; Knudsen et al., 1987; Camhi, 1988; Schildberger, 1988; Stein, 1989; Anderson, 1989; Sparks and Mays, 1990), and some common principles of sensorimotor processing are beginning to emerge (cf. Grobstein, 1988, 1991 on activity-gated divergence; Grobstein, 1991 on motor equivalence). Here I want to focus on the particular issue of how spatial location is represented in the sensorimotor interface. I will review recent findings in the frog which indicate the existence in the sensorimotor interface of a form of representation of spatial location which is not only experimentally distinguishable from those present on either the the sensory or motor sides of the nervous system but also distinctive in its organization. A related picture of one or more intermediate spatial representations, some with properties quite similar to those present in the frog, is emerging in other systems, suggesting that with regard to forms of representation of spatial information, as with other matters, it is becoming possible to make some useful general statements about how nervous systems deal with the problem of directed, ballistic movements.

INTERMEDIATE SPATIAL REPRESENTATIONS HAVE A DISTRIBUTED (PARCELLATED) CHARACTER

Our evidence for an intermediate spatial representation derives from a series of studies of behavior following selected central nervous system lesions (Kostyk and Grobstein, 1982, 1987a,b,c; Grobstein et al., 1985, 1988, 1990; Masino and Grobstein, 1989a,b; Grobstein and Staradub, 1989). The logic and appropriateness of lesion studies for an analysis of this kind have been discussed elsewhere (Grobstein, 1990a,b), and the basic findings reviewed from several different perspectives (Grobstein et al., 1983; Grobstein, 1988, 1989, 1991). Here I will only briefly sketch our reasons for believing that there exists in the frog a form of spatial representation distinct from that on either the sensory or motor sides of the nervous system, as background for a more extensive discussion of the details of the organization of this representation, and of its relation to representations seen in other systems.

Unilateral lesions of the retinal projections or of the optic tectum, the stages of initial sensory processing in the frog, produce characteristic deficits in visually triggered prey orienting behavior, the generation of rapid, ballistic movements directed towards a prey item. Such deficits consist of a failure to respond to prey items at any location within an area of visual space which is defined by a retinal coordinate frame. Lesioned frogs respond normally to stimuli at all locations within the visual field of one eye (which includes one visual hemifield and an additional 45 degrees across the midline in front; see Figure 1). They fail to respond to stimuli at remaining locations. In contrast, unilateral lesions of a descending tectofugal pathway (see Masino and Grobstein, 1989b, 1990, for relevant anatomical studies), in either the caudal midbrain or the caudal medulla, produce a deficit defined in a head or body centered coordinate frame. Lesioned frogs respond normally to stimuli presented at any location within the contralateral visual hemifield, and abnormally to stimuli at locations within the *entire* ipsilateral visual hemifield. The deficit region is defined by the mid-sagittal plane, rather than by the edge of the visual field of either eye (Figure 1). In short, there exists a spatial representation subsequent to the tectum which is distinct from that in the retinotectal projection by virtue of being in a head or body centered coordinate frame, rather than in a retinal coordinate frame.

The form of representation of spatial information in the caudal midbrain and medulla differs from that in the retina and tectum not only in being in a head or body centered coordinate frame (and hence in being more strictly lateralized) but also in being "parcellated" (Grobstein, 1988; Masino and Grobstein, 1989a). Following retinal or tectal lesions, the deficit involves a failure to respond to stimuli. Following more caudal lesions, the deficit consists instead of abnormally directed movements, with the abnormality relating specifically to one component of three dimensional stimulus location. Lesioned frogs respond to stimuli at all locations in the ipsilateral hemifield with a forwardly directed movement. The movements do not vary with the horizontal eccentricity of the stimulus, but do vary with its distance and elevation. What this indicates is that information about the horizontal eccentricity of a stimulus exists in the intermediate representation in a form which is independent of information about its elevation and distance. This is not the case at the level of the retinotectal projection, whose map-like organization means that activity at a given location represents simultaneously information about horizontal eccentricity and at least one of the other two variables.

The lateralized and parcellated intermediate spatial representation can be operationally distinguished not only from initial sensory representations of target location but also from representations closer to the motor side of the nervous system. This distinction is based on the finding that the lesions studied affect not only the directedness of movements in response to stimuli in the ipsilateral hemifield but their qualitative character as well. The forwardly directed responses observed in lesioned frogs are the normal movement patterns appropriate for frontal stimuli, as opposed to being the movement patterns appropriate for more eccentric stimuli degraded by the absence of a turn component. What this implies is that the lesions are not disturbing the production of motor patterns *per se*, but are instead disturbing a prior act of motor choice (see Grobstein, 1991 for a discussion of this phenomenon in a larger context). For the purposes of comparing findings in the frog with those in other organisms, as I will do briefly below, it is worth emphasizing that the logic of the distinction between the observed intermediate spatial representation and others closer to the motor side of the nervous system relates to the existence of an intervening step of motor choice, rather than depending on any knowledge or presumptions about the organization or number of representations either centrally or closer to the motor periphery.

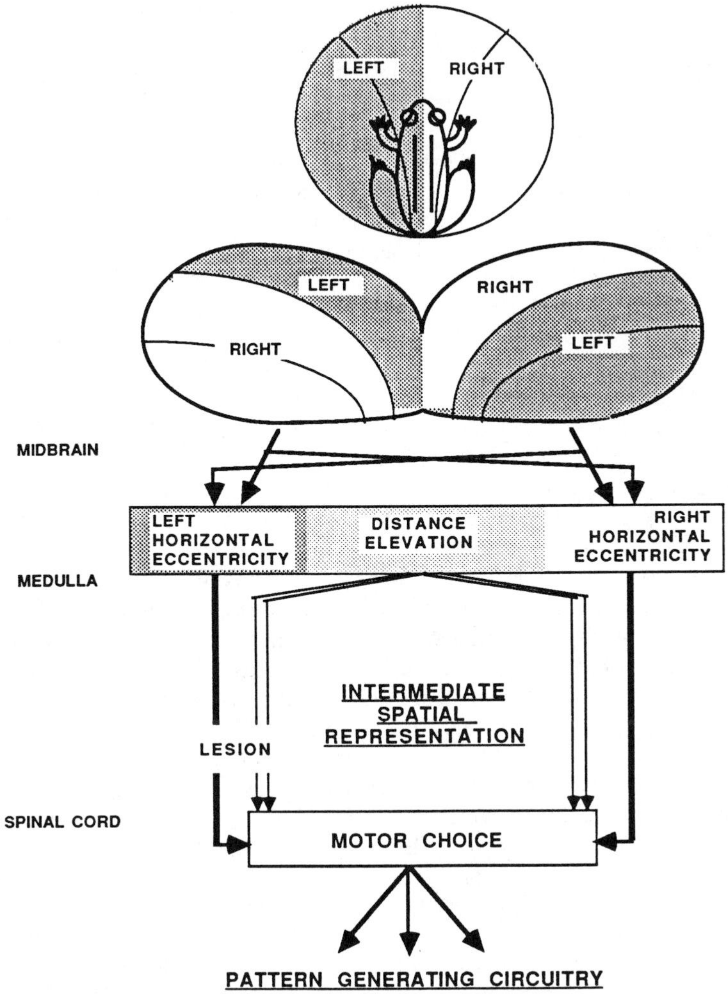

FIGURE 1. A model of the neuronal organization underlying directed prey orienting movements in the frog, based largely on observations of behavior following selected central nervous system lesions. The visual fields of the two eyes of a frog, and their representations on the two lobes of the optic tectum, are shown above. The visual field of each eye includes the entire ipsilateral hemifield, and extends across the midsaggital plane to include a portion of the contralateral hemifield (up to the curved lines). The visual field of each eye is topographically mapped, in its entirety, on the contralateral tectal lobe, so that each tectal lobe contains a representation of the midsaggital plane and of portions of both the left and right hemifields (shaded and clear areas, respectively).

This two-dimensional map-like representation is subsequently transformed into a three-dimensional, parcellated "intermediate spatial representation." The transformation apparently takes place in the ventral midbrain (elongated box in middle of figure) and yields several descending pathways (arrows from box). These carry discrete components of the intermediate spatial representation to the spinal cord where they are combined to yield a choice of one or another kind of movement ("motor choice"), after which appropriate signals are distributed to circuitry responsible for creating particular motoneuron discharge patterns ("pattern generating circuitry).

As illustrated, one discrete component of the intermediate spatial representation is a signal related to the horizontal eccentricity of a stimulus, defined in terms of angle to the left or right of the midsaggital plane. Left and right horizontal eccentricity signals descend unilaterally, on the ipsilateral side of the brain (heavier arrows). In contrast, signals related to stimulus elevation and distance descend bilaterally. The pathways carrying these two signals (parallel thinner arrows) are distinct from those carrying the horizontal eccentricity signals. Whether they are also distinct from one another remains to be determined. A unilateral lesion, as illustrated, affects turning toward stimuli at all locations in the entire ipsilateral visual hemifield, while sparing information about the distance and elevation of such stimuli since such information descends bilaterally.

Figure 1 provides a schematic summary of some of the most basic organizational features of the intermediate spatial representation in the frog, as we currently understand them, and the relation of this representation to some known or suspected earlier and subsequent information processing steps. The representation appears to be created in the ventral midbrain (Masino and Grobstein, 1989b; 1990), based on inputs from the optic tectum and, presumably, other structures. Signals related to the horizontal eccentricity of a stimulus to the left or right of the mid-sagittal plane in front descend on the left and right sides of the brain respectively. It is unilateral interruption of one of these pathways which results in animals responding to stimuli at all locations in one hemifield with forwardly directed movements. Signals related to the elevation and distance of a stimulus, in contrast, descend bilaterally (probably each along separate pathways but we do not as yet know this for sure), and so are spared for all locations in space by unilateral lesions Why the system is organized such that information about distance and elevation descends equally well on either side of the brain, while information about horizontal eccentricity is lateralized, is an interesting question deserving of further investigation. Since our earlier studies involved only unilateral lesions, they left open the possibility that the bilateral pathway for elevation and distance consisted of the combination of the two lateralized pathways which carry horizontal eccentricity information. More recently, we have found that bilateral lesions can markedly affect the distance signal without significant effect on horizontal eccentricity signals (Grobstein and Staradub, 1989; see Figure 2), suggesting that the horizontal eccentricity and distance pathways are, as illustrated, physically distinct.

The parcellated form of spatial representation seen in the frog is of interest both in its contrast to better known forms of spatial representation, and as an instance of an emerging generalization that parcellated spatial representations are characteristic of sensorimotor processing in a wide variety of organisms (Grobstein, 1989, and below). Maps, in which spatial location is represented by the location of activity in a particular neural structure and adjacent locations within the structure correspond to adjacent locations in space, certainly represent the most widely documented form of spatial representation. In all known cases, maps are associated with laminar neural structures and accordingly represent spatial location in terms of no more than two dimensions, corresponding to the two dimensions of the sheet-like structure in which the map occurs. Three dimensional maps have not been observed in any organism, making it at least an entertainable possibility that parcellated representations are a more common form of representation of three-dimensional spatial location.

Consistent with the notion that representations of three-dimensional space typically involve parcellated representations are older clinical studies (Beevor, 1909; see also Iannone and Gerber, 1983) suggesting the existence of a parcellated representation underlying head movements in humans (where the three dimensions correspond to roll, pitch and yaw), as well as recent psychophysical studies showing the existence of parcellated representations underlying human limb movements involved in pointing to targets in three dimensional space (Soechting and Flanders, 1989a,b; Flanders and Soechting, 1990; see also Favia et al., 1990). At the same time, parcellated representations have long been known to exist in the circuitry involved in primate saccade control (see discussion in Grobstein, 1988), and have recently been documented for head orientation movements in owls (Masino and Knudsen, 1990), both being situations in which there is no obvious reason why stimulus localization needs to be done in three dimensions rather than in two. Hence, while it may well prove to be true that parcellated representations are the most common way of representing three-dimensional spatial information, it is also possible that there is some more general reason for the existence of such representations.

An interesting and perhaps significant feature of the parcellated, intermediate representation in the frog is that it is apparently an abstract representation, in several senses (Grobstein, 1989, 1991). A variety of different input signals, each based on a different coding scheme and using a different coordinate frame, provide input to the intermediate representation. Distance, for example, may reflect binocular cues, presumably deriving from the place-coded retinocentric maps, or a totally different kind of signal, some kind of corollary discharge associated with lens accommodation (see House, 1989; Grobstein, 1991 for references and discussion). In addition to visual signals reflecting target

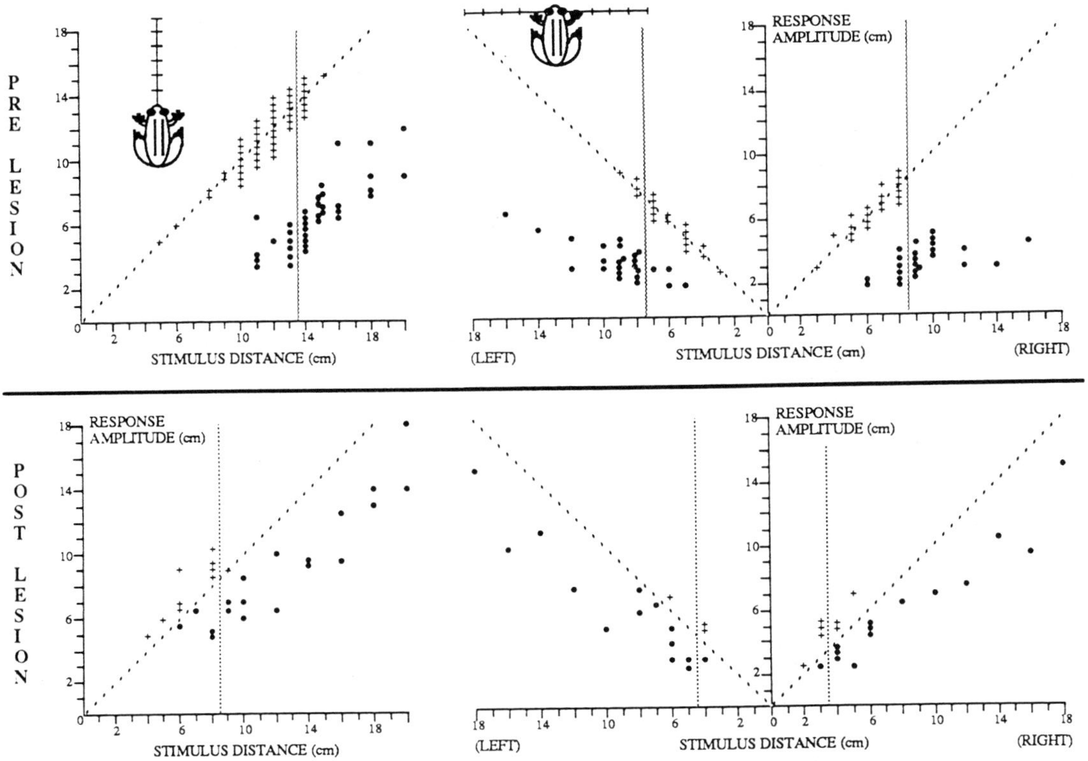

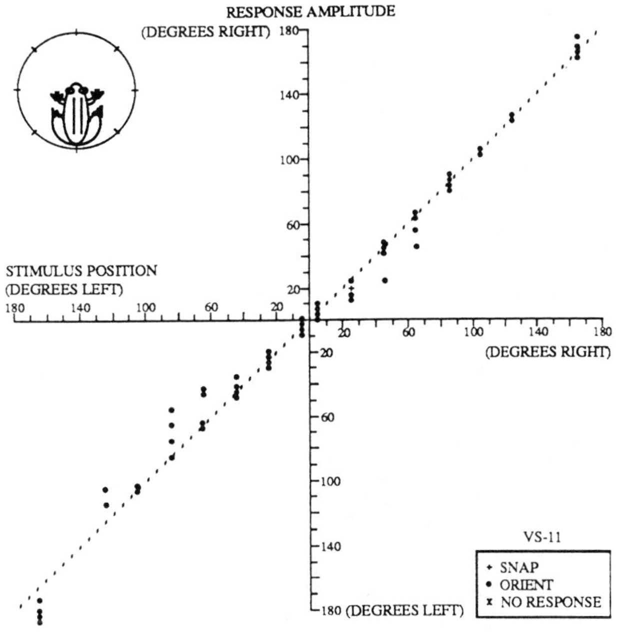

FIGURE 2. Effects of a bilateral lesion on distance and horizontal eccentricity signals. The upper two panels of the figure show the pre-lesion responses of a frog to stimuli presented at varying distances on the mid-saggital plane in front (left panel) and on the frontal plane to each side (right panel). The middle two panels show the post-lesion responses to the same stimuli. Each symbol corresponds to a single trial and shows the amplitude of the response (ordinate) for given stimulus distances (abscissa). Crosses correspond to snapping movements and dots to orienting movements lacking a direct snap. Note that the former occur for nearer distances and the latter for greater ones, with a definable distance at which orienting movements predominate (vertical line).

Following the lesion, the movements made by the frog are those appropriate for stimuli at distances greater than the ones at which the stimuli were actually presented. This can be seen both in the amplitudes of the responses to stimuli at particular distances as well as by the fact that the lesioned frog orients to stimuli at distances where it had previously snapped (vertical lines are shifted to shorter distances).

The lesion in this case affected the distance signal without having detectable effect on the horizontal eccentricity signal, as shown in the panel to the lower right. This panel shows the post-lesion responses of the frog in terms of the observed angle of turn (ordinate) for stimuli presented at a series of different horizontal angles (abscissa). Each dot corresponds again to a single trial. Turn angles are similar to the angles of stimulus presentation, as in normal frogs. An example of disturbed horizontal eccentricity signals can be seen in Figure 3. Data from Grobstein and Staradub (1989).

location, tactile signals related to target location are also represented in the intermediate representation (Grobstein et al., 1988). Just as a given intermediate signal may correspond to any of a variety of different inputs, a given intermediate signal may be associated with a number of different body movements, involving different amounts of head, limb, and body movement and different combinations of body rotation and translation (Grobstein et al., 1990). What this suggests is the possibility that parcellated spatial representations are perhaps the computationally most convenient way to relate different spatial representations one to another (Grobstein, 1989). A similar suggestion has been made in connection with the owl (Masino and Knudsen, 1990). In the case of limb movements, the parcellated representation is believed to be a representation of target location, one which must be combined with kinesthetic information about initial limb position prior to construction of a motor command, and so the same argument might apply here. It remains, however, to be shown exactly how parcellated representations might serve to simplify computational problems, and one would like more information about additional examples of such representations before trying to come to a firm conclusion. Modelling studies (cf House, 1989; Arbib and Cobas, 1991; Cobas and Arbib, 1991a,b; Mittlestaedt and Eggert, 1990) could also play a significant role in helping to understand the advantages of parcellated representations (see Anastasio, 1991, for discussion of a related problem in distributed coding in a different context).

In thinking about such questions, some additional possible generalizations about parcellated representations may be worth bearing in mind. One might in principle imagine such representations to be defined in terms of any of a variety of different component axes. In fact though, there is a surprising similarity in the components of parcellated representations in most systems in which they have been described. In particular, a variable like horizontal eccentricity is present in all systems. While this might be imagined to reflect no more than a similarity in the conventions used by experimenters to describe spatial locations, it is in fact intrinsic to findings in the frog, at least, that the relevant component is a radial rather than a Cartesian one (Masino and Grobstein, 1989a). An even more interesting, because apparently arbitrary, similarity is that horizontal eccentricity seems to be not only lateralized in a variety of quite different situations, but lateralized in the same sense as in the frog, with information about locations to one side of the mid-saggital plane represented on the ipsilateral side of the nervous system. This is true not only of vertebrates, but of at least some invertebrates as well (Comer and Dowd, 1987; Schildberger, 1988), suggesting that there may be something generally significant about this particular kind of parcellated representation.

The discovery of an intermediate spatial representation in the frog was significant because investigators working in that system had earlier presumed that there was a fairly direct linkage between sensory maps and pattern-generating circuitry. In other systems, it seems likely that there exists not one but several different spatial representations between the sensory maps and the circuitry responsible for generating particular motoneuron discharge patterns, a reality which Soechting and his colleagues have called particular attention to in the case of human pointing movements. This may well be true for the frog as well, and it is, in any case, of interest to try and relate particular intermediate representations seen in one system with one or more of those seen in others, both from the perspective of trying to understand the significance of parcellated representations and that of approaching further generalizations about the nature of sensorimotor processing.

Based on findings in both motor cortex and mammalian colliculus, Soechting and Flanders (1990) make a useful distinction between representations of target location and representations of the movement necessary to reach the target given the initial starting position (a "motor error" representation, which has been documented for both cortex and mammalian colliculus). Soechting and Flanders suggest that the latter is derived from the former (by the addition of kinesthetic information), that the parcellated representation they observe is of the target representation sort, that if collicular function is similar in all vertebrates the representations seen in frog and owl are probably "motor error" representations, and hence that there may exist at least two kinds of parcellated representations, one more toward the sensory and the other more toward the motor side of the sensorimotor interface.

That the representation in the frog is some distance from the motor side of the nervous system is evident from the arguments emphasized above, and it is for this reason that we have referred to it as a representation of target location without further specification. At the same time, there is some published work whose significance we did not previously appreciate which now seems relevant in attempting a more precise characterization of this representation. Kostyk and Grobstein (1987a) reported studies of the responses of normal and lesioned animals to stimuli presented near the mid-saggital plane behind the animals. Normal frogs may respond to such stimuli with either leftward or rightward turns, the amplitude of which can exceed 180 degrees. Interruption of the horizontal eccentricity tract on one side of the brain results in animals responding to ipsilateral caudal stimuli primarily with forwardly directed movements. These animals however also exhibited occasional contralateral turns of an amplitude of greater than 180 degrees. It would appear from this that what has been disturbed is not strictly information about target location in one hemifield, but rather information about how to get there using either rightward or leftward turns. While this may seem at first to push the parcellated representation more toward the motor side of the interface, the observations could equally well be described in terms of disturbance of one of two representations of stimulus location which use respectively clockwise and counterclockwise conventions for measuring eccentricity.

This leaves open the issue of whether the representation is of a "motor error" sort, and the issue is actually more complex still. There are two somewhat different ideas combined in the concept of a "motor error" representation. One is the notion of a signal which takes into account the location of parts of the organism with respect to other parts. Whether the representation in the frog has this character can be explored by comparing the orienting movements of normal and lesioned frogs made from a standard posture with those made from other initial postures, with the head initially yawed into the deficit hemifield for example. If the deficit border were to prove to be aligned with the mid-sagittal plane of the body, rather than remaining aligned with the head, one could reasonably conclude that kinesthetic signals have contributed to the observed representation. Such experiments are in progress, but will not, however, for a significant reason, fully resolve the issue of whether one is dealing with a "motor error" representation. This concept, as it is currently used, includes the additional idea that a given signal corresponds to a particular direction and amplitude of movement irrespective of the initial position from which a movement is made (and hence does not correspond to a particular location in space). What is at issue here is the use of kinesthetic information to generate a signal related to target location in some abstract coordinate frame, as opposed to the use of kinesthetic information to translate such a target location signal into appropriate movement of a body part whose position relative to the rest of the body is variable. The latter may be a problem specific to eye and limb movements, and not relevant in situations like frog orienting in which the directed movements are primarily of the entire body. On the other hand, recent findings raise the possiblity that intermediate representation in the frog may persist after inaccurate movements and be compared with a kinesthetic signal so as to generate further movements (Grobstein et al., 1990; Grobstein, 1991). A signal reflecting such a comparison might well qualify as a "motor error" signal, and this possibility is worth further exploration.

The issue of whether what leaves tectum is a "motor error signal," as opposed to "an input to circuitry establishing a representation of spatial location in an abstract head- or body centered coordinate frame (Grobstein, 1989)," is, it seems to me, still unresolved, not only in the frog but in most organisms (Grobstein, 1988). This does not, however, detract from my main concerns in this section, and in some ways serves to highlight them. The sensorimotor interface, in a variety of animals, clearly contains idiosyncratic spatial representations, many of them having a parcellated character. As such, they are different from many known spatial representations in being distributed, rather than map-like. Understanding the origins, significance, and relations among such representations within particular animals, and in comparisons among animals, is a problem which deserves the attention of both experimentalists and modellers, and one which is probably significant in the context of robotics and artifical intelligence as well.

POPULATION (ACTIVITY) CODING IN AN INTERMEDIATE SPATIAL REPRESENTATION

The intermediate spatial representation in the frog, as described in the previous section, is a distributed representation in the sense that it is activity in several different neuronal structures which represents target location, rather than the location of activity within one neural structure. In this section, I want to describe evidence that it is a distributed representation in a second sense as well: that the value of components of the parcellated representation are coded in terms of the level of activity across a population of neurons, rather than in terms of which particular elements of a population of neurons are active. This evidence also suggests that there exist defaults against which spatial representation values are defined, and implies certain characteristics of the processes which decode the intermediate spatial representation (see Lehke and Sejnowski, 1990) for an interestingly related discussion in a quite different context).

As described earlier, complete lesions of a defined descending tectofugal pathway result in animals responding to stimuli at all locations in the ipsilateral hemifield as if they were located on the mid-sagittal plane in front of the frog. If the horizontal eccentricity information being carried in this pathway was coded in terms of the presence or absence of activity in particular neurons, one would expect that partial lesions would result in accurate turning toward stimuli at some locations in the ipsilateral hemifield, and failure to turn toward others. What is instead observed is one or another degree of more or less systematic disturbance of turn angle for all locations in the affected hemifields (see Figure 3; a large number of additional cases are illustrated in Kostyk and Grobstein, 1987a; Masino and Grobstein, 1989a). These invariably consist of a reduction of turn amplitude for all stimulus locations, with a persistance of the monotonic increase of turn amplitude with increasing stimulus eccentricity which is seen in normal animals. Though we have not systematically analyzed such partial lesion effects at a high degree of resolution, the observed reduction in turn amplitude is at least grossly correlated with the amount of damage to the identified descending tract, with no obvious dependence on which particular part of the tract is damaged

These observations make it highly unlikely that individual neurons contributing axons to the descending tract code for particular horizontal eccentricities, and suggest instead that horizontal eccentricity is coded by some variable such as the sum total of activity in all of the axons of the tract. Complete interruption of the tract results in forwardly directed movements, implying that zero activity codes a location on the mid-sagittal plane in front, and that subsequent circuitry is organized so as to equivilently interpret an absent signal (as in the case of a complete lesion). Increasing levels of activity code for increasing horizontal eccentricities. Partial damage to the tract somewhat reduces the level of population activity for all stimulus locations and hence results in reductions of turn amplitude for stimuli throughout the hemifield.

More recently, we have obtained evidence that a similar coding scheme is used in the pathways that carry information about target distance (Grobstein and Staradub, 1989). As shown in Figures 2 and 3, partial bilateral damage produces changes in the distance component of orienting movements not for particular stimulus distances but for all distances within the range tested. What is perhaps unexpected is that the change resulting from partial damage is not a reduction in the distance component of the movements, but rather an increase. This can be seen in the values for individual trials, but more strikingly in the change in the location of the snap/hop border (Ingle, 1972; Grobstein et al., 1985; Grobstein, 1988), the distance at which animals shift from a motor pattern in which a tongue flick is directed to the target to an approach motor pattern lacking a tongue flick (vertical dotted lines in Figures 2 and 3). The borders are clearly closer to the animals after the lesion than they are before. This is to be expected if the movements being made for stimuli at a given location are the movements normally made for stimuli at more distant locations (hops rather than snaps are being made for an intermediate range of distances). What this implies is that increasing levels of activity in the distance pathways code

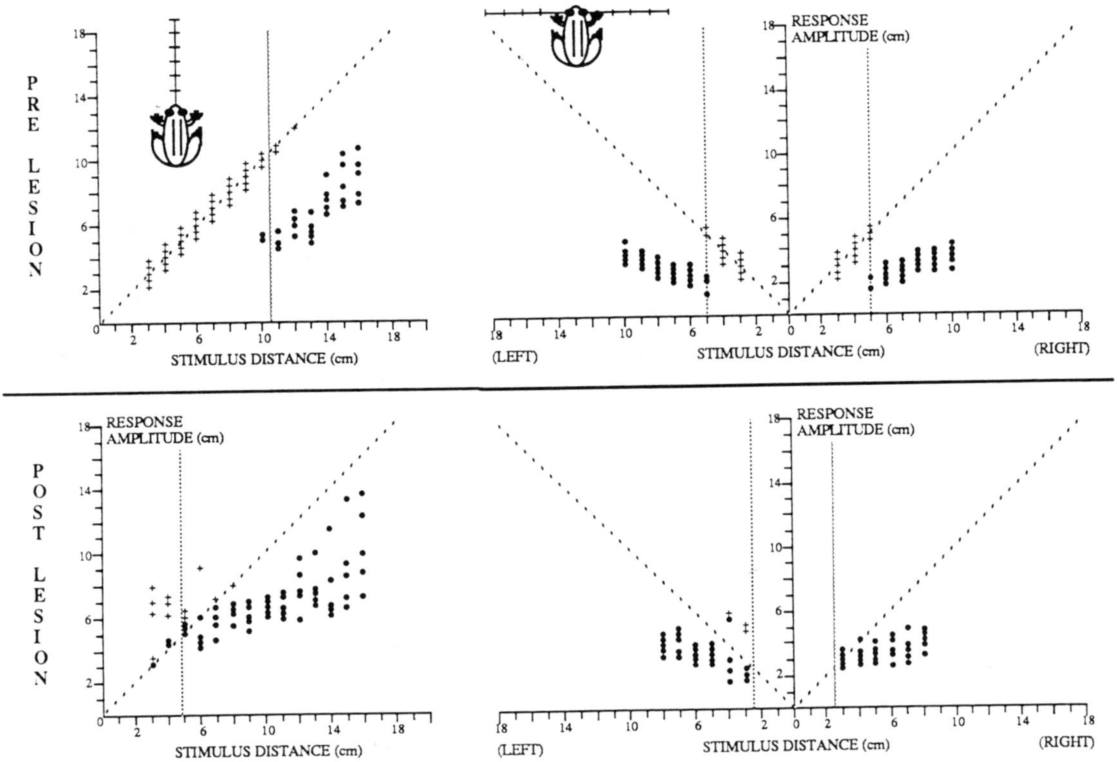

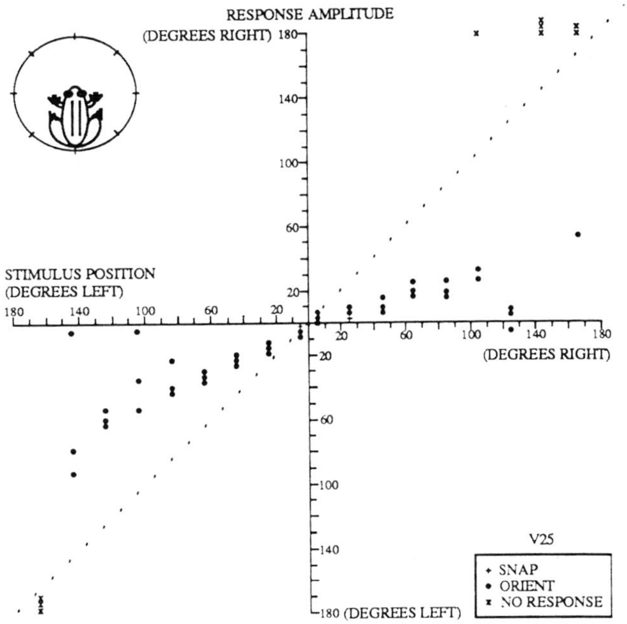

FIGURE 3. Effects of a bilateral lesion on distance and horizontal eccentricity signals. Conventions as in Figure 2. The lesion in this case affected the distance signal as well as the horizontal eccentricity pathways bilaterally. The data is presented to illustrate that disturbances in the distance and horizontal eccentricity pathways are similar in their effects, implying that the two pathways both represent signals in terms of a population activity code. The disturbance in responses to stimuli presented at varying distances is similar to that in the case illustrated in Figure 2, again involving a systematic alteration in response amplitude for all stimulus distances. There is in this case, in addition, a systematic reduction in the angle of turn observable at all horizontal eccentricities of the stimulus (lower right panel). The effect is present in both hemifields, but more dramatic in one, presumably due to different amounts of damage in the two horizontal eccentricity pathways. Grobstein and Taglianetti, unpublished observations.

for stimulus locations nearer to the frog, rather than more distant, and that the zero activity default value is presumably at some distant location rather than at zero distance measured from the frog.

We have as yet substantially fewer cases of partial lesions in the distance pathways and there are some variations in these effects which we have yet to understand. In some cases, the relation between stimulus distance and response distance is not only shifted vertically (as illustrated) but flatter in slope, an effect which may provide important clues to the underlying coding scheme at a finer level of detail and which also makes us for the moment somewhat uncertain about what distance corresponds to zero activity. The basic partial lesion phenomena as described is, however, robust and, as noted, strikingly similar to what is seen in the horizontal eccentricity pathways, suggesting that a coding scheme based on population activity may be a general characteristic of the intermediate spatial representation in the frog.

Population coding has been increasingly recognized as an important aspect of sensorimotor processing in a variety of organisms and situations (McIlwain, 1972; Georgopoulos et al., 1988; Dowd and Comer, 1988; Sparks and Mays, 1990), so that in this regard too the frog provides an illustrative example of an emerging generalization about sensorimotor processing. At the same time, the history of the recognition of the population coding phenomena in the frog is different from that in other sensorimotor systems, and there may be significant differences in the phenomena themselves. By and large, the history in other systems has involved situations in which it was first apparent that there was a prominent map-like aspect to the coding, and it only subsequently emerged there was significant information about spatial location in the simultaneous activity of large numbers of neurons in the map-like structure. In the tectum, for example, it was first noted that individual neurons are active in relation to movements whose direction and amplitude relate to their location in the plane of the tectum, and only later fully appreciated that the associated movement fields were quite broad, that large numbers of neurons were active in association with movements of a particular direction and amplitude, and that this was a significant factor in interpretation of the coded message (Lee et al., 1988).

Population coding in the intermediate spatial representation in the frog, in contrast, was immediately apparent because of the lesion methodology employed to explore this system. An obvious question, at this point, is whether, as in other systems, there exists in this representation a place-coding aspect as well. Our findings, as described, could be accounted for on the assumption that all axons in the relevant tracts carry a similar signal, whose frequency increases with increases in the relevant variable (horizontal eccentricity and nearness to the animal). Removal of any subset of the axons in such a case (by a partial lesion) would reduce the total activity for all stimulus locations. It is, however, also possible that there exists some range fractionation in the coding of these signals, as is known for example to exist in the coding of stimulus intensity in the auditory system and that for intensity of muscle activation in the motor periphery. In both cases, increases in intensity are represented by some degree of increase in firing frequency of active neurons as well as by recruitment of additional, higher threshhold elements. Assuming elements with different threshholds were randomly distributed in the descending pathways in the frog, partial lesions would yield in this case, as well as in that of pure activity coding, the same effect, and so we cannot, without electrophysiological observations, say whether individual elements are distinctive in their responses to stimuli at different locations. At the same time, there is a clear noteworthy difference between the character of population coding in the frog and that in the case of tectal efferent projections involved in saccade control. In the latter, what is significant is not the total activity in the population but rather the distribution of activity around some most active elements which (by their place coding properties) represent particular values of the signal. In such a system, removal of active elements on one side of the distribution will lead to a decrease in the value of the population coded signal, and removal of active elements on the other side of the distribution will lead to an increase in this value (Lee et al., 1988). In the frog, this seems not to be the case; partial lesions always lead to a decrease in the value of the population coded signal. What this implies is that, while range fractionation may exist, the population coding in the frog has less place-coding character than that in the case of saccade control. Individual elements may differ in the relation between input signal and firing frequency but must all have firing frequencies which are a monotonic function of the input signal, rather than having peak firing frequencies at some particular value of the input signal. Presumably the different kinds of population coding require different kinds of decoding circuitry.

It is also worth stressing that the coding in the intermediate spatial repesentation appears to be based on total population activity, rather than on frequency. The latter is presumably significant for individual elements but what represents the behaviorally meaningful signal is the total activity across the population. In the case of primate saccade control, a system which has until recently dominated thinking about sensorimotor processes, frequency coding is quite significant in motoneuron activity, and a sense has grown up that the major problem in sensorimotor processing is to achieve a transformation from a place-coded to a frequency coded signal. Such transformations are included in models of a variety of sensorimotor systems, including of the frog (Cobas and Arbib, 1991). In actuality, the saccade system is very much a special case. Even in this system, increases in muscle activation involve recruitment of motoneurons as well as increases in firing frequency. In many sensorimotor systems, perhaps most, it is the former effect rather than the latter that dominates, so that the population activity coding scheme of the intermediate representation in the frog is probably a closer match than frequency to most output representations. This suggests that, in general terms, it may make sense to think of sensorimotor transforms in terms of a transformation from place coding to population activity coding, rather than as a transformation from place coding to frequency coding.

Though it is not yet entirely clear whether the particular kind of population coding seen in the frog will prove to be a common property of sensorimotor systems, the considerations just described suggest this may be so and provide, in any case, a possible rationale for its existence. It is becoming increasingly clear in a variety of systems that effective sensorimotor performance depends on feedback systems that relate intended motor performance to signals about actual motor commands and their effects (see Grobstein, 1991 for discussion and references). Given that the latter are largely represented by population activity coded signals, having the former in a similar coding scheme may facilitate needed comparisons and calculations. Whether this proves true or not, it would seem appropriate to give increased attention to the problem of how one builds neural circuits so as to create (and decode) signals in a form where it is not the firing of individual neurons but rather the ensemble activity which is significant. Of interest as well is the recognition that there exist various forms of population coding and the associated questions, like those for different kinds of spatial representation, of which forms occur in which circumstances and why.

SUMMARY AND CONCLUSIONS

In an earlier paper (Grobstein, 1988), I characterized the sensorimotor interface as "a poorly understood terra incognita whose very dimensions are still uncertain." As described in this chapter, there is much yet to learn about its organization in even the relatively simple cases of ballistic, directed movements. At the same time, important aspects of interface organization are becoming clearer, and this in turn calls attention to new and interesting problems for future exploration. In frog orienting, and in a significant number of other organisms and behaviors where relevant observations have been made, it has been possible to identify in the interface at least one representation of spatial location, and perhaps several, which are experimentally distinguishable from those located more toward the sensory and motor sides of the nervous system. The representation, in the frog, is distributed, rather than map-like, in a particular sense. It is "parcellated," with different components of the representation coded in distinct neural structures. Parcellated representations, with independent coding of distinct components, are, as discussed, emerging as a general characteristic of the sensorimotor interface. In the frog, each component of the parcellated representation is apparently itself represented in a population coded fashion, again in a particular sense, as the total activity across a population of neurons. Whether this particular form is an idiosyncracy of the frog remains to be seen but population coding generally seems also to be significant in a variety of sensorimotor systems.

Each of the three phenomena discussed in this chapter (a distinctive central spatial representation, its distributed (parcellated) character, and population (activity) coding) represented unexpected findings, certainly in the frog, and to one extent or another in other systems as well. Recent findings thus not only contribute to better understanding sensorimotor processing but also call attention to the need to alter the kinds of expectations many neurobiologists have of nervous system organization (see also Grobstein, 1989, 1990a), and point toward previously unthought of problems worthy of exploration. That space would inevitably be represented in the nervous system in a map-like fashion is one of the expectations that clearly needs to be altered. Indeed, the still more general issue of whether any kind of central representation isomorphic to three dimensional space should be expected in the nervous system, and if so why, is among those which deserve more attention (Grobstein, 1989; 1990b, 1991). There exist as well other intriguing problems at both conceptual and neuronal levels. What advantages are gained by one or another kind of distributed representation? by one or another kind of population coding? What kinds of neuronal network will transform a map-like representation into a distributed one, and deal as well with the added problem, discussed elsewhere (Grobstein, 1991), that the dimensionality of the input is higher than that of the central representation? What kinds of neuronal networks will generate and translate population coded signals, in each of the forms in which these may exist? These sorts of questions clearly lie on, and are significant for, the "common ground of brains, machines, and mathematics" (Arbib, 1964). Their existence is testimony to the fruitfulness of an effective interaction between those exploring neuronal models and those exploring the brain itself, and their answers will inevitably come from the continued cross fertilization provided by the combination of insights gained from the brain and from efforts to model it.

Acknowledgements. My research program is supported by NIH Grant R15 NS 24968, and a grant from the Whitehall Foundation. My thanks to a large number of colleagues and collaborators without whose thoughts and efforts the present chapter would not exist, to Jeff Carr and Vivian Taglianetti for careful readings of drafts of this chapter, and to Vivian Taglianetti for help with the figures.

REFERENCES

Anastasio, T.J. (1991) Distributed processing in vestibulo-ocular and other oculomotor subsystems in monkeys and cats. In: Visual Structures and Integrated Functions (Arbib, M.A. and Ewert, J.-P., eds.) Springer-Verlag, in press.

Anderson, R.A. (1989) Visual and eye movement functions of the posterior pariental cortex. Ann. Rev. Neurosci. 12: 377-403.

Arbib, M.A. (1964) Brains, Machines, and Mathematics. Mc-Graw Hill.

Arbib, M.A. and Cobas, A. (1991) From prey-recognition to predator avoidance 1: maps and schemas. In: Visual Structures and Integrated Functions (Arbib, M.A. and Ewert, J.-P., eds.) Springer-Verlag, in press.

Beevor, C.E. (1909) Remarks on paralysis of the movement of the trunk in hemiplegia. Brit. Med. J. 1: 881-885.

Berkinblitt, M.B., Feldman, A.G., and Fukson, O.I. (1986) Adaptability of innate motor patterns and motor control mechanisms. Behav. Brain Sci. 9: 585-638.

Camhi, J. M. (1988) Escape behavior in the cockroach: distributed neural processing. Experientia 44: 401-408.

Cobas, A. and Arbib, M.A. (1991) From prey-recognition to predator avoidance 2: modelling the medullary deficit. In: Visual Structures and Integrated Functions (Arbib, M.A. and Ewert, J.-P., eds.) Springer-Verlag, in press.

Cobas, A. and Arbib, M.A. (1991) Prey-catching and predator-avoidance in frog and toad: defining the schemas. Manuscript under review.

Comer, C.M. and Dowd, J.P. (1987) Escape turning behavior of the cockroach. J. Comp. Physiol. 160: 571-583.

Dowd, J.P.and Comer, C.M. (1988) The neural basis of orienting behavior: a computational approach to the escape turn of the cockroach. Biol. Cybern. 60: 37-48.

Ewert, J.-P. (1987) Neuroethology of releasing mechanisms: prey-catching in toads. Behav. Brain Sci. 10: 337-387.

Favia, M., Gordon, J., Hening, W., and Ghez, C. (1990) Trajectory control in targeted force impulses. VII. Independent setting of amplitude and direction in response preparation. Exp. Brain Res. 79: 530-538.

Flanders, M. and Soechting, J.F. (1990) Parcellation of sensorimotor transformations for arm movements. J. Neurosci. 10: 2420-2427.

Georgopoulos, A.P. (1986) On reaching. Ann. Rev. Neurosci. 9: 147-70.

Georgopoulos, A.P., Kettner, R.E., and Schwartz, A.B. (1988) Primate motor cortex and free arm movements to visual targets in three-dimensional space. II.Coding the direction by a neuronal population. J. Neurosci. 8: 2928-2937.

Grobstein, P. (1988) Between the retinotectal projection and directed movement: topography of a sensorimotor interface. Brain, Behav. Evol. 31: 34-48.

Grobstein, P. (1989) Organization in the sensorimotor interface: a case study with increased resolution. In: Visuomotor Coordination: Amphibians, Comparisons, Models, and Robots (Ewert, J.-P. and Arbib, M.A., eds), Plenum, pp 537-568.

Grobstein, P. (1990a) Strategies for analyzing complex organization in the nervous system. I. Lesion experiments, the old rediscovered. In: Computational Neuroscience. (Schwartz, E., ed.), MIT Press, pp 19-37

Grobstein, P. (1990b) Strategies for analyzing complex organization in the nervous system. II. A case study: directed movement and spatial representation in the frog. In: Computational Neuroscience. (Schwartz, E., ed.), MIT Press, pp 242-255.

Grobstein, P. (1991) Directed movement in the frog: motor choice, spatial representation, free will? In: Neurobiology of Motor Programme Selection: New Approaches to Mechanisms of Behavioral Choice (Kien, J., McCrohan, C., Winlow, B., eds.) Manchester University Press, in press.

Grobstein, P. and Staradub, V. (1989) Frog orienting behavior: the descending distance signal. Soc. Neurosci. Abstr. 15: 54.

Grobstein, P., Comer, C., and Kostyk, S.K. (1983) Frog prey capture behavior: betwen sensory maps and directed motor output. In: Advances in Vertebrate Neuoroethology. (Ewert, J.-P., Capranica, R.R., and Ingle, D.J., eds.), Plenum, pp 331-347.

Grobstein, P., Reyes, A., Zwanziger, L., and Kostyk, S.K. (1985) Prey orienting in the frog: accounting for variations in output with stimulus distance. J. Comp. Physiol. 156: 775-785.

Grobstein, P., Crowley, K., and Spiro, J. (1988) Neuronal organization for directed movement in the frog: similarities in visual and tactile prey orienting. Soc. Neurosci. Abstr. 14: 1236.

Grobstein, P., Meyer, J., and Egnor, R. (1990) Directed movement in the frog: motor equivalence, multi-dimensionality, internal feedback? Abstr. Soc. Neurosci., in press.

House, D. (1989) Depth Perception in Frogs and Toads. Springer-Verlag.

Iannonne, A.M., and Gerber, G.M. (1983) Brown-Sequard syndrome with paralysis of head turning. Ann. Neurol. 12: 166.

Ingle, D.J. (1972) Depth vision in monocular toads. Psychonom. Sci. 29: 37-38.

Knudsen, E.I., du Lac, S, and Esterly, S.D. (1987) Computational maps in the brain. Ann. Rev. Neurosci. 10: 41-66.

Kostyk, S.K. and Grobstein, P. (1982) Visual orienting deficits in frogs with various unilateral lesions. Behavioral Brain Research 6: 379-388.

Kostyk, S.K. and Grobstein, P. (1987a) Neuronal organization underlying visually elicited prey orienting in the frog. I. Effects of various unilateral lesions. Neuroscience 21: 41-55.

Kostyk, S.K. and Grobstein, P. (1987b) Neuronal organization underlying visually elicited prey orienting in the frog. II. Anatomical studies on the laterality of central projections. Neuroscience 21: 57-82.

Kostyk, S.K. and Grobstein, P. (1987c) Neuronal organization underlying visually elicited prey orienting in the frog. III. Evidence for the existence of an uncrossed descending tectofugal projection. Neuroscience 21: 83-96.

Lee, C., Rohrer, W.H., and Sparks, D.L. (1988) Population coding of saccadic eye movements by neurons in the superior colliculus. Nature 332: 357-360.

Lehky, S.R. and Sejnowski, T.J. (1990) Neural model of stereoacuity and depth interpolation based on a distributed representation of stereo disparity. J. Neurosci. 10: 2281-2299.

Masino, T. and Grobstein, P. (1989a) The organization of descending tectofugal pathways underlying orienting in the frog, Rana pipiens. I. Lateralization, parcellation, and an intermediate spatial representation. Exp. Brain Res. 75: 227-244.

Masino, T. and Grobstein, P. (1989b) The organization of descending tectofugal pathways underlying orienting in the frog, Rana pipiens. II. Evidence for the involvement of a tecto-tegmento-spinal pathway. Exp. Brain Res. 75: 245-264.

Masino, T. and Grobstein, P. (1990) Tectal connectivity in the frog, Rana pipiens: tecto-tegmental projections and a general analysis of topographic organization. J. Comp. Neurol. 291: 103-127.

Masino, T. and Knudsen, E.I. (1990) Horizontal and vertical components of head movement are controlled by distinct neural circuits in the barn owl. Nature 345: 434-437.

McIlwain, J.T. (1972) Lateral spread of neural excitation during microstimulation in intermediate grey layer of cat's superior colliculus. J. Neurophysiol. 55: 97-112.

Mittelstaedt, H. and Eggert, T. (1990) How to transform topographically ordered spatial information into motor commands. In: Visuomotor Coordination: Amphibians, Comparisons, Models, and Robots (Ewert, J.-P. and Arbib, M.A., eds), Plenum, pp 569-585.

Schildberger, K. (1988) Behavioral and neuronal mechanisms of cricket phonotaxis. Experientia 44: 408-415.

Sparks, D.L. and Mays, L.E. (1990) Signal transformations required for the generation of saccadic eye movements. Ann. Rev. Neurosci. 13: 309-336.

Soechting, J.F. and Flanders, M. (1989a) Sensorimotor representations for pointing to targets in three-dimensional space. J. Neurophysiol. 62: 582-594.

Soechting, J.F. and Flanders, M. (1989b) Errors in pointing are due to approximations in sensorimotor transformations. J. Neurophysiol. 62: 595-608.

Stein, P.S.G. (1989) Spinal cord circuits for motor pattern selection in the turtle. Ann. N. Y. Acad. Sci. 563: 1-10.

Prey-catching and predator avoidance 1: Maps and Schemas [1]

Michael Arbib and Alberto Cobas [2]
Center for Neural Engineering
University of Southern California
Los Angeles, CA 90089-2520, U.S.A.

Abstract

We model the construction of motor actions through the interaction of different motor schemas via a process of competition and cooperation wherein there is no need for a unique schema to win the competition (although that might well be the result) since two or more schemas may simultaneously be active and cooperate to yield a more complicated motor pattern. Based on lesion data, our model is structured on the principles of segregation of coordinate systems and participation of maps intermediate between sensory and motor schemas. The motor schemas are driven by specific internal maps which between them constitute a distributed internal representation of the world. These maps collectively provide the transition from topographically-coded sensory information to frequency-coded inputs to the diverse motor schemas that drive muscle activity. As a challenge to further comparative analysis of the prey-catching and predator-avoidance systems, we argue that the Positional Heading Map hypothesis, (that the heading map codes the position of the object) should be rejected in favor of the Motor Heading Map hypothesis. This holds that each system has a separate projection pathway that converges in a different way onto the heading map, coding the required motor response, which has a single connection pattern to those motor schemas common to both systems.

The Role of Schemas in *Rana computatrix*

The complexity of data on neural mechanisms is such that an account organized solely at the cellular level overburdens comprehension. We thus use schemas to present models which clarify the relation of behavior to the interaction of brain regions. Such schemas serve both to summarize much data on behavior and lesions and, perhaps surprisingly, on neural connectivity. More explicitly, schemas are programs developed to represent, at least, perceptual structures and distributed motor control. The brain can support many concurrent activities for recognition of different objects, and the planning and control of different activities. Thus schema theory views the use, representation, and recall of knowledge as mediated through

---

1 The research described in this paper was supported in part by grant no. 1RO1 NS 24926 from the National Institutes of Health (M.A.Arbib, Principal Investigator) and Fulbright/ MEC fellowship FU88-35011116 (Spain) to A.C.

2 Present address: Center for the Neurobiology of Learning and Memory, University of California at Irvine, Irvine, CA 92717, USA.

the activity of a network of interacting computing agents, schema instances. The activity level of an instance of a perceptual schema represents a "confidence level" that the object represented by the schema is indeed present; while that of a motor schema may signal its "degree of readiness" to control some course of action. A schema network does not, in general, need a top-level executor since schema instances can combine their effects by distributed processes of competition and cooperation (i.e. interactions which, respectively, decrease and increase the activity levels of these instances).

This paper and its companion (Cobas and Arbib 1991) present a model of prey-catching and predator-avoidance which exemplifies the schema level of our work on modeling visuomotor coordination in frog and toad (see Arbib 1987 for a review of this *Rana computatrix* project). A model of behavior is phrased in terms of interacting schemas. A given schema may be distributed across more than one brain region; a given brain region may be involved in many schemas; hypotheses about neural localization of subschemas may thus be tested by lesion experiments.

A schema-level model explores the interactions of schemas with specified functions, irrespective of the neural network or other method used for implementation. The function is defined in terms of the way in which patterns of activity arriving at certain input ports are related to activity observed on output ports, and may in general depend upon some internal state. RS (the Robot Schema language: Lyons and Arbib 1989) formalizes each schema as a port automaton with a set of input and output ports through which it can communicate with other instances. Instantiation and deinstantiation operations capture the notion that, as action and perception progress, certain schema instances need no longer be active (they are deinstantiated), while new ones are added as new objects are perceived and new plans of action are elaborated (schemas are instantiated as new schema instances which are linked into the network). A basic schema definition includes a behavior specification (in RS, this is a C-like program) which makes explicit how an instance of the schema will read inputs, write outputs, change internal state, and execute (de)instantiations. An assemblage also has ports, but its behavior is defined through interactions in a network of schema instances.

Schemas may be formed hierarchically. A schema assemblage may itself be considered a schema for further processes of assemblage formation; and the network itself will be dynamic, growing and shrinking as various instantiations and deinstantiations occur. The RS syntax for an assemblage tells us how to put schemas together in a way which does not depend on how the behavior specification is given, whether directly as in a basic schema, or indirectly, as when the schemas constituting an assemblage are themselves schema assemblages. (It is thus simple to extend RS to include neural net specifications by making it possible, in the basic schemas, to define the behavior directly in terms of a neural network.) We speak of *encapsulation*: the internal workings of a schema are hidden (they are encapsulated) — all that matters for overall behavior is that each schema passes messages to and from other schemas in concordance with some overall specification.

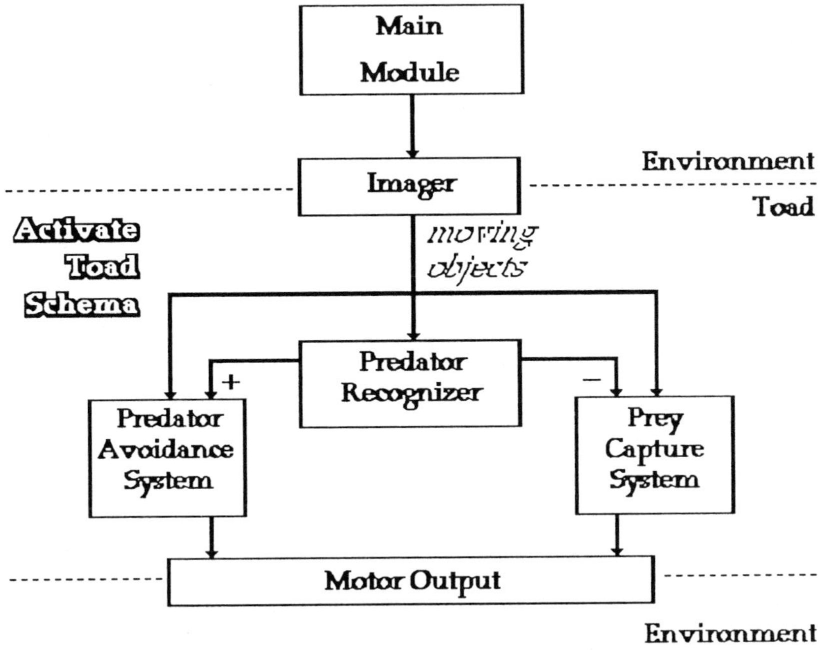

Figure 1: The ActivateToad schema assemblage integrates all the toad's schemas for one cycle of predator-avoidance and prey-catching. The Main Module maintains the environment representation, and, via the Imager, provides the toad model with a current list of moving predators and prey.

The Overall Schema Assemblage

In this paper we describe schema models for prey capture and predator avoidance behaviors. The simulation of these models was written in C to run on a Sun workstation. It represents an environment in which a variable number of predators, prey and a toad move independently and interact with each other. Each object (toad, prey, or predator) is represented as a data structure encoding the object's position in the environment, its dimensions and other properties. In the simulation, toads only react to moving stimuli, following their natural behavior. The toad is modeled by schema programs explicating the neural interactions underlying behavior; while predators and prey move following simple rules giving, for each step, a movement distance and heading at random within a preprogrammed range. In the simulation, we thus find two classes of functions: (i) environment functions that are in charge of the maintenance of the simulated environment and its interface with some of the schema functions, and (ii) schemas that represent functions of the toad brain, and that communicate with each other through specific "input/output ports".

The Main Module (Fig. 1) is the "environment function" that is executed first. It starts the actual simulation by executing a loop in which each cycle consists in: (i) activation of all prey objects via the schema ActivatePrey, (ii) activation of all predator objects via the schema ActivatePredator, and (iii) activation of the toad object, via the schema ActivateToad which is the heart of the model. Each object performs one 'ballistic' behavior episode and returns the distance and heading moved, relative to the object's own position and orientation. Additionally, the Activate Toad reports whether it has attempted a snap at a given prey. The Main function then repositions the objects and removes prey accordingly, and executes a new cycle if prey remain. The simulation ends when all prey have been snapped.

The Imager schema picks up from the environment all predator and prey objects that have moved since the last time frame, without regard for their identity. The positions of all these objects are transformed so they are relative to the toad's position and orientation, and the list is returned in the Imager's output port so that it may serve as input to various of the toad's schemas. If no moving objects are detected by the Imager, the ActivateToad schema finishes and no movement is elaborated for the toad on this cycle. Otherwise, the PredatorRecognizer schema is activated, and if it detects a predator from the objects passed by the Imager schema, the predator-avoidance system is enabled. If not, the prey-capture system is activated. More specifically, when the PredatorRecognizer schema identifies a predator_object from the Imager output, the immediate action is to activate the Predator Avoidance System and block the Prey Capture system (cf. the + and - signs in Fig. 1). To relate this to the neural data, we note that when the pretectum is lesioned the toad responds with prey acquisition behavior to predator stimuli moving in the affected part of the visual field, as if there were a lack of inhibition and specificity of the PreySelector schema operating in the Prey Capture system (Ewert 1976). We thus localize the PredatorRecognizer schema in the pretectal region, and hypothesize that it is able to inhibit the corresponding retinotopic locus of the PreySelector schema in the tectum. In this way, the activity produced in this locus, due to the lack of stimulus specificity characteristic of the prey capture system, is effectively blocked by the PredatorRecognizer.[3] The activity evoked by prey in tectum would be blocked by a strong inhibition to the Approach and Snap schemas. This inhibition would prevent the activation of these two schemas by the DepthMap output. In addition, the detection of a predator stimulus by the PredatorRecognizer schema would prime the Escape and Defend schemas, making them sensitive to the commands of their respective Map schemas.

The use of the Imager as a mere processor of lists of moving objects is part of the encapsulation strategy of schema theory. We do not need to probe here the interior details of the Retina, PreyRecognizer, Depth-mapper or PredatorRecognizer which provide input to the schemas whose interactions represent the neural interactions revealed by the behavioral data on normal and lesioned animals to be reviewed here and in Cobas and Arbib (1991). Detailed models of the four named systems do exist, however (see Teeters 1989, Cervantes-Perez et al. 1985, House 1989, and Liaw

---

[3] Note that, once we start talking about "retinotopic locus" (as we shall do in greater detail in the next section), we provide a refined description of the schema which can bridge toward models implemented in terms of interactions in layered neural networks.

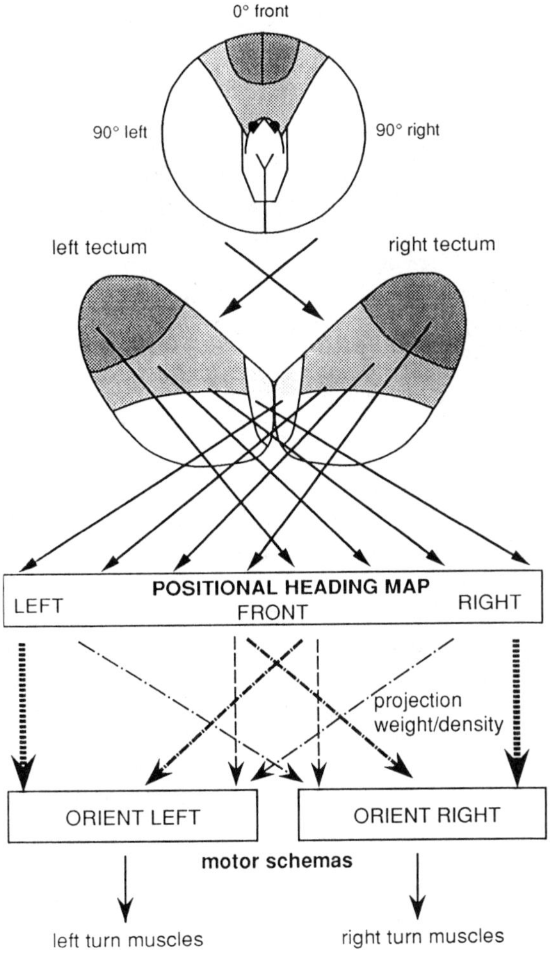

Figure 2: Layout of a positional heading map used for both prey acquisition and predator avoidance. For the sake of clarity, only the connections associated with the contralateral visual field in the binocular regions of each tectal hemisphere are represented. The connections corresponding to the ipsilateral binocular field would terminate in the same loci as the homologous ones in the other tectum. The map codes the position of the stimulus in the world, no matter whether it be prey or predator. The systems diverge between the map and the Orient motor schemas, as shown by the - - → arrows for Prey vs. -·-·→ arrows for Predators.

and Arbib 1991, respectively), and the refinement and reconciliation of these models on the basis of the insightsgained from the present schema analysis is a challenge for current research. Of course, what we learn at the neural level may well refine the specification of the schemas, in which case we can use the schema model to assess the implications of this change *without necessarily refining other schemas to the neural network level.*

Neural Maps for Prey Capture and Predator Avoidance

Neural maps can provide an internal representation of the external world onto which several sensory channels may converge. We are particularly interested in neural maps which are relative to the toad's body rather than any particular sensory framework. This idea of sensory convergence has been discussed by Grobstein (e.g., 1989) for the Prey Capture system, and we here extend such considerations to the Predator Avoidance system, but we first wish to consider how these neural maps could serve in the transformation of spatially-coded information gathered by the sensory organs into population-coded information used by the motor schemas. This can be accomplished based on a differential projection density from different regions of the map onto the same motor schema (cf. Scudder 1988 for control of eye movements). Grobstein has shown that the representation of prey location is parcellated in the tectal output, with a contralateral pathway carrying information about the orientation of prey anywhere in the contralateral hemifield. We shall call this the Heading Map. Unfortunately, there are few data on the transition from spatial maps to population codes (but see Grobstein & Staradub 1989, Grobstein 1991). Since stimuli in a variety of spatial positions will trigger innervation of the same musculature though with different intensity, the change from one system to another has to take place somewhere, and we postulate that it occurs in the transition from Map schemas to motor schemas. In the present case, if the toad locates a prey at 90° in the left hemifield the stimulus would end up represented in the lateral left part of the Heading Map, as determined by the incoming retinotopic projections from the contralateral tectum. One way to turn this into an appropriate command to orient the animal is to have this region in the Heading Map form a dense projection onto the left Orient schema, so that this stimulus will evoke in the latter schema a high level of activation (see the ---→ arrows [not the -··-→ arrows] from the Heading Map to the Orient schemas in Fig. 2) to command the left turn muscles to produce an adequate left orienting movement toward the prey. The central portion of the Heading Map has a very light projection onto the Orient schema, and thus a prey falling into that region will only elicit a weak activation and consequently a very small turning movement or perhaps no turn at all.

However, in the case of avoidance behavior, the escape route chosen by the animal is rarely directed just opposite to the threat. Ingle's (1976) frog data suggest that the escape direction is a compromise between the forward direction and that away from the threat. In any case, the crucial distinction is this: In the Prey Capture system, the heading of the stimulus and the heading of the response are the *same;* in the Predator Avoidance system, they are *different.* This raises the key question: where does the divergence between these two systems in the mapping from sensory input to motor output arise? Two different approaches will be considered here. In each, we shall assume a common heading map, and assume that both systems share a common motor pattern generator for the Orient schema. However, experiments must also be conducted to test the possibility that each system has its own heading map and/or Orient schema:

i) *The Positional Heading Map Hypothesis* (Fig. 2): Both systems have the same projection to the heading map, which codes *the position of the object* relative to the toad's body, so that the required motor behavior arises from a different connection to the motor schemas depending on which system is active at the moment.

ii) *The Motor Heading Map Hypothesis* (Fig. 3): Each system has a separate projection pathway that converges in a different way onto the heading map, coding *the required motor response,* which has a single connection pattern to those motor schemas common to both systems. This is the hypothesis that we favor, and that is employed in the schema models described below.

The Positional Heading Map (or internal representation) approach is favored by Grobstein (1989), who speculates that the same pathways signal target position for prey and predator but that a different pathway triggers the motor pattern generators (MPGs) for approach and avoidance, with these MPGs transforming the spatial code differently. In this case, both systems have a heading map with similar tectal input but a notably different output projection both in destination and in density as well (cf. Fig. 2) which must be gated by the dominant system. In the case of the prey capture system, the motor schema activated by the left Heading Map is the (ipsilateral)

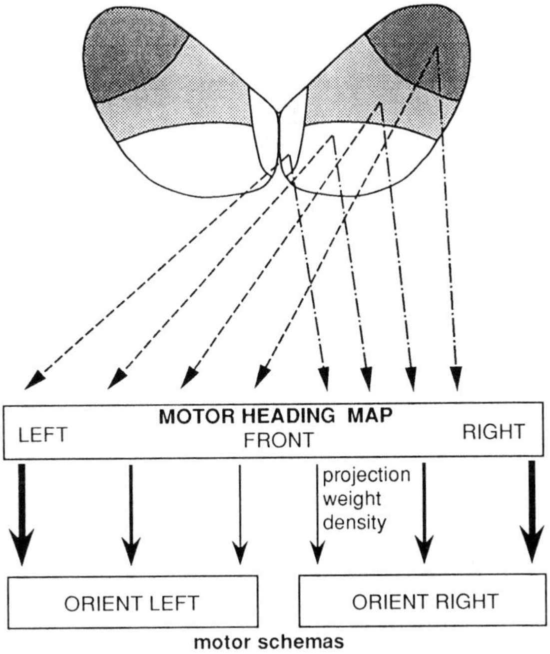

Figure 3: Layout of a motor heading map applicable for both prey acquisition and predator avoidance. For the sake of clarity, only the connections associated with the contralateral visual field of one tectal hemisphere are represented. Those of the contralateral tectum would terminate in symmetrical locations. The connections corresponding to the ipsilateral binocular field would terminate in the same loci as the homologous ones in the other tectum. The map codes the direction of the appropriate turn for prey acquisition or predator avoidance. The divergence between both systems occurs between the tectum and the motor heading map (as shown by the - - - → arrows for Prey vs. the -·-·→ arrows for Predators) and so its connections with the Orient motor schemas can be the same for both systems.

left Orient schema, while in the predator-avoidance system the output target of the same map region is the (contralateral) right Orient schema. Symmetrical relations hold for the right portion of the Heading Map. Additionally, regions on or near the "front" or "ahead" location in the map have less influence on the turning command in approach, while in avoidance behavior they have the most, and the reverse occurs with regions located farther to the left or to the right of the map.

If the Heading Map is indeed coding the position of the object, then from each locus there are two outputs, one directed to the ipsilateral Orient motor schema and another to the contralateral one, each with the corresponding weight or density of innervation. The mechanism of action would be as follows. The tectum would stimulate equally both outputs, but only one of them would be able to find an adequately primed connections to an Orient schema to elicit the turning movement.

In the alternate Motor Heading Map hypothesis (Fig. 3), the output that connects the map with the common motor schemas is the same in destination and density for both behavior systems. The projection pathway from the tectum thus differs depending on whether the object is identified as a predator or not, which seems reasonable in view of the fact that prey and predator are signalled by different tectal populations (e.g., T5.2 and T3, respectively [Ewert 1984]; cf. Cervantes-Perez et al. 1985 and Liaw and Arbib 1991, respectively). The projection from tectum to HeadingMap for prey acquisition is the same as that shown in Fig. 2. However, for avoidance behavior, we postulate that for visual input corresponding to the contra-

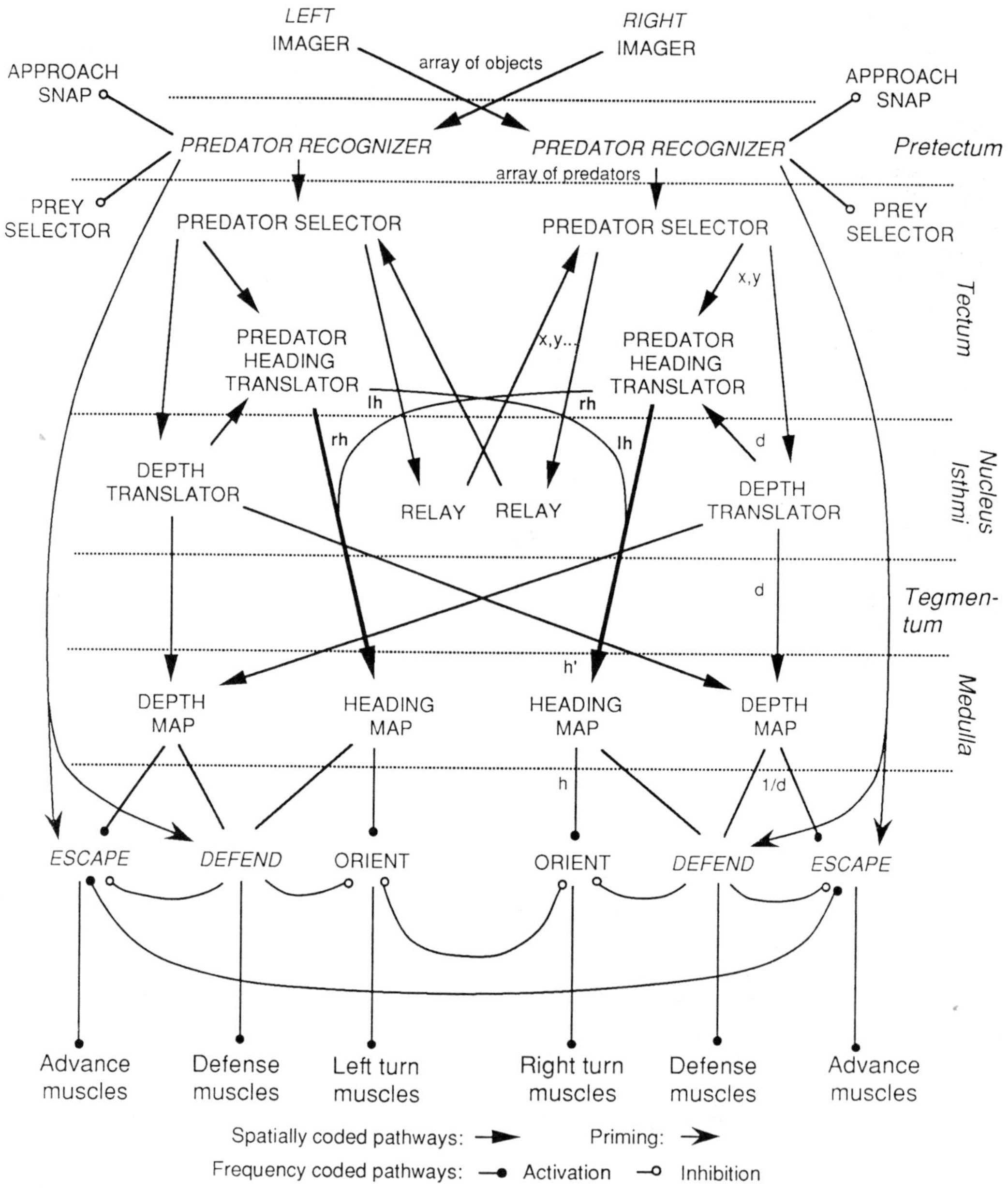

Figure 4: The overall schema model for predator-avoidance. To simplify comprehension, first examine the left half of the figure to see the overall relation between perceptual and motor schemas for input from the right eye and right visual hemifield, then look at the bilateral connections involved in depth mapping and coordination of motor schemas. The output of the PreyHeadingTranslator schema is primarily ipsilateral.

lateral hemifield (the most common situation), the tectal projection is ipsilateral, leading to the part of the motor heading map coding for turns directed to the contralateral side, guiding avoidance to the side away from the stimulus, triggering a sharper turn the closer the predator is to the forward direction.

In light of these considerations, we now develop our schema models for Predator Avoidance and Prey Capture using the Motor Map Hypothesis, i.e., the two systems use the HeadingMap to code direction of response, and they thus share a common connection pattern from the Heading Map to the Orient Schema. Each model is characterized by a symmetrically organized system of schemas showing a basic top-to-bottom flow of information but also bilateral connections at different levels[4]. We tentatively assign the various schemas to brain regions, based on several anatomical findings such as those on the relation of tectum and pretectum mentioned above. This is in principle independent of the performance of the model from an algorithmic point of view, but is important for simulating lesion experiments.

A Schema Model for Predator Avoidance

The PredatorSelector (Fig. 4) is the first schema to be activated after the Predator Avoidance assemblage has itself been activated by the PredatorRecognizer schema (Fig. 1). The Predator Recognizer extracts a unique predator object from the list of moving objects passed by the Imager schema. In the present implementation, it picks the nearest. (The reader may note an apparent paradox here, for it is the output of the Predator Selector that provides the input to the DepthTranslator that computes the distance to the predator stimulus [its output is actually the "closeness" of the selected object].) However, the schema level of description simply guarantees that if the Predator Selector chooses a predator then other schemas will use that information correctly to compete and cooperate in yielding the overall behavior of the animal. The internal algorithm of the schema is irrelevant *when we work at the schema level.* When we proceed to the neural net level, we only require that the network conform to the schema's input/output specification, not that the structure of the network match the structure of the "placeholder" algorithm.

The animal approaches the prey but avoids the predator — and so the heading is *away from* the selected object, not towards it. The PredatorHeadingTranslator (PredatorHT) schema obtains the heading *for escaping from* from the toad to the selected predator. If h is the direction (in degrees) of the predator relative to the toad (where $h = 0$ is straight ahead, $0 < h \leq 180$ means to the left and $0 > h > -180$ means the predator is to the right), we use the formula

$$\text{heading} = \begin{cases} 0 & \text{if } |h| \geq 135 \\ (h - 135)/2 \text{ modulo } 360 & \text{otherwise} \end{cases}$$

---

[4] In our <u>simulations</u>, homologous schemas were lumped into a single bilateral schema function handling both sides (note that homologous schemas are actually duplicates of the same process that act upon data from different origin). This technique allowed us to more easily implement the models without losing control over each side of the representation by changing the inflow or outflow of any given side to represent the effects of unilateral lesions.

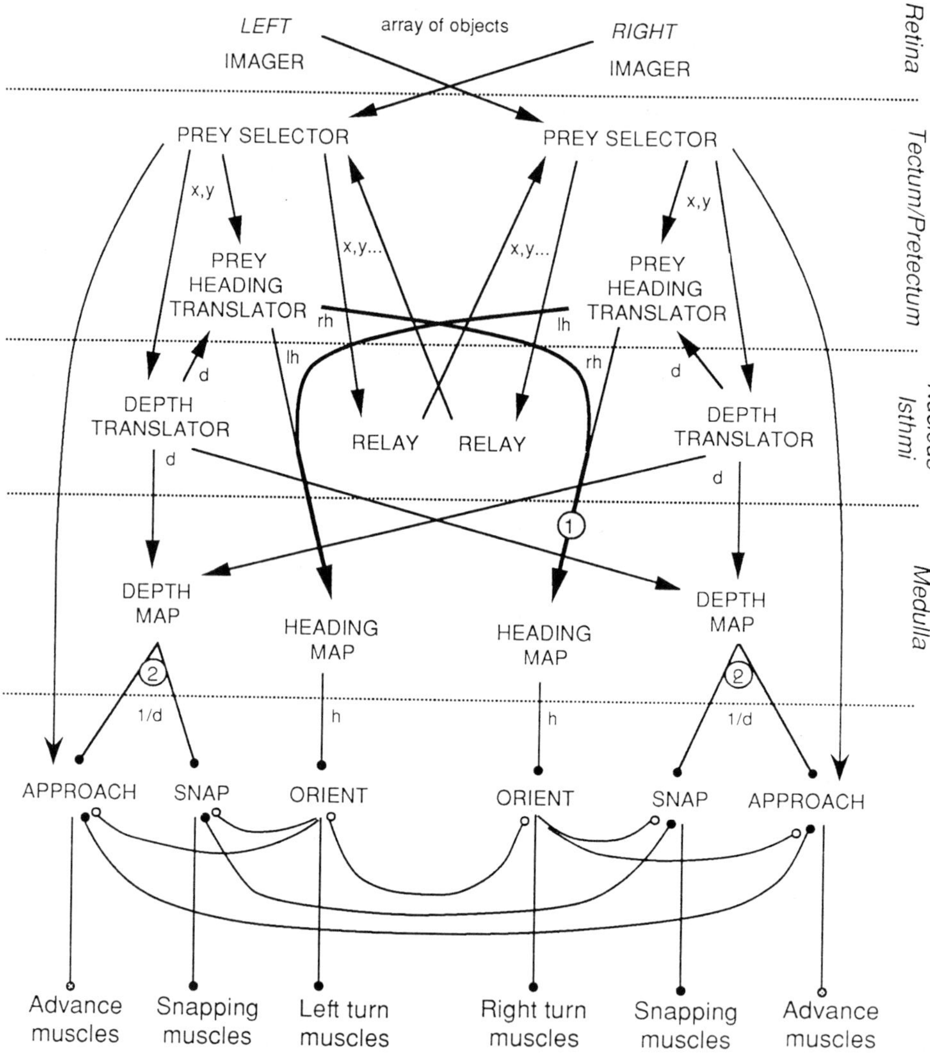

Figure 5: The overall schema model for prey capture. For conventions in labelling connections, see Fig. 4. Note that the main output pathway for the PreyHeadingTranslator schema is contralateral.

If the computed heading corresponds to the ipsilateral hemifield, the heading parameter is fed to the contralateral HeadingMap schema, otherwise it is transmitted to the ipsilateral HeadingMap schema. The HeadingMap then stimulates the Orient schema, so that a predator situated in the lateral visual field would trigger a frontal escape direction while one in the frontal visual field would produce an escape directed to the opposite side.

The size and distance to the predator determine the activation of the Defend schema. The Orient schema is modulated by the action of the Defend schema. The

Escape schema outputs a distance to jump, also modulated by the Defend schema. The actual specification of the motor schemas Orient (which is the same as for prey capture), Defend and Escape is presented in Cobas and Arbib (1991), which also predicts effects of lesion experiments. Since there are few data on Predator Avoidance (as distinct from Prey Capture) to constrain most of the hypotheses used to construct the model, a major goal of this effort is to stimulate new experiments.

A Schema Model for Prey Capture

If there is a moving object, but no moving predator is recognized, the Prey Capture assemblage will be activated (Fig. 5), starting with the PreySelector schema which assumes that all moving objects passed to it by the Imager schema are prey and so extracts a unique prey object. The DepthTranslator schema then obtains the distance. The retinotopic position of the selected prey from both sides also goes to the PreyHeadingTranslator which obtains the heading from the toad to the selected prey which is used by the Orient schema to compute and return the toad's new heading. If the prey's position corresponds to, say, a left contralateral heading, this is sent to the left HeadingMap schema, while if it corresponds to the right side (as when coming from the part of the binocular visual field corresponding to the ipsilateral side) its output is sent to the right HeadingMap schema in the medulla. In the simulation, the PreyHeadingTranslator computes the angle to the prey_object from the toad and puts this value in its output port. A positive value represents a left heading, while a negative value represents a right heading. The Snap schema uses the closeness to the prey and the toad's heading to decide whether to aim a snap at the prey, while the Approach schema returns the distance to advance. The final motor pattern of the toad may thus result from the cooperative activation of several individual motor schemas, rather than through sequential activation of a short sequence of such schemas.

Ablation of nucleus isthmi: Caine and Gruberg (1985) found that the observed effect of ablation of nucleus isthmi on prey-catching is very similar to that obtained for tectal ablation, resulting in a scotoma in the monocular field of the affected side. The nucleus isthmi is also involved in predator-avoidance behavior, for its lesion abolishes response to threat stimuli in certain parts of the visual field, depending on the extent of the damage. Nevertheless, this influence may not be specific since the prey capture behavior is also abolished in the same affected regions. Interestingly, the predator-avoidance responses, but not prey capture, are preserved when the nucleus isthmi is injected with ibotenic acid, an excitatory neurotoxin.

In the model, the result that unilateral ablation of nucleus isthmi yields a scotoma in the monocular field of the affected side would be obtained by blocking the Relay schema and thus avoiding completion of the selection of the prey after the link with the other hemisphere has been eliminated. In our simulation, the removal of, say, the left Relay schema and the consequent inactivation of the right PreySelector, right PreyHT and right DepthTranslator, is emulated by resetting to 0 the output of the DepthTranslator function and that of the PreyHT when the heading reported by the latter schema is between 45° and 180° (left monocular field). The result would be no activity when the stimulus is within 45° and 180° to the left of the midline. We have to note here, however, that if damage to the isthmic region also damages the DepthTranslator schema, only orientation movements could be observed in the rest of the visual field due to the lack of distance estimates (recall that the contralateral DepthTranslator has been rendered ineffective because of the inoperativity of the PreySelector contralateral to the lesion).

Modeling of the effects of other lesions depends upon our particular approach to the motor schemas for Orient, Snap and Approach. These are described in the companion paper (Cobas and Arbib 1991).

Discussion

We have offered a model in which each motor schema is driven by specific internal maps which between them constitute a distributed internal representation of the world. These maps collectively provide the transition from topographically-coded sensory information to frequency-coded inputs to the diverse motor schemas that drive muscle activity. In favor of the Motor Heading Map hypothesis, we suggest that the *activity-gated* divergence postulated by Grobstein (1989) is not required for orienting responses, because the Orient schema commanded by the Heading Map works in the same fashion for both prey acquisition and predator-avoidance. Nevertheless, in the case of the Depth map, on which schemas specific for each system depend, the need for a gated divergence exists. "Priming" (enacted by the PreySelector or PredatorRecognizer) sensitizes the schemas specific to one of the systems, leaving the others insensible to the output produced by the Depth Map (avoiding, for example, the trigger of the Snapping command during predator-avoidance behavior).

Unfortunately, most of the experimental data regarding the internal representation have been obtained from the study of prey acquisition, and one cannot guarantee that the threat avoidance system is organized in a fashion comparable to that of the prey system. A map coding for the position of the prey is indistinguishable from one coding for the point to which movement should be aimed, whereas in the predator-avoidance system the location of the object is clearly different from the desired target for the turning movement. The Predator Avoidance model thus makes assumptions that require experimental testing, such as:

1. Predator Avoidance always takes priority over Prey Capture. We need data on response to various patterns combining prey and predator stimuli.
2. If several predators are present, one is chosen to avoid — the PreySelector and PredatorSelector schemas use the same mechanism (although they are made to work with inputs of different identity). There are indications that toads will choose one prey out of a group of similar ones, even if presented with two virtually identical stimuli in the symmetrical positions of the visual field (Cobas, unpublished observations). We need data which study the response to multiple predators, including cases in which one predator lies in the preferred escape direction defined by the others.

References

Arbib, M. A., 1975, Artificial intelligence and brain theory: unities and diversities. *Ann. Biomed. Eng.* 3: 238-274.

Arbib, M.A., 1981, Perceptual structures and distributed motor control. In *Handbook of Physiology — The Nervous System II. Motor Control* (V.B. Brooks, Ed.), Bethesda, MD: Amer. Physiological Society. pp. 1449-1480.

Caine, H. S., and Gruberg, E.R., 1985, Ablation of nucleus isthmi leads to loss of specific visually guided behavior in the frog *Rana pipiens, Neuroscience Letters*, 54:307-312.

Cervantes-Perez, F., Lara, R., and Arbib, M.A., 1985, A Neural Model of Interactions Subserving Prey-Predator Discrimination and Size Preference in Anuran Amphibia", *J. Theor. Biol.*, 113:117-152.

Cobas, A., and Arbib, M.A., 1991, Prey-catching and predator avoidance 2: Modeling the medullary hemifield deficit, in *Visual Structures and Integrated Functions*, (M.A. Arbib and J.-P. Ewert, Eds.), *Research Notes in Neural Computing*, Springer-Verlag.

Didday, R.L., 1976, A model of visuomotor mechanisms in the frog optic tectum. *Math. Biosci.* 30: 169-180.

Ewert, J.-P.,1976, The visual system of the toad: behavioural and physiological studies on a pattern recognition system in *The Amphibian Visual System*, (Fite, K., Ed) Academic Press, pp. 142-202.

Ewert, J.-P., 1984, Tectal mechanism that underlies prey-catching and avoidance behavior in toads, in *Comparative Neurology of the Optic Tectum*, (H. Vanegas, Ed.), Plenum Press, NY.

Grobstein, P., 1989, Organization in the sensorimotor interface: A case study with increased resolution, in *Visuomotor Coordination: Amphibians, Comparisons, Models, and Robots*, (J.-P. Ewert and M.A.Arbib, Eds.), Plenum Press, pp.537-568.

Grobstein, P., 1991, Directed movement in the frog: a closer look at a central spatial representation, in *Visual Structures and Integrated Functions*, (M.A. Arbib and J.-P. Ewert, Eds.), *Research Notes in Neural Computing*, Springer-Verlag.

Grobstein, P., Reyes, A., Zwanziger, L., and Kostyk, S.K., 1985, Prey orienting in frogs: Accounting for variations in output with stimulus distance, *J. Comp. Physiol. A*, 156:775-785.

Grobstein, P. & Staradub, V., 1989, Frog orienting behavior: the descending distance signal, *Soc. Neurosci. Abstr.* 15:54

Gruberg, E.R., 1989, Nucleus isthmi and optic tectum in frogs, in *Visuomotor Coordination: Amphibians, Comparisons, Models, and Robots*, (J.-P. Ewert and M.A.Arbib, Eds.), Plenum Press, pp. 341-356.

Gruberg, E. R. and Udin, S. B., 1978, Topographic projections between the nucleus isthmi and the tectum of the frog *Rana pipiens* , *J. Comp. Neurol.* 179: 487-500.

House, D., 1989, *Depth Perception in Frogs and Toads: A Study in Neural Computing, Lecture Notes in Biomathematics 80*, Springer-Verlag, Berlin.

Ingle, D., 1976, Spatial visions in anurans. In *The amphibian visual system*, (K. Fite, Ed.), Academic Press: New York, pp 119-140.

Liaw, J.-S., and Arbib, M.A., 1991, A Neural Network Model for Response to Looming Objects by Frog and Toad, in *Visual Structures and Integrated Functions*,

(M.A. Arbib and J.-P. Ewert, Eds.), *Research Notes in Neural Computing*, Springer-Verlag.

Lyons D.M., and Arbib, M.A., 1989, A Formal Model of Computation for Sensory-Based Robotics, *IEEE Trans. on Robotics and Automation*, 5:280-293.

Minsky, M.L., 1985, *The Society of Mind*, Simon and Schuster.

Scudder, C.A., 1988, A new local feedback model of the saccadic burst generator, *J. Neurophysiol.*, 59, 1455-1475.

Teeters, J.L., 1989, A simulation system for neural networks and model for the anuran retina. Technical Report 89-01, Center for Neural Engineering, University of Southern California, Los Angeles.

Prey-catching and predator avoidance 2: Modeling the medullary hemifield deficit[1]

Alberto Cobas [2] *and Michael Arbib*
Center for Neural Engineering
University of Southern California
Los Angeles, CA 90089-2520, U.S.A.

Abstract

We provide a biological and behavioral analysis of our schema-theoretic model of prey-catching and predator-avoidance. Based on lesion data, our model is structured on the principles of segregation of coordinate systems and participation of maps intermediate between sensory and motor schemas: The motor schemas are driven by specific internal maps which between them constitute a distributed internal representation of the world. These maps collectively provide the transition from topographically-coded sensory information to frequency-coded inputs to the diverse motor schemas that drive muscle activity. We simulate data on approach and avoidance behavior of the frog or toad under normal conditions and under lesion of different brain centers. The model postulates the construction of motor actions through the interaction of different motor schemas via a process of competition and cooperation wherein there is no need for a unique schema to win the competition (although that might well be the result) since two or more schemas may simultaneously be active and cooperate to yield a more complicated motor pattern. The model generates different motor zones for prey-catching behavior which match those observed experimentally in normal conditions and in the medullary hemifield deficit, and offers predictions for new experiments on both approach and avoidance behaviors.

1. Introduction

In the companion paper (Arbib and Cobas 1991), we have introduced a schema-level analysis of prey-catching and predator-avoidance in frog and toad. Here, we specify the motor schemas and show how they enable us to simulate a wide range of behavior in normal and lesioned animals. Perhaps the most important features not considered in previous models are independent processing of the different parameters that define the stimulus position, interaction among different procedures for parameter evaluation, and implementation of intermediate maps using coordinates relative to the animal's body.

---

[1] The research described in this paper was supported in part by grant no. 1RO1 NS 24926 from the National Institutes of Health (M.A.Arbib, Principal Investigator) and Fulbright/ MEC fellowship FU88-35011116 (Spain) to A.C.

[2] Present address: Center for the Neurobiology of Learning and Memory, University of California at Irvine, Irvine, CA 92717, USA.

Frogs and toads show a variety of responses to presentation of a predator-like stimulus (Ingle, 1976a,b, Ewert, 1984). For example, toads may duck, puff, turn away, jump or run from an enemy, depending on specific characteristics of the stimulus like its position, movement direction, speed, shape, size, etc. Escape takes the form of jumping away from the threat in a more or less directed manner. Ingle's (1976a) data suggest that, at least in frog, the escape direction is a compromise between the forward direction and that away from the threat. Our goal will be to address the following body of data:

Ablation of one tectum: After ablation of the entire optic tectum, both visual prey-catching behavior and predator-avoidance behavior fail to occur to objects in the affected monocular field (Ewert 1968; Ingle 1977), although some particular avoidance patterns, such as ducking or sidestepping, can still be observed when the pretectal region is stimulated (Ingle, 1983). The lesion experiments suggest that the pretectal region has an inhibitory action on the tectum, blocking prey catching and activating avoidance responses when a predator stimulus is recognized.

The "midbrain hemifield deficit": In contrast with unilateral tectal ablation, unilateral lesions of a defined tract caudal to the tectum in the medulla (Kostyk & Grobstein 1982; Grobstein & Masino 1986) do not produce any scotoma, although the animal responds abnormally to stimuli located in the *entire half of the visual field* (including the binocular field) ipsilateral to the lesion. The effects evoked by this kind of lesion have been termed a "hemifield medullary deficit". For a left hemifield deficit, the animal is incapable of executing a visually elicited left turning movement when the prey is in the left hemifield and, in its place, it would either advance or snap to the front. These movements are independent of the stimulus horizontal position (eccentricity), but change depending on the stimulus elevation and distance. One conclusion is that the representation in the medulla is different from that in the tectum, in the sense that this representation is relative to the toad's body and no longer to the toad's eye. It can also be concluded that different components of the stimulus position are handled though different pathways that can be disturbed independently.

The Snap-Hop Transition: In normal conditions, the frog snaps when a prey is located nearby, but reorients, and hops, toward a stimulus that is further (but not too far) away. The transition between the two patterns occurs at a distance of about one body length for a lateral stimulus and two body lengths for a frontal stimulus (Ingle 1970; Grobstein et al. 1985). In frogs with the midbrain hemifield deficit, the transition boundary is invariably at twice the body length when the stimulus is located anywhere in the affected hemifield (Kostyk & Grobstein 1987), the same as if the stimulus had been located in front, which is also the same direction to which the animal always advances. This shows that the interpretation of certain sensory cues will depend on the interpretation of others, e.g., the distance to the prey stimulus is interpreted after the horizontal position of that stimulus is evaluated and a turning angle has been chosen, in order to decide whether to snap at the stimulus.

The "snapping" and "turn/snapping" deficit: A snapping deficit is observed with small bilateral lesions performed in the ventral part of the medulla that seem to selectively interrupt the pathway that carries depth information (Grobstein & Staradub 1989). In this circumstance the frog turns more or less appropriately but fails to snap and instead advances an incorrect distance, overestimated in relation to the extent of the lesion. The "turn/snapping" deficit is observed when the pathways that carry the depth and eccentricity information are both interrupted by a large bilateral lesion in the ventral part of the medulla (Grobstein & Staradub 1989). The result is that the frog will always respond with a small advancing movement, no matter the actual prey distance, without ever snapping or turning.

If the tectum's input map defines the frame in which avoidance is determined, we must still ask "Where in the brain is the position of the threat remapped into the direction of escape?" Grobstein (1988) indicates that the tectum does not contain a global motor map, but instead is a relatively early stage within the visual information-processing sequence. This opens the possibility that remapping occurs somewhere after the tectum, and we have evaluated two alternatives in Arbib and Cobas 1990. The definition of the Orient schema in this paper is explicitly based on the *Motor Heading Map* hypothesis, which we favor, whereas Grobstein appears to favor the *Positional Heading Map* hypothesis.

2. Motor Schemas for Prey Capture

A flow-chart representation of the prey-capture model was given in Arbib and Cobas 1991. Our task is now to look in detail at *Orient*, *Snap*, and *Approach*, the three motor schemas involved in the animal's response to prey.

```
Orient:
threshold=0, activation_level=0, heading=0, defend_command=0, orient_command=0
get heading, defend_command
if heading < 0 then left_turn = FALSE else left_turn = TRUE
increment activation_level by |heading|
decrement activation_level by 1000 * defend_command
if activation_level ≥ threshold then:
   if left_turn = FALSE then new_heading = -activation_level
   if left_turn = TRUE then new_heading = activation_level
return new_heading
```

Figure 1: Summary algorithm for the Orient motor schema. This schema is shared between the Prey Capture and Predator Avoidance system. In the Prey Capture system, the defend_command parameter defaults to 0 and has no effect. Note that resetting to 0 the positive (negative) new_heading input from the HeadingTranslator schema simulates the lesion of the pathway carrying left (right) heading in the prey capture system (medullary deficit).

The input to the Orient schema is the heading toward the target stimulus, which is provided by the HeadingMap schema. The connections from the HeadingMap are such that the more lateral the stimulus is, the more strongly the Orient schema is activated. If this activation level surpasses a threshold, the pool of motoneurons controlling the turn muscles are activated in proportion. Also, the activity of the Orient schema influences that of the Approach and the Snap schemas. In our simulation (Fig. 1) the activation level of the Orient schema increases by the product of the absolute value of the angle (0 heading is straight ahead for the toad, while positive and negative values represent left and right headings, respectively) and a predetermined weight factor, which in the current simulation implementation is equal to 1. If the final activation is reaches or surpasses the threshold, then the orient command is enabled, the toad's new heading is made equal to the prey angle (positive to the left, negative to the right) and that heading is returned in the output port.

The DepthMap is connected to both the Snap and Approach schemas in such a way that the activity level of these two motor schemas will increase more vigorously in response to close objects than to objects located farther away. Both Snap and Approach schemas need to be primed by the PreySelector schema in order to be able to respond to the stimulation by the Depth Map schema. Also, both schemas are inhibited by the activity of the Orient schema. The Snap schema determines if the prey is within the snapping zone. This evaluation is an increasing function of the closeness of the prey (inverse of the distance) and a decreasing function of the activity of the Orient schema. The snapping command is made effective if and only if the final activation is no smaller than its threshold (Fig. 2).

```
Snap:
threshold=1, activation_level=0, closeness=0, orient_command=0,
snapping_command=0
get closeness, orient_command
increment activation_level by 2 * closeness
decrement activation_level by 0.01* orient_command
if activation_level ≥ threshold then:
   snapping_command = activation_level
return snapping_command
```

Figure 2: Summary algorithm for the Snap motor schema of the Prey Capture model. Note that dividing the closeness input from the DepthTranslator schema simulates the bilateral partial lesion of the distance pathway (snapping deficit).

In a similar way, the Approach schema determines the distance for the toad to advance as a positive function of the closeness to the selected prey and a negative function of the activation level of the Orient schema. The presence of the prey by itself contributes to raising the activation level to a default value through a direct priming action from the PreySelector schema (Fig. 3). If the approach command is finally effective, the actual moving distance is made no greater than the maximum move the toad is estimated to make in a single ballistic episode.

The schemas of the motor pattern generator circuitry are bilaterally interconnected, in such a way that both Snap schemas would activate each other, as would the Approach schemas, while both Orient schemas would inhibit each other. The action would be practically identical for both bilateral counterparts, with the

```
Approach:
threshold=1, activation_level=0, moving_distance=0, default_distance=4,
   closeness=0, orient_command=0
get closeness, orient_command
if closeness and orient_command = 0 then closeness = 1/default_distance
increment activation_level by 10 * closeness
decrement activation_level by 0.1* orient_command
if activation_level ≥ threshold then:
   moving_distance = min (1/closeness, toad_max_move)
return moving_distance
```

Figure 3: Summary algorithm for the Approach motor schema of the Prey Capture model.

exception of the Orient schemas, which would trigger turning movements in opposite directions under the same level of activation. The Snap schema triggers the complete snapping movements, while the Approach and Orient schemas trigger an action proportional to the amount of activation over their respective threshold. So, for example, the more the Orient schema is activated (over the threshold) the more acute the resulting turn would be. The activation of the Snap and Orient schemas would trigger an immediate snap or turn, while the activation of the Approach schema could sometimes evoke a longer lasting action depending on whether a single step or jump or a sequence of steps or jumps are performed, as regulated by the distance to be covered and/or the "motivational" state of the animal. The mechanism of action of the Approach schema would be such that the higher its activity, the shorter the distance to the target and the less the advance muscles would have to be stimulated, and the lowest, the longer the distance required and the more intense the advance muscles would need to be driven. Also, no activity over the threshold for the Approach schema would correspond to no stimulation of the advance muscles.

The final motor behavior may result from the combined activity of these motor schemas. In this way, we can obtain a snapping movement directed to a prey located at a short distance at the right side of the toad by the activation of the Approach and the Snap schemas due to the short distance (high closeness), and the Orient schema using the required right heading value. Their coordination would be regulated by the appropriate transmission delays from one schema to another and from each schema to the controlled musculature. The same kind of ballistic movement obtained in the Lara et al. (1984) model by a sequence of several schema activations is achieved here by basically the same schemas, activated with a variable intensity and performing concurrently. This can yield more flexible and varied behavioral patterns than the purely sequential approach. After the appropriate motor action has been performed, if there are prey that have not yet been captured (i.e., snapped at), a following evaluation of the environment would trigger the next response.

2.1 Normal function

We now examine the inputs to the three motor schemas and their interactions under several situations in normal conditions. The weight of the interaction among these schemas will determine the shape of a series of "motor regions" (Fig. 4), which would delimit those areas of space where a prey would elicit an orienting movement, an approach movement, a snapping movement, or some combination of these. For example, the weight of the depth input to the Snap schema determines the extension of the snapping zone to the front, and the weight on the input from the Orient schema determines the width of the zone laterally (see Fig. 2).

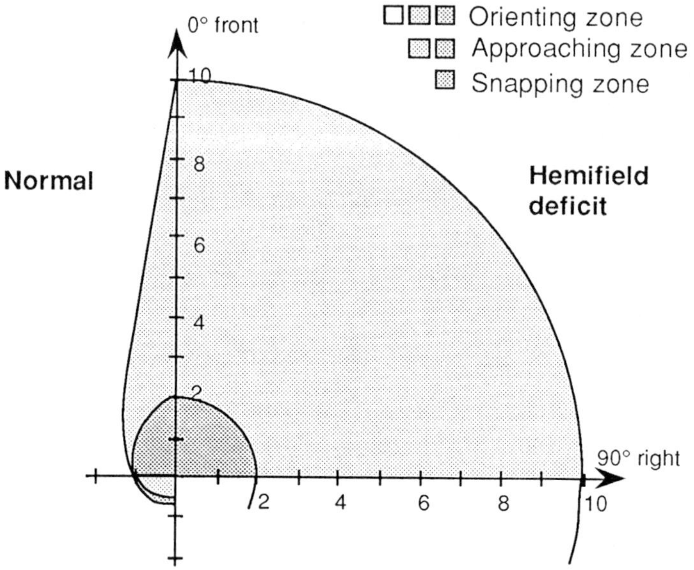

Figure 4: The model yields motor zones in prey acquisition which differ between the normal animal (left) and one with a medullary hemifield deficit (right). Each unit corresponds to 1 body length.

In Fig. 4 we see that the snapping zone resulting from our simulation extends 2 body lengths to the front and 1 body length 90° to one side, for the normal situation. These values more or less correspond with those reported in the literature (Ingle, 1970; Grobstein at al., 1985). No similar experimental values for the Approaching zone exist in the literature, and thus we estimated 10 body lengths in the frontal direction and 1 in the lateral direction to be a reasonable approximation. Note that the precise shape of the boundary line results just from the simple linear interaction of the heading and closeness to the prey (see Figs. 2 and 3). Since the exact actual shapes of the fields are ill-undefined, especially for the approach zone and for the posterior field of the toad, further experiments are necessary to reveal whether the current approach is precise enough. In the following paragraphs we analyze how the model would resolve several hypothetical prey configurations (refer to Fig. 4).

Prey at front within the snapping zone: The small distance to the prey would be enough to activate the Snap schema over its threshold and increase the activity of the Approach schema in such a way that a short frontal advancing movement combined with a snapping movement would be obtained. The Orient schema would not reach the threshold and would not have any effect. The activity of both sides would be equivalent.

Prey at right outside the snapping zone: The conditions would be the same as in the previous case with the addition that a strong right heading value would intensely activate the Orient schema, which in turn would decrease the activity of the Snap and Approach schemas below their respective threshold. The result would be a simple right turning movement.

Prey at right inside the snapping zone: The increase of the closeness value (decrease of distance to the prey) over the preceding situation would suffice to activate the Approach and Snap schemas over their threshold to yield a snapping movement aimed to the right. In this way, the action of a pronounced right heading requires a higher closeness value to obtain a snapping movement as compared to a frontal situation, yielding a snapping zone wider at the front than at the sides, as observed in behavioral experiments for the snapping zone (Ingle 1970; Grobstein et al. 1985).

Prey at front-left outside the snapping zone: The closeness parameter will be too weak in this case to activate the Snap schema over its threshold, and a mild heading to the left would be enough to activate the left Orient schema over its threshold. The Approach schema would be sufficiently activated to remain over the threshold after the small inhibition by the Orient schema, and so an advance movement directed a little bit to the left would be obtained.

2.2 "Lesion" effects

The "midbrain hemifield deficit": In our model, the corresponding "simulated" lesion in the right side would leave the right HeadingMap schema without the heading input and would therefore produce no stimulation of the right Orient schema for all the prey stimuli falling onto the right visual field (to the right of the animal's midline). The immediate consequence is that the animal would be incapable of executing visually elicited right turning movements when the prey is in the right hemifield and, in its place, it would always either advance or snap as if the prey was at the same actual distance but at the front.

In our simulation, the same effect can be replicated just by resetting any *negative* heading value returned by the PreyHT to 0. Since, in our simulation, negative heading values correspond to the right and positive to the left, whenever the prey is located in any part of the right hemifield, the activity of the Orient schema will invariably also be 0. This results in no right turning movement and cancels the influence of the Orient schema on the Snap and Approach schemas (see Figs. 2 and 3), which now always perform as when the prey is located at the front in the normal situation.

In the hemifield deficit, the snapping zone in the affected side appears equally wide laterally as at the front; and this is replicated in the model since what narrows the width of the snapping zone at the sides is the inhibitory action of the Orient schema output. The elimination of that output makes the width of the snapping zone (and that of the approaching zone) dependent only on the distance to the prey and independent of its angle within the affected field (Fig. 4).

The "snapping deficit": In our model, this effect is produced by the inability to excite the Snap schema above its threshold due to the diminished input from the DepthMap. The Approach schema is also affected by the same reason although not completely due to the stronger influence the DepthMap has on it. The predicted reaction would be as if the prey were always outside the snapping zone (which after the lesion would be very much reduced or nonexistent), and therefore would always yield an advance movement perhaps combined with a turn.

This lesion is reproduced in the simulation by proportionately reducing the output of the DepthTranslator[3] schema and therefore the input to both the Snapping and Approach schemas. When the closeness value is decreased enough, the Snap schema is not able to raise its activity level over the threshold and thus fails to produce a snapping movement, while the Approach schema is still able to pass its threshold, but computes a larger than necessary distance and overshoots the prey. The Orient schema would remain unaffected.

The "turn/snapping deficit": In our model, the elimination of the DepthMap and HeadingMap schemas leave the Orient and Snap schemas inactivated. The Approach schema, having been aroused by the priming of the PreySelector and without any further stimulation, will increase its own activity and constitute in this way the default response of the system. This property would be specific of the Approach schema, and not applicable to the Snap schema. The predicted reaction would be as if the prey were always straight outside the snapping zone, and therefore would always be an advance movement.

This effect is simulated by resetting the input to all three motor schemas to 0. Both Orient and Snap fail to increase their activation level, and therefore remain inactive. The Approach schema, nevertheless, gets a default distance[4] that is effective when no input is received (this symbolizes the priming effect from the PreySelector schema) and commands a fixed approach irrespective of the actual location of the stimulus.

Ablation of one tectum: The animal would not respond to prey objects located in the affected monocular field. The information on the other most medial part of the affected visual field would be provided by the binocular region of the other tectum

---

[3] Recall that in the simulation the Translator schemas, in addition to their own computations, perform also the function of the corresponding Map schemas.

[4] Grobstein does not mention any figure for the distance that the toads advance in this situation. We use 4 body lengths as a reasonable placeholder until the necessary data are available.

through the connection of the affected HeadingMap with the ipsilateral PreyHT and the bilateral projection of the DepthTranslator in the contralateral side of the lesion (the other DepthTranslator schema would be rendered inoperable for lack of input from its corresponding tectum).

In our simulation, the removal of, say, the left tectum and consequently the elimination of the left PreySelector and PreyHT, would be emulated by resetting to 0 the depth value reported by DepthTransator and the heading reported by PreyHT when the later ranges from -45° to -180°. (The distance and heading estimates for angles between 0 and -45°, being part of the binocular field, would still be provided by the right PreyHT and Depth Translator.) The result would be no reaction for prey located from 45° to 180° to the right of the toad's midline.

3. A schema model for Predator Avoidance

The fundamental idea for the predator-avoidance model is quite similar to that for the prey capture model. In fact, many of the schemas in the Predator Avoidance system are hypothesized to be the same or equivalent to those of the Prey Acquisition system (Fig. 7 of Arbib and Cobas 1990). Since there is little experimental evidence to constrain most of the hypotheses used to construct the model, a major goal of this effort is to stimulate the necessary experiments. With this, we turn to a specification of Defend, Orient, and Escape, the motor schemas of the predator avoidance model.

```
Defend:
 activation_level = 0, threshold = 1/(2*toad_size), defend_command = 0
 get predator_size
 increment activation_level by 1/predator_size
 if activation_level ≥ threshold then:
    defend_command = 1
 return defend_command
```

Figure 5: Summary algorithm for the Defend schema of the predator-avoidance model.

The Defend schema is a compound schema in the sense that it incorporates a variety of motor responses such as sidestepping, ducking, poison gland secretion, etc., each one of which would be triggered as a function of certain predator features like its size, position (distance), shape, movement direction, speed, etc. In the current model, however, the size of the predator stimulus as transmitted from the pretectum by the PredatorRecognizer schema as part of the priming effect, is the only parameter we use to control the activation level of the Defend schema. Smaller objects tend to stimulate the Defend schema more intensely than larger objects, and so, only predators smaller than a certain maximum size are able to activate this schema over the threshold and trigger a defensive response. In the absence of experimental data, we assumed as a reasonable limit for making the Defense schema effective when the size of the predator was equal or smaller than twice the size of the toad.

In the simulation, the Defend schema (Fig. 5) increases its activity level as a function of the inverse of the size of the predator. The Defend schema outputs its activation level, or 0 if that is less than the threshold. The 0 indicates that no defense mechanism is to be performed. As a default (compare the priming of the Approach schema in the prey capture system), the presence of a predator by itself can raise the activation of the Defend schema to the threshold level by priming from the PredatorRecognizer schema though the predator size parameter. This does not occur with the Escape schema.

The Orient schema in predator avoidance is the same as that in the prey capture system (see Fig. 1). In our simulation, the PredatorHeading Translator (PredatorHT) schema directly provides the Orient schema with a heading value transformed from the direct retinotopic position of the stimulus. This transformation emulates the main ipsilateral projection that the PredatorHT has onto the HeadingMap schema in the model, which in turn feeds the Orient schema. In this way, the nearer the threat is to the forward direction, the higher the activation of the Orient schema is. However, when the Defend schema is effective, the Orient schema is completely inhibited. Otherwise, evaluation of the activation level results in a turn that is a compromise between the forward direction and that directly away from the predator (in the simulation the resulting heading ranges between 0 and 70° to both sides).

```
Escape:
threshold=tan(30/2), activation_level=0, closeness=0, defend_command=0,
   moving_distance=0
get predator_size, closeness from DepthTranslator, defend_command from Defend
increment activation_level by (predator_size/2) * closeness
decrement activation_level by 1000 * defend_command
if activation_level ≥ threshold then:
   moving_distance = toad_max_move
return moving_distance
```

Figure 6: Summary algorithm for the Escape schema of the predator-avoidance model.

The Escape schema (Fig. 6) gets the size of the predator through the priming effect by the Predator Recognizer schema and, in our current simulation, an estimate of the closeness to the predator directly from the DepthTranslator schema (in the model this is done via the DepthMap schema). Its level of activation then increases in proportion to the predator size and distance, getting higher the nearer the predator is. The threshold is set so the escape reaction is triggered only when the angular size of the stimulus[5] exceeds a preset value. In our simulation we used an angular size limit of 30°, which has been experimentally determined for frogs (Ingle & Hoff, 1990). When the conditions are met, the escape distance is made equal to a

[5] In our simulation, we use the predator dimension and distance from the toad to calculate its angular size.

preset jump distance value. As with the Orient schema, an effective defense command completely inhibits the escape.

3.1 Normal function

In a general situation in which a predator is spotted, say, coming at 90° to the left, the Escape schemas would be stimulated because of the priming by the PredatorRecognizer schema, the excitation from the DepthMap, and their mutual excitatory interaction. At the same time, the right Orient schema would be activated to command a turn to that side. The Defend schemas would also be influenced by the priming effect of PredatorRecognizer. If the predator is large enough (e.g, larger than twice the size of the toad, according to the current settings of our simulation — see Fig. 5), the Defend schema would not be effective and so a normal escape reaction would be obtained when the stimulus angular size is larger than the preset value (in our case, 50° — see Fig. 6). The resulting turning orientation would be some compromise between the forward direction and the contrary to the threat direction which, in this example, would correspond to a jump of some 22° to the right. If the predator is smaller, the Defend schema would become active, and the actions of the Escape and Orient schemas would be inhibited, blocking a normal escape turn and jump. The action of the Defend schema could be one or several of the described defensive reactions described in the literature. For example, it could include the rotation of the toad at 45° to the left, in a perpendicular direction to the threat, so that the animal could lower the right side of his body to present the back to the threat as a protective strategy.

3.2 "Lesion" effects

Little is known about the effect of lesions on predator-avoidance behavior, and thus the constraints imposed on our model are rather few. It is known that the elimination of the posterior thalamus-pretectal region eliminates escape behavior and evokes a prey capture response to abnormally large objects that would have been avoided otherwise (Ewert 1976). If we suppose that the localization of the Predator Recognizer schema is in that zone, its elimination would consequently block the rest of the predator-avoidance behavior and leave the PreySelector schema in the prey acquisition system completely disinhibited. Given the nonspecificity of the latter system in relation to prey-predator discrimination, this would thus allow the trigger of an approach response toward any moving stimulus, even if it has predator characteristics. Neither the Defend nor the Escape schemas would be activated by the respective Map schemas, due to the lack of priming from the PredatorRecognizer schema.

4. Discussion

We have offered a model, consistent with a large body of behavioral data, which shows how the toad's brain may construct behavior through the interaction of different perceptual and motor schemas via a process of competition and cooperation where, although a unique motor schema might win the competition, in general two or more motor schemas may simultaneously be active and cooperate to yield a complex motor pattern. Each one of these motor schemas is driven by specific internal maps which between them constitute a distributed internal representation of the world. These maps collectively provide the transition from topographically-coded sensory information to frequency-coded inputs to the diverse motor schemas that drive muscle activity.

Recent experiments by Ingle & Hoff (1990) indicate that the avoidance response to predator stimuli in frogs is in fact controlled by the direction in which the stimulus is moving, rather than merely by its position. When the predator trajectory crosses in front of the frog, the animal jumps to the hemifield that the stimulus is coming from, in a sort of cut-back maneuver, instead of away from it as it does when the predator is on a collision course with the frog. This response is similar to that observed in the prey acquisition system, because in both cases the frog moves toward the stimulus. After studying several lesion effects, Ingle & Hoff conclude that the frog uses predominantly an ipsilateral tectofugal pathway (as in classical predator avoidance) to direct movement away from the stimulus position, and a contralateral tectofugal pathway (as in classical prey acquisition or non-colliding predator avoidance) to move toward the stimulus. This favors the hypothesis of a motor heading map, since they point to a divergence of the two kinds of responses already at the level of tectal output.

To account for this new evidence, our predator avoidance model would have to be modified: the selection of a predator would activate both the PredatorHT schema projecting ipsilaterally, which directs movement away from the stimulus, and the PreyHT schema projecting contralaterally, which directs movement toward the stimulus. The pathways would compete with each other, in such a way that only the most active one would eventually command the motor schemas. The trajectory of the predator stimulus would determine the bias: when the predator is on a collision course with the toad, the PredatorHT schema would be activated more intensely, while the PreyHT schema would be more activated when the predator trajectory crosses in front of the toad. Experimental evidence shows that transection of the ipsilateral tectal pathway abolishes the escape away from predator stimuli located in the opposite visual field and instead triggers a response directed to the same hemifield (Ingle & Hoff, 1990). According to the hypothesis described above, the effect would be produced by the activation of the contralateral tectal pathway which, in the absence of competition, would invariably command movement toward a stimulus, prey or predator, placed in the opposite hemifield. We did not simulate this effect, however, because the detection of the stimulus' direction of movement, on which this new evidence is based, is not supported in the present simulation.

We have designed our predator avoidance schema model to conform as much as possible to the principles of segregation of coordinate systems and participation of maps intermediate between sensory and motor schemas that were considered in our prey acquisition model. We have made assumptions about the transitions between different modes of predator avoidance which need to be tested in normal and lesioned animals. It would be interesting to know the effects on predator avoidance produced by lesions of known consequence in the prey capture system, like the medullary lesion of the distance pathway carrying depth information (Grobstein & Staradub, 1989).

Even with such data available to refine the model proposed here, still more data are required to show how the coordination of motor responses can be obtained based on the information processed by the thalamus and the tectum where "planning" has to occur, as when the animal needs to detour around obstacles and construct a path to get to the prey (e.g., Collett 1982, Lara et al. 1984). Arbib and House (1987) showed how depth perception could be used to yield maps whose interaction provided the data for inferring the proper motor response for detour behavior, but the model, while highly parallel, is non-neural, and is not structured in light of the separation of pathways documented and modeled in this paper. We look to future research to show how a combination of all these theories will combine with an expanding data base to clarify the neural basis of visuomotor coordination.

5. References

Arbib, M.A., and Cobas, A., 1991, Schemas and neurons: A multi-level perspective on visuomotor coordination, this volume.

Arbib, M.A., and House, D.H., 1987, Depth and Detours: An Essay on Visually Guided Behavior, in *Vision, Brain and Cooperative Computation*, (M.A. Arbib and A.R. Hanson, Eds.), A Bradford Book/The MIT Press, pp. 129-163.

Collett, T., 1982, Do toads plan routes? A study of the detour behaviour of *Bufo Viridis*, *J. Comp. Physiol.* 146: 261-271

Ewert J.-P., 1968, Der Einfluss von Zwischenhirndefekten auf die Visuomotorik im Beute- und Fluchtverhalten der Erdkröte (*Bufo bufo* L.) *Z. Vergl. Physiol.* 61: 41-70

Ewert, J.-P.,1976, The visual system of the toad: behavioural and physiological studies on a pattern recognition system in *The Amphibian Visual System*, (K. Fite, Ed) Academic Press, pp. 142-202.

Ewert J.-P., 1984, Tectal mechanisms that underlie prey-catching and avoidance behaviors in toads. In *Comparative neurology of the optic tectum*, (Vanegas, H., Ed.), Plenum, New York, pp 247-416

Grobstein, P., 1988, Between the retinotectal projection and directed movement: topography of a sensorimotor interface, *Brain Behav. Evol.* 31:34-48.

Grobstein, P., Reyes, A., Zwanziger, L., and Kostyk, S.K., 1985, Prey orienting in frogs: Accounting for variations in output with stimulus distance, *J. Comp. Physiol. A*, 156:775-785.

Grobstein, P. & Masino, T., 1986, Sensorimotor circuitry underlying directed movement if the frog: evidence for an intermediate representation of space in the tectofugal pathways, *Soc. Neurosci. Abstr.* 12:684

Grobstein, P. & Staradub, V., 1989, Frog orienting behavior: the descending distance signal, *Soc. Neurosci. Abstr.* 15:54

Ingle, D., 1970, Visuomotor functions of the frog optic tectum, *Brain Behav. Evol.* 3: 57-71.

Ingle, D., 1976a, Spatial visions in anurans. In *The Amphibian Visual System*, (K. Fite, Ed), Academic Press: New York, pp 119-140

Ingle, D., 1976b, Behavioral Correlates of Central Visual Function in Anurans, in *Frog Neurobiology* (R. Llinás and W. Precht, Eds.) Springer-Verlag, pp. 435 - 451.

Ingle, D., 1977, Detection of stationary Objects by Frogs After Ablation of Optic Tectum. *J. Comp. Physiol. Psych.* 91:1359-1364.

Ingle, D., 1983, Brain mechanisms of visual localization by frogs and toads, in *Advances in Vertebrate Neuroethology* (J.-P.Ewert, R.R. Capranica and D.J. Ingle, Eds.) Plenum Press, pp. 177-226.

Ingle, D. J., and Hoff, K. vS., 1990, Visually elicited evasive behavior in frogs, *Bioscience*, 40:284-291.

Kostyk, S.K. and Grobstein, P., 1982, Visual Orienting deficits in Frogs with Various Unilateral lesions, *Behavioural Brain Res.* 6:379-388.

Kostyk, S.K., and Grobstein, P., 1987, Neuronal organization underlying visually elicited prey orienting in the frog. I. Effects of various unilateral lesions, *Neuroscience* 21:41-55.

Lara, R., Carmona,M., Daza, F., and Cruz, A. (1984) A global model of the neural mechanisms responsible for visuomotor coordination in toads, *J. Theor. Biol.* 110:587-618.

A Neural Network Model for Response to Looming Objects by Frog and Toad [1]

Jim-Shih Liaw
Michael A. Arbib
Center for Neural Engineering
University of Southern California
Los Angeles, CA 90089-2520, U.S.A.

Abstract. *Toads exhibit a wide variety of avoidance patterns depending on the stimulus situation. The analysis of this situation is achieved through interaction between the optic tectum and pretectum. The retinal signal is first received and processed by tectal neurons and their outputs then converge onto pretectal neurons. Based on these converging inputs, neurons in thalamic pretectum are able to analyze the stimulus situation and determine an appropriate avoidance action. The spatial location of the stimulus is encoded in the topography of tectal neurons. This signal is projected onto a motor heading map which specifies the direction of the avoidance movement. We develop a neural network model to account for the toad's detection of and response to a looming stimulus.*

1. Introduction

While prey-catching behavior requires accurate motor output (to catch a relatively small prey), predator-avoidance behavior requires analysis of a complicated stimulus situation to determine a crude but quick estimate of the course of action (to move anywhere in a relatively large "escape zone"). Different avoidance responses are observed in frogs depending on the stimulus situation (Ingle 1976, Ingle and Hoff 1990, Ewert, 1984) but here we concentrate on the response to a looming object: If the approaching object is small and from the upper visual field, frogs would duck. If the approaching object is large and on a collision course, frogs jump away from it. The direction of such a jump is a compromise

---

1 The research described in this paper was supported in part by grant no. 1RO1 NS 24926 from the National Institutes of Health.

between the forward direction and that away from the threat, but when the object is not on a collision course, frogs may cut behind it (Fig. 1).

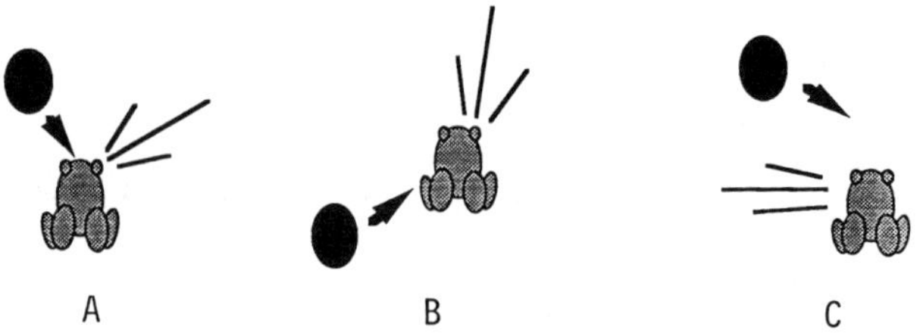

Figure 1: The escape direction. When the looming stimulus is on a collision course with a frog, the escape direction of the frog is a compromise between the forward direction and that away from a looming object (A,B). If the stimulus is not on a colliding trajectory, the frog will jump in such a direction as to "cut back" behind the looming object (C) (adapted from Ingle and Hoff, 1990).

We develop a neural network model to account for the toad's detection of and response to a looming stimulus, and for the translation of that recognition into an appropriate command. The spatial location of the stimulus is encoded in the topography of tectal neurons. This signal is projected onto a motor heading map which specifies the direction of the avoidance movement (Cobas and Arbib, 1990). The avoidance action is initiated by the signals from tectum and pretectum. The model is implemented in NSL, a Neural Simulation Language (Weitzenfeld, 1989).

2. Neural network model for avoiding looming objects

A. The basic cell types

Two types of neuron are sensitive to a stimulus moving toward the eye:

T3 neurons: Grüsser and Grüsser-Cornhels (1970) reported that T3 neurons in the optic tectum respond vigorously to stimuli moving toward the frog's eye, in contrast to their low sensitivity to movement away from the eye or around the body at a constant distance. The excitatory receptive field (ERF) of these neurons ranges from 20° to 30°. To elicit a T3 neuron response, the stimulus must have an angular size of at least 3° (Grüsser and Grüsser-Cornhels, 1973).

TH6 neurons: Ewert (1971) categorized neurons from caudal thalamus into 10 classes based on their receptive fields and response characteristics. Among these pretectal neurons, TH6 neurons are sensitive to a looming stimulus, especially that from the upper visual field. Moreover, the closer the object, the stronger the TH6 response it can elicit. TH6 neurons have ERF of either 180° or 360°. The input to "total field" TH6 neurons comes from the contralateral eye via the ipsilateral optic tectum (Brown and Ingle, 1973) — small tectal lesions produce scotoma in these neurons.

Other type of neuron relevant to detection of a looming stimulus include:
T6 neurons: The ERF of T6 neurons lies in the upper visual field, with fronto-caudal range at least 120° and left-right range greater than 90°. They can be elicited by stimuli larger than 8° diameter and moving faster than 5°/sec.
TH3 neurons: The ERF of TH3 neurons is about 30°. These neurons receives input from the contralateral eye. They respond best to moving stimuli that fill their ERF. Based on their response characteristic to configurational stimuli, Ewert and Wietersheim (1974) postulated that these neurons are critical in the recognition of a predator.

Given these data, we postulate that the optic tectum contains a spatial map of looming stimulus locations encoded by a population of T3 neurons, while TH6 neurons carry out the recognition of a looming stimulus based on converging tectal inputs and depth information. Once a predator is recognized by the pretectal system, the prey-catching system in the optic tectum will be shut off by inhibitory modulation from pretectum.[2] If the looming stimulus is large (signaled by the size-sensitive TH3 neurons), the jumping motor schema will be activated. On the other hand, if the stimulus is recognized as a small air-borne object (signaled by T6 neurons), the ducking motor schema will be elicited. In summary, the topography of the T3 neurons provides the base for the spatial map of the stimulus location whereas the activation of the tectal and pretectal neurons serves as the triggering command of the predator-avoidance behavior (as shown in Fig. 6).

The retinal input to our model will comprise R3 and R4 neurons. This is based on the observation that the activation of R3 and R4 retinal neurons has a high correlation with the flight and hiding response (Grüsser and Grüsser-Cornhels 1976). They are more sensitive to larger stimuli than other retinal neurons. Ewert (1984) reported that caudal thalamus receives inputs from R3 and R4 neurons but not R2 neurons. R3 neurons are sensitive to the leading edge of a visual stimulus while R4 neurons exhibit prolonged firing activity to the dimming of light (Tsai and Ewert, 1987; Grüsser and Grüsser-Cornhels, 1976).

---

[2] It has been postulated that the prey-catching system involves inhibitory modulation of the optic tectum by the pretectum (Ewert and von Seelen, 1974). Ewert (1974) found that the activity of T5.2 cells is highly correlated to the prey-catching response. Lara, Arbib & Cromarty (1982) proposed a tectal column model of prey-catching. However, the detailed study of neural interactions between prey-catching and predator-avoidance is beyond the scope of the present paper.

B. The looming-detection network

In our model, the visual stimulus is transmitted to the avoidance circuitry via R3 and R4 ganglion cells. These are modeled by a version of the Teeters (1989) model of frog and toad retina implemented in NSL (following Wang & Arbib, 1990). T3 neurons detect movement toward the eye and T6 neurons monitor stimulus activity in the upper visual field. Based on Brown and Ingle's (1973) finding on pretectal "total field" neurons, there is no direct retinal projection to the TH6 neurons. Instead, these tectal signals, along with depth information, converge onto the TH6 neurons which determines whether the visual stimulus is a looming threat (Fig. 2). The topography of the T3 neurons provides the spatial map of the stimulus location and the activity of the TH6 neurons indicates the presence of a looming stimulus.

The detection of movement toward the eye by T3 neurons is achieved by arranging (Fig. 3) the mask for the projection from the retina to T3 neurons so that the weights increase as one moves away from the center of the mask. Through such a mask, the sensitivity of R3 cells to the leading edges of a visual stimulus (Tsai and Ewert, 1987) provides the information of expanding edges in all directions. This gives the signal for an approaching object. The prolonged excitation of R4 neurons to the dimming of light (Grüsser and Grüsser-Cornhels, 1976) ensures the continuity inside the expanding edges. Thus, the T3 neuron is able to detect looming objects and yet would not be fooled by expanding edges alone (such as two vertical bars moving away from each other).

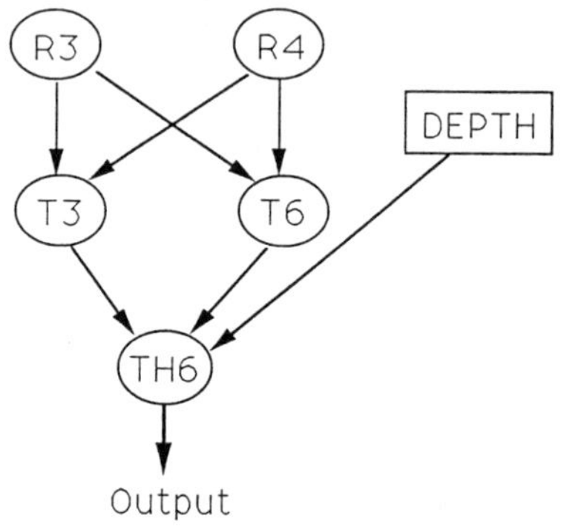

Figure 2: The looming-detection model. The visual stimulus is transmitted to the network via R3 and R4 ganglion cells. T3 neurons detect movement of a looming object and T6 neurons monitor stimulus activity in the upper visual field. These tectal signals, along with depth information, converge onto the TH6 neurons which determine whether the visual stimulus is a looming threat.

How can we localize a stimulus which is expanding in all directions? We propose a method of population encoding as a solution to this problem: The position of a looming object can be determined by localizing its focus of expansion (FOE). In our model, the T3 neuron with its center of mask aligned with the FOE of a looming stimulus would have the highest activation. Neighboring neurons would have less activation which decays as one moves away from the FOE (Fig. 4). Thus the spatial position of a looming stimulus is encoded in a population of T3 neurons where the more central neurons have higher activations.

The ERF of a T6 neuron is in the upper visual field. This effect is achieved by projecting retinal input to T6 neurons through a mask which has larger weights for higher position. Input from T6 neurons to TH6 neurons makes them more responsive to an approaching aerial threat. TH6 neurons also receive depth information from some depth perception circuitry. In our current model, such depth information is obtained from a schema which is not analyzed further. A candidate for neural implementation is the Cue Interaction Model of House (1989), but the present schema computes the depth map of each point on retina at each time step on the basis of the initial position, speed, and final position of a stimulus. More precisely, the output of the schema is "closeness", inversely proportional to the distance of the stimulus. By properly combining the inputs from T3 neurons, which signal the presence of expanding edges, and the depth signal from the depth perception system, the TH6 neurons will respond more strongly to a smaller but closer one.

Figure 3: Radiating mask for T3 neurons. The retinal signals are projected through a radiating mask to T3 neurons. R3 cells signal the leading edges of a stimulus while the prolonged excitation of R4 neurons to the dimming of light ensures the continuity inside the expanding edges.

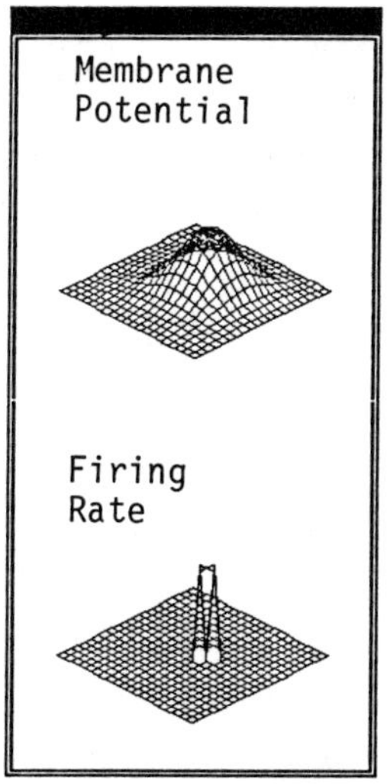

Figure 4: Population encoding of stimulus position. The position of a looming object is determined by localizing its focus of expansion. The T3 neuron with its center of mask aligned with the FOE would have the highest activation. Neighboring neurons would have decaying activation as one moves away from the FOE. Thus the spatial position of a looming stimulus is encoded in a population of T3 neurons where the more central neurons have higher activations.

C. The motor heading map

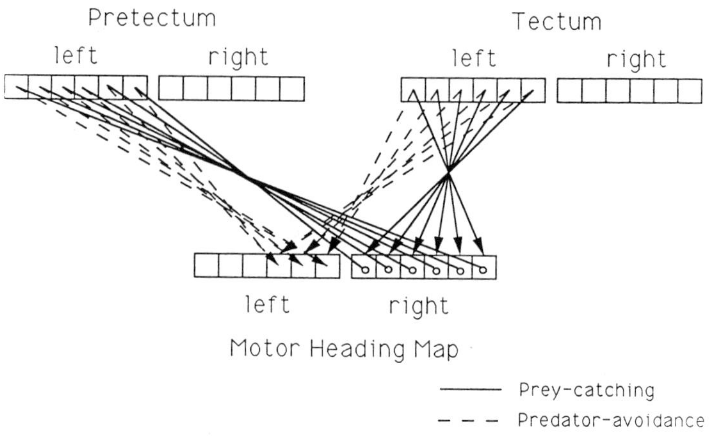

Figure 5: Gating of tectal projection onto motor heading map. The optic tectum projects bilaterally to the heading map with differential weights. The pretectum gates the tectal signal by projecting excitatory signals to the ipsilateral heading map and inhibitory signals to the contralateral one.

Once a looming object is detected, toads must determine which direction to jump to avoid it. Cobas and Arbib (1990) propose the Motor Heading Map hypothesis for the determination of the direction to jump: prey-catching and predator-avoidance systems share a common map for the heading (coded in body-coordinates) of the Orient motor schema, as distinct from a common tectal map for the direction of the stimulus.[3] The projection from optic tectum to the heading map thus must differ depending on whether a visual stimulus is identified as prey or predator.

In our present model, we only consider situations involving approaching prey and predator, without the "cut back" maneuver. The correct projection from the optic tectum onto the heading map can be achieved via gating by the pretectal predator system: The optic tectum projects bilaterally to the heading map with differential weights. The pretectum projects excitatory signals to the ipsilateral heading map and inhibitory signals to the contralateral one (Fig. 5). In this way, the pretectum can gate the appropriate heading information to the heading map based on the presence or absence of a predator.

---

[3] The direction of prey and the direction of prey-catching are the same, but the direction of a predator and the direction of escape are different. Thus, in the latter case, the sensory map and the motor map must be distinguished.

D. Motor schema selection: ducking vs. jumping

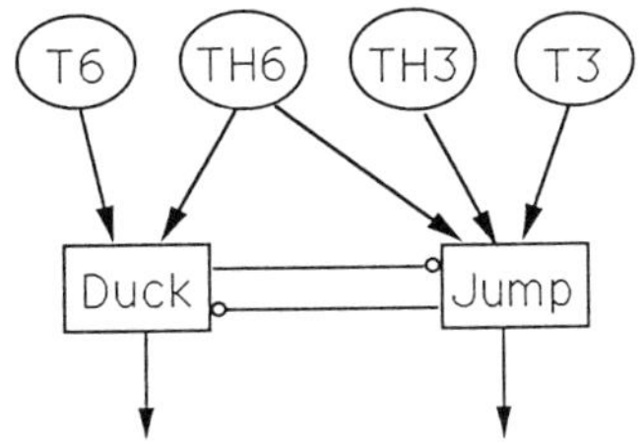

Figure 6: Selection of motor schema. The selection of the proper motor schema is achieved through competition between the ducking motor schema and the jumping motor schema, depending on the size and position of a looming stimulus. TH3 and T3 neurons favor a larger stimulus while T6 neurons prefer stimuli from higher elevations.

Toads respond to different stimulus situations with different movement patterns. Cobas and Arbib (1990) proposed a general mechanism of motor pattern selection through the interaction of motor schemas. In our present model, we are concerned with the toad's response to a looming stimulus. If the approaching object is small and from the upper visual field, the toad would duck down, while if the object is large or ground-borne, the toad tends to jump away (Ewert, 1984). We postulate a specific mechanism for the selection of proper action (ducking vs. jumping) depending on the size and position of a looming stimulus.

Our approach is to utilize the sensitivity of TH3 neurons to stimulus size and the sensitivity of T6 neurons to stimuli in the upper visual field. In our model, each TH3 neuron integrates visual signals from 49 R3 and R4 cells (through a 7x7 mask). In this arrangement, the larger the stimulus, the stronger the activity of, as well as the more the number of, TH3 neurons it can elicit. T6 neurons, on the other hand, are more sensitive to the elevation of the looming stimulus. By connecting TH3, TH6, and T3 neurons to the jumping motor schema, and TH6 and T6 neurons to the ducking motor schema, an appropriate action could be chosen through their competition, depending on the size and elevation of the stimulus (Fig. 6).

3. Results of the simulation

The predator-avoidance model is implemented in NSL (Weitzenfeld, 1989) running on a Sun workstation. Visual stimuli are represented as black squares in a 2 dimensional array. In our current simulation, the dimension of this array is 21x21 and each square corresponds to about 17° of visual angle. This is a rather coarse grain in comparison to the resolution of toad's retina, and we wish to increase the resolution of our model in the near future so that precise quantitative simulation can be carried out. A program is used to generate visual stimuli with different positions (in the x-y plane), initial distances, and speeds. The distance (closeness) of the stimulus at each time step is also calculated by this program. The visual stimulus is then projected to the retinal network and on to the predator-avoidance model.

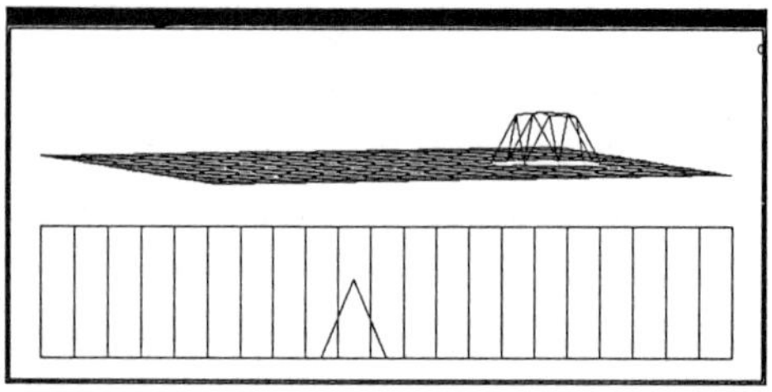

Figure 7: Escape direction. The top row represents a two dimensional array of T3 neurons which encode the position of a stimulus. The bottom row represents a one-dimensional heading map which signals the escape direction.

A. Stimulus position vs. direction to jump

Fig. 7 shows the activity of T3 neurons (top row) and the directions to jump (bottom row) in response to looming stimuli from different positions. In this figure, the stimulus approaches from the right and the jump is directed to the leftward front. In other simulations, when the stimulus approaches from the front, the jump is directed to the far right and when the stimulus approaches from the right, the jump is directed to the leftward front. All these directions are roughly a compromise between the forward direction and that opposite to the incoming stimulus. This result is achieved by bilateral projection from the optic tectum and ipsilateral inhibitory projection from the pretectum onto the heading map.

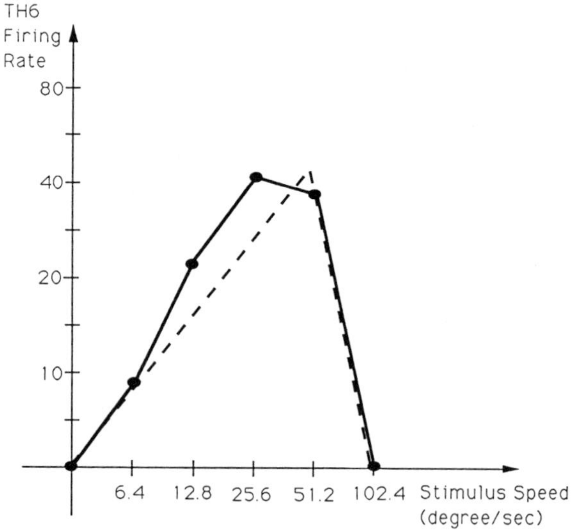

Figure 8: Stimulus speed and avoidance response rate. The solid line is the projection obtained from computer simulation. The dashed line represents the data from an experiment in which a disc is moved around over the toad's head (Ewert and Rehn, 1969).

B. The effect of stimulus speed and size

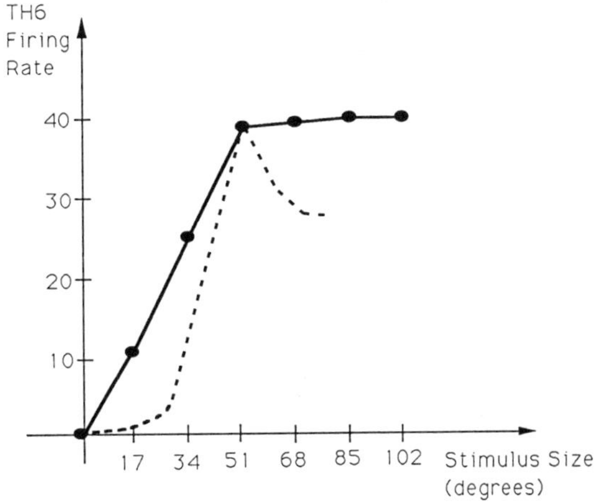

Figure 9: Stimulus size and avoidance response rate. The solid line is the projection obtained from computer simulation. The dashed line represents the corresponding experimental data.

Neuronal response is dependent upon the speed of the stimulus. The average impulse frequency of R3 and R4 ganglion cells increases as the stimulus speeds up (Grüsser and Grüsser-Cornhels, 1976). Avoidance activity of toads increases almost linearly with the speed of the stimulus, within limits (Ewert and Rehn, 1969): Response = $KV^{0.9}$ for $1<V<50$ (^{o}/sec), where V is the speed of the stimulus and K is some constant. When the stimulus moves too slowly (< 1°/sec) or too quickly (> 100°/sec), toads would not respond at all. The simulation results of our model with different stimulus speeds is shown in Fig. 8. Without tuning the parameters specifically for this simulation, the result fits the experimental data well: The response increases almost linearly with the speed of the stimulus and the lower and upper limits to elicit an avoidance response match those observed.

The toad's avoidance response rate is also subject to the size of the visual stimulus. Ewert and Rehn (1969) found that toad's escape activity increases with the size of the stimulus, with a size of about 50° diameter being optimal. By generating stimuli with varying sizes and projecting them to our model, we obtain the effect of stimulus size on avoidance activity plotted in Fig. 9. The model behaves much like the observed animal behavior except that it shows saturation of response activity when the stimulus is larger than the optimal size, while the observed data show a slight decrease in this situation.

C. Ambiguous stimulus situations

One of the clues in determining if an object is moving toward the eye is the expansion of image size on the retina. The expansion results in leading edges moving away from each other. Such movement of the edges can also be produced by several objects moving away from each other. Ambiguity thus arises if this is the only clue available. In our model, this ambiguity is resolved by integrating the signal from R4 neurons. The R4 neurons exhibit prolonged activation to the dimming of light and, in effect, can ensure the continuity inside the expanding edges. Computer simulation shows that the T3 neuron is not fooled by some tricky stimulus situations such as four squares (or two vertical bars) moving away from each other. But, if two vertical bars are overlapped at first, then for a short period of time after they start to move away from each other, they appear to be a single expanding bar. In this case, the simulated T3 neuron is fooled.

D. Stimulus/background contrast

Schiff (1965) reported that frogs do not respond to a white expanding stimulus on a black background and Ingle (personal communication) observed that frogs are hit by a white looming object on a black background. We carried out a related simulation and observed that our model did not respond to a white looming stimulus on a black background.

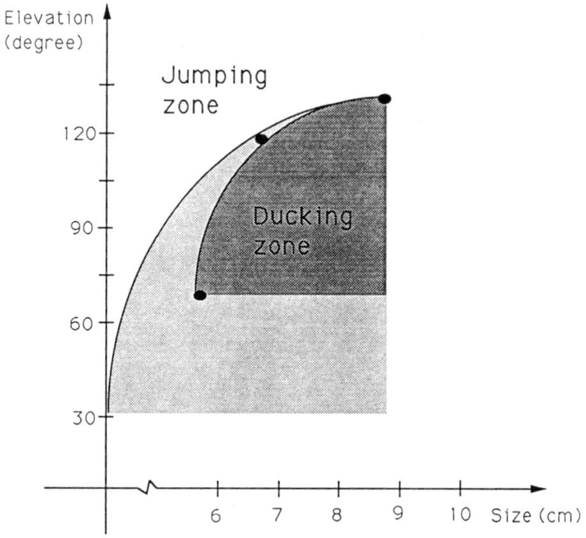

Figure 10: Motor schema selection. The darker grey area represents the ducking zone predicted by the computer simulation. This area is relatively small due to the crude resolution of the simulation. The lighter grey region is a postulated ducking zone.

E. Motor schema selection

There are no quantitative data regarding the stimulus parameters which discriminate the ducking and jumping responses. Hence, we carry out a series of simulations to determine the effect of varying the size and position of a looming stimulus on the choice of avoidance actions. Our simulation results show that the space formed by the size and elevation of a stimulus is divided into a ducking zone and a jumping zone (Fig. 10). The lighter grey area represents the postulated ducking zone whereas the darker area is drawn from our simulation result.

4. Discussion

In developing the predator-avoidance model, we have tried to make it as consistent as possible with known biological data. All the functional units employed have been physiologically identified, except the schema for depth perception. However, many of the details of the connections between different layers of neurons are unknown and, therefore, assumptions are made in developing the model. Several hypotheses are postulated for the underlying mechanisms of toad's detection of a looming stimulus, determination of direction to jump, and selection of proper avoidance actions. Some predictions are also made by our model.

1) The T3 neuron receives retinal input with weights that favor signals away from the center of its ERF (Fig. 3). This is a rather unique property in contrast to the commonly observed Gaussian distribution in the ERF of neurons.

2) We propose a mechanism to localize the position of a looming object which appears to expand in all directions. The position of a looming object is determined by localizing its focus of expansion (FOE) and is encoded in a population of T3 neurons where the more central neurons have higher activations. Different arrangements of the mask for T3 neurons are used in coding the position of a looming stimulus and the results are shown in Fig. 11. In this simulation, the radiating mask used in our model shows better performance.

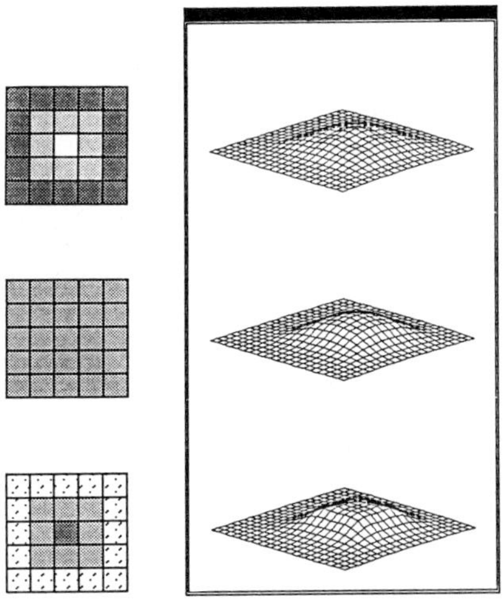

Figure 11: Various masks for T3 neurons. Different masks for T3 neurons are shown in the left column while the corresponding simulation results of the firing rate of a T3 neuron array are shown in the right. Note the more precise localization of the peak in the top row, and the splitting into "subpeaks" in the bottom row.

3) We postulate the existence of a depth perception system whose firing rate encodes the "closeness" of a stimulus, independent of its visual angle on the retina. We also assume that T3 neurons do not receive depth information and hence are not sensitive to the distance of a

stimulus, acting only as an expanding-edge detector. The depth system projects to TH6 neurons to make them more responsive to smaller, closer stimuli than to larger but further ones. This property is the major difference between these two z-axis movement sensitive neurons (T3 and TH6).

4) Cobas and Arbib (1990) postulated a motor heading map as a common interface between the spatially coded sensory map and the frequency-coded motor output for both prey-catching and predator-avoidance behavior. In our present model, we proposed a way for gating tectal signals onto the motor heading map by pretectal activity based on the presence or absence of a predator.

5) The origin of the triggering command for prey-catching and predator-avoidance behavior is still controversial (Ingle, 1983, Ewert, 1987). We postulate that T3, T6, TH3 and TH6 neurons all contribute to the selection of motor schemas, through competition between the ducking motor schema and the jumping motor schema which depends on the size and position of a looming stimulus (Fig. 6).

6) We predict that input from R4 ganglion cells signals the continuity inside the expanding edges. The continuity requirement in our model provides an interesting contrast to Ewert's finding that continuity of stimulus is not necessary for the recognition of prey or predator (Ewert & Traud, 1979). Schiff (1965), on the other hand, reported that frogs do not respond to discontinuous expanding surfaces. It would be interesting to present the above ambiguous looming stimulus to toads and compare the results.

7) We have reproduced, without tuning any parameters, the observed phenomena in which the avoidance activity increases with the size and speed of a looming stimulus (Fig. 8 and 9). The simulation results fit the experimental data well except for stimuli larger than 51°: our model shows saturated response rate in contrast to the experimental findings which show slight decrease.

8) We predict that the space formed by the size and elevation of a stimulus is divided into a ducking zone and a jumping zone (Fig. 10). Although the result obtained is crude, it shows that the higher a stimulus is, the more likely it is that a ducking response will be elicited. On the other hand, the larger a stimulus is, the more likely is it that a jumping response will be elicited.

Our present model does not account for Ingle's new finding (Ingle and Hoff, 1990) on the frog's ability to "cut back" behind a crossing stimulus. This requires the analysis of the direction of stimulus movement to determine if it is on a collision course.

5. Summary

The predator-avoidance behavior requires analysis of a complicated stimulus situation to determine the corresponding avoidance action. We developed a neural network model to account for the detection of a looming stimulus, the determination of the direction to jump and the selection of a proper avoidance movement. In our model, the retinal signal is first received and processed by tectal neurons and their outputs then converge onto pretectal neurons. Based on these converging inputs, neurons in pretectum are able to analyze the stimulus situation and determine an appropriate avoidance action.

Future research will address the analysis of more complicated stimulus situations and the integration of the corresponding responses with detour behavior when barriers are present along with predators. Many interesting and challenging issues raised by lesion experiments also remain to be addressed in our modeling. These include rewired retinotectal projection, and recovery following pretectal lesions. Their study requires a mechanism for synaptic plasticity and for remapping among populations of neurons.

References

Brown, W.T., and Ingle, D., 1973, Receptive field changes produced in frog thalamic units by lesions of the optic tectum, *Brain Res.* **59**: 405-409.

Cobas, A., and Arbib, M. A., 1990, Prey-Catching and Predator-Avoidance in Frog and Toad: Defining the Schemas, (in press).

Ewert, J.-P., 1971, Single unit response of the toad (*Bufo americanus*) caudal thalamus to visual objects, *Z. Verg. Physiol.* **74**:81-102.

Ewert, J.-P., 1984, Tectal mechanism that underlies prey-catching and avoidance behavior in toads, in: *Comparative Neurology of the Optic Tectum*, H. Vanegas, ed., Plenum Press, NY.

Ewert, J.-P., 1987, Neuroethology of releasing mechanisms: prey-catching in toads, *Behav. and Brain Sci.*, **10**: 337-405.

Ewert, J.-P., and Rehn, B., 1969, Quantitative Analyse der Reiz-Reaktionsbeziehungen bei visuellen Auslosen des Fluchtverhaltens der Wechselkrote (*Bufo bufo* Laur.), *Behaviour* **35**: 212-234.

Ewert, J.-P., and Seelen, W. v., 1974, Neurobiologie und System-Theorie eines visuellen Muster-Erkennungsmechanismus bei Kroten, *Kybernetik* **14**: 167-183.

Ewert, J.-P., and Wietersheim, A.v., 1974, Musterauswertung durch tectale und thalamus/ praetectale Nervennetze im visuellen System der Krote (*Bufo bufo* L.), *J. Comp. Physiol.* **92**: 131-148.

Ewert, J.-P., and Traud, R., 1979, Releasing stimuli for antipredator behavior in the common toad *Bufo bufo* (L.), *Behaviour* **68**: 170-180.

Grüsser, O.-J., and Grüsser-Cornhels, U., 1970, Die Neurophysiologie visuell gesteuerter Verhaltensweisen bei Anuren, *Verhandlungsbericht Dtsch. Zool. Ges.*, **64**: 201-218.

Grüsser, O.-J., and Grüsser-Cornhels, U., 1973, Neuronal mechanisms of visual movement perception and some psychophysical and behavioral correlations, in: *Handbook of Sensory Physiology*, R. Jung, ed., vol VII/3A, Springer-Verlag, NY.

Grüsser, O.-J., and Grüsser-Cornhels, U., 1976, Neurophysiology of the anuran visual system, in: *Frog Neurobiology*, R.Llinas and W. Precht, eds., Springer-Verlag, NY.

House, D., 1989, Depth Perception in Frogs and Toads: A Study in Neural Computing, in: *Lecture Notes in Biomathematics 80*, Springer-Verlag, Berlin.

Ingle, D., 1976, Spatial vision in anurans, in: *The Amphibian Visual System*, K.V. Fite, ed., Academic Press, NY.

Ingle, D., 1983, Brain mechanism of visual localization by frogs and toads, in: *Advances in Vertebrate Neuroethology*, J.-P Ewert, R. Capranica and D. Ingle, eds, Plenum Press, NY.

Ingle, D., and Hoff, K. vS., 1990, Visually Elicited Evasive Behavior in Frogs: Giving memory research an ethological context, *BioScience* Vol. 40, No. 4.

Lara, R., Arbib, M.A., and Cromarty, A.S., 1982, The role of the tectal column in facilitation of amphibian prey-catching behavior: a neural model, *J. Neurosci.* **2**: 521-530.

Schiff, W., 1965, Perception of impending collision, *Psychological Monographs*, **79**, No. 604.

Teeters, J.L., 1989, A Simulation System for Neural Networks and Model for the Anuran Retina, Technical Report 89-01, Center for Neural Engineering, Univ. of Southern California.

Tsai, H.J., and Ewert, J.-P., 1987, Edge preference of retinal and tectal neurons in common toads (*Bufo bufo*) in response to worm-like moving stripes: the question of behaviorally relevant 'position indicators,' *J. Comp. Physiol. A*, **161**: 295-304.

Wang , D.L. and Arbib, M.A., 1990, How does the toad's visual system discriminate different worm-like stimuli? *Biol. Cybern.* (in press).

Weitzenfeld, A., 1989, NSL, Neural Simulation Language, Technical Report 89-02, Center for Neural Engineering, Univ. of Southern California.

Control of Frog Evasive Direction: Triggering and Biasing Systems

David Ingle
Northeastern University
Marine Science Center
East Point
Nahant, MA 01908

Abstract

Midbrain systems for triggering either feeding or evasive behaviors will be reviewed. These systems interact with modulating or "biasing" systems in telencephalon or diencephalon in determining the final choice of directions taken in either feeding or evasive sequences. Some biasing effects seem to act directly upon the tectum (e.g., those from pretectum and from n. isthmi) while others (including pretectum and striatum) may converge in the brainstem with tectal efferents to guide orientation. New data on the role of cerebellum in biasing the direction of avoidance behavior on a tilted surface leads to the suggestion that both cerebellum and striatum bias avoidance directions via a common biasing center in the tegmentum. Effects of tegmental lesions support this model.

Dual Avoidance Strategies

It is common knowledge among zoologists and small boys that frogs respond to the approach of a would-be predator or captor by turning and leaping away. In nature, frogs such as *Rana pipiens* or *Rana climitans* are strongly biased to jump towards water. This is usually the home pond, but frogs along a mountain trail will jump unerringly into mud puddles and disappear into the mud. Escaped frogs in the laboratory will usually leap and dive under a cabinet or into a crack between furniture: a deliberate "hole-seeking" strategy which is easily demonstrated in a visually controlled test arena (Ingle, 1983). However, in a large empty white arena (free from obstacles or hiding places) the frog's direction is mainly determined by the angle of approach of a threatening dark object (Ingle, 1976). The direction of escape elicited by a stimulus looming directly at the frog on a collision course is a compromise between leaping straight ahead to obtain maximum distance, and turning directly away from the threat. Since the initiation of escape jumps depends upon an intact optic tectum (but not upon any rostralwards thalamic or telencephalic structure) we propose that the well-known topography of the retinotectal projections mediates the mapping of escape directions depending upon

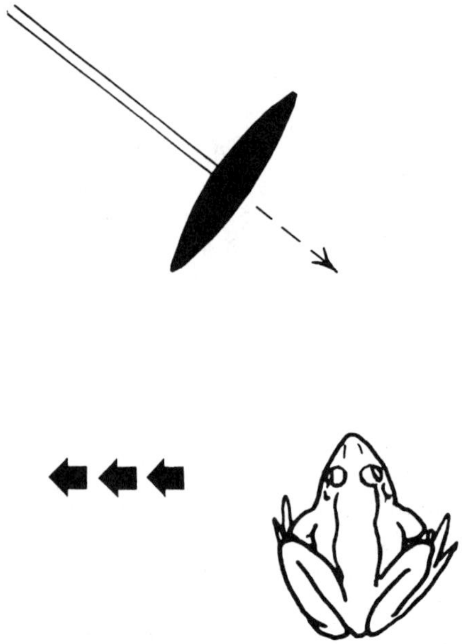

Figure 1. The trajectory of a crossing black disk typically elicits an avoidance leap into the hemifield from which the threat originates.

the region of tectum stimulated by looming threat. However, this tectal efferent system differs from that by which prey-catching directions are mapped out. While severing the crossed tectal projections in ventral tegmentum abolishes turning towards prey, it has no obvious effect on escape behavior.

We have repeatedly confirmed that the ability of frogs to turn consistently away from looming threat depends upon an ipsilateral tectum to brainstem projection, presumably the large bundle projecting from tectum to the medulla. Yet, most of these "hemisected" frogs could still escape by jumping forwards, or by turning in the general direction of the threat. We wondered what might be the natural function of a system which turned the animal *towards* approaching threat, and we soon discovered that such ipsiversive turns are normally elicited by a threat which approaches obliquely within the frontal field, on a trajectory which crosses the rostral midline (Fig. 1). Film analysis revealed that frogs usually initiate ipsiversive jumps when the *crossing* stimuli are still on the same side of the midline where they first appeared, and at locations which overlap with those of *looming* stimuli which elicit contraversive jumps. We conclude that the location of threat as it triggers escape does not by itself determine evasive direction, but that the stimulus trajectory (where it is headed) is of critical importance in the frog's choice between contraversive and ipsiversive jumps.

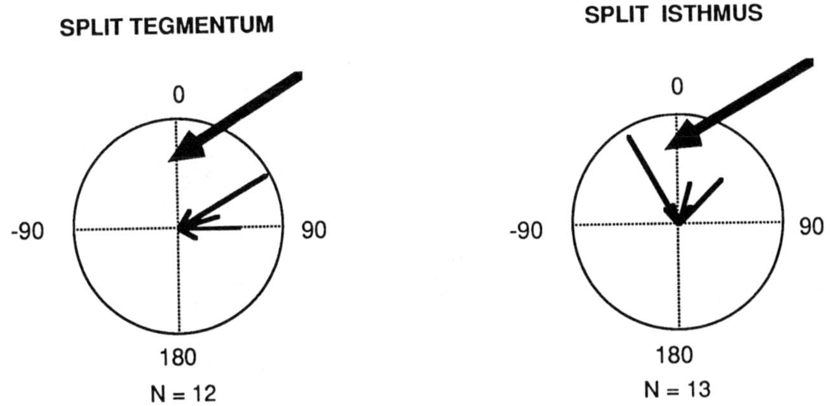

Figure 2. Frogs with tegmentum split jump ipsiversively as do normals in response to crossing threat. This tendency is lost after splitting the isthmus.

Dual Efferent Mechanisms for Avoidance

We have noted that interruption of ipsilateral tectal efferents (including non-tectal fibers) abolishes the typical contraversive escape pattern. We next discovered that splitting the midline at the isthmus level (below the cerebellum and caudal to the ansulate commissure) would abolish any consistent tendency to turn ipsilaterally in response to crossing stimuli (Fig. 2). Indeed, many of these split-isthmus frogs made no ipsiversive jumps at all, but leapt into the path of the crossing stimulus and sometimes collided with it. Such maladaptive behavior shows that the "cut-back" manoeuvre is as useful for a frog as it is for a football carrier evading an approaching cornerback.

How might tectal efferent cells access the opposite brainstem, in order to guide the cut-back turn? Since there are no direct projections crossing the midline at that level, we presume that the tectum contacts one of the three large tegmental cell groups (anterodorsal, anteroventral or posteroventral) whose fibers cross the isthmic decussations. In pilot studies we found that injections of the cell-toxin, ibotenic acid, into the posteroventral region at one side abolished cut-backs unilaterally, as our model predicts. An unconfirmed prediction from this model is that cells within the critical relay nucleus (e.g. PV) would demonstrate greater sensitivity to crossing threat stimuli than to colliding stimuli.

Figure 3. The distribution of avoidance directions elicited by lateral threat destined to hit the frog (H) is mainly within the opposite frontal field while caudalward (C) moving threat produces a similar range of avoidance directions. However, threat moving rostrally (R) mainly elicits jumps into the read field quadrants.

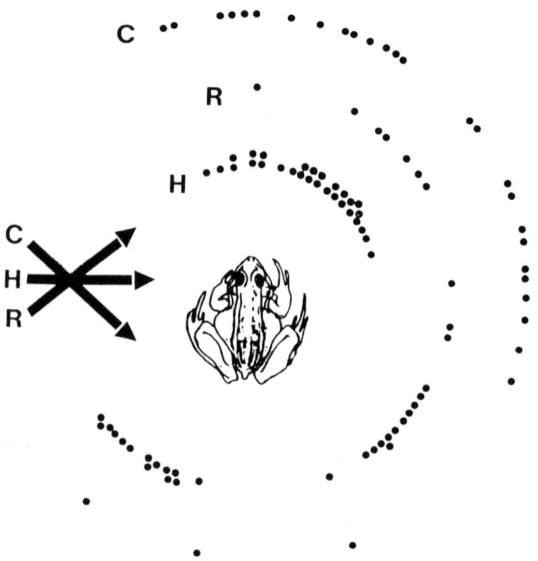

Note that the stimuli designated as "crossing" contain two motion vectors: looming + forward motion. It is likely to be more than coincidence that Grüsser & Grüsser-Cornhels (1976) described "large field" tectal cells as including both "loom-detectors" and "forward-motion detectors" but did not notice a population of caudalward motion-detectors. We tested the idea that forwards motion had a special significance for the frog's avoidance system by comparing escape directions in response to lateral-field threat approaching the frog either (a) directly, (b) obliquely forwards, of (c) obliquely backwards. These results, summarized in Fig. 3 show that frogs jump mainly forwards when threatened by either caudalward or looming stimuli (i.e. they do not discriminate these trajectories). Yet, they jump mainly into one of the rear quadrants in response to rostralwards threat. The caudal-ipsiversive turns can be accounted for by a retinotopic mapping of crossed tectal output cells sensitive to forwards-motion. We suggest that the caudal-contraversive jumps could reflect a vector interaction between the contribution of crossing efferents and the forward tendency elicited by L-cells. It remains to explain why, on some trials, the L-cells projecting ipsilaterally (Fig. 4) influence the jump direction (i.e., contralateral) while on other trials the F-cells dominate the decision to turn ipsilaterally. Does "vector interaction" result from the independent outputs of the two sides of the brainstem, or from an integration of the two "turn generators" via commissural interactions at the medullary level? Such questions may be approached by detailed study of frogs with (a) the medullary-spinal junction hemi-

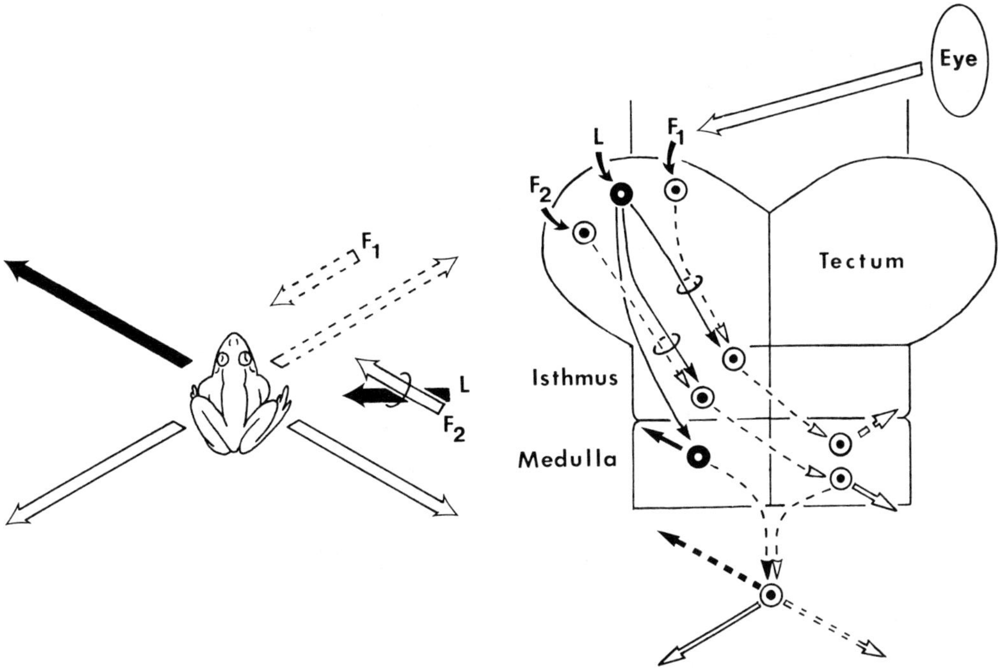

Figure 4. The main avoidance directions shown in Fig. 3 are schematically depicted on the left side together with effects of crossing stimuli (F2). The dark arrows (L) refer to looming threat, while F arrows refer to forward moving threat. The two F stimuli activate different tectal regions which map onto regions of a crossed isthmo-medullary pathway, and there activate different turn amplitudes via the right medulla. Each pathway also receives an input from loom-sensitive tectal cells (L). The frequent occurrence of turns to the lower left in response to F2 stimuli is here explained as a vector interaction of the crossing F2 output and the ipsilateral L-cell output (both dashed arrows).

sected, or (b) the medulla split down the midline. Also useful in sorting out loci of F-cell interaction will be unit recordings in tectum, tegmentum and medulla in order to see at which stage in the visuomotor pathway the precise trajectory of threat is encoded.

Avoidance Direction is Strongly Modified by Location of Obstacles

In an earlier review, Ingle (1976) noted that an opaque obstacle placed within the frog's jump path will divert escape jumps in either direction past the object. We assume that the frog's pretectum is involved in barrier modification of escape routes since it is critical for detours around barriers during escape from a tail-pinch (Ingle, 1980) and for barrier-detours to approach prey (Ingle, 1983). Our further unpublished studies revealed that the frog's ability to detour past a lateral barrier-edge on the side opposite an approaching threat (Fig. 5A) was abolished by pretectal lesion opposite to the critical barrier-edge.

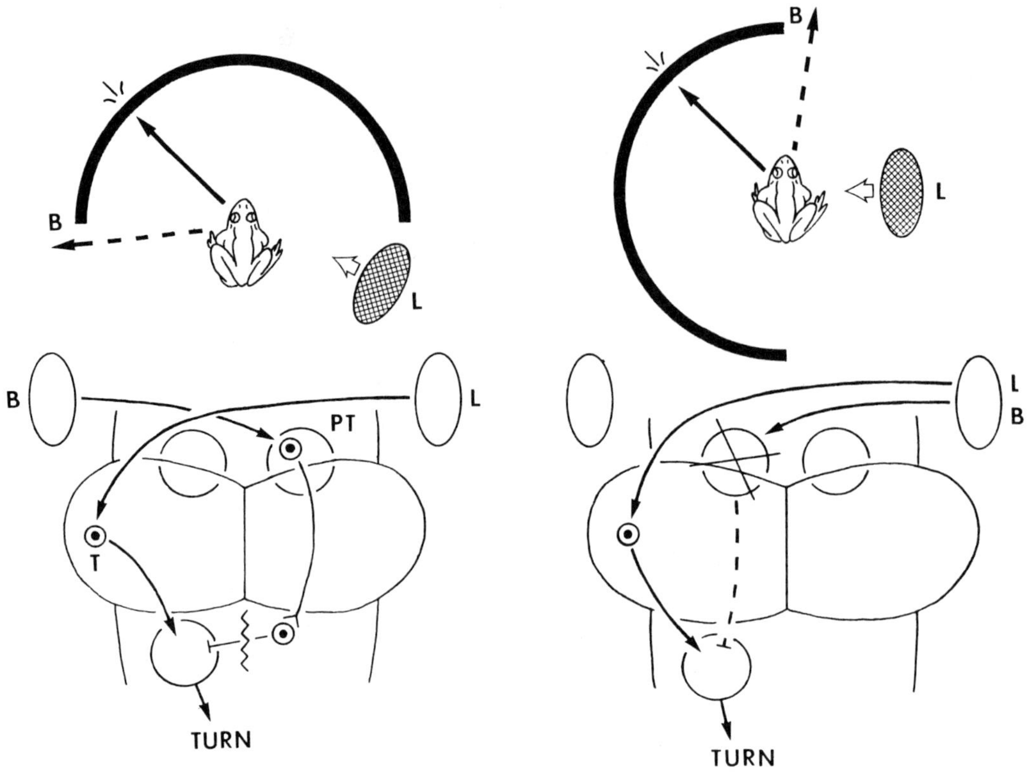

Figure 5. The left side shows the effect of splitting the isthmus on the frog's ability to avoid a solid barrier while escaping from a looming disk (L) in the rear right field. Instead of clearing the barrier edge (B) as a normal frog would do (dashed line) these frogs jump as if no barrier were seen and collide with it on each test. We explain this deficit as a disconnection of a pretectofugal inhibitory system (PT) which crosses in the isthmus from the ipsilateral descending tectofugal system (T). The right shows a task which the split-isthmus frog can still perform, in which the looming stimulus and the barrier-edge are seen via the same eye. This result implies that pretectum also modulates the tectofugal system via ipsilateral descending projections. Ablation of the left pretectum produces the expected deficit in avoiding this barrier location, although the same frog would perform normally were the looming threat (L) and barrier locations reversed.

The frog's ability to circumvent lateral barrier edges during prey catching is also abolished by splitting the isthmus region (Ingle, 1983), probably because pretectum connects to a tegmental cell group with decussating fibers. In any case, I have proposed (Ingle, 1983) that this crossing efferent system derived from pretectum converges in the medulla with crossed tectal efferents which control turn amplitude. This model may work for threat-elicited detours as well because splitting the isthmus abolishes the frog's ability to avoid a left-side barrier while escaping from right side threat (Fig. 5 Left). However, since the same frog can still avoid a barrier-edge seen on the same side as that of looming threat (Fig. 5 Right), it can presumably use ipsilateral pretectal efferent pathways descending to the medulla. While Ingle (1983) has described this ipsilateral pretectofugal bundle as descending along a similar route as the ipsilateral tectofugal system into anterior

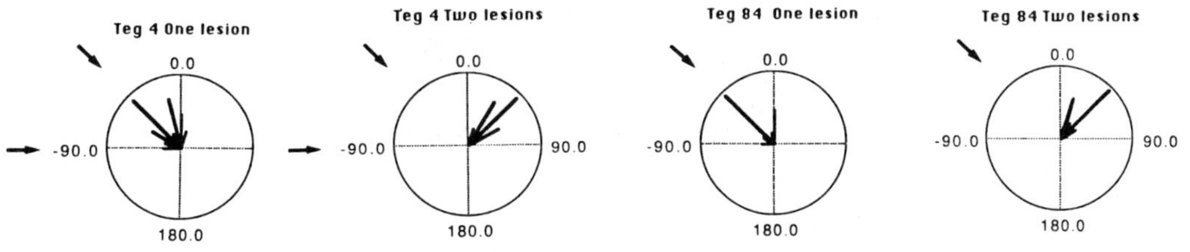

Figure 6. The effect of a second symmetrical tegmental lesion in reversing the maladaptive leftwards turning tendency produced by a right-side tegmental lesion. Not shown is the ability of the double lesioned frog to jump leftwards to avoid a right-side threat.

medulla, nothing is known concerning the precise terminations of either ipsilateral or contralateral output systems. Here is a problem where unit recordings of loom-sensitive cells in the medulla can supplement anatomy.

Avoidance Directions are also Biased by Tilting the Substrate

Another way that we can readily bias the frog's choice of escape directions from a threatening stimulus is to tilt the floor of the test arena by 10° in either forwards or sidewards directions. Fig. 6 shows that a frog tilted downhill will avoid a threat moving uphill mainly by lateral jumps, while frogs tilted uphill evade a threat moving downwards toward them by turning and jumping downhill. Clearly the frog avoids making short uphill jumps and favors long downhill leaps which better evade the predator. When tilted 10° to one side, the frog jumps downhill on 75-80% of trials. Frogs tested on tilted panes of glass where the visible floor was flat showed the full biasing effect, guided presumably by proprioception. On the other hand, frogs sitting on a flat glass with the visible floor tilted by 10 degrees showed symmetry in avoiding frontal threat. Finally we found that frogs tested for several months after removal of the cerebellum lost and never recovered the biasing effect of tilting the floor, although their avoidance vigor remained high. It is of great interest in light of these results, that the deep nucleus of the frog's cerebellum projects rostrally into the ventral tegmentum (Montgomery, 1988), an area which can strongly bias a frog's preferred escape direction, as we shall next review.

The Tegmental Biasing System

About two years ago we attempted to test the proposal of Grobstein (1988) that tegmental nuclei played a critical role in programming the amplitude of turning during the frog's prey-orienting response. We first made large unilateral suction ablations in the area of the nucleus ruber, and found that while these lesions always caused the frog to tilt conspicuously to one side, they produced neither inaccuracy of

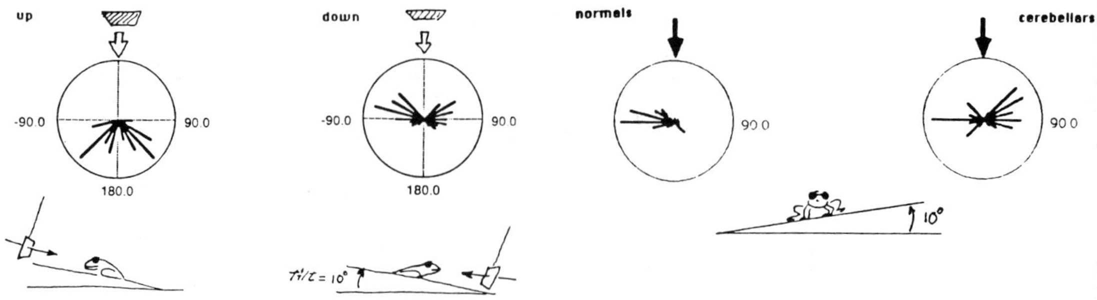

Figure 7. On the left are shown the tendency of frogs tilted upwards to escape a frontal threat by turning into the rear quadrants, and the tendency of a downward tilted frog to escape into rostral quadrants. On the right are shown responses of normal and cerebellum-ablated frogs to frontal threat while sitting on a surface tilted sidewards by 10°. Each of these four histograms represents 100 jumps.

prey-catching nor distortion of evasive behavior. However, when unilateral ablations invaded the posterior tegmentum, we noticed gross modifications of evasive behavior, even in absence of postural distortions. After a unilateral lesion which invaded the AV cell group, frogs seldom were able to turn away from contralateral threat but instead jumped ipsilaterally in the general direction of approaching threat. This wrong-way turning resembled that of frogs with post-tectal hemisections but was more adaptive in that these animals seldom collided with the looming disks.

Unlike the hemisected animals, unitegmental lesions produced an equally strong tendency to escape *into* barriers placed in the field contralateral to the lesioned side. In fact, some frogs jumped *farther* into the wrong field (hitting the barrier) when the barrier was present than when it was absent. Thus it seemed that the sight of the barrier induced a strong effort to jump to one side of the midline, but for these lesioned frogs the added effort was applied in the wrong direction.

Two other remarkable features of this side-bias effect need be added to make sense out of it. First, despite the persisting tendency to avoid threat (visual or tactile) by jumping away from the lesion-side, these frogs could turn and strike accurately at prey. Even when compulsively jumping 45° to one side in escaping threat or a tail pinch, they could accurately catch prey anywhere within the visual fields. We conclude that the biasing effect was not due to an overt postural asymmetry nor did it apply to prey-catching movement; in our tests it only distorted avoidance turns. Secondly, despite the persistence of such distorted avoidance jumping for 3-4 months post surgery, we "cured" the deficit in each of 5 frogs subjected to a second symmetrical tegmental ablation. Although these frogs were less vigorous after the double lesion, their jumps and body orientation were now normally directed away from the threat approaching on either side (Fig. 7). Similarly barrier avoidance was now carried out efficiently clearing the barrier-edges up to 90° on either side. Such normalized avoidance behavior was seen as soon as the animals had recovered from anesthetic.

We conclude that the ablated regions (including the AV zone) are not necessary for achieving normal avoidance directions away from looming threat, or around frontal barriers. Yet following a unilateral removal of the AV region, the opposite intact region has a powerful biasing effect on subsequent avoidance behavior. One natural use of this biasing mechanism may be the tilt effect, where unilateral activation of the AV region by the cerebellum follows a postural asymmetry. However, effects of forward or backward tilting (Fig. 6) must involve more subtle modulation of tegmentum than that produced by asymmetry. It is possible that this tegmental zone harbors a rostrocaudal "motor map" whose descending influence on a medullary turning center can either enlarge or decrease the amplitude of avoidance turns. The suggestion of a tegmental motor map was also made above in our discussion of the mechanism underlying ipsiversive turns (Fig. 4). A further use of the tegmental biasing center will be proposed in the second article ("The Striatum and Short-Term Spatial Memory: From Frog to Man", this volume), which describes the role of visual memory in biasing avoidance behavior.

References

Grobstein, P., 1988, Between the retinotectal projection and directed movements: topography of a sensorimotor interface. *Brain Behav. Evol.* 31: 34-48.

Grüsser, O.J., and Grüsser-Cornhels, U., 1976, Neurophysiology of the anuran visual system. In: R.Llinás and W.Precht, eds. *Frog Neurobiology*, Springer-Verlag, New York, pp. 297-385.

Ingle, D., 1976, Spatial vision in anurans. In: K.Fite, ed. *The Amphibian Visual System*, Academic Press, New York, pp. 119-141.

Ingle, D., 1980, Some effects of pretectum lesions on the frog's detection of stationary objects. *Behav.. Brain Res.* 1: 139-163.

Ingle, D., 1983, Brain mechanisms of localization in frogs and toads. In: J.P. Ewert, RR.Capranica and D.Ingle, eds. *Advances in Vertebrate Neuroethology*. Plenum Press, New York, pp.177-226.

Montgomery, N., 1988, Projections of the vestibular and cerebellar nuclei in *Rana pipiens*. *Brain Behav. Evol.* 31: 82-95.

Approach and Avoidance Systems in the Rat

PAUL DEAN and PETER REDGRAVE

Department of Psychology
University of Sheffield
Sheffield S10 2TN, England

Abstract. Current models of the mammalian superior colliculus focus on its role in the production of saccadic eye movement. In these models a single command, related to desired change in eye position, is passed down a single output channel to the brainstem, where it is transformed into an appropriate signal for the oculomotor neurons. In rodents, however, both anatomical and behavioural findings point to the existence of multiple functional output channels. These appear able to command a wide range of responses appropriate to an unexpected sensory stimulus, including not only orienting and approach, but a set of defensive reactions that vary from freezing to rapid flight, together with appropriate physiological changes in cortical EEG, blood pressure and heart rate. It may therefore be possible to use the mammalian superior colliculus as a preparation for studying and modelling a number of basic issues in sensorimotor control. These new models need not be concerned only with avoidance or defence. For example, in models of saccade generation, the trajectory of the movement is formed downstream of the superior colliculus: in rodents, however, it appears that the colliculus can control head-movement trajectory directly. Data concerning this collicular control system might therefore be of use in testing models of how the brain computes movement trajectories.

Introduction

Since the purpose of the workshop is to relate models and experimental data, this chapter is organized as follows. First, a brief look at current models of the mammalian superior colliculus. Second, an account of recent experimental work on the rodent superior colliculus, indicating its role in both approach and avoidance behaviours. Finally, some speculations concerning how current models of the superior colliculus might be refined by these empirical discoveries.

1 Current Modelling of Mammalian Superior Colliculus

To date, models involving the mammalian superior colliculus have focused on its role in the production of saccadic eye-movements (e.g. Grossberg and Kuperstein, 1989: Sparks and Mays, 1990: Van Gisbergen and Van Opstal 1989). The function of these movements is to change the direction of gaze from one position to another as rapidly as possible. The basic ingredients of models of saccade-production are shown in Fig.1

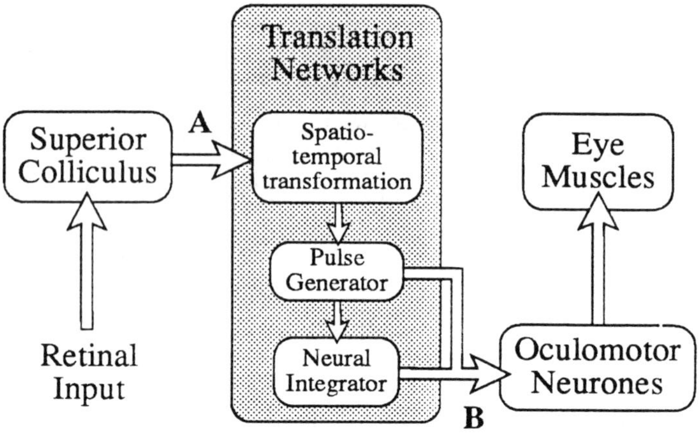

Figure 1 Elements common to most models of primate saccadic eye-movements. **A.** Collicular output specifies desired change in eye position coded spatially in terms of the centre of gravity of activity within the superior colliculus. **B.** Summed output of the translation networks provides a temporally coded train of impulses to the oculomotor neurones.

The superior colliculus issues a command signal which corresponds to something like desired change in eye-position. This command is coded spatially, by the location of active cells within the collicular output array. Neuronal machinery between the superior colliculus and the oculomotor neurons translates the collicular command into language the latter can understand. This is generally thought to mean (i) transforming the position signal from a spatial to a temporal code; (ii) providing the appropriate velocity pulse; and (iii) integrating the pulse to give a tonic position command, which holds the eyes in place after the phasic 'change-position' command is finished.

The key feature of these models, as far as the superior colliculus is concerned, is that in general there is only one important anatomical link between it and the brainstem (where the rest of the saccadic machinery is mainly located), and only one important collicular command signal.

2 Rodent Approach and Avoidance Systems

Recent work on the functional anatomy of the superior colliculus in rodents suggests that the scope for fruitful modelling of collicular-based sensorimotor processing need not be restricted to the production of saccadic movements. This work is described here in three sections:

(i) Anatomical findings concerning collicular output channels.

(ii) Behavioural findings concerning responses mediated by the superior colliculus.

(iii) Attempts to relate (ii) and (i).

2.1 Anatomy of Collicular Output Channels in Rat

Injections of the orthograde tracer WGA-HRP (wheatgerm agglutinin conjugated with horseradish peroxidase) into the rat superior colliculus reveal massive projections streaming to diencephalon, midbrain, pons and sites further caudal in brainstem and spinal cord. We have concentrated on the descending projections (Redgrave et al. 1987a) which are shown in schematic form in Fig.2

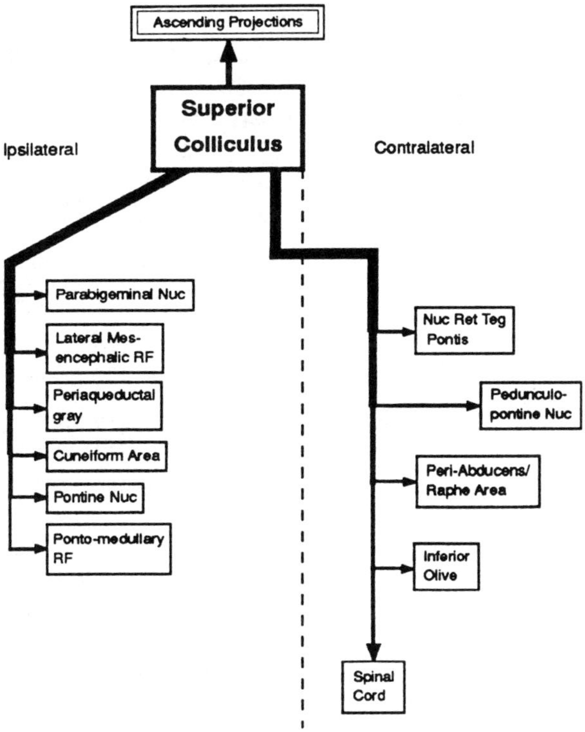

Figure 2 Schematic representation of the major descending projections of the mammalian superior colliculus. Nuc = nucleus; RF = reticular formation; Ret teg = reticularis tegmentis.

These are split into two main bundles.
(1)An ipsilateral projection that travels caudally and laterally to innervate target areas in the midbrain and lateral pons.
(2) A contralateral projection that travels ventrally and medially, crosses midline in the dorsal tegmental decussation, then turns caudally to run in the predorsal bundle. From there it innervates a succession of targets mainly in medial pons and medulla, and eventually reaches the cervical spinal cord (and is therefore sometimes referred to as the tectospinal or tecto-reticulo-spinal tract).

From the point of view of the models outlined in Section 1, the important feature of these projections is the number of targets that they innervate. There seem to be at least 10 distinguishable target areas. Some of these targets are known to be directly involved in the production of head and eye movements (e.g. medial pontomedullary reticular formation, periabducens area); but others are precerebellar nuclei (dorsolateral pons, nucleus tegmentis reticularis pontis, inferior olive) whose precise functions are the topic of intense debate; and yet others have been so little studied that debate about their function has scarcely arisen (cuneiform area, caudolateral midbrain reticular formation). Nor is it the case that collicular descending projections in the rat are wildly aberrant - rather the reverse, in that the pattern of projections we found in rat was already well known from studies in other mammals including cat and monkey (Huerta and Harting 1984).

Maybe, though, collicular outputs are functionally homogeneous, and distribute their single message to many different parts of the brain via axon collaterals. We investigated this possibility by making large injections of

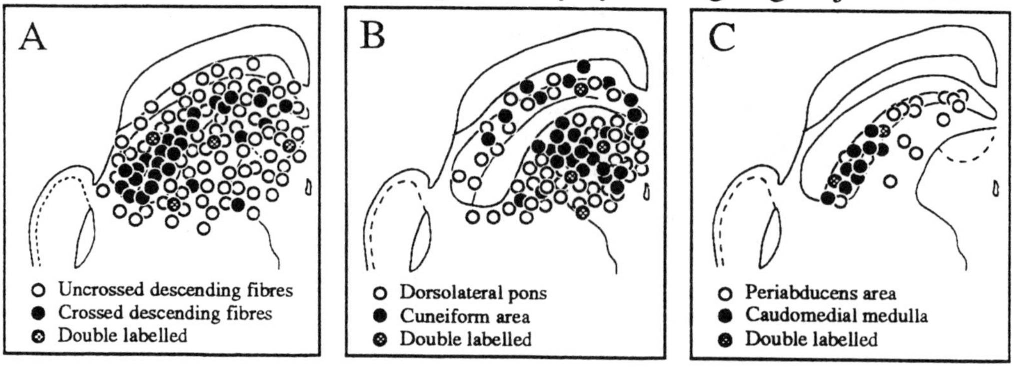

Figure 3 Schematic representation of retrogradely labelled cells from three experiments in which a double labelling technique was used to investigate the organization of tectal outputs. **A.** Combined injections of fluorescent tracers were made into contralateral predorsal bundle fibres and ipsilateral terminal zones at the level of the rostral pons. **B.** Combined ipsilateral injections of tracers were made into the dorsolateral pontine nuclei and the cuneiform area. **C.** Combined contralateral injections of tracers were made into the periabducens area and the ventromedial caudal medulla. In all cases fewer than 10% of labelled cells contained both tracers and cells singly labelled with different dyes were often regionally segregated.

one fluorescent tracer (True Blue) into the fibres of the crossed descending projection, and even larger injections of another (Diamidino Yellow) into the terminal areas of the ipsilateral descending pathway (Redgrave et al. 1986).

These substances are transported backwards down axons to the parent cell body, and after the injections thousands of labelled cells were visible in the superior colliculus. However, almost all these cells were labelled with a single tracer: fewer than 10% of counted cells were labelled with both (Fig 3A).

It therefore seems unlikely that the terminal areas of the uncrossed and crossed projections receive identical messages from the superior colliculus. Moreover, a similar fragmentation appears to occur *within* each of the major descending pathways (Fig 3B, C). The ipsilateral projections to dorsolateral pons and the cuneiform area arise from separate cells (Redgrave et al. 1987b) as do the contralateral projections to the periabducens area and caudal medulla/spinal cord (Redgrave et al. 1990; Keay et al. 1990b). There are at least four separate channels from superior colliculus to brainstem in rat, and possible several more: the organisation of the remaining projections has yet to be investigated.

These anatomical findings raise the possibility that the models of section 1, insofar as they aim to represent a full account of collicular outputs, are seriously incomplete. Again it is worth saying that multiple output channels are not a peculiarity of the rat. Results from experiments using orthograde transport, or retrograde transport of single label, suggest that a similar arrangement is present in other animals, including garter snake, squirrel, cat and monkey (e.g. Edwards, 1980; Holcombe and Hall, 1981a,b; Huerta and Harting, 1984; Dacey and Ulinski, 1986; Moschovakis et al. 1988; May and Porter, 1989).

2.2 Responses Mediated by Rat Superior Colliculus

What do these multiple output channels do? Surely they cannot all be involved in the production of saccades? Early hints came from studies of the effects of lesions of the superior colliculus. For example, Goodale and Murison (1975) showed that rats with such lesions failed to react in any way to novel stimuli that produced a wide range of behaviour in normal animals. This included interrupting what they were already doing (which was running towards a target); orienting, rearing and investigating; and freezing, retreat and flight. Analysis of this and other studies suggested the hypothesis that the superior colliculus was part of the circuitry for many of the reactions that common sense indicates a transient unexpected visual stimulus can evoke (Dean and Redgrave 1984).

One source of support for this position comes from dramatic demonstrations of the effects of collicular lesions in particular situations. Thus, wild rats with damage to the visual collicular layers will allow people to approach within touching distance (Blanchard et al. 1981: for related effects in hamsters and gerbils see Merker 1980; Ellard and Goodale 1988). Laboratory rats with similar lesions cannot be woken, either behaviourally or as assessed by the cortical EEG, by even very intense light flashes (Dean et al. 1984). Nor will such animals interrupt what they are doing to approach a suddenly appearing peripheral light, although extensively trained before operation to do so (Overton et al. 1985).

A second source of support is provided by stimulation studies. Stimulation of the rodent superior colliculus can produce contralaterally-directed movements of the head and eyes resembling naturally occurring orienting or approach. But it can produce other kinds of response as well. For example, we serendipitously discovered that injections of the GABA-blocker picrotoxin into the superior colliculus can cause laboratory rats to treat sensory stimuli (eg fingers) that they normally explore as if they were highly threatening; as the drug takes effects, active orienting and sniffing are replaced by immobility, retreat and eventually violent flight (Redgrave et al. 1981). The production of defensive-like responses by appropriate activation of the superior colliculus has subsequently been confirmed in number of laboratories, with electrical and chemical stimulation, using rats, hamsters and gerbils (for references see Dean et al. 1989).

The range of responses produced can be thought of in relation to naturally occurring threats (Fig 4).

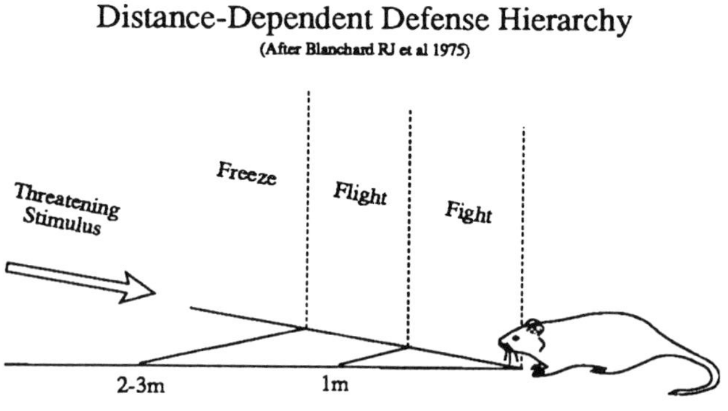

Figure 4 An example of the way in which defensive strategies of the wild rat can be affected by the imminence of visual threat the the form of an experimenter (Blanchard et al 1975).

Remote threats produce freezing (often accompanied by orienting); as the threat becomes more immediate, freezing is replaced by flight, if an escape route is available, or attack if it is not; and finally, if all else fails, the animal 'plays dead' (tonic immobility) (Blanchard et al. 1986; Rodgers and Randall 1987). Moreover, in natural circumstances these response patterns are accompanied by appropriate physiological responses, and so too can stimulation of the rat superior colliculus produce changes in cortical EEG, changes in blood pressure and heart rate, and analgesia (Dean et al. 1989).

2.3 Relating Behaviour to Anatomy

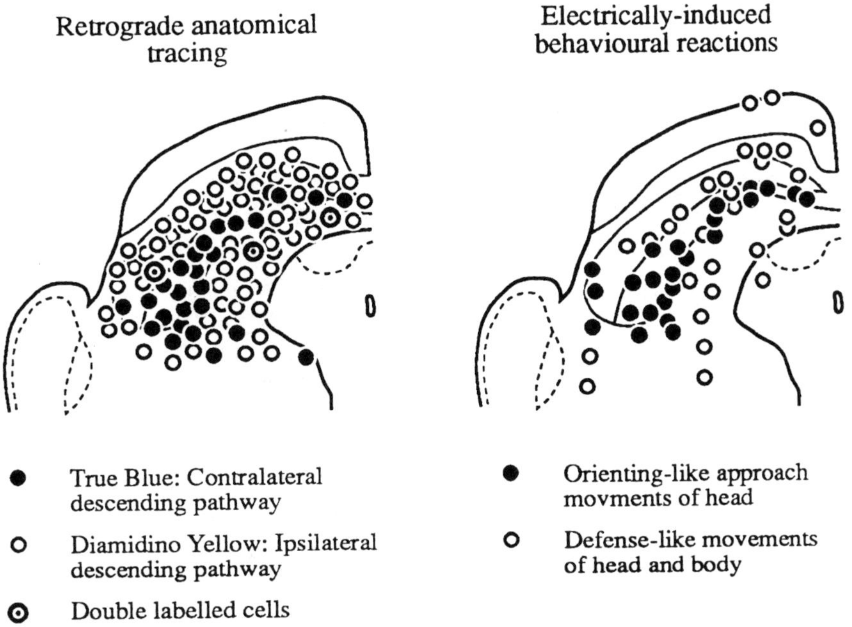

Figure 5 Schematic illustrations of the regional distribution of (**A**) tectal cells retrogradely labelled after combined injections into the ipsilateral and contralateral descending pathways (Redgrave et al 1987) ; and (**B**) electrical stimulation sites from which approach-like and defense-like movements were elicited at threshold (Sahibzada et al 1986).

It is an obvious inference that the multiple anatomical outputs from the rodent superior colliculus mediate the various responses to unexpected stimuli with which the colliculus is involved. We and others have attempted to test this idea using three main techniques. One is to stimulate the superior colliculus electrically or chemically in many different locations, and compare the resultant map with the anatomical distribution of cells of origin.
of particular output pathways (Redgrave and Dean 1985; Sahibzada et al 1986; Keay et al. 1988, 1990; Dean et al. 1988a, b). An example is shown in Fig 5. The second technique is to look at the behavioural effects of stimulating or interrupting the output pathway itself (Dean et al. 1986, 1988; Ellard and Goodale 1986, 1988; Mitchell et al. 1988; Tehovnik and Yeomans 1986). Finally, it is possible to record electrophysiologically from cells identified as contributing to a given output projection (Keay et al. 1990, Westby et al. 1990).

At present, the following main conclusions have been drawn from such experiments.

(i) *Avoidance System.* Defence-like responses appear to be mediated, at least in part, by the ipsilateral projection to the cuneiform nucleus, adjacent tegmentum and periaqueductal grey. Some of the cells that project to the cuneiform area respond to moving visual stimuli, in some cases apparently selectively to movement towards the animal (loom). There is probably response segregation within this general area: stimulation close to the cuneiform nucleus produces either freezing or flight, whereas sites further rostrolateral produce jumping and biting.

(ii) *Approach System.* Responses resembling orienting, approach and pursuit appear to be mediated by multiple output pathways. For example, section of the crossed descending pathway does not eliminate such movements, though it affects them severely. Moreover, within the crossed descending pathway, cells projecting to the periabducens area respond primarily to auditory stimuli, whereas cells projecting to the caudal medulla and spinal cord respond mainly to stimulation of the vibrissae. The functional significance of this multiple representation is not completely clear: however, there seem to be two distinguishable kinds of contralateral head movement produced by collicular stimulation, one of which resembles saccadic movements, and the other of which is clearly non-saccadic. The non-saccadic system is described further in Section 3.1.

Thus, as far as current evidence goes, the commonsense conjecture that anatomically distinct output pathways from the rodent superior colliculus carry functionally distinct messages seems to be supported. In addition, there is some suggestion that different output channels may receive particular subsets of the sensory input that reaches the superior colliculus (see also McHaffie et al. 1989). It has been argued that such an organisation for the superior colliculus would be consistent with the "mosaic" structure of the colliculus - some inputs to the intermediate and deep layers of the superior colliculus are distributed very patchily, as are certain histochemical markers - and with the marked heterogeneity of electrophysiological responses of single units in these layers (Dean et al 1989). Although the details of this organisation are only beginning to be explored, it is nonetheless worth asking what its implications are for future models of collicular function.

3 Implications for Modelling

The experimental data indicate that the rodent superior colliculus can send a number of different command signals to the brainstem, so the restricted models represented generically in Fig.1 (which were devised primarily to explain the production of saccadic eye movements in primates) clearly require expansion. This, however, raises the issue of how important it is to have a more accurate model of the rodent superior colliculus.

3.1 Modelling Structures vs. Processes

The problem with modelling specific brain structures in particular species is that there are many brain structures and many species. Consequently some measure of interest or importance is needed to decide which to model (or at least which to model first). This is an extremely complex issue (cf Arbib 1987 1989a), but one particular worry surfaces clearly in relation to the present topic: that the collicular organisation described here might be a rodent eccentricity, so that attempts to model it would tell us only about the little tricks that rats use, for example, to keep out of the clutches of cats. Rather than explore whether the rat superior colliculus has cat-detector units it might be preferable to study problems of more direct relevance to human behaviour, such as how the primate superior colliculus controls saccadic eye movements. This kind of worry can easily lead to primate envy.

In fact there are reasons for suspecting that collicular organization described here is not a rodent eccentricity, but is instead common to many mammals (Dean et al. 1989). For example, the major anatomical features described in Section 2.1, and the mosaic-like distribution of afferent input and of histochemical markers, are found in a number of species (e.g. Illing and Graybiel 1986; Wallace 1988). However, here we wish to pursue a separate point, which concerns the difference between modelling particular processes as opposed to particular structures.

In process-centred modelling the focus is on understanding the principles that inform the component processes of intelligent behaviour. This seems in fact to have been a major reason for investigating and modelling the saccadic system (Fig 1). It has been argued that eye-movement control systems function "at a level of complexity somewhere between the banality of the spinal reflex and the inscrutability of the voluntary act" and that it is "now becoming apparent that many of what were formerly thought to be unique properties of the eye movement control system have direct parallels even in a system as complex as the control of the hand" (Carpenter 1988). Eye-movement systems are thus simple enough to be soluble, but also interesting in that they embody principles which help to explain more complex examples of sensorimotor control.

From this perspective, the discovery that the rodent superior colliculus does more than produce saccades may be a welcome one. It may be useful as a preparation to study a wider range of individual processes (section 3.2) and also how integrated control of such processes might be implemented (section 3.3).

3.2 Individual Processes

The data reviewed in Section 2 suggest it will be possible to use the rodent superior colliculus to investigate control of particular components of defensive reactions, both somatomotor and physiological. Movement-triggered defence is a widespread and basic competence, clearly of great importance - life or death, indeed - yet we know surprisingly little about its organisation (Dean et al. 1989).

Perhaps less obviously, the rodent superior colliculus may also be useful for studying the control of movements directed *towards* stimuli rather than away from them. In the class of model illustrated in Fig 1, the colliculus controls only the destination of the movement: how the eye reaches that destination is determined by the downstream machinery, in particular the pulse generator. The characteristics of the pulse generator account for the strong relationship between the amplitude of a saccade and its velocity (the 'main sequence'), and also for the inability of collicular stimulation in primates to alter that relationship (eg Robinson 1972, 1981).

In contrast, stimulation at certain sites within the rat superior colliculus, particularly in the lateral intermediate layers, can affect head-movement amplitude and velocity independently. Thus, at a given current intensity, movement amplitude is related to the number of pulses within a stimulating train, whereas movement velocity is related to the frequency of the pulses (Fig 6) (King et al. 1990). The rat superior colliculus is therefore potentially able to control the trajectory of head movements, ie it may possess a competence delegated to the pulse -generator in Fig 1. If so, studies of the rat superior colliculus may help to refine current models of trajectory formation, even though these were formulated primarily in the context of arm movements (eg Bullock and Grossberg 1988; Houk et al, 1989).

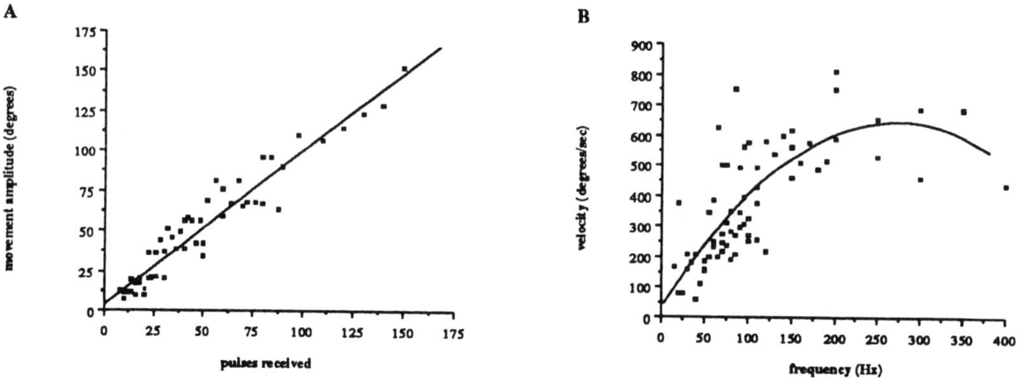

Figure 6 The relationship between number of stimulating pulses and head-movement amplitude (**A**), and between stimulating frequency and movement velocity (**B**), elicited from a site in lateral superior colliculus.

3.3 How Individual Processes are Coordinated

Because the rodent superior colliculus has more than one output system, we can ask how the decision is taken to activate one output rather than another. This issue is related to the general underlying idea of 'unexpected stimulus' (Section 2-2), and to the way that individual output channels are controlled (Section 2-3). Examples of possible decision rules for the rat, based on position of a stimulus in the visual field, and the way in which it moves, have been discussed elsewhere (Dean et al. 1989). Here we consider briefly the more general problem of how control can be smoothly passed from individual process to individual process in such a way that an adaptive, 'intelligent' stream of behaviour is produced.

This problem has been addressed in a variety of contexts by Arbib and his collaborators (eg Arbib 1987, 1989a,b), with particular emphasis placed on the construction of a working frog - *Rana computatrix*. In a related endeavour, Brooks' group have produced insect-like robots (eg Brooks 1989). In fact the advent of more sophisticated robots and autonomous vehicles has made the design of flexible control architectures an increasingly topical question. It is possible that the rodent superior colliculus, mediating as it does a number of competences relevant to robot survival, could be useful for testing out ideas about particular control architectures - e.g. "schemas" (Arbib) or "subsumption architecture" (Brooks). If so, *Rana computatrix* may be shortly joined by *Rattus computator*.

4 References

Arbib MA (1987) Levels of modeling of mechanisms of visually guided behavior. *Behav Brain Sci* 10: 407-465.

Arbib MA (1989a) Visuomotor coordination; Neural models and perceptual robotics. In Ewert J-P, Arbib MA (eds) *Visuomotor Coordination*. Plenum New York, pp 121-171.

Arbib MA (1989b) *The metaphorical brain 2: neural networks and beyond*. Wiley Interscience, New York.

Blanchard DC, Williams G, Lee EMC, Blanchard RJ (1981) Taming of wild *Rattus norvegicus* by lesions of the mesencephalic central grey. *Physiol Psych* 9: 157-163.

Blanchard RJ, Flannelly KJ, Blanchard DC (1986) Defensive behaviors of laboratory and wild *Rattus norvegicus*. *J Comp Psychol* 100: 101-107.

Brooks RA (1989) A robot that walks: emergent behaviors from a carefully evolved network. *Neural Comp* 1 253-262.

Bullock D, Grossberg S (1988) Neural dynamics of planned arm movements: emergent invariants and speed-accuracy properties during trajectory formation. *Psych Rev* 95; 49-90.

Carpenter RHS (1988) Movements of the eyes, 2nd edition. Pion London.

Dacey DM, Ulinski PS (1986) Optic tectum of the eastern garter snake, *Thamnophis sirtalis*. II. Morphology of efferent cells. *J Comp Neurol* 245: 198-237.

Dean P, Mitchell IJ, Redgrave P (1988a) Contralateral head movements produced by microinjection of glutamate into superior colliculus of rats: evidence for mediation by multiple output pathways. *Neuroscience* 24: 491-500.

Dean P, Mitchell IJ Redgrave P (1988b) Responses resembling defensive behaviour produced by microinjection of glutamate into superior colliculus of rats. *Neuroscience* 24: 501-510.

Dean P, Redgrave P (1984) The superior colliculus and visual neglect in rat and hamster. *Brain Res Rev* 8: 129-163.

Dean P, Redgrave P, Mitchell IJ (1988c) Organisation of efferent projections from superior colliculus to brainstem in rat: evidence for functional output channels. In Hicks TP, Benedek G (eds) *Vision within extrageniculo-striate systems, Prog Brain Res* 75: 27-36. Elsevier Amsterdam.

Dean P, Redgrave P, Molton L (1984) Visual desynchronization of cortical EEG impaired by lesions of superior colliculus in rats. *J Neurophysiol* 52: 625-637.

Dean P, Redgrave P, Sahibzada N, Tsuji K (1986) Head and body movements produced by electrical stimulation of superior colliculus in rats: Effects of interruption of crossed tectoreticulospinal pathway. *Neuroscience* 19: 367-380.

Dean P, Redgrave P, Westby GWM (1989) Event or emergency? Two response systems in the mammalian superior colliculus. *Trends Neurosci* 12: 137-147.

Edwards SB (1980). The deep cells of the superior colliculus: Their reticular characteristics and structural organisation. In Hobson JA, Brazier MAR (eds.) *The Reticular Formation Revisited.* Raven Press New York, pp 193-209.

Ellard CG, Goodale MA (1986) The role of the predorsal bundle in head and body movements elicited by electrical stimulation of the superior colliculus in the Mongolian gerbil. *Exp Brain Res* 64: 421-433.

Ellard CG, Goodale MA (1988) A functional analysis of the collicular output pathways: a dissociation of deficits following lesions of the dorsal tegmental decussation and the ipsilateral collicular efferent bundle in the Mongolian gerbil. *Exp Brain Res* 71: 307-319.

Goodale MA, Murison RCC (1975) The effects of lesions of the superior colliculus on locomotor orientation and the orienting reflex in the rat. *Brain Res* 88: 243-261.

Grossberg S, Kuperstein, M (1989) *Neural dynamics of sensorimotor control, expanded edition.* Pergamon Press New York.

Holcombe V, Hall WC (1981a) Laminar origin of ipsilateral tectopontine pathways. *Neurosci* 6: 255-260.

Holcombe V, Hall WC (1981b) The laminar origin and distribution of the crossed tectoreticular pathways. *J. Neurosci* 1: 1103-1112.

Houk JC, Singh SP, Fisher C, Barto AG (1989) An adaptive sensorimotor network inspired by the anatomy and physiology of the cerebellum. *COINS Technical Report* 89-108, Univ of Massachusetts at Amherst MA.

Huerta MF, Harting JK (1984) The mammalian superior colliculus: Studies of its morphology and connections. In Vanegas H (ed),*The Comparative Neurology of the Optic Tectum.* Plenum Press New York, pp 687-773.

Illing R-B, Graybiel AM (1986) Complementary and non-matching afferent compartments in the cat's superior colliculus: innervation of the acetylcholinesterase-poor domain of the intermediate gray layer. *Neuroscience* 18: 373-394.

Keay KA, Dean P, Redgrave P (1990) N-methyl D-aspartate (NMDA) evoked changes in blood pressure and heart rate from the rat superior colliculus. *Exp Brain Res* 80: 148-156.

Keay KA, Redgrave P, Dean P (1988) Cardiovascular and respiratory changes elicited by stimulation of rat superior colliculus. *Brain Res Bull* 20: 13-26.

Keay KA, Westby GWM, Frankland P, Dean P, Redgrave P (1990b) Organization of the crossed tecto-reticulo-spinal projection in rat - II. Electrophysiological evidence for separate output channels to the periabducens area and caudal medulla. *Neuroscience* in press.

King SM, Dean P, Redgrave P (1990) Collicular control of rat head movement: further evidence for position and velocity guided control mechanisms. *Neurosci Lett* Supplement 38: S50.

May PJ, Porter JD (1989) Laminar origins of the descending tectal pathways in the macaque monkey. *Soc Neurosci Abstr* 15: 239.

McHaffie JG, Kao C-Q, Stein BE (1989) Nociceptive neurons in rat superior colliculus: response properties, topography, and functional implications. *J Neurophys* 62: 510-525.

Merker B (1980) *The sentinel hypothesis: A role for the mammalian superior colliculus.* Unpublished doctoral thesis, Massachusetts Institute of Technology.

Moschovakis AK, Karabelas AB, Highstein SM (1988) Structure-function relationships in the primate superior colliculus. I. Morphological classification of efferent neurons. *J Neurophys* 60: 232-262.

Overton P, Dean P, Redgrave P (1985) Detection of visual stimuli in far periphery by rats: Possible role of superior colliculus. *Exp Brain Res* 59: 559-569.

Redgrave P, Dean P (1985) Tonic desynchronisation of cortical EEG by electrical stimulation of superior colliculus and surrounding structures in urethane-anaesthetised rats. *Neuroscience* 16: 659-671.

Redgrave P, Dean P, Souki W, and Lewis G (1981) Gnawing and changes in reactivity produced by microinjections of picrotoxin into the superior colliculus of rats. *Psychopharmacology* 75: 198-203.

Redgrave P, Dean P, Westby GWM (1990) Organization of the crossed tecto-reticulo-spinal projection in rat - 1. Anatomical evidence for separate output channels to the periabducens area and caudal medulla. *Neuroscience* in press.

Redgrave P, Mitchell I, Dean P (1987a) Descending projections from the superior colliculus in rat: a study using orthograde transport of wheatgerm-agglutinin conjugated horseradish peroxidase. *Exp Brain Res* 68: 147-167.

Redgrave P, Mitchell I, Dean P (1987b) Further evidence for segregated output channels from superior colliculus in rat: ipsilateral tecto-pontine and tecto-cuneiform projections have different cells of origin. *Brain Res* 413: 170-174.

Redgrave P, Odekunle A, Dean P (1986) Tectal cells of origin of predorsal bundle in rats: Location and segregation from ipsilateral descending pathway. *Exp Brain Res* 63: 279-293.

Robinson DA (1972) Eye movements evoked by collicular stimulation in the alert monkey. *Vision Res* 12: 1795-1808.

Robinson DA (1981) The use of control systems analysis in the neurophysiology of eye movements. *Ann Rev Neurosci* 4: 463-503.

Rodgers RJ, Randall JI (1987) Defensive analgesia in rats and mice. *Psych Rec* 37: 335-347.

Sahibzada N, Dean P, Redgrave P (1986) Movements resembling orientation or avoidance elicited by electrical stimulation of the superior colliculus in rats. *J Neurosci* 6: 723-733.

Sparks DL, Mays LE (1990) Signal transformations required for the generation of saccadic eye movements. *Ann Rev Neurosci* 13: 309-336.

Tehovnik EJ, Yeomans JS (1986) Two converging brainstem pathways mediating circling behavior. *Brain Res* 385: 329-342.

Van Gisbergen JAM, Van Opstal AJ (1989) Models. In: Wurtz RH, Goldberg ME (eds) *The neurobiology of saccadic eye movements.* Elsevier Science Publishers North Holland, pp 69-101.

Wallace MN (1988) Lattices of high histochemical activity occur in the human, monkey, and cat superior colliculus. *Neuroscience,* 25: 569-583.

Westby GWM, Keay KA, Redgrave P, Dean P, Bannister M (1990) Output pathways from the rat superior colliculus mediating approach and avoidance have different sensory properties. *Exp Brain Res* in press.

Computation of Absolute Distance in the Mongolian Gerbil (*Meriones unguiculatus*): Depth Algorithms and Neural Substrates

COLIN G. ELLARD[1] *and MELVYN A. GOODALE*[2]

Departments of Psychology,

[1] *Mount Allison University, Sackville, N.B., Canada, E0A 3C0*

[2] *The University of Western Ontario, Ontario, Canada, N6A 5C2*

Abstract. *Mongolian gerbils (Meriones unguiculatus) can be trained to estimate the absolute distance of a visual target in order to make a ballistic jump to that target. We have conducted a number of experiments designed to discover the sources of depth information that gerbils use, the ways that they combine independent depth estimates, and the neural pathways underlying these distance computations. Before carrying out a jump, gerbils often execute a series of vertical oscillations of the head. These "head bobs" increase in number, size, and velocity as the distance to be jumped increases. Furthermore, there is a weak correlation between an animal's overall accuracy and its propensity to bob its head. These findings suggest that gerbils produce retinal motion in order to compute the distance to the target. When gerbils are trained to jump to a target of consistent width, insertion of targets of differing widths will produce errors in jumping performance, indicating that retinal image size (RIS) is being used as a cue to distance. Errors that gerbils make are smaller than would be predicted if RIS were the only source of distance information, suggesting that information from disparate sources is blended to arrive at the final distance estimate. By manipulating the reliability of RIS information, it is possible to influence this blending process.*

Neither lesions of posterior neocortex, superior colliculus nor pretectal nuclei completely abolish the ability to compute distance in this task (although lesions of at least the former two structures produce degradations in accuracy). The use of RIS for distance computation survives large cortical ablations, even though gerbils can no longer distinguish between target sizes in a discrimination learning task. This finding suggests that there may be specialized circuitry for distance computation that is independent of the cortical mechanisms responsible for recognizing the properties of objects.

In a series of experiments conducted with Mongolian gerbils, we set out to study the estimation of absolute distance in a visuomotor task. By design, the experimental apparatus (borrowed from Richardson (1909) and Russell (1932)) has been kept very simple and unconstrained. Using operant methods, animals are trained to jump across a gap from one raised platform to another (Figure 1). The size of the gap is varied randomly from trial to trial such that some estimate of distance is required if an animal is to perform well. This simple test of depth vision has two attractive qualities. First, it is possible to obtain a quantitative measure of the distance estimate that a gerbil has used to carry out a jump. By measuring the landing position of a gerbil from a videotape record, we have access to the result of the distance computation that was used to program the movement. This estimate of distance must be an absolute one, and therefore differs from estimates that are required in most other experimental procedures designed to study depth processing in mammals other than humans (Walk and Gibson, 1961; Mitchell and Timney, 1982).

Second, in many of our experiments we do not take extreme measures to constrain the type of depth information that is available. By providing access to a rich variety of distance cues, we can entice the gerbils to take advantage of the situation either by selecting the most appropriate data for the distance computation or by combining a set of independent distance estimates.

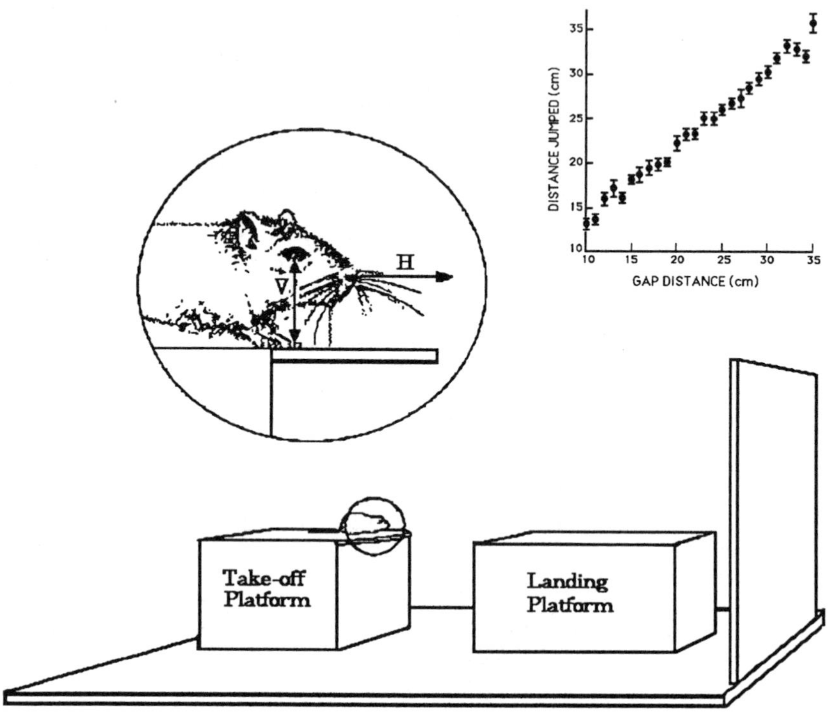

Figure 1: A schematic diagram of the apparatus used to test absolute distance estimation in the gerbil. The inset figure shows the field of view of the close-up camera that was used to measure head and body movements that preceded the jump. The small graph illustrates the relationship between the size of the gap and the distance jumped in ten normal animals. Data points represent means and error bars are standard error of the mean (Adapted from Ellard et al., 1984).

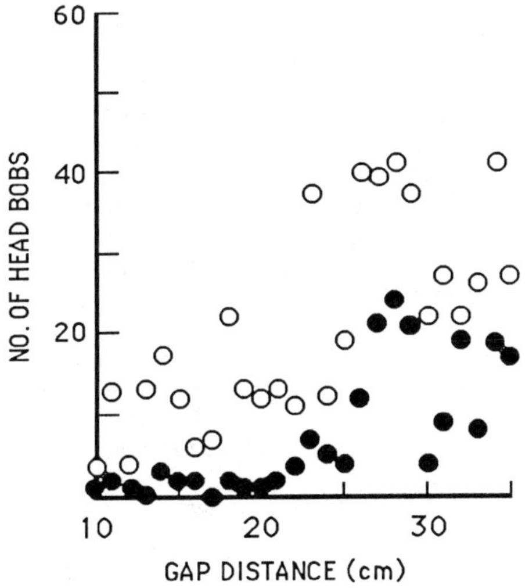

Figure 2: Graph showing frequency of vertical head movements as a function of gap distance. Data points represent sums of head bobs in ten animals. Open circles represent head bobs smaller than 1.5 cm (the interocular distance) and filled circles represent head bobs larger than 1.5 cm (and therefore potentially more informative than retinal disparity). Both large and small head bobs tended to increase with increasing gap distance, but the increase was proportionately greater for the large head bobs.
(Adapted from Ellard et al., 1984)

In an initial experiment (Ellard et al., 1984), ten animals were trained to jump distances of up to 36 cm (about 2.5 -3 body lengths). As figure 1 shows, gerbils were able to calibrate accurately the magnitude of the jump as a function of the size of the gap. Having satisfied ourselves that gerbils were carrying out distance computations with some degree of accuracy, we devoted much of our subsequent attention to questions concerning the sources of distance information that the gerbils were using.

Retinal Motion Information

In the initial experiment, we noticed that gerbils often carried out a series of vertical oscillations of the head ('head bobs') just prior to initiating a jump. Such head movements might be useful for generating retinal motion information that could be used to compute the distance to the target. Head movements of this type have been observed in other species, and a similar suggestion has been made in these cases (Walk and Gibson, 1961; Wallace, 1959). Only in the case of the locust, however, has a substantive case been made for the use of retinal motion as a cue to depth (Wallace, 1959; Collett, 1978). If head movements were involved in distance estimation, one seemingly straightforward prediction might be that the more an animal bobbed its head, the more accurate a jumper it would be. We examined the relationship between an individual animal's accuracy (variance in landing position) and its propensity to bob its head. When only large head bobs were included in the analysis, there was a weak but significant correlation between the incidence of head bobbing and an animal's overall accuracy ($r = 0.33$). In retrospect, the small amount of variance in performance that was accounted for by head bobbing should not have been as surprising to us as it was at the time. On many trials, animals jumped without bobbing their heads at all. It would be more surprising if such a feat were attempted in the complete absence of reliable distance information. It now seems more likely, especially in light of our later work, that gerbils use more than one source of information to solve the distance problem in this task, and that variability in the degree to which head bobbing is employed is correlated with the use of other sources of information rather than with jumping performance.

In light of our failure to find a convincing relationship between accuracy and head bobbing, we turned our attention to a more detailed analysis of the circumstances in which head bobs were produced and the dynamics of these movements. One of the most striking findings was that the incidence of head movements was strongly related to the size of the gap; larger gap distances usually elicited more head bobs (see Figure 2). One interpretation of this relationship that is in accord with what has just been suggested--that retinal motion might be one member of a constellation of independent sources of depth information--is that retinal motion information becomes more useful at longer distances, while other cues suffice at shorter distances. We have thought of two possible reasons for this state of affairs. First, it is possible that other cues to distance are rather limited in the range of distances over which they can operate effectively. Collett and Harkness (1982) have made convincing arguments that many distance cues are limited by physical constraints of various kinds. Binocular stereopsis, for example, is limited by interocular distance. Accomodation is limited by the motility of the lens. Because gerbils are capable of fairly large head movements (4.5-5 cm or roughly 3 times the interocular distance), some of the limitations of other distance cues might apply more weakly to retinal motion. The second reason why retinal motion might become more important at longer distances has more to do with the biomechanics of the task than with depth vision per se. As the distance to the target increases, the amount of impulse power that is required of the muscles increases in an exponential fashion (Alexander, 1968). Miscalculations, therefore, become increasingly costly in energy terms. Also, since longer jumps require higher velocities, the consequences of a missed jump become increasingly aversive. Therefore, not only is depth information degraded as target distance increases, but the magnitude of the tolerable performance error may decrease as well.

When they occur, head bobs can be produced singly, or can be chained together into series of up to about a dozen discrete movements. By measuring the kinematic properties of these head movements, we have tried to find systematic trends in bouts of head bobs and to relate these trends to both task and performance variables. These analyses yielded two pieces of information, at least one of which strengthens the case for the use of retinal motion. Measurements of the amplitudes and both peak and mean velocities of head movements revealed that there was some tendency to produce larger and faster head movements as the size of the gap was increased. This finding, we believe, is in accord with the retinal motion interpretation. As target distance increases, the amplitude and velocity of head movement that is required to maintain some minimal amount or velocity of retinal motion will also increase.

We also found a strong relationship between the amplitude and velocity of head movements. Larger head bobs were also faster. The significance of this for retinal motion may be that the duration of a head bob tends to be fairly constant. If the head bobs do represent something like triangulation in the temporal domain, then there may be constraints on the time over which information can be integrated.

Our attempts to find structure within bouts of head bobs have been unsuccessful so far. Upward and downward movements are about equally frequent and tend to alternate, but there do not appear to be very strong trends in the velocities or amplitudes of successive head bobs. Some gerbils tend to execute a series of quite similar head bobs while others produce both large and small head bobs in an apparently random fashion.

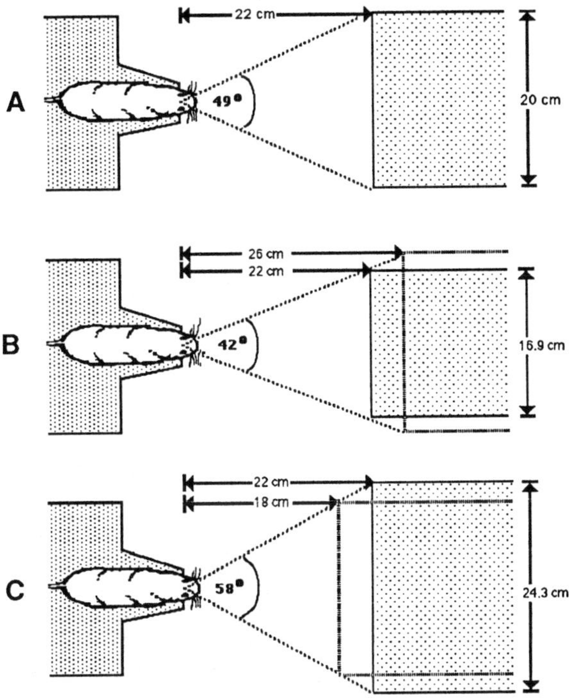

Figure 3: A demonstration of the manipulation used to demonstrate the use of retinal image size information by gerbils. **A.** A representation of a gerbil facing a 20 cm wide target that is located 22 cm away. The retinal angle subtended by the target is 49°. **B.** A gerbil facing a target whose true distance is still 22 cm but whose width has been decreased to 16.9 cm. A gerbil using RIS will interpret the retinal angle of 42° as representing a target that is 26 cm away. **C.** A gerbil facing a target that is wider than the standard target. In this case, the gerbil will interpret the retinal angle as a target at 18 cm. In the experiment, another set of probe platforms was also used in which the size of error predicted if RIS were used would be either = or - 8 cm (Adapted from Goodale et al., 1990).

Retinal Image Size Information

Our observation that many gerbils jumped without producing a single head bob, along with the strong correlation between the incidence of head bobbing and gap distance, implied that gerbils must be using at least one other source of distance information in order to determine whether head bobs are required for completion of the jump. Another prominent source of information that is available to gerbils is the size of the retinal image subtended by the target platform. Gerbils are trained over the course of many weeks and hundreds of trials using a platform of constant dimensions. It is possible that this constancy might be used by calibrating the retinal image size of the target with the force that

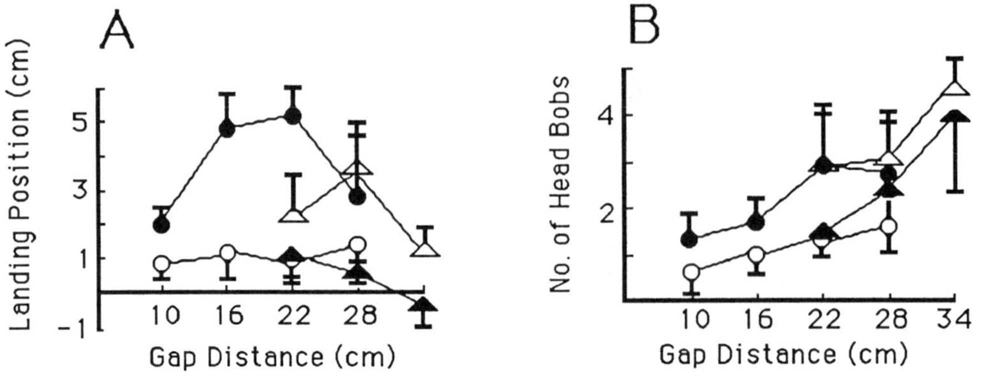

Figure 4: Results of changes in width of landing platform on performance in the jumping task. The target was made to appear either 8 cm closer or 8 cm farther away (according to RIS information). **A.** Each point represents the mean landing position beyond the leading edge of the landing platform for six animals. Circles represent sets of trials in which the landing platform was made narrower (Open circles - control performance; filled circles - narrower probe platforms). Triangles represent sets of trials in which the landing platform was made wider (Open triangles - control performance; filled triangles - wider probe platforms). **B.** Each point represents the mean number of head bobs preceding a jump for six animals. All conventions are the same as for **A.** The differences between control performance in the two conditions are accounted for by the fact that the condition with the narrower probes was carried out before the condition with wider probes. There was a tendency for gerbils to compensate for the errors produced by the probe platforms during the course of testing. Hence, in the condition producing underjumps, gerbils tended to overjump slightly on control trials (see Goodale et al. (1990) for discussion).

is required for the jump. In order to test this idea, we first trained gerbils to jump in the standard task, and then presented them with unpredictable probe trials in which the standard target platform was replaced with a new platform that was either wider or narrower than the one used for training (see Figure 3). When we measured the landing positions of animals on the probe trials, we discovered that we had indeed been successful in manipulating the landing position of the animals by changing the width of the target (see Figure 4). Hence, we have good evidence that retinal image size is used as a source of distance information in the task. It is important to note that the sizes of the errors that were produced by gerbils were always smaller than would be predicted if RIS was the only cue that was being used. For instance, when the target platform was changed such that an RIS estimate yielded a distance value that was 8 cm greater than the real target distance, the average overjump was less than 4 cm. This finding suggests that RIS information is being blended with other information to arrive at the final estimate. We will return to this point later on.

In more recent work (Ellard et al., in press) we have been able to show that gerbils are capable of forming more than one RIS calibration at a time. Gerbils were trained to form two different RIS calibrations in two distinct contexts (the context was formed by the locations in the testing room and the unique backgrounds behind the target platforms). In each of the two contexts, a different sized target platform was used during training (narrow in one context and wide in the other). In test sessions, probe trials were presented in which the probe target platform was midway in size between the two training targets. The same probe platform was presented to the same animals in both contexts. Depending only on the context in which the probe target was presented, we were able to produce either overjumps or underjumps. This finding suggests that gerbils are able to form separate calibrations for objects that depend more upon the surroundings that they associate with the object than with properties of the object itself.

One of the reasons for our interest in other sources of depth information was the finding that head bobbing was strongly correlated with distance--the dimension that it was purportedly helping to measure. Our hope was that the resolution to this paradox lay in the fact that other depth cues were providing information that helped to determine whether head bobbing was required. If RIS is such a cue, then one would expect the insertion of probe targets to result not only in changes in depth estimates, but to also affect the correlation between head bobbing and target distance. As Figure 4b shows, this was indeed the case. Not only did a wider target, for instance, result in shorter jumps, but it also reduced the frequency of head bobbing prior to these jumps. This suggests that at least some of the information that is used to determine whether head bobbing is necessary is derived from retinal image size.

Other Distance Algorithms

In experiments performed so far, we have concentrated on retinal motion and retinal image size. We do not ourselves believe, and do not wish to convey the impression that these two algorithms constitute an exhaustive set of distance algorithms used by gerbils. In some of the experiments described below, we mention the use of other algorithms such as stereopsis and loom. Strong direct evidence for the importance of these sources of distance information is lacking so far, but there are many suggestive indications that these sources may play a role in the jumping task.

Interactive Use of Multiple Sources of Depth Information

One of the reasons for our interest in this particular task is that large amounts of information from a number of independent sources are available to the gerbils. As such, they are free to use whatever information might be available and to combine independent estimates to arrive at a single distance value. Undoubtedly, this situation is closely analogous to the ways that distance problems present themselves to animals in natural contexts. A testament to the effectiveness of the algorithms that gerbils use in our own experiments is the great difficulty that we have had in causing an animal to perform poorly despite a variety of interventions into both the constraints that are presented in the task and even into the nervous system of the animals (see next section). Indeed, we have most often convinced ourselves that our manipulations have had an effect by the presence of subtle changes in the gerbils' behaviour (such as changes in the frequency of head bobbing or in the latency to execute a jump) rather than by gross degradations in jumping performance. The manipulations that we have employed thus far fall into two categories: manipulations of the availability of information or manipulations of the reliability of this information.

Information availability

One of our first manipulations of availability was conducted by forcing gerbils to complete the jumping task using only one eye. In the experiment described at the beginning of this paper, the ten animals that had learned the depth task using both eyes had one eye sutured closed without lid resection (this allowed us to re-open the eye later in the experiment). The initial response to the monocular condition was a reluctance to jump among some animals (this was quantified as the failure to initiate a jump within a one minute time interval). Interestingly enough, there was considerable variability among animals in this regard. When we examined the relationship between failure to jump and behavior during the earlier binocular performance, we discovered that those animals that bobbed their heads most frequently in the binocular condition were least likely to refuse to jump in the monocular condition. Frequent binocular head bobbers were also willing monocular jumpers. Overall, the frequency of head bobbing was considerably increased in the monocular condition, especially at shorter distances where head bobs are fairly infrequent in normal animals. Among those animals that jumped in the monocular condition, there was no apparent decrement in the accuracy of performance, as measured by the variance in landing position on the target platform. Finally, with repeated testing, both of the effects described above (inflation of head bob frequencies and reluctance to jump) disappeared. This pattern of results is typical of the effects of a number of different types of manipulation. It would appear that the initial response to unavailability of a source of depth information is to boost the contribution from one or more other sources. It should not be surprising that some animals are better at this than others, and that less skilled animals are unwilling to jump in the absence of sufficiently precise information.

Another of our early observations was that there were considerable differences between animals in the style of approach to the edge of the take-off platform. Some animals approached the gap with fairly large and smooth forward movements while others used very small shuffling movements. We observed that those animals in the former category were less likely to produce head bobs than the latter animals. One hypothesis is that large, smooth forward movement produces a different kind of retinal motion information--retinal expansion or loom. If an animal is an effective generator of loom information, it is less likely to need to bob its head to generate translational retinal motion. In order to test this idea, we retrained six of the original ten animals to jump from a very small take-off platform that precluded any type of forward movement prior to the jump. The effect of this manipulation was again to inflate head bobbing frequencies but to have no effect on accuracy.

In other experiments, animals have been tested in a situation devoid of retinal image size information. This was done by randomly varying the width of the target platform from trial to trial. Whether gerbils are trained and tested using this scheme or trained in the conventional manner (with a constant platform) and then tested with a randomly varying target platform, the results are similar: there is no apparent loss of accuracy but there is an increase in the frequency of head bobs (Goodale et al., 1990; Ellard, unpublished results).

Information reliability

In the experiment described earlier in which gerbils formed two RIS calibrations simultaneously, we have been able to show that gerbils take into account the reliability of distance information when arriving at an estimate using multiple sources of information (Ellard et al., in press). In that experiment, we included two testing conditions. In the MIXED condition, gerbils were moved on a random schedule from one context to another. On each trial, therefore, an animal had to determine which location it was in before it could use RIS. In the SEPARATE condition, gerbils carried out 15 consecutive trials in one context and then a further 15 trials in the other. Testing in each context occurred on a daily basis and the order of testing was reversed from day to day. One of the differences between these conditions is that gerbils in the SEPARATE condition would have the opportunity to sample their surroundings on a number of consecutive occasions. Gerbils in the MIXED condition would not have such an opportunity. On each trial, they would have to make a new determination of context before completing the jump. It is likely, therefore, that gerbils in the MIXED condition would be less certain about their location than those in the SEPARATE condition. Hence, the RIS information might be considered less reliable in the MIXED condition. Results showed that although probe effects occurred in both conditions, the size of these effects was somewhat smaller in the MIXED condition. This finding, and the one described previously in which RIS information was severely degraded during acquisition by randomly varying the size of the target platform (Goodale et al., 1990) are all that we know about the influence of cue reliability.

Computation of a single distance value from several independent sources

At present, we have only a fragmentary knowledge of how independent estimates of distance from different depth algorithms are combined to arrive at a single value. The information that we have, however, is sufficient to rule out several classes of models and to weakly suggest some alternatives. One scheme might consist of a "winner-take-all" strategy in which information that is both readily available and reliable is employed for the estimate, and other less reliable information is discarded. Collett and Harkness (1982) have argued against this scheme on the basis of the fact that it is probably no less computationally expensive to arrive at an estimate and then discard it than it is to use it, even if the final contribution of the less reliable information is only marginal. As well, the evidence from our retinal image size studies, in which the landing position was intermediate between the true distance and that which would be derived from RIS argues against this type of process.

Without any very firm evidence, we have always suspected that separate estimates of distance are averaged in some fashion in order to arrive at the final distance value that is used to propel the jump. Collett and Harkness (1982) have presented many arguments based on optics and geometry, in combination with limiting factors inherent in neural systems (primarily resolution acuity or some variant thereof) that all distance algorithms are better used in some circumstances than in others. Factors such as ambient illumination, motion of the target, familiarity of the target and so on are all likely to have an effect on the utility of a particular distance algorithm. If a mechanism for estimating distance is to use multiple distance algorithms effectively, then it ought to be able to take into account this type of variability. There are at least two types of mechanisms that might be employed. A very sophisticated system might use a form of weighted average in which each algorithm yields up its best estimate and the contributions of each of these estimates are adjusted by some external weighting function in accord with their reliability. One can imagine, for instance that in the context-bound RIS experiment described above, the "gain" of the RIS estimate is decreased in the MIXED condition to reflect the increased uncertainty about which particular calibration to use. This type of model accounts for a good number of the RIS effects that we have seen, but seems rather unwieldy and unnecessarily complicated. If the system were to take into account all of the important attributes of the target and context that it should, a good deal of preliminary computation would be required in order to arrive at an accurate set of weights.

A simpler way of producing something akin to a weighting function might be to adopt a statistical approach to the problem. For the sake of argument, let us assume that a gerbil faces the edge of the gap in the apparatus with some set value of tolerable performance error. The problem for the distance-sensing system then reduces to arriving at an estimate whose associated error is less than or equal to the tolerable performance error. An effective way of reducing the error of an estimate is to increase the size of the sample. A gerbil could do this either by using as many depth algorithms as possible or by increasing the number of estimates contributed by each algorithm. If a single algorithm contributed a number of estimates, the appropriate weight would be generated automatically by the variability among the estimates. There are some suggestions in our data that gerbils might act to decrease uncertainty by repeated sampling. In the experiments where some information was made unavailable, the result was always an increase in the incidence of head movements. Even when a normal animal is presented with the task without constraints, not only does the incidence of head bobbing increase with distance, but the number of head bobs that precede a jump rises sharply as well. Although it is easy to imagine that a series of head bobs consists of a set of independent depth estimates, it is less obvious that other algorithms such as loom or retinal image size are used repeatedly during a trial. It is not at all uncommon, however, for a gerbil in our task to run to the edge of the take-off platform, pause, and then retreat for a few seconds before making another run. This pattern of behavior sometimes occurs five or six times before the jump is executed. Perhaps this is not just "vacillation", but is an attempt to use repeated samples. A simple extension of this theory might also account for the differences between performance in the MIXED and SEPARATE conditions. Perhaps the RIS effect on probe trials is smaller in the MIXED condition because repeated RIS estimates are made using both of the available contextually cued calibrations. In some cases, therefore, the correct calibration will be used and in other cases, the incorrect one will be used.

One experimental finding that does not fit well with this theory is the decrease in the incidence of head bobbing in RIS probe trials where the platform was made to appear closer than it really was. Since this manipulation would produce two sets of samples with a 4 or 8 cm difference between their means, one would expect to see increases in sampling in order to reduce error, and therefore higher rates of head bobbing. The implication of this finding is that although repeated sampling may be used to decrease variability (either within or across information domains), the mean distance value that results from RIS computations helps to determine whether or not to bob the head. We do not know whether the mean value also affects the sampling frequencies of other depth algorithms, but there are reasons to suspect that there might be special constraints on the use of head bobbing. Gerbils are preyed upon by a number of birds with acute vision (Agren et al., 1989). Head bobbing is a slow, cumbersome and conspicuous behavior that is incompatible with rapid locomotion. This is likely to draw the attention of a predator and its use may be a last resort when all other algorithms have failed to produce a sufficiently precise estimate.

Neural Circuitry Involved in Distance Estimation

Just as we have had little success in preventing gerbils from performing well in the jumping task by depriving them of visual information of one kind or another, we have also found that distance estimation algorithms are quite resistant to the effects of removal of parts of the central visual system. In one study (Ellard et al., 1986) we removed either posterior neocortex, superior colliculus, or pretectum in gerbils that had been trained to jump preoperatively. Only removal of the neocortex produced significant effects on accuracy and even in these animals there was clear evidence that the size of the jump was strongly influenced by the size of the gap. In a more recent study in which longer periods of preoperative training were employed (Carey et al., 1990), performance close to that of shams was obtained from animals with large visual cortical lesions extending beyond primary visual cortex to include most or all of extrastriate areas 18a and 18b. In the earlier study, at least, the performance of these animals was far from normal in that head bobs were produced on all trials regardless of the size of the gap. One possible reason for this is that retinal motion information was used to fill in information that was unavailable because of the lesions. Another possibility is that the neural circuitry that is responsible for initiating head bobbing is separate to that which interprets the self-generated motion information. In accord with the repeated sampling hypothesis, one might expect such circumstances to lead to very large numbers of head bobs in an attempt to overcome increased errors of estimation.

Lesions of the superior colliculus resulted in slightly increased rates of head bobbing, but no decrease in accuracy. Curiously, animals in this group had a tendency to overjump consistently compared to shams. One possible explanation of this finding is that collicular animals were jumping conservatively. It is difficult to understand why they should do so, however, since their distance estimates (as judged by variance in landing position) were as accurate as those of control animals. Another explanation is that the lesions might have given rise to a consistent error in distance estimation, perhaps even of non-neural origin (collicular lesions in rats have reportedly produced exophthalmus which might have an effect on optics (Pope and Dean, 1979)).

Lesions of pretectal nuclei also produced a small increase in head bobbing, but no changes in either accuracy or mean landing position.

In a more recent lesion study, the effects of cortical lesions on the RIS algorithm have been examined. In this experiment, animals with either lesions of primary visual cortex or both primary and secondary visual cortices were presented with probe trials of the type described above (Carey et al., 1990). The main finding in this experiment was that the use of RIS survived even the largest of cortical lesions (see Figure 5). Animals with such lesions showed effects on probe trials that were indistinguishable from those that have been observed in shams. This finding suggests that the RIS system is independent of at least visual cortical areas and is perhaps completely subcortical. What makes this finding especially intriguing is that in Lashley's (1939) studies of the effects of cortical lesions on visual discrimination in rats, similar lesions to those carried out by Carey et al., abolished the ability to discriminate objects on the basis of size. In unpublished work conducted in Goodale's laboratory using a similar discrimination paradigm, this finding was replicated with gerbils. This striking dissociation is somewhat reminiscent of the phenomenon of "blindsight" in humans in which visual information that is unavailable to consciousness is able to guide visuomotor activity (Perenin and Jeannerod, 1975; Weiskrantz et al. 1974). The implication of the finding with gerbils is that the neural circuitry that is

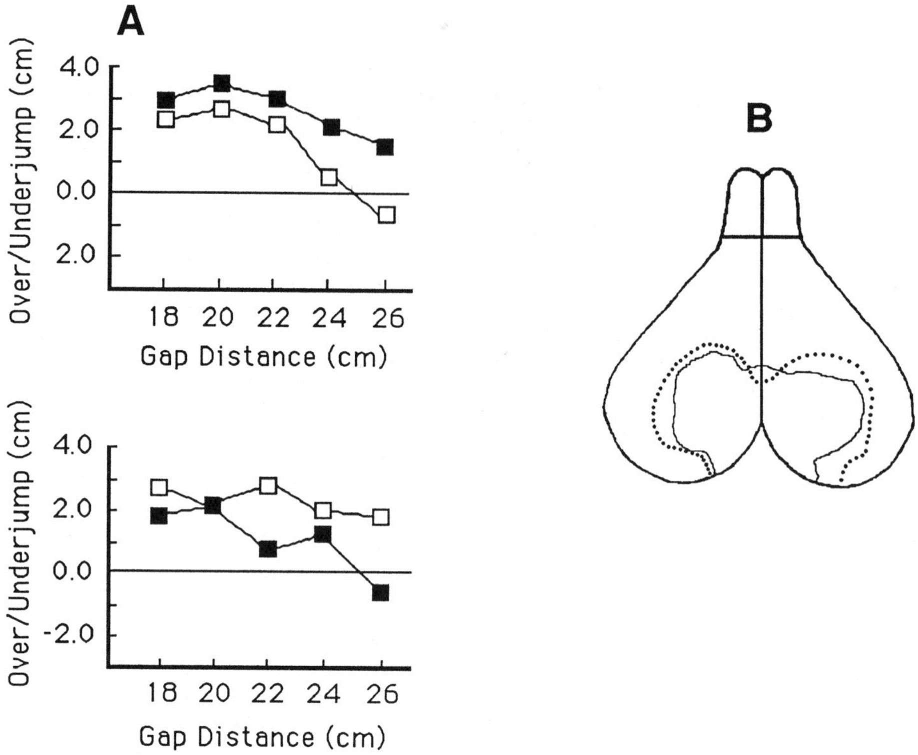

Figure 5: Effects of posterior cortical lesions on use of the retinal image size algorithm. **A.** Graphs showing the mean landing position of animals with lesions on both standard and probe trials. Open symbols - trials using the standard platform. Filled symbols - trials with probe platforms. Upper figure - a narrower probe platform was used. Lower figure - a wider probe platform was used. **B.** Dorsal view reconstruction of the lesions. The solid line is the boundary of the smallest lesion in the group. The dotted line outlines the largest of the lesions. (Adapted from Carey et al., 1990)

involved in object recognition does not have access to the information that must be present somewhere else in the nervous system in order for RIS to be used. This could be either because the proper anatomical connections do not exist or because the information is not in a suitable form to be used in other types of computations. It might be, for instance, that the RIS information that is used for depth computation is gathered while the animal is moving, hence reducing what we have called RIS to a form of loom information. Perhaps the task of the depth system is to extract distance from invariants present in the moving optic array while the cortical mechanisms involved in size discrimination are required to "factor out" contributions from image movement to arrive at a measurement of a property of the static object.

Future Directions

Work in both of our laboratories is currently driven by two objectives. First, we are pursuing the questions concerning the ways in which separate estimates of distance are combined. In order for this work to proceed, it seems necessary that we find ways to manipulate landing position by interfering with a depth algorithm other than RIS. If this were possible, we could examine the relative weights attached to two different types of estimates at the same time and also examine the influence of changes in task contingencies on these two weights. So far, our attempts to do this using translational retinal motion have been frustrated by the finding that gerbils appear to calibrate the retinal motion of the target with the amplitude or velocity of head bobbing, rather than the differential movement of target against background. Even very large changes in the distance between the target and the background are without effect (Goodale et al., 1990). In order to manipulate the use of this cue, therefore, we will have to find a way to actually move the target in close temporal relationship with movements of the animal's head. An attempt is underway in Goodale's laboratory to build an apparatus that will accomplish this.

Future work will also include an effort to track down the neural pathways involved in both RIS and retinal motion. This work will include lesions of subcortical nuclei and physiological recording. Before this work bears fruit, however, it is likely that we will have to learn more about the form of information that is used in RIS. Does RIS information consist of the retinal distance between two of the vertices of the target platform, for instance, or is the velocity of divergence of these two points the critical quantity?

A third avenue of enquiry concerns the relationship between the methods that we have observed that gerbils use to measure distance in a ballistic jumping task with methods used in other types of visuomotor tasks. Does a gerbil running toward a target, for instance, use RIS to determine when to decelerate? Does a gerbil that is being pursued by a threat use similar algorithms to determine the fastest route to shelter, or does it refer to an internalized map of the layout of its environment, only using visual means to fine-tune its trajectory as it nears the target?

In all of this work, the emphasis will be placed on attempts to understand the use of depth in tasks requiring movement towards or away from targets in relatively unconstrained settings. We believe that such tasks are likely to bear strong resemblances to the problems for which depth-sensing systems evolved in the gerbil and that an understanding of the methods used to solve these problems is likely to illluminate the general principles involved in visuomotor behavior.

References

Agren G, Zhou Q, Zhong W (1989) Ecology and social behaviour of Mongolian gerbils, *Meriones unguiculatus,* at Xilinhot, Inner Mongolia, China. *Animal Behaviour* 37: 11-27

Alexander RM (1968) *Animal Mechanics,* Sidgwick and Jackson, London

Carey DP, Goodale MA, Sprowl EG (1990) Blindsight in rodents: The use of a 'high level' distance cue in gerbils with lesions of primary visual cortex. *Behav Brain Res* 38: 283-289

Collett T (1978) Peering: A locust behaviour pattern for obtaining motion parallax information. *J Exp Biol* 76: 237-241

Collett T, Harkness L (1982) Depth vision in animals. In: Ingle DJ, Goodale MA, Mansfield RJW (eds) *The analysis of visual behavior.* MIT Press, Cambridge MA, pp 111-176

Ellard CG, Chapman DG, Cameron KA (in press) Calibration of retinal image size with distance in the Mongolian gerbil: Rapid adjustment of calibrations in different contexts. *Perception and Psychophysics*

Ellard CG, Goodale MA, MacLaren Scorfield D, Lawrence C (1986) Visual cortical lesions abolish the use of motion parallax in the Mongolian gerbil. *Exp Brain Res* 64: 599-602

Ellard CG, Goodale MA, Timney B (1984) Distance estimation in the Mongolian gerbil: The role of dynamic depth cues. *Behav Brain Res* 14: 29-39.

Goodale MA, Ellard CG, Booth L (1990) The role of image size and retinal motion in the computation of absolute distance by the Mongolian gerbil (*Meriones unguiculatus*). *Vision Res* 30: 399-413

Lashley KS (1939) The mechanism of vision: XVI. The functioning of small remnants of the visual cortex. *J Comp Neurol* 70: 45-67

Mitchell DE, Timney B (1982) Behavioral measurement of normal and abnormal development of vision in the cat. In Ingle DJ, Goodale MA, Mansfield RJW (eds) *The analysis of visual behavior*, MIT Press, Cambridge MA pp 483-523

Perenin MT, Jeannerod M (1975) Residual vision in cortically blind hemifields. *Neuropsychologia,* 13: 1-7

Pope SG, Dean P (1979) Hyperactivity, aphagia and motor disturbance following lesions of superior colliculus and underlying tegmentum in rats. *Behav Neural Biol* 27: 433-453

Richardson F (1909) A study of sensory control in the rat. *Psychol Monogr* 12: 1-124.

Russell JT (1932) Depth discrimination in the rat. *J Gen Psychol* 40: 136-159.

Walk RD, Gibson EJ (1961) A comparative and analytical study of visual depth perception. *Psychol Monogr* 75: 1-44

Wallace GK (1959) Visual scanning in the desert locust *Schistocerca gregaria forskal*. *J Exp Biol* 36: 512-525

Weiskrantz L, Warrington E, Sanders MD, Marshall J (1974) Visual capacity in the hemianopic field following restricted occipital ablation. *Brain* 97: 709-729

Generating Motor Trajectories

Equilibrium point mechanisms in the spinal frog

Simon Giszter, Ferdinando A. Mussa-Ivaldi and Emilio Bizzi.
Department of Brain and Cognitive Science, Massachusetts Institute of Technology, 77 Massachusetts Ave., Cambridge, MA 02139

1 Introduction

1.1 Capabilities of the frog spinal cord

The frog spinal cord has been shown to be capable of precise positioning and complex multi-joint trajectory generation (Fukson et al. 1980, Giszter et al. 1989, Schotland et al. 1989). It is well known, for example, that the spinal frog is capable of generating a coordinated sequence of multi-joint hindlimb movements directed to the removal of a noxious stimulus from the skin. This "wiping reflex" requires complex information-processing. Thus, the spinal cord must contain circuitry that coordinates the motion of multiple limb segments (see figure 1A). While this ability has been revealed utilizing wiping behaviors, we hypothesize that these and other behaviors may use a general limb positioning mechanism residing in the spinal cord. Such a mechanism might be expected to combine information from several sources: spinal pattern generators, higher centers in the central nervous system, and peripheral feedback. Computation in the positioning mechanism must depend on the form of final output pattern utilized. An ongoing framework for both theoretical and experimental investigation in motor control has been the concept that the viscoelastic properties of the muscles determine the type of computations needed within the central nervous system in order to plan and execute movements. This has led in particular to the equilibrium point hypothesis (Feldman, 1974, Bizzi, 1976, Bizzi et al.,1984, Hogan, 1984, Flash, 1987). We will examine the idea that general positioning by the spinal cord utilizes such a mechanism.

1.2 The equilibrium point hypothesis

Muscles are arranged about the joints in an agonist-antagonist configuration. The elastic behavior of the muscles implies that when a cocontracted limb is perturbed by an external force, the limb is displaced by an amount that varies with both the external force and the stiffness of the muscles. When the external force is removed, the limb should return to its original position. Experimental studies of arm movements in monkeys have shown that forearm posture can be seen as an equilibrium point between opposing elastic forces (Bizzi et al., 1976). The equilibrium point hypothesis posits transitions among such equilibrium postures as a fundamental mechanism for trajectory generation. In contrast other viewpoints consider postural control and trajectory generation as separate processes. An inverse dynamics computation of necessary torques to achieve a trajectory might be performed for example. These different hypotheses have different implications for the way in which a general positioning mechanism would be constructed and utilized by the spinal cord.

1.3 Experimental approach

To search for a general positioning mechanism in the frog spinal cord we first supposed that perhaps local activation of some premotor areas of the spinal cord, either together or in isolation, might cause the development of specific and reproducible postures or trajectories in the limb of the frog. We therefore developed a preparation which allowed us to record the bulk of the limb musculature (90% by mass), and a full set of forces and torques generated in the limb before, during and after microstimulation of a small volume of neuropil in the premotor regions of the spinal cord gray matter. We might envisage several possible outcomes for such an activation. In view of different hypotheses for control of posture and movement these might include: (1) A spinal cord locus could specify a constant direction and magnitude of end-point force regardless of limb configuration. In this way an acceleration or velocity of movement might be specified, but not a position. (2) A spinal cord locus could specify a pattern of joint torques underlying a class of trajectories to a single point. Accelerating and decelerating patterns of torques moving the limb would be found.(3) A spinal cord locus could specify a stable limb configuration and position (a posture) with a surrounding field of restoring forces (an equilibrium point or configuration mechanism). The development of this new posture could evolve with smooth transitions between postures in joint space or end-point space, or in a more complex manner.

We believe it is possible to assess the pattern of use of muscles, and the types of joint torques and endpoint forces generated by the spinal cord by local activation of premotor regions. This paradigm enables us to determine the trajectory and postural generating mechanisms which the isolated frog spinal cord's neural organization can best support.

2 Effects of microstimulation in premotor neuropil

2.1 Experimental preparation

The work reported here is based on microstimulation experiments on 20 spinalized bullfrogs. The spinal cord was transected at the level of the calamus scriptorius. The tectal and other anterior areas were

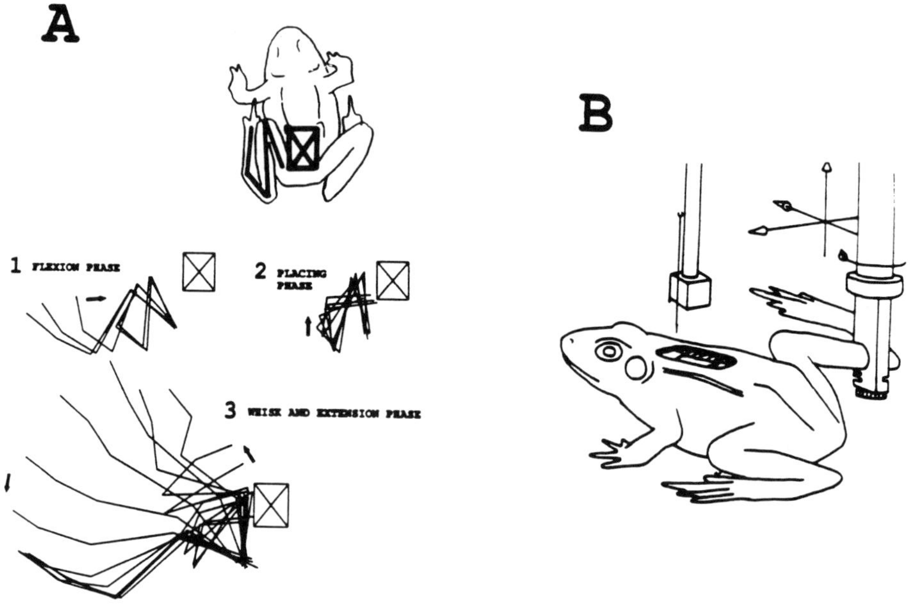

Figure 1. A: An example of fine positioning and trajectory generation in the spinal frog. The frog wipes a noxious stimulus from the back using the hindlimb (from Giszter et al. 1989). This involves several phases and intermediate postures.
B: The preparation used for microstimulation. A stimulating electrode or array is inserted in the lumbar spinal cord of the spinalized frog. The right ankle of the frog is attached to a six-axes force transducer measuring force and torque about three orthogonal axes.

destroyed. Then, the lumbar spinal cord was exposed by removing the spinal arches of the fourth, fifth and six vertebrae. In the same surgical session, we also implanted electromyographic (EMG) electrodes with bipolar leads in 11 leg muscles.

2.2 Stimulation Technique

We elicited motor responses by microstimulating the premotor layers in the grey matter of the spinal cord. We used fine stainless steel stimulating electrodes with impedances ranging from 1 to 10 Megaohms. The electrode was placed at depths ranging from 800 to 1400 microns from the dorsal surface of the spinal cord. Each stimulus consisted of a train of biphasic current impulses lasting 300 milliseconds (duration of 1 millisecond and a frequency of 40 Hz). The peak current amplitude ranged between 1 and 6 microamperes. We chose the current amplitude so as to elicit measurable mechanical responses while remaining as close as possible to the threshold of EMG activations. Marked electrode locations (using electrolytic lesions) were visualized post-mortem in 30-micron, frozen sections stained with Cressyl violet.

2.3 Data Recording

For each spinal stimulation, we collected the EMG signals from the implanted muscles and the mechanical force at the ankle. The stimulation train was set to begin after 200 milliseconds from the onset of data collection. The raw EMG signals were sampled at a rate of 1000 samples per second and rectified and filtered off-line with a time constant of 20 milliseconds.

To measure the mechanical responses seen at the right ankle of the frog we used a six-axes force transducer (Lord LTS-210F), as shown in Figure 1B. The output of the force transducer was a set of three force and three torque components with respect to the x,y and z axes. The resolution was 0.01 lb (approximately 5 gm) for the force components and 0.01 inch-lb for the torque components. The sampling rate for the force signal was 70 Hz. Although we sampled and stored all six force/torque channels, our subsequent analysis was limited to the x and y components of the force vector.

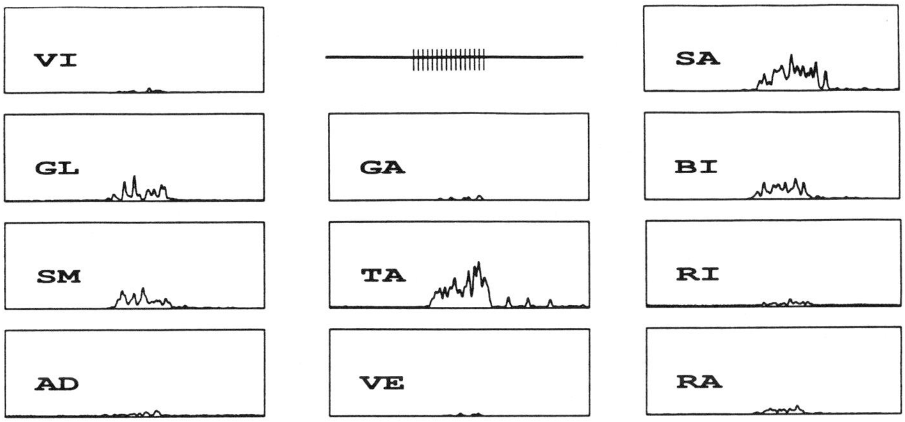

Figure 2. EMG responses to a microstimulation of lateral neuropil measured in 11 muscles. The trace top center is a stimulus marker of duration 300ms. Muscles recorded were as follows: GL, gluteus; VE, vastus externus; VI, vastus internus; RA, rectus anticus; BI, biceps; SM, semimembranosus; GA, gastrocnemius; TA, tibialis anticus; SA, sartorius; AD, adductor magnus.

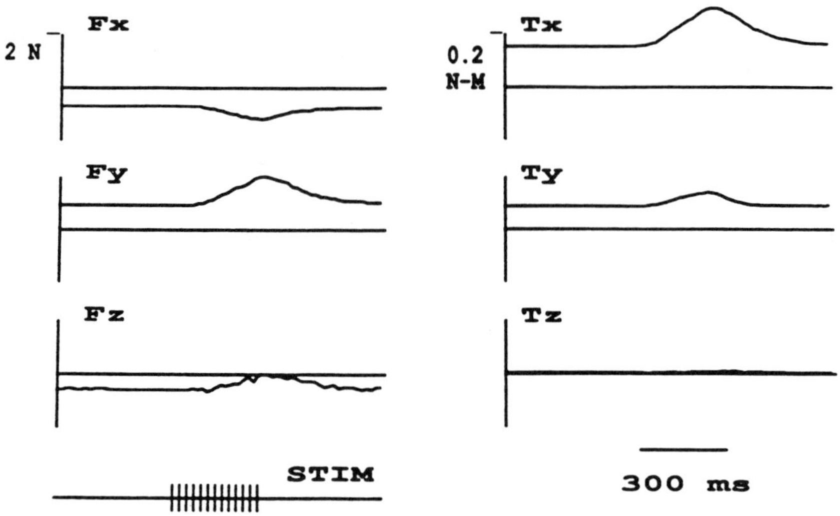

Figure 3. Force (Fx,Fy,Fz) and torque (Tx,Ty,Tz) signals measured by the six axes force transducer before, during and after a stimulation (STIM). Each signal represents a force along or a torque about the relevant axis (for example Fx : force along x axis). Scales are from -2 to 2 Newtons for force and from -0.2 Newton-meters to 0.2 Newton-meters for torque.

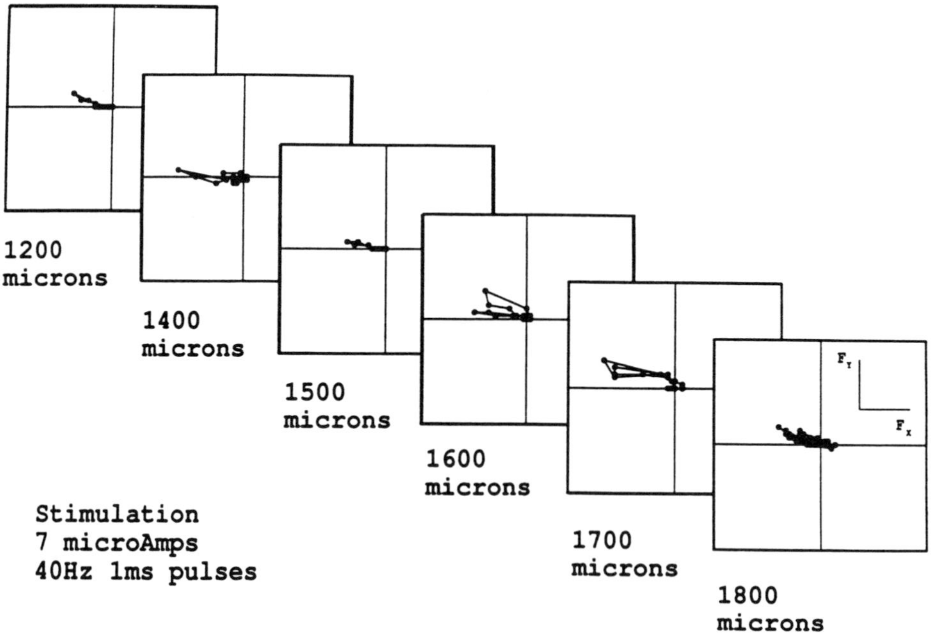

Figure 4. The effect of electrode depth in the spinal cord lateral neuropil on orientation of the stimulus generated force vector. It can be seen that over time the vector grows smoothly and somewhat directly in a particular orientation which is maintained as the electrode penetrates the spinal cord.

2.4 Force and EMG Responses in the lateral neuropil fields at a single limb configuration

2.4.1 EMG patterning

When our electrodes were placed within the lateral interneuron areas we routinely recruited a collection of muscles at threshold. This same group of muscles remained activated through a range of currents and as our electrodes moved from about 600 to 1400 microns. At still higher currents new muscles were recruited. A typical record is shown in figure 2. Small rostrocaudal movements of the electrode produced penetrations eliciting the same muscles. The observations indicate a local homogeneity of response

2.4.2 Force responses

Accompanying the recruitment of muscles, forces would be generated (see figure 3). Before the onset of the stimulus, each force component was at a resting value that depended upon the configuration of the limb. The mechanical response to the stimulation appeared as a change from this resting value after a latency of about 30-50 milliseconds. Force following stimulation would rise to a plateau level, and then some variable period after termination of stimulation decline back to the baseline level. The proportions of the above-baseline components chosen for detailed study (Fx Fy) seemed to be characteristic of the particular site in the cord. Variation of current short of levels recruiting new muscles led to 'actively generated' force vectors of greater magnitude but of constant orientation.

2.4.3 Effect of depth on Force orientation

As the electrode penetrated deeper in the cord, the orientation of the 'active' above-baseline force vector remained constant, though the magnitude might be varied. Usually, so long as the electrode remained in the lateral neuropil this orientation was maintained. An example is seen in figure 4.

3 Estimation of a limb Force Field

The preceding section showed that a consistent set of effects resulted from microstimulation of a region of premotor gray matter. A specific collection of muscles and orientation of forces was found. How this collection of muscle activations and force magnitude and orientations is varied as the configuration of the limb is altered and the nominal end-point at which measurements are made is moved about the workspace was the next question we addressed. In fact, the major goal of these experiments was to determine the field of static forces associated with the stimulation of a spinal-cord site.

3.1 Measurement procedure

To assess configuration effects we measured the mechanical response to the same spinal stimulation at different ankle locations. Typically, we recorded the force vectors in a set from 9 to 16 locations forming a regular grid over the ankle's workspace. At each grid location, we recorded the force vector elicited by the stimulation of the same spinal-cord locus.

After the mechanical responses had been measured in all grid locations, the last stimulation was delivered with the ankle placed in the first tested grid location as a control. Then we compared the outcome of the last stimulation with the first stimulation. Any visible difference between either the EMG records or the force records in these two stimulations was attributed to some unwanted change in the electrode location or in the state of the preparation. We only used for further analysis the data obtained from a stable preparation.

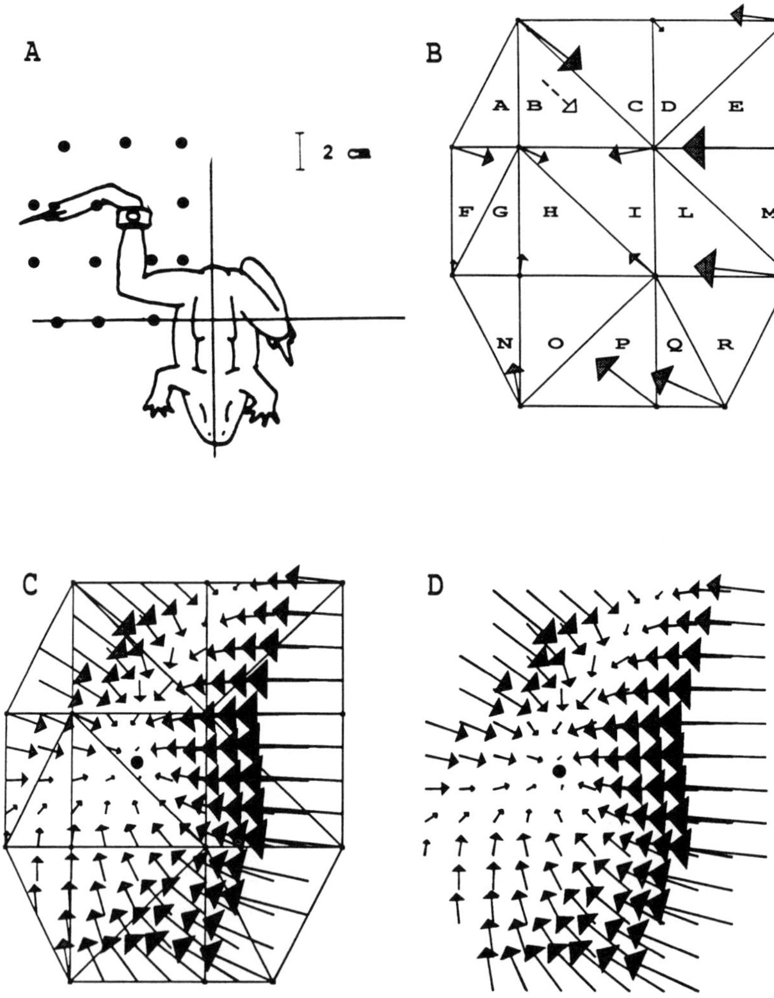

Figure 5. The procedure used to estimate a force field. A: The tested locations of the workspace are shown in reference to the frog's body. All fields shown in subsequent figures can be referenced to the frog in a similar way. B: The partitioning of the tested workspace into a set of non overlapping triangles. Each vertex of a triangle is a tested point. Force vectors measured at each point at the same latency are displayed as solid arrows. The dashed line in triangle B shows an interpolated vector, obtained linearly from the vectors at the three vertices of triangle B. C: Interpolated field . D: the interpolated field with tesselation triangles removed for clarity of display.

3.2 Force field reconstruction

For a single stimulation site, the force vectors measured at the different force locations were considered as samples of a continuous force field. By definition, a force field is a function relating every workspace location to a corresponding force vector. To estimate the force field in a large region of the ankle's workspace we implemented a piecewise-linear interpolation procedure (see Figure 5).

1. First, we partitioned the ankle's workspace into a set of non-overlapping triangles (A,B,C,... in Figure 5B) using a Delaunay triangulation algorithm (Preparata and Shamos, 1985). The vertices of each triangle were tested grid points.

2.Within each triangle, we applied a linear interpolation to the force vectors measured at the corners. Thus, within each interpolation triangle the force components were given as:

$$F_x = a_{1,1}x + a_{1,2}y + a_{1,3}$$
$$F_y = a_{2,1}x + a_{2,2}y + a_{2,3}$$

Note that the above expression had a unique solution for the $a_{i,j}$ when supplied with three data vectors.

3. The above interpolation procedure was applied to the data vector collected at a given latency from the onset of the stimulus. Using data sets obtained at two subsequent latencies we derived two estimates (or "frames") of the force field generated by a single spinal stimulation. The presence of one or more equilibria was tested for by searching within each interpolation triangle for a location (x_0,y_0) at which F_x and F_y were both zero.

3.3 Separation into quiescent and active fields

Because we recorded the pre stimulus (or baseline) forces, we have a measure of the quiescent postural force field of the frog's limb. The effect of the stimulation is a new force field. This could arbitrarily be decomposed into the quiescent postural field and a collection of additional forces resulting from the applied stimulation by a vector subtraction of two of the the pre and post stimulus force fields. This operation is highly significant in assessing contributions of stimulus triggered activation of muscles in shaping the properties of the force fields observed (see below).

4 Experimentally determined force fields

4.1 Convergent fields in the lateral neuropil areas

Prior to stimulation we measured a force field associated with the quiescent resting leg. This consisted of an equilibrium position and a convergent force field which stabilized this resting posture of the limb's nominal end-point (see figure 6, part 1A). Remarkably, we also found that the force fields associated with stimulation of lateral neuropil regions (see Fig 6) produced convergent force fields in the limb. These fields were of a very simple structure, with only a single stable equilibrium. Often the equilibrium of such a convergent field lay within the area of the workspace tested. In other instances the equilibrium position lay outside the tested region. Such equilibrium points were distributed throughout the tested area of workspace. Different spinal cord loci generated different equilibrium fields. An example of such fields from different loci is seen in figure 6. The span of final equilibria seen in a small sampling of spinal cord loci in a single frog is shown in figure 6 part 2. It can be seen equilibria lie both in extension and in flexion. For this reason we do not believe we are simply eliciting wiping or flexion withdrawal.

4.2 Divergent fields among motoneurons in the motor nuclei

When the electrode was lower in the gray matter, among the motoneurons, or in the ventral roots, the result was a set of parallel or divergent forces with no equilibria in the tested region of the workspace. The motor nuclei of the frog are frequently commingled, and often the same collection of muscles were activated in the motoneuron region as in the lateral neuropil areas. The balance was, however, sufficiently different to cause these sizeable field differences. These differences are shown in figure 7.

4.3 The convergence of fields is actively generated

The question arises as to whether the equilibrium fields of lateral neuropil are (1) a result of stimulus generated active contributions which possess an equilibrium, imposed on top of the quiescent postural field, or are (2) a result of a field without an equilibrium in the tested region, imposed over the quiescent postural field. Figure 8 shows that these stimulus generated fields possess an actively generated equilibrium even when the quiescent postural forces are subtracted. An equilibrium is specified by the stimulus generated activation of muscles and this results in a stiffer total force field. The results therefore suggest that the stimulation of premotor areas lead to a balanced recruitment of muscles.

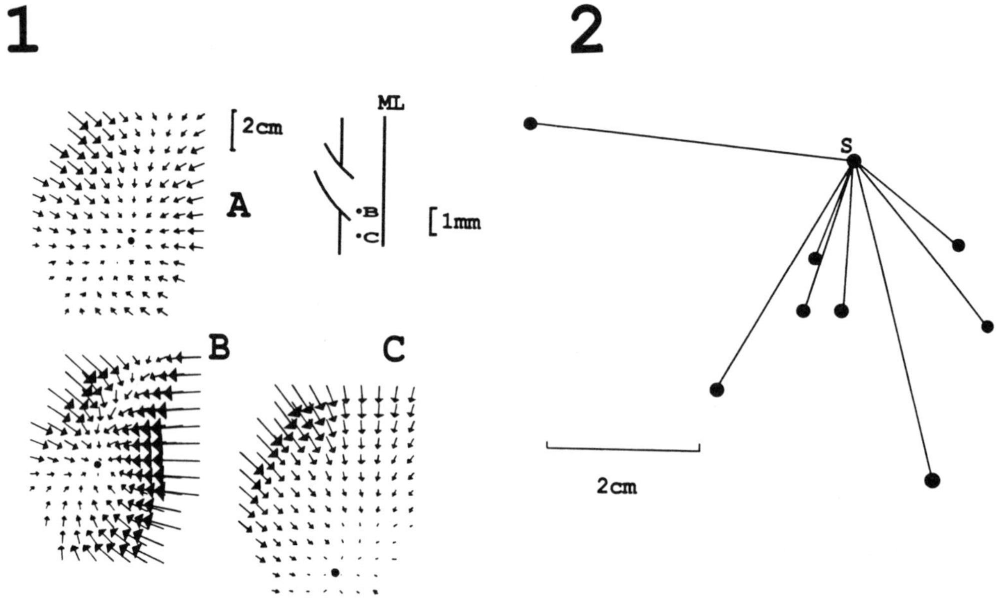

Figure 6. Lateral neuropil force fields. 1: Inset: schematic of spinal cord dorsal view. A: pre stimulus postural field B: post stimulus peak field obtained at a location (B in the inset) C: a second peak post stimulus field obtained at a location 1mm rostral to location B (C in the inset). 2: S pre stimulus equilibrium location. The end of each ray from S is a new equilibrium location generated in the peak post stimulus field at a different electrode location in a single frog.

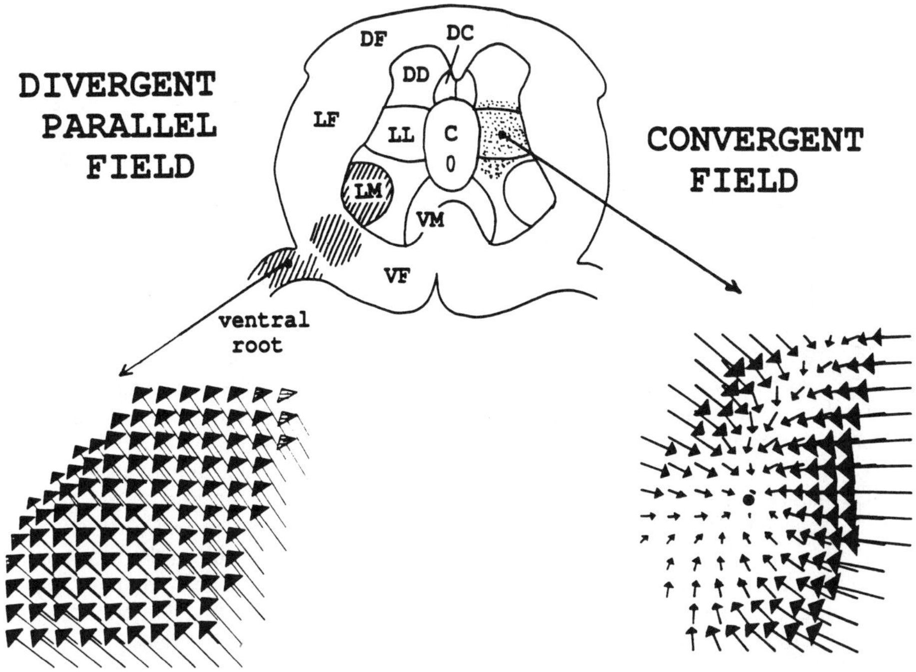

Figure 7. Examples of the two field types found in different spinal cord locations. Center: Diagram of a transverse section of the spinal cord in the lumbar enlargement. DF: dorsal funiculus, DC: dorsal column nucleus, DD dorsal neuropil region, LF: lateral funiculus, LL: lateral neuropil region, C: central neuropil region, VM: ventromedial neuropil region, VF: ventral funiculus. Hatched regions are areas in which parallel or divergent force fields were found. Stippled regions are areas in which convergent force fields were observed.

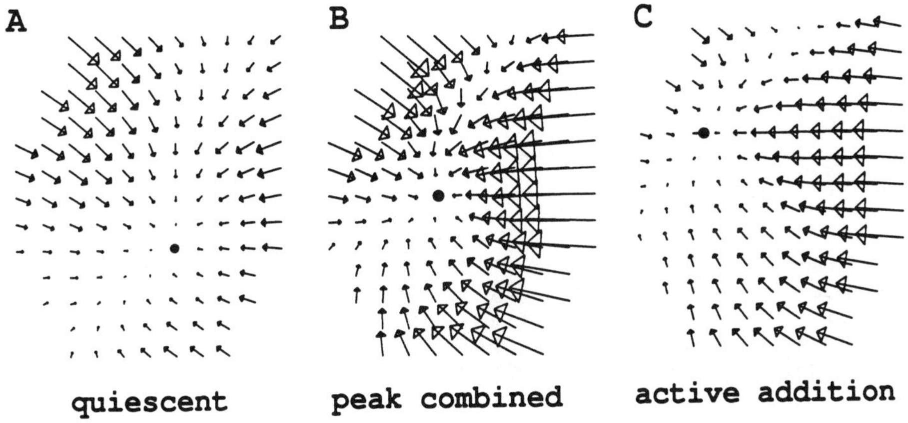

Figure 8. An example of the separation of pre and post stimulus contributions to force field organization. Field C was obtained by a vector subtraction of A from B. The field in C demonstrates that the convergence of the field is an actively specified property and not a result of a pre-stimulus postural field.

4.4 The equilibrium trajectory

After the delivery of a stimulation, the force field as a whole changed with time. The dependency of the force field upon time is graphically captured by a sequence of "frames" (Figure 9A). Each frame represents the force field measured at a given latency from the onset of the stimulus. In the first frame (latency = 0), we have the resting force field, that is the field as it was before the stimulus had produced any effect. The subsequent frames are separated by intervals of 86 milliseconds. They show the effect of the stimulus as a smooth change in the overall pattern of forces. In several instances, we have observed the following sequence of events (as indicated in figure 9A):

1.After a brief delay from the onset of the stimulus (about 50 milliseconds), the pattern of forces began to change and the equilibrium position started to "move" in a given direction (Figure 9, frames 1 to 3).

2.Then (Figure 9A, frame 4) the equilibrium position reached a point of maximum displacement within the workspace. This point was maintained for a time interval depending upon the stimulation parameters (current, train duration, etc.) At the same time, the field forces reached a maximum amplitude around the equilibrium position, corresponding to a maximum in end-point stiffness.

3. Finally (Figure 9A, frames 5 and 6) the field forces started to decrease in amplitude and rotate toward their original directions. At the same time, the equilibrium moved back to its resting location.

Figure 9 also displays a summary of the temporal evolution of the equilibrium point corresponding to the full sequence of fields at the highest temporal resolution measurement possible This sequence of static equilibria is by definition an "equilibrium trajectory": as the neuromuscular activity changes gradually in time (as shown in Figure 9B), the equilibrium undergoes a gradual shift. Furthermore, after the EMG activities have returned to their resting value, the equilibrium returns to the resting location.

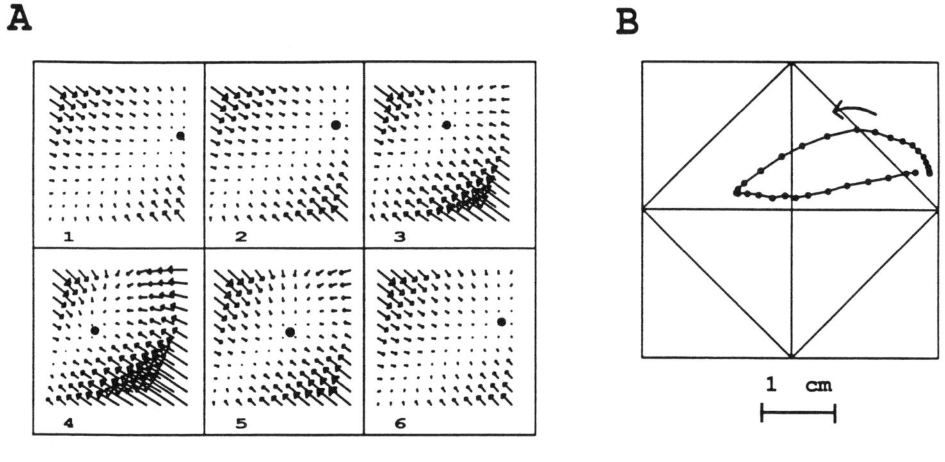

Figure 9. The equilibrium trajectory. Left panel: A temporal sequence following a stimulation. The six frames are ordered left to right, and separated by 86 milliseconds. The filled circle represents the equilibrium point. Right panel: The trajectory of the equilibrium point at maximum resolution superimposed on the tesselating triangles used. Note that the equilibrium moves smoothly across several triangles. As it crosses the edges different vectors are used to interpolate.

4.5 EMG variation with configuration

We have examined the variation of activation (as indicated by EMG) as a function of configuration. An example for three muscles is shown in figure 10. Qualitatively the collection of muscles activated was almost always invariant across the tested range of configurations and workspace locations. Indeed the *quantitative* variations were also usually modest except in extreme positions. In a few instances, we failed to observe EMG responses in some muscles at a location, especially at locations corresponding to extreme limb flexion or extension. However most frequently a modest afferent/configuration based component could be detected. Taken together this suggests a strong feedforward component in the organization of the equilibrium force fields with a modest afferent modification.

5 The patterns of muscle activation underlying equilibria

To address the question of how much of the patterning of forces could be accounted for using a feedforward model of muscle activation (i.e. a configuration insensitive activation), we measured the force field contributions of individual muscles. A force field such as we measured is produced by some balance of activation of a collection of muscles acting through skeletal mechanisms.

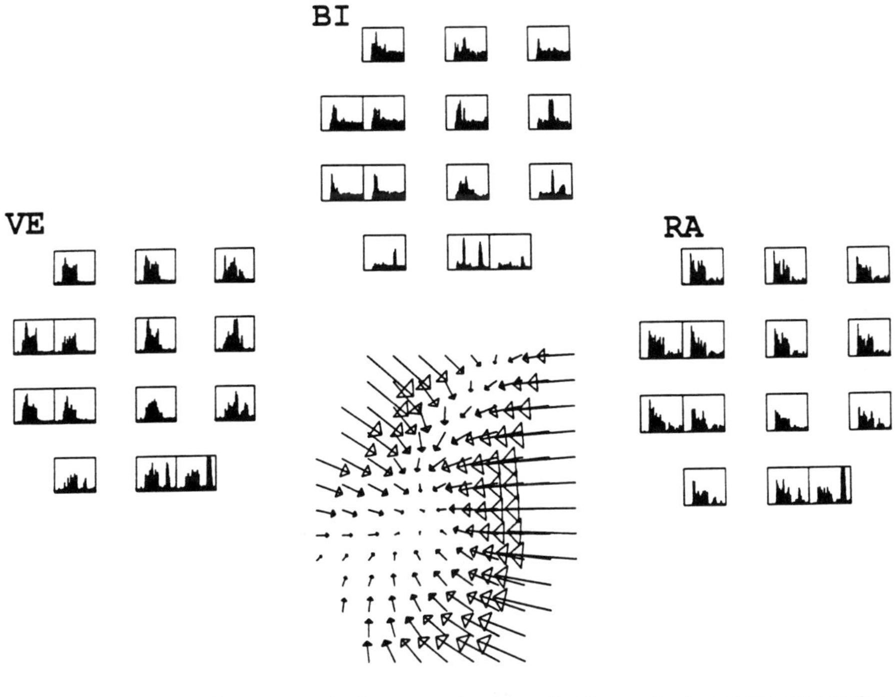

Figure 10. Variation of EMG response in three muscles recruited in a spinal cord site generating a convergent force field. VE: vastus externus. BI: biceps. RA: rectus anticus. Each small square is an EMG record. The lower left corner of each square is anchored to the grid point at which the record was collected. Note that the records are at first examination quite similar. However the biceps shows some inhibition in extreme flexion (lower right area of grid). Also note the configuration dependent reflex activation of muscles towards the end of each record and after the cessation of stimulation.

5.1 Effects of individual muscles in generating a force field

The muscles we have recorded can only interact with one another through the skeleton and not directly. Thus their contributions can be examined in isolation. Direct electrical stimulation of a single muscle was used to produce a near maximal activation. This was used together with positioning of the limb at each of a grid of points to measure a muscle generated field. The end-point force measured results from torques produced by this muscle alone. The muscle is acting through the skeleton and therefore the force produced depends on configuration. Since the measurements are static, velocity can be neglected as a factor in the muscles contraction. The torques generated thus depend only on muscle length and moment arm (at a particular configuration), and the level of muscle activation. However, the muscle's length and its moment arm are are also functions of configuration. Thus only two sets of variables, configuration parameters and levels of activation, fully determine the end-point force generated by a given muscle. In the experiments described here configuration was fully constrained and determined by end-point position. Thus a unique field can be measured and associated with each muscle. This is the muscle's contribution after removal of the baseline force variations seen in the quiescent leg (see above). This individual muscle field can then be modulated by a coefficient (c) representing the level of activation of the muscle:

$\mathbf{f}(\mathbf{r},t) = c^*\mathbf{F}(\mathbf{r},t)$ where $\mathbf{F}(\mathbf{r},t)$ is the muscle force field, and $\mathbf{r}$ the configuration

Examples of such individual muscle fields are shown in figure 11 for the three muscles shown in figure 10. It should be noted that no individual muscle generated an active field contribution with any equilibria in the regions of measurement. Convergent or divergent fields were seen which were very repeatable and characteristic of each muscle.

5.2 Combination of muscles to generate a force field

A collection of individual muscle fields could be combined by a simple vector sum to produce a composite field representing the result of some balance of feedforward activation of muscles, that is, an activation unmodulated by configuration feedback. In this way we could assess the possibility (or otherwise) of obtaining particular force fields without afferent modulation, and under any particular recruitment hypothesis. Note that if we include feedback modulation of activation the parameter c above would also become a function of configuration.

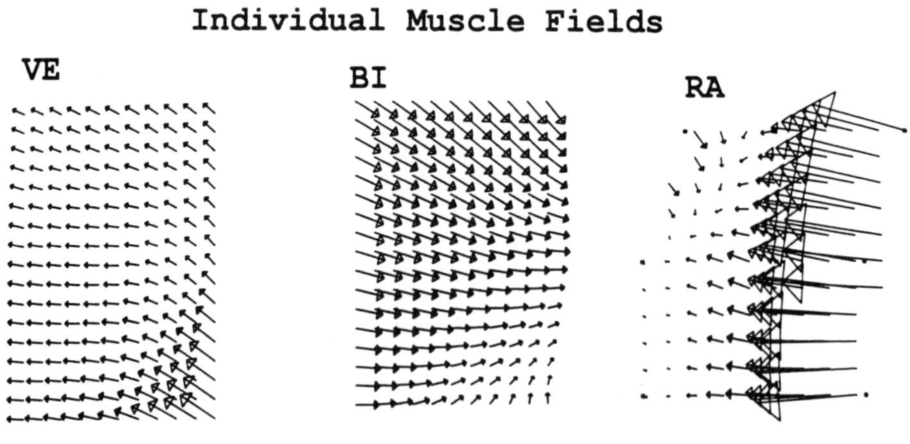

Figure 11. The individual force fields generated by direct electrical stimulation of three muscles. The pre stimulus postural fields have been removed. Note the configuration effects on force as a result of muscle length, moment arm and linkage kinematics. VE: vastus externus. BI: biceps. RA: rectus anticus.

5.3 An anatomically based feedforward model of force field generation

We used this feedforward model of equilibrium force field generation to assess a simple hypothesis of spinal cord organization. We examined the hypothesis that the balanced recruitment seen experimentally was due to recruitment of motoneurons in proportion to their frequency in a locus of spinal cord. We used the data of Cruce 1974 on the organization of motor nuclei in the lumbar cord of the frog to predict motoneuron frequency at a point in the cord. Additionally we assumed our electrode recruited cells in a gaussian distribution about the electrode location. Thus depending on current spread (the variance of the electrode gaussian) and motoneuron distributions in the cord, different balances of muscle activation will result from stimulation of different spinal cord loci. These factors determined a coefficient for each measured muscle field, which were then combined. An example of results from such a simulation is shown in figure 12. It can be seen that in the simulation some regions of cord can be expected to produce equilibrium configurations (based solely on balance of motoneuron frequencies in the region), however other areas can produce divergent areas in fields similar to those we observed experimentally when stimulating among motor nuclei. It should be noted that the precise spinal cord anatomy and form of the stimulus pulse spread are quite critical in the exact fields produced in these simulations. From these simulations it appears that the balance of muscles recruited by premotor stimulation in the lateral regions of the cord cannot be account for simply by motoneuron frequency using the anatomical and stimulation parameters we have tested. We conclude this since we have never seen divergent fields in premotor areas experimentally. We therefore believe additional mechanisms must be invoked. Among these might be motoneuron dendritic processing, operation of the size principle, modulation and recruitment through interneuronal structures, and the afferent modulation of activation observed experimentally.

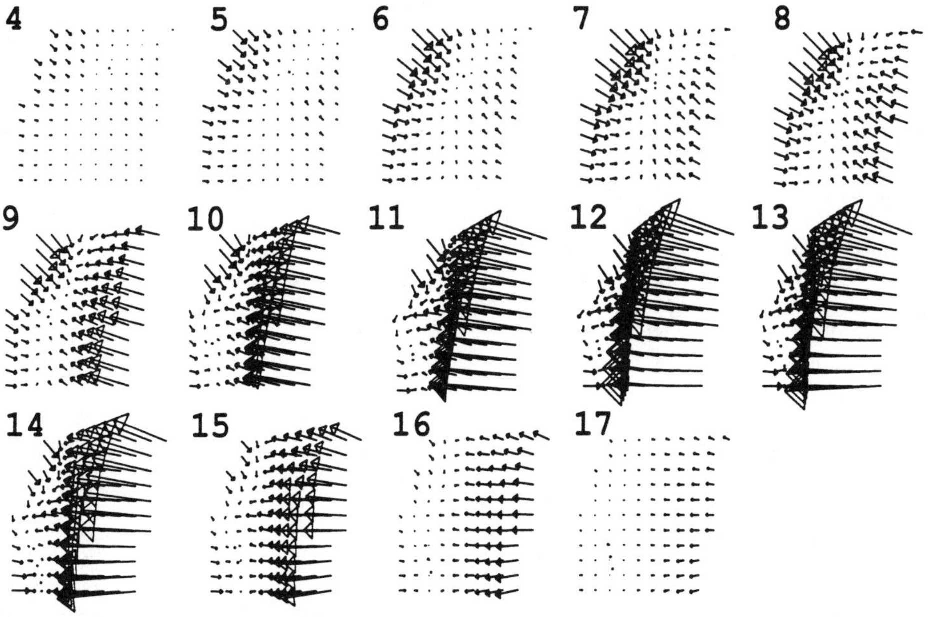

Figure 12. Force fields generated by a simulation combining the individual force fields of twelve muscles under the assumption of configuration independent activation of muscles. The strength of each muscle's field is modulated in accordance with the hypothesized spread of excitation and the estimated motoneuron frequency in an area of the lumbar spinal cord. The numbers indicate distance in millimeters along the lumbar enlargement. It can be seen that there is a systematic variation of the fields along the cord, becoming dominated by vectors diverging into extension in one region of the cord. Such divergence was never observed in experimental results from lateral neuropil.

6 Discussion and Conclusions

The experiments described in this paper have demonstrated that the stimulation of a site in the spinal cord generates a particular pattern of EMG activities and a field of elastic forces with a single equilibrium point. The field of elastic forces is a complete characterization of limb posture: it specifies an equilibrium location together with the restoring forces generated by the muscles when the limb is away from the equilibrium. Stimulation of different spinal sites produced different EMG patterns as well as different displacements of the equilibrium point. Equilibria were found both in extension and flexion which is not consistent with simply eliciting wiping patterns. In this context it is worthy of note however that extensive cocontraction ***has*** been demonstrated to occur during wiping (Schotland et al. 1989).

Using a simple model it has been shown that an equilibrium limb configuration can be achieved simply by a balanced use of muscles. This balanced recruitment could operate in a strictly feedforward way so as to specify equilibria at various locations in the limb workspace. However, afferent effects are also observed experimentally and some fine details of the force field might rely on such afference.

We have explored the hypothesis that simple recruitment of motoneurons based on their frequency in lumbar spinal cord underlies our observations and found that it is likely additional mechanisms must be invoked to account for the complete absence of divergent fields in the microstimulation of premotor areas. These mechanisms might be dendritic processing within the motoneurons coupled with operation of the size principle, the combined use of the interneuron-motoneuron assemblies, and the recruitment and interaction of afferent fibers with these other mechanisms. We are currently unable to distinguish these alternatives as possible sites of action and mechanisms underlying our microstimulation results. Regardless of the exact basis however, we believe these experiments may give extensive further insight into the organization spinal mechanisms.

Our results show a clear-cut relationship to the equilibrium-point hypothesis. This hypothesis was first proposed by Feldman (1974). The gist of this hypothesis is that the CNS generates movements as a shift of the limb's equilibrium posture -- that posture at which the net torque generated by the muscles at the joints is zero. The equilibrium-point hypothesis is strongly rooted into the biomechanics of muscles and in particular in their tunable spring-like behavior (Rack and Westbury, 1969; Hogan, 1984, McIntyre, 1990): the isometric force generated by a muscle depends on the level of neuromuscular activity as well as upon the length of the muscle. In other words, the state of activation of a muscle does not determine tension alone but a whole length-tension curve. Experimental work on single-joint movements (Bizzi et al. 1984) has supported the idea that postural control is achieved by the CNS through the choice of agonist and antagonist length-tension curves. These curves determine the equilibrium position for the limb and the stiffness about the joints.

The equilibrium-point hypothesis has important implications for the central representation of movement and posture. According to the views based on the explicit representation of inverse dynamics, posture and movements are separate issues: complex computations are necessary for generating movements, whereas posture is a separate feedback control problem with practically no computational content. In contrast, according to the equilibrium-point hypothesis, the major task for a biological system is that of establishing a map between patterns of neuromuscular activities and limb postures. This map may be considered the primal form of motor coordination upon which rests the ability to generate limb movements. The work presented here suggests that major parts of such a primal map may reside entirely in the spinal cord.

In summary, we have observed a clear correspondence between neuromuscular activations and limb postures. We observed that movements generated by spinal neural activation are accomplished by a gradual change of the equilibrium position and of the surrounding field of restoring forces. We have shown these equilibrium postures can be generated by a balanced co-activation of muscles, even in a simple feedforward and configuration independent manner. Our experiments thus provide evidence for the role of equilibrium-point mechanisms in the generation of posture and movement by the frog spinal cord.

Acknowledgements

This work was supported by NIH grants NS09343 and AR26710 and ONR grant N00014/88/K/0372.

7 References

1.Bizzi, E., A. Polit and P. Morasso (1976) Mechanisms underlying achievement of final head position. Journal of Neurophysiology, 39: 435.
2.Bizzi, E., N. Accornero, W. Chapple and N. Hogan (1984) Posture control and trajectory formation during arm movement. Journal of Neuroscience, 4: 2738.
3. Cruce, L.E. (1974) The anatomical organization of hindlimb motoneurons in the lumbar spinal cord of the frog, rana catesbiana. J. Comp. Neurol, 153:59.
4.Feldman, A. G. (1974) Change of muscle length due to shift of the equilibrium point of the muscle-load system. Biofizika, 19: 534.
5.Flash, T. (1987) The control of hand equilibrium trajectories in multi-joint arm movements. Biological Cybernetics, 57: 257.
6.Fukson, O. I., M. B. Berkinblitt and A. G. Feldman (1980) The spinal frog takes into account the scheme of its body during the wiping reflex. Science, 209: 1291.
7.Giszter, S. F., J. McIntyre and E. Bizzi (1989) Kinematic strategies and sensorimotor transformations in the wiping movements of frogs. Journal of Neurophysiology, 62: 750.
8.Hogan, N. (1984) An organizing principle for a class of voluntary movements. Journal of Neuroscience, 4: 2745.
9.Hollerbach, J. M. and T. Flash (1982) Dynamic interactions between limb segments during planar arm movement. Biological Cybernetics, 44: 67.
10.Kelso, J. A. S. (1977) Motor control mechanisms underlying human movement reproduction. Journal of Experimental Psychology, 3: 529.
11.McIntyre, J. (1990) Utilizing elastic system properties for the control of posture and movement. PhD. Thesis, MIT, Department of Brain and Cognitive Sciences, Cambridge, MA.
12.Mussa-Ivaldi, F. A., N. Hogan and E. Bizzi (1985) Neural , mechanical and geometric factors subserving arm posture in humans. Journal of Neuroscience, 5: 2732.
13.Preparata, F. P. and M. I. Shamos (1985) Computational Geometry. An Introduction. Springer-Verlag, Heidelberg.
14.Rack, P. M. H. and D. R. Westbury (1969) The effects of length and stimulus rate on tension in the isometric cat soleus muscle. Journal of Physiology (London), 217: 419.
15.Schotland J. L., W. A. Lee and W. Z. Rymer (1989) Wiping reflex and flexion withdrawal reflexes display different EMG patterns prior to movement onset in the spinalized frog. Experimental Brain Research, 78: 649.

Actuator and Kinematic Redundancy in Biological Motor Control*

Reza Shadmehr

Center for Neural Engineering
University of Southern California
Los Angeles, CA 90089-2520 USA

Abstract

A task must be specified in terms of both the position and stiffness of the limb before muscle forces and activations are unambiguously assigned. To illustrate this, we begin with the problem of how to control an inverted pendulum with a pair of muscles. An active state model of the frog's gastrocnemius is used to derive three criteria for the stiffness characteristics of the system during posture and movement. The differential equation representing this model is solved to indicate the relationship between force and stimulation frequency. This result leads to an interesting prediction of muscle forces in a minimum stiffness equilibrium point control scheme: neural activity in the agonist muscle should decrease as the joint rotates the limb against gravity. For the case where the number of joints exceeds the task's degrees of freedom, an algorithm for mapping end-effector position and stiffness to the lengths of the muscles is considered. We show that a previously proposed algorithm for control of multi-joint limbs (Berkinblit et al. 1986a, Hinton 1984) is in fact a special case of this mapping. We contend that these kinematic maps must be augmented by a mechanism that takes into account the dynamics of the muscle-load-feedback system. We suggest an adaptive control scheme where the derived kinematic relationships are used to set the bias of the stretch reflex feedback loop, while a learning mechanism produces a virtual equilibrium trajectory that compensates for the second order dynamics of the load, as well as the dynamics of the muscles.

1 Introduction

The process of generating a movement may be viewed as a series of transformations from an overall movement objective (e.g., hitting a target with a baseball) to a plan specifying the desired behavior of the end-effector (e.g., the path that the hand should follow before the ball is released), and finally to a pattern of muscle activations. There is a wealth of data that describes the patterns of muscle activity and behavior of proprioceptive feedbacks during execution of

*This research was supported in part by NIH grant R01-NS24926-05 to Prof. Michael Arbib. The author is recipient of an IBM Graduate Fellowship in Computer Science. He can be reached by E-Mail at reza@rana.usc.edu.

particular trajectories, yet there has been relatively little progress made in describing algorithms that the CNS might be using to learn the dynamics of the muscle-load-feedback structures, and to actually perform the planned trajectory. There has been some progress in understanding the movement planning process in the CNS: By looking at the kinematics of reaching, Hogan (1984) has suggested that it is the position of the hand, rather than the joint angles, that is being planned. Yet it is not apparent how a task planned in terms of hand coordinates can be executed by muscles that are inherently joint-based: There are generally more degrees of freedom in a limb than there are degrees of freedom in the movement, so producing a trajectory of muscle lengths for a desired hand trajectory is an ill-posed problem. This is the issue of kinematic redundancy.

The situation is further complicated by the fact that at least two muscles act on a single joint, and in many cases, a muscle spans more than one joint. The problem is that the map from joint torques to muscle forces is not one-to-one, i.e., many levels of co-contraction can lead to the same effective joint torque. How does one decide on the amount of co-contraction? This is the issue of actuator redundancy.

Aside from the fact that kinematic redundancies in the system make the transformation from a planned hand trajectory to muscle activations an ill-posed problem, there is also the issue that the skeleton has non-linear and coupled dynamics: Forces produced by a muscle at a given joint affect the position of all the joints in the limb. Therefore, to execute a task, one must take into account the dynamics of the muscles, the skeleton, and the load. The controller is further restricted by a relatively slow afferent system, prohibiting implementation of a host of techniques that are commonly used in robotics in order to avoid compensation for dynamics of the limb.

In this paper we begin with the problem of how to control a single joint (an inverted pendulum) with a pair of antagonistic muscles. The muscles are modeled to faithfully reproduce the mechanical behavior of a frog's gastrocnemius, and are based on the active state theory (Gasser and Hill 1924) as parameterized by Inbar and Adam (1976). Our objective is to be able to activate these muscles so that the joint moves along a desired trajectory, with a particular stiffness profile. Based on stability requirements of the limb and the intrinsic mechanical limitations of muscles, we will derive three criteria for joint stiffness during posture and movement. A minimum stiffness strategy is used to calculate forces that should be produced by the muscles if an equilibrium point model (Feldman 1966) is used for control of the limb. This approach is contrasted with a model that takes into account the dynamics of the task. We will then explore algorithms for the problem of learning trajectory control in multi-joint biological limbs, with particular emphasis on representing and regularizing motor redundancy. We will show that the mechanical characteristics of the muscles and the feedback systems can be used by the CNS to not only solve the redundancy issues (as has been suggested by Mussa Ivaldi et al. 1988), but also to provide a means for building an adaptive internal model of the muscle-load-feedback system for precise execution of desired tasks.

2 Actuator Redundancy

One of the fundamental differences between the way one programs robots to perform tasks and the way biological systems move is in the information that is needed before a command can be sent to the actuators. In robots, once appropriate joint torques are determined, the robot will move more-or-less along the desired trajectory. In this section we will see that this information alone will not be sufficient for assignment of muscle activation rates.

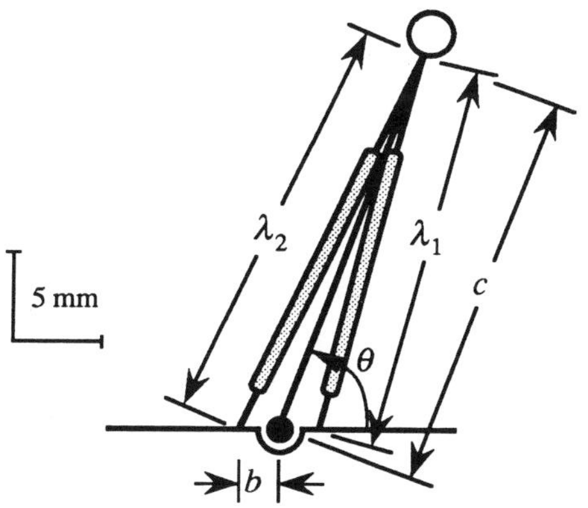

Figure 1: A ball-and-socket joint with two muscle-like actuators. Length of each muscle is denoted by λ_i.

Consider the ball-and-socket joint of Fig. 1, where the dynamics of the skeleton are described by the following:

$$\tau = mc^2\ddot{\theta} + \nu\dot{\theta} + mcg\cos(\theta) \tag{1}$$

where m is the point mass at the end of the limb, c is the length of the limb, ν is the joint's viscous parameter, and g is the gravitational constant. The lengths of the muscles, λ_1 and λ_2, are related to the joint angle θ by the functions:

$$\lambda_1 = \sqrt{b^2 + c^2 - 2bc\cos(\theta)} \tag{2}$$

$$\lambda_2 = \sqrt{b^2 + c^2 + 2bc\cos(\theta)} \tag{3}$$

Differentiating the muscle lengths with respect to the joint angle yields the following:

$$\frac{d\lambda_1}{d\theta} = \frac{bc\sin(\theta)}{\lambda_1} \tag{4}$$

$$\frac{d\lambda_2}{d\theta} = \frac{-bc\sin(\theta)}{\lambda_2} \tag{5}$$

The relationship between torque τ at the joint, and muscle forces ϕ_1 and ϕ_2 is:

$$\tau = -\frac{d\lambda_1}{d\theta}\phi_1 - \frac{d\lambda_2}{d\theta}\phi_2 = -\frac{bc\sin(\theta)}{\lambda_1}\phi_1 + \frac{bc\sin(\theta)}{\lambda_2}\phi_2 \tag{6}$$

Note that in Eq. (6), for a given joint torque trajectory $\tau(t)$, a trajectory in terms of muscle forces $\phi(t)$ cannot be calculated because there is an infinite set of antagonistic muscle forces that can lead to generation of the same joint torque. A simple example of this is evident in postural control: one can hold a limb at the same position with varying degrees of stiffness. For the limb in Fig. 1, joint stiffness, K_J, is defined as:

$$K_J = \frac{d\tau}{d\theta} = -\frac{d^2\lambda_1}{d\theta^2}\phi_1 - \left(\frac{d\lambda_1}{d\theta}\right)^2\frac{d\phi_1}{d\lambda_1} - \frac{d^2\lambda_2}{d\theta^2}\phi_2 - \left(\frac{d\lambda_2}{d\theta}\right)^2\frac{d\phi_2}{d\lambda_2} \tag{7}$$

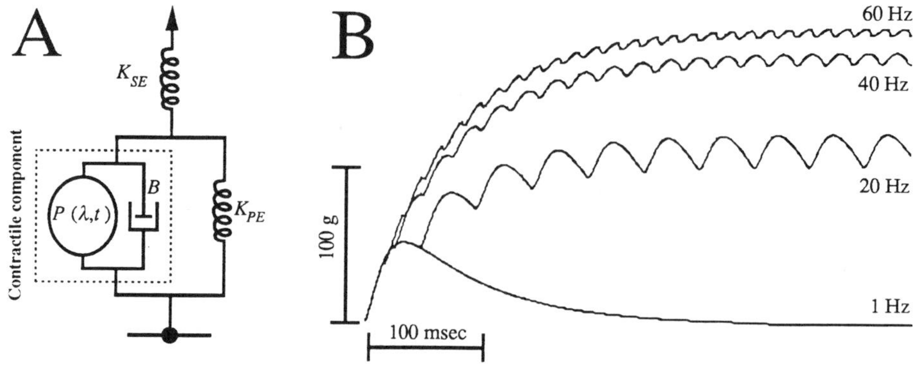

Figure 2: An active state muscle model. A) The active state $P(\lambda, t)$ is the tension developed by the force generator. B) Isometric tension produced by the muscle model at various stimulation rates.

By specifying joint stiffness, one is describing, for example, how the limb should react in case any part of it comes in contact with the environment. In the next section we will describe a muscle model and show that a unique set of muscle forces will result when joint torque and stiffness are specified.

From Fig. 1, we can see that K_J should always be negative: if a disturbance displaces the limb, the resulting change in torque should be in the opposite direction. Setting K_J depends on the stability requirements of the system: The equilibrium of the system is at $\theta_E = \cos^{-1}(\tau/mcg)$, and in order to guarantee stability, the stiffness must be:

$$K_J < -mcg\sin(\theta_E) \tag{8}$$

The fact that muscles cannot push against the skeleton will also limit the possible range of K_J during motion. In order to explain this, we need to fully specify (6) and (7) by formulating a muscle model.

2.1 A Muscle Model

Here we describe a two component active state muscle model: an elastic component in series with a contractile component (Fig. 2A). This is the model proposed by Gasser and Hill (1922) to explain the tension dynamics of various frog muscles. The following differential equation describes force development in this model:

$$\dot{\phi} = \frac{K_{SE}}{B}(K_{PE}\Delta\lambda + B\dot{\lambda} - (1 + \frac{K_{PE}}{K_{SE}})\phi + P(\lambda, t)) \tag{9}$$

where K_{SE} is the series elastic component in the tendons, K_{PE} is the parallel elastic component which with K_{SE} accounts for the passive tension properties of the muscle, B represents the viscous resistance opposing force development during shortening, λ_0 is the rest-length of the muscle beyond which passive force is developed (Aubert et al. 1951), and $\Delta\lambda = \lambda - \lambda_0$ when

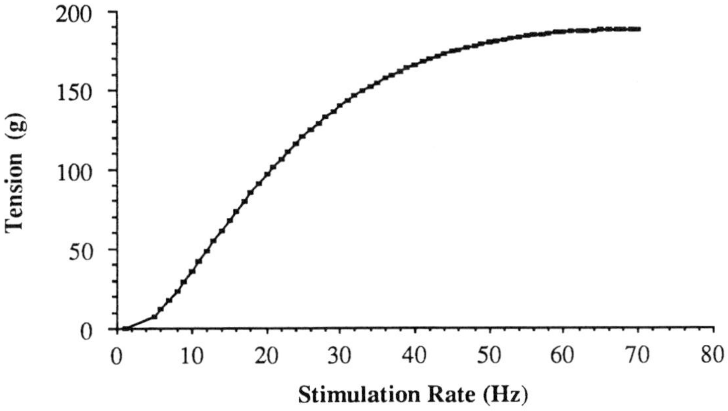

Figure 3: Tension output of the muscle model during isometric conditions at resting length after $1/\Delta t$ impulses, and just before the arrival of the next impulse.

$\lambda_0 < \lambda$, else $\Delta\lambda = 0$. $P(\lambda, t)$ is the active force produced by the contractile mechanism in the muscle. It can be represented by the product of a length dependent function $S(\lambda)$, and a function $Q(f)$ which depends on the history of muscle stimulation $f(t')$ for $t' \leq t$. Using a linear systems approach to $Q(f)$, the input to the contractile mechanism, $f(t)$, is a series of impulses, while the output is force:

$$P(\lambda, t) = S(\lambda) \int_{-\infty}^{t} f(t')h(t - t')dt'$$

Using a parameter estimation technique, Inbar and Adam (1976) have calculated $h(t)$, i.e., the impulse response of Q. We approximated their results in Fig. 5A by a sum of two exponentials, and derived the active tension (in grams) when the input to the system is a series of impulses of frequency $1/\Delta t$:

$$P = S(\lambda) \sum_{n=0}^{\infty} (\exp(-70(t - n\Delta t)) - \exp(-210(t - n\Delta t))(u(t - n\Delta t) - u(t - (n+1)\Delta t)) \quad (10)$$

where $u(t)$ is a unit step function at $t = 0$, and $S(\lambda) = 1200(2\lambda/\lambda_0 - 1)$ when $\lambda \leq \lambda_0$, else $S(\lambda) = 0$. Fig. 2B is the simulation results of (9) and (10) during isometric conditions at $\lambda = \lambda_0$.

By formally defining a muscle model we have specified the dynamics of the system in Fig. 1. However, control of this system requires understanding how much activation the muscles should receive in order to produce a particular force profile (the inverse dynamic model of the muscles). Our approach is as follows: We assumed that λ_0 for the muscles in Fig. 1 is at the point of maximal extension for each muscle in the physiological workspace (this means that the force produced by each muscle is at a maximum when the muscle has its greatest length in the workspace). We then solved the differential equation in (9) for an isometric muscle preparation in order to indicate the amount of tension in the entire muscle after n impulses, and just before the $(n+1)$st impulse:

$$\phi = S(\lambda)a_1 \left(\frac{\exp(-210\Delta t)}{210 - a_1 a_2} - \frac{\exp(-70\Delta t)}{70 - a_1 a_2} + a_3 \exp(-a_1 a_2 \Delta t) \right) \frac{\exp(-a_1 a_2 n \Delta t) - 1}{\exp(-a_1 a_2 \Delta t) - 1} \quad (11)$$

where $a_1 = K_{SE}/B$, $a_2 = 1 + K_{PE}/K_{SE}$, and $a_3 = 1/(70 - a_1a_2) - 1/(210 - a_1a_2)$. Eq. (11) is an approximation of muscle dynamics in (9). In Fig. 3 we've plotted muscle tension at $\lambda = \lambda_0$, and $n\Delta t = 1$. Using this figure we can now map a desired muscle force at a given muscle length into a frequency of activation for that muscle.

Let us now return to the claim in the previous section that specification of joint torque and joint stiffness will allow for a unique solution to the muscle forces. In analogy with eq. (7), $d\phi/d\lambda$ is muscle stiffness, and by differentiating eq. (11) and using the linear expression for $S(\lambda)$, we see that muscle stiffness is linear with respect to muscle force when $\lambda \leq \lambda_0$:

$$d\phi/d\lambda = \frac{\phi}{\lambda - \lambda_0/2}$$

By using the above relation for $d\phi_1/d\lambda_1$ and $d\phi_2/d\lambda_2$ in eq. (7), K_J now becomes a linear function of the muscle forces, and since it is linearly independent of eq. (6), we can find a unqiue set of muscle forces for a given set of joint torque and stiffness.

2.2 Experiments

Let us begin with the question of how to maintain posture with the muscle-skeleton system of Fig. 1, i.e., how to assign muscle forces so the limb stays at a desired position θ_d. The procedure is to find the set of muscle forces which position the equilibrium of the system at θ_d, and to ensure that this position is stable. By using the relationship between joint equilibrium and torque: $\tau = mcg\cos(\theta_E)$, and by setting $\theta_E = \theta_d$ and using eq. (8), we can solve (6) and (7) for the muscle forces. For the minimum joint stiffness that satisfies (8), we've plotted the muscle forces as a function of joint equilibrium position in Fig. 4A. Increasing joint stiffness (i.e., making it more negative) simply scales this approximately parabolic relationship between muscle forces and equilibrium joint angle, i.e., the ratio of muscle forces is independent of joint stiffness at equilibrium.

An elegant model of motor control (Feldman 1966, Flash 1987) suggests that movement may be thought of as a shift in equilibrium position of the system. Assume that we wish the limb in Fig. 1 to rotate from θ_1 to θ_2 in Δt seconds and follow a desired trajectory $\theta_d(t)$ which minimizes the time derivative of acceleration, i.e., a minimum jerk trajectory (Hogan 1984):

$$\theta_d(t) = \theta_1 + (\theta_1 - \theta_2)\left(15(t/\Delta t)^4 - 6(t/\Delta t)^5 - 10(t/\Delta t)^3\right) \tag{12}$$

As an example, we considered a movement from 45 to 135 degrees in 0.5 seconds. By shifting the equilibrium position of the system along $\theta_d(t)$ and using, for example, a minimum stiffness protocol, we can solve for the muscle forces $\phi_1(t)$, and $\phi_2(t)$. The resulting force trajectory for each muscle is plotted in the "Equilibrium model" of Fig. 4B.

A second approach to programming muscle activation is to consider the dynamics of the moving limb when we assign muscle forces. This means that the torque trajectory should include the influence of joint velocity and acceleration along the desired trajectory θ_d:

$$\tau(t) = mc^2\ddot{\theta}_d + \nu\dot{\theta}_d + mcg\cos(\theta_d) \tag{13}$$

Since this means that the torques experienced by the system will be much higher than when the equilibrium position of the system is changed, it may not be possible for the muscles to produce a high joint torque while maintaining the stiffness requirements of (8). Stiffness of a joint in motion is constrained by the fact that muscles cannot push against a load. We implemented

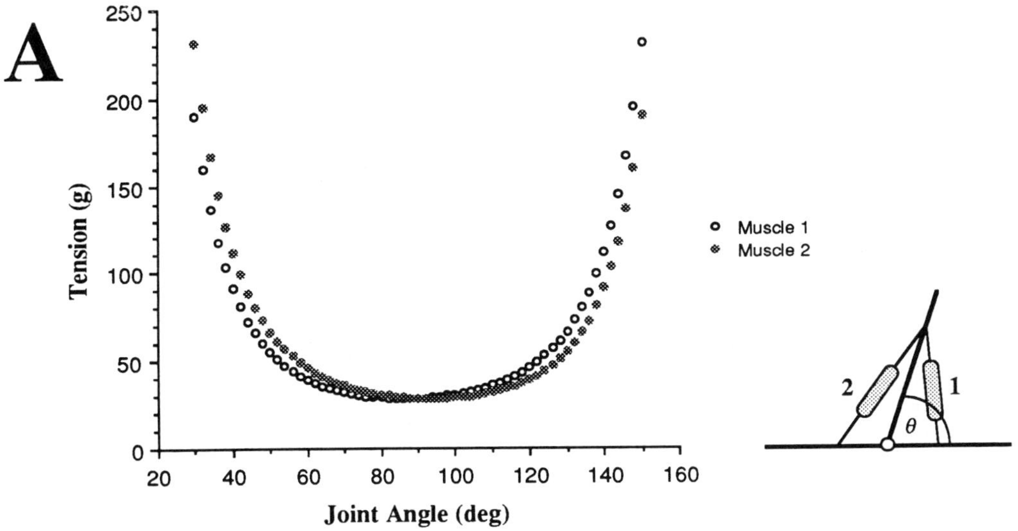

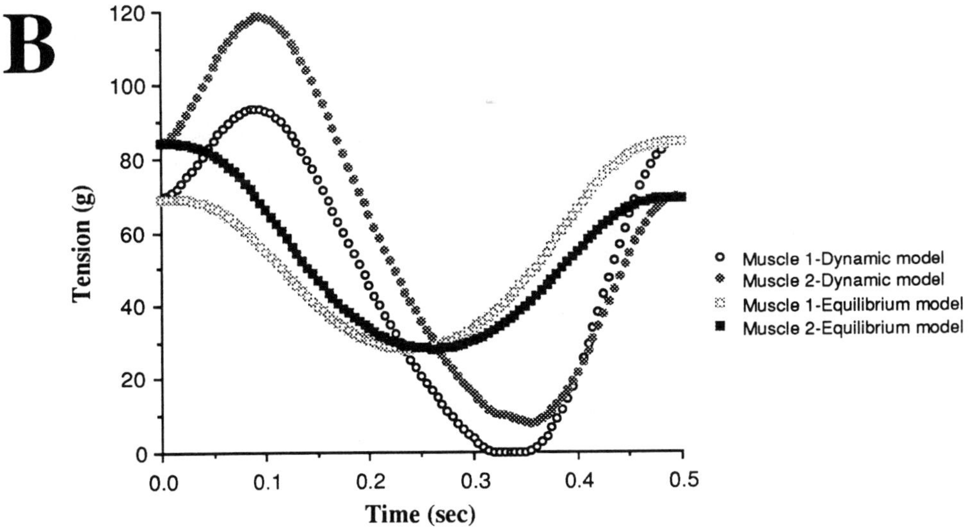

Figure 4: A) Muscle tension required to keep the joint at stable equilibrium, and at minimum joint stiffness. B) Muscle tension trajectory for a minimum stiffness movement from 45 to 135 degrees. The results for a Dynamic Model and an Equilibrium Model strategy are illustrated.

this constraint on the system by limiting the range of K_J so that a muscle is never asked to produce a negative active force (this criterion is a weaker condition than (8) when the limb is at rest, however during motion, it becomes the limiting factor). The procedure is to solve for ϕ_1 and ϕ_2 in terms of K_J and τ, and then find the conditions for which neither muscle force can become negative:

$$K_J < -\frac{\tau(\lambda_0 b^2 c^2 \sin^2(\theta) + \lambda_0 \lambda_2^2 bc \cos(\theta) - 2\lambda_2^3 bc \cos(\theta))}{\lambda_2^2 bc \sin(\theta)(2\lambda_2 - \lambda_0)} \tag{14}$$

$$K_J < -\frac{\tau(\lambda_0 \lambda_1^2 bc \cos(\theta) - \lambda_0 b^2 c^2 \sin^2(\theta) - 2\lambda_1^3 bc \cos(\theta))}{\lambda_1^2 bc \sin(\theta)(2\lambda_1 - \lambda_0)} \tag{15}$$

We repeated the original movement with the minimum stiffness that met the criteria of (8), (14), and (15), using the torque trajectory of (13). The force trajectory for each muscle is plotted in Fig. 4B (labeled Dynamic model).

Fig. 4B illustrates an essential property of the equilibrium model: It predicts that in order to move the limb in Fig. 1 from a small joint angle to a larger one, the force (and neural activity) for both the agonist and antagonist should *decrease* (this is also observed in Fig. 4A). The reason for this is that it takes less torque to hold the mass at, for example, 80 degrees, than 45 degrees. Never the less, the equilibrium model accomplishes the joint rotation due to the stiffness requirement of (8): when the current position is not at the desired equilibrium position, the joint stiffness is large enough to produce a correcting torque that exceeds the effect of the gravitational pull on the load and moves the limb toward the equilibrium position. It appears to us that this property of the equilibrium hypothesis can be directly tested if the effect of the spinal reflexes can be eliminated or explicitly accounted for in a separate model. Referring to Fig. 4B, in contrast to the results of the equilibrium strategy, using the dynamic strategy suggests muscle forces that require an increasing burst to begin moving the limb, then both muscles nearly become quiet, and finally the antagonist is strongly activated to brake the movement.

The point of this example was to show that even for a single joint movement, the nature of the biological actuators is such that a task must be specified in terms of both position and stiffness before muscle activations can be programmed. The notion of an equilibrium point, as introduced by Feldman (1966), and demonstrated as a control algorithm by Flash (1987), suggests that we can begin with the spring-like characteristics of the antagonistic muscles, ignore the dynamics of the skeleton, shift the minimum of a potential energy surface along the desired path, and the limb will more-or-less follow. We have specified the stiffness conditions for which this hypothesis holds true, as well as the expected muscle forces for a typical movement. Our results provide an easy test of the equilibrium hypothesis as a control paradigm for movement generation: the force and neural activity in the agonist should decrease as a joint rotates a load against gravity.

3 Kinematic Redundancy

The issue of kinematic redundancy arises when the degrees of freedom in a limb exceed the degrees of freedom in the movement, e.g., a planar three-joint arm. Formally, the definition is as follows: Consider a multi-joint limb with an end-effector attached to the distal link. If the position of the end-effector is denoted by vector $\mathbf{x} = [x_1, x_2, \ldots, x_m]^T$, and $\boldsymbol{\theta} = [\theta_1, \theta_2, \ldots, \theta_n]^T$ specifies the joint angles, then there exists some continuous non-linear function $h(\boldsymbol{\theta})$, such that

$\mathbf{x} = h(\boldsymbol{\theta})$, i.e., the forward kinematic mapping. If $m < n$, then the arm is kinematically redundant, meaning that there is a unique $\mathbf{x}$ associated with each $\boldsymbol{\theta}$, but there may be many $\boldsymbol{\theta}$ associated with each $\mathbf{x}$. When there are muscles attached to the joints, the issue becomes how to assign muscle lengths for a given end-effector position: Assume that $\boldsymbol{\lambda} = [\lambda_1, \lambda_2, \ldots, \lambda_q]^T$ is a vector of muscle lengths where λ_i is the length of muscle i, and $n < q$. Then there exists some continuous non-linear function $g(\boldsymbol{\theta})$, where $\boldsymbol{\lambda} = g(\boldsymbol{\theta})$. Given a planned end-effector trajectory $\mathbf{x}(t)$, describing an appropriate trajectory in terms of muscle lengths $\boldsymbol{\lambda}(t)$ requires finding $g(\ h^{-1}(\mathbf{x})\ )$. A trajectory in terms of the muscle lengths may be required for setting the bias of the stretch reflex loop (activation of the γ-motoneurons), and for assigning muscle activations as in the discussion in the previous section.

To find $h^{-1}(\mathbf{x})$, the most direct approach is certainly to find an analytical expression, an approach that has proved to be very difficult in robotics, due to the complexity of $h(\boldsymbol{\theta})$ (Sciavicco and Siciliano 1988). A novel approach has recently been suggested by Mussa Ivaldi et al. (1988): taking advantage of the elastic properties of the neuromuscular system, they have proposed an algorithm that allows one to map small changes in the position of the end-effector into changes in joint angles. We have sketched the derivation of the algorithm below:

$$d\boldsymbol{\theta} = \mathbf{C}_J \, d\boldsymbol{\tau} \tag{16}$$

$$d\boldsymbol{\tau} = \mathbf{J}_S^T \, d\mathbf{f} \tag{17}$$

$$d\mathbf{f} = \mathbf{K}_S \, d\mathbf{x} \tag{18}$$

$$\text{by substitution:} \quad d\boldsymbol{\theta} = \mathbf{C}_J \, \mathbf{J}_S^T \, \mathbf{K}_S \, d\mathbf{x} \tag{19}$$

where $\mathbf{C}_J$ is the limb's joint compliance, $\mathbf{J}_S$ is the Jacobian at the end-effector: $\mathbf{J}_S = \partial\mathbf{x}/\partial\boldsymbol{\theta}$, $\mathbf{f}$ is a force vector at the end-effector, and $\mathbf{K}_S$ is the limb's stiffness at the end-effector. Given a limb's joint compliance $\mathbf{C}_J$, by using the principle of virtual work one can calculate the end-point stiffness $\mathbf{K}_S$:

$$\mathbf{K}_S = (\mathbf{J}_S \, \mathbf{C}_J \, \mathbf{J}_S^T)^{-1} \tag{20}$$

which basically shows how to go from impedance at joint coordinates to impedance at end-point coordinates. The inverse in (20) always exists because from the change in potential energy of the system, it can be shown that $\mathbf{C}_J$ is a positive definite matrix (Mussa Ivaldi et al. 1988), and $\mathbf{J}_S$ is full rank by construction. The idea of the algorithm in (19) is that a particular pattern of changes in joint angles will occur if an external agent forced the end-effector to make small displacements $d\mathbf{x}$. The resulting changes in the limb's configuration can be fully specified if the limb's impedance is known. So to perform a movement with the algorithm in (19), one would need *a priori* knowledge regarding the impedance of the limb during the movement.

Assume that a task is specified in terms of an end-effector trajectory $\mathbf{x}(t)$ for a kinematically redundant biological limb. What (19) implies is that the task must also specify joint stiffness during motion (or that the CNS can assume a minimum stiffness value that is appropriate for the task). Since the task needs to be performed by muscles, we should rewrite the algorithm in (19) in terms of muscle lengths:

$$d\boldsymbol{\lambda} = \mathbf{J}_M \, d\boldsymbol{\theta} \tag{21}$$

$$d\boldsymbol{\lambda} = \mathbf{J}_M \, \mathbf{C}_J \, \mathbf{J}_S^T \, \mathbf{K}_S \, d\mathbf{x} \tag{22}$$

where $\mathbf{J}_M = \partial\boldsymbol{\lambda}/\partial\boldsymbol{\theta}$ (for example, see (4) and (5)). The algorithm in (22) suggests a method by which the CNS can set the bias of the stretch reflex circuitry for each muscle, given an arbitrary end-effector trajectory and joint impedance. The contribution of Mussa Ivaldi et al.'s

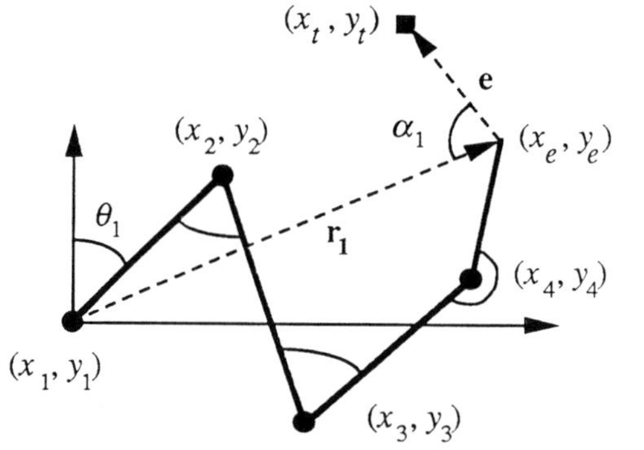

Figure 5: Model robot nomenclature and vector definition for the Error Vector Algorithm.

(1988) work has been to exploit the fact that motor impedance provides a unique solution to the configuration of a mechanism given an externally imposed motion to its end-effector. In the next section we will review a previously proposed algorithm, referred to as the *Error Vector Algorithm*, for solving kinematic redundancies (independently developed by Berkinblit et al. 1986a, and Hinton 1984), and we will show that this algorithm is a special case of the relation derived in (19).

4 Error Vector Algorithm

A current hypothesis in motor control is that motor behaviors are organized of compartments or segments that can be combined in different ways to form new movement patterns (e.g., Berkinblit et al. 1986b, Fentress 1987, Viviani and Terzuolo 1982). The essential component of this hypothesis is that "every limb joint is subserved by a set of individual control systems which interact in the process of solving a common motor task" (Berkinblit et al. 1986b). These organizations are said to exist in locomotion of cats—where it has been argued that limb joints are controlled by a set of generators which interact with each other to produce an overall locomotor pattern (Grillner 1975)—as well as in reaching movements (Hinton 1984), and the wiping reflex in the frog (Berkinblit et al. 1986a).

In this hypothesis, control of each joint is accomplished in parallel while information about the position of each joint relative to the end-effector is shared between all controllers in order to perform a common motor task. Each controller "produces an individual movement in the corresponding joint, based on the information on the position of the target and on the result of collective work of all the limb's joints, in particular, the knowledge of the position of the limb's tip relative to the target" (Berkinblit et al. 1986a). We refer to this as the *Error Vector Algorithm.* In this section we will generalize the algorithm for a robot with n joints, and show that the algorithm uses the transpose of the robot's Jacobian at the end-effector to simulate attachment of an imaginary spring between the target and the end-effector. Our discussion will indicate that is a special case of the relationship in (19).

Consider a planar, multi-joint limb such as the one in Fig. 5. Assume that each joint has one degree of freedom (along the axis perpendicular to the plane of motion) and joints are connected by a rigid link of some length l_i. Point (x_i, y_i) is the location of joint i, (x_e, y_e) is the position of the end-effector, (x_t, y_t) is the position of the target, and a limb of length l_i connects to joints $i-1$ and i. Assuming that $(x_1, y_1) = (0,0)$, then the forward kinematics are described by the following:

$$\begin{aligned} x_i &= x_{i-1} + l_{i-1}\sin(\theta_{i-1} - \theta_{i-2} + \theta_{i-3} - \ldots + (-1)^i\theta_1) & (23) \\ y_i &= y_{i-1} + (-1)^i l_{i-1}\cos(\theta_{i-1} - \theta_{i-2} + \theta_{i-3} - \ldots + (-1)^i\theta_1) & (24) \end{aligned}$$

The iterative algorithm proposed by Hinton (1984) and Berkinblit et al. (1986a) describes how a particular target can be reached when the number of joints exceed the number of coordinates that define the target. The idea is to plan changes in joint angles as calculated by an error vector that points from the tip of the limb to the target position, as in Fig. 5: For each joint i, a vector $\mathbf{r}_i$ points to the current position of the end-effector. There is also a vector $\mathbf{e}$ which points from the end-effector to the target position. If c_i is a constant, and α_i is the angle between $\mathbf{r}_i$ and $\mathbf{e}$, then the change in the angle of joint i is a vector that points along an axis perpendicular to the plane of movement, and is defined by Berkinblit et al. (1986a) as:

$$\Delta\theta_i = c_i\,|\mathbf{r}_i|\,|\mathbf{e}|\,\sin(\alpha_i) \tag{25}$$

where $|\mathbf{x}|$ is the magnitude of the vector $\mathbf{x}$.

We will examine the rationale for this algorithm first intuitively, and then rigorously by finding the Jacobian of the robot at the end-effector position. In (25), a rotation from $\mathbf{r}_i$ to $\mathbf{e}$ through an angle α_i will change θ_i about an axis perpendicular to the plane of motion. It is desirable to make the change in θ_i proportional to the magnitude of the error vector $\mathbf{e}$. Also, $\Delta\theta_i$ is reasoned to be proportional to $|\mathbf{r}_i|$ since the longer this vector is, the more effectively the position of the end-effector can approach the desired target.

To show the reasoning behind (25) in a rigorous fashion, we need to initially express $\Delta\theta_i$ in terms of the end-effector coordinates. By use of geometry (Law of Cosines), α_i can be eliminated from (25). In general, for the notation that was introduced in Fig. 5, and using equations (23) and (24), we can express (25) as follows:

$$\begin{aligned} \text{if } i \text{ is even:}\quad \Delta\theta_i &= c_i((x_e - x_i)(y_t - y_e) - (y_e - y_i)(x_t - x_e)) & (26) \\ \text{if } i \text{ is odd:}\quad \Delta\theta_i &= c_i((y_e - y_i)(x_t - x_e) - (x_e - x_i)(y_t - y_e)) & (27) \end{aligned}$$

For the 4-joint limb of Fig. 5, let us derive the Jacobian matrix $\mathbf{J}_S$, where $\mathbf{J}_S = \partial\mathbf{x}/\partial\boldsymbol{\theta}$. From (23) and (24), substituting (x_e, y_e) for (x_5, y_5) and differentiating with respect to $\boldsymbol{\theta}$ we have:

$$\begin{aligned} dx_e &= l_1\cos(\theta_1)d\theta_1 + l_2\cos(\psi_1)d\psi_1 + l_3\cos(\psi_2 + \theta_1)(d\psi_2 + d\theta_1) \\ &\quad + l_4\cos(\psi_3 + \psi_1)(d\psi_3 + d\psi_1) & (28) \\ dy_e &= -l_1\sin(\theta_1)d\theta_1 + l_2\sin(\psi_1)d\psi_1 + l_3\sin(\psi_2 + \theta_1)(d\psi_2 + d\theta_1) \\ &\quad + l_4\sin(\psi_3 + \psi_1)(d\psi_3 + d\psi_1) & (29) \end{aligned}$$

where $\psi_i = \theta_{i+1} - \theta_i$. For example, the element in the first row of the first column in the Jacobian matrix $\mathbf{J}_S$ will be:

$$J_{S\,(1,1)} = l_1\cos(\theta_1) - l_2\cos(\theta_2 - \theta_1) + l_3\cos(\theta_3 - \theta_2 + \theta_1) - l_4\cos(\theta_4 - \theta_3 + \theta_2 - \theta_1) \tag{30}$$

By comparing (24) with (30), we see that the (30) is in fact y_e. Similarly, we can show that:

$$\mathbf{J}_S = \begin{pmatrix} y_e & y_2 - y_e & y_3 - y_e & y_4 - y_e \\ -x_e & x_e - x_2 & x_e - x_3 & x_e - x_4 \end{pmatrix} \tag{31}$$

By substituting the above Jacobian in (26) and (27), we can rewrite (26) and (27) in the following format:

$$\Delta\boldsymbol{\theta} = \mathbf{C}\,\mathbf{J}_S^T\,\mathbf{e} \tag{32}$$

where $\mathbf{C}$ is a diagonal matrix made up of c_i (one constant for each joint), and $\mathbf{e}$ is the error vector, $\mathbf{e} = [(y_t - y_e), (x_t - x_e)]^T$. The relation in (32) is another way to write the Error Vector Algorithm that has been suggested by Berkinblit et al. (1986a) and Hinton (1984).

Now compare the relation in (32) with (19). In (19), $\mathbf{C}_J$ is the joint compliance matrix, and $\mathbf{K}_S$ is the end-point stiffness. The relation in (19) assigns changes in joint angles as a function of small displacements in the end-effector and limb impedance. In (32), the error vector $\mathbf{e}$ represents the affect of the displacement after it interacts with end-point stiffness (to become a force acting on the end-effector).

It turns out that when we simulated movements with a kinematic model, the algorithm in (32) worked only if the matrix C had elements that were all very small (on the order of 1% of $|\mathbf{e}|$ before onset of movement), otherwise, the limb would either oscillate about the target position, or become unstable. Therefore we conclude that the algorithm proposed by Hinton (1984) and Berkinblit et al. (1986a) in equation (25), is a special case of the relation in (19)—the special case being that (25) assumes an identity matrix for end-point stiffness $\mathbf{K}_S$.

The idea that every joint has a controller which interacts with other joint-controllers in the process of solving a common motor task can be described by the relation in (19) when one realizes that (19) is a precise formulation of (25). Each controller "works" by imitating the effect of a displacement at the tip of the limb on the joint that it controls. To do this, the controller needs to know three kinds of information: (1) where the tip of the limb is with respect to the target, (2) where the joint is with respect to the tip of the limb, and (3) what the impedance of the limb is. Since the algorithm is iterative and highly dependent on initial conditions, it is likely that quite different joint angles may be observed when the end-effector cycles through a given trajectory. Another reason for this phenomenon is that the mapping in (19) is not integrable, meaning that after tracing a circle, the joint angles do not return to their original position at the start of the trace. A final reason is that we have only mapped the kinematics of movement here. The actual trajectory of limb is affected by forces linked to its motion.

In the next section we will consider some of the dynamic forces inherent in motion. Our goal will be to show how to rapidly learn to compensate for muscle and limb dynamics in order to produce a precise end-effector trajectory.

5 Learning System Dynamics

Assume that we have a task that requires the end-effector to follow a trajectory $\mathbf{x}(t)$, with a stiffness profile $\mathbf{K}_J(t)$. Based on our discussion in the previous two sections, we have an algorithm in (22) which specifies what the muscle lengths should be during this task. This algorithm's purpose is to specify the kinematics of the task in the same coordinate system as the actuators. The kinematics of the task are, however, only a static representation of the dynamics of the system. For example, consider that the skeleton has non-linear and coupled dynamics, meaning that the relation in (16) is only valid for small disturbances from equilibrium since it has not taken into account the effect of centripetal, Coriolis, or gravitational forces. In order for the end-effector to produce the desired trajectory, a trajectory of muscle activations will have to be arrived at which compensates for these forces.

We propose that the dynamics of the muscle-load-feedback system can be learned with a

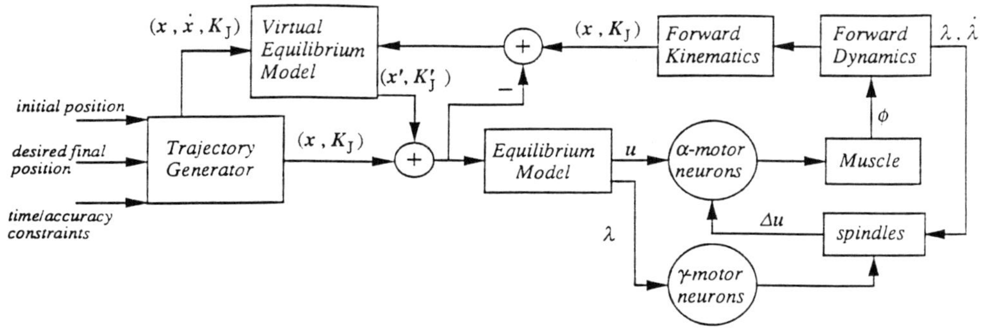

Figure 6: A schematic for learning dynamics of a biological limb. The trajectory generator specifies end-effector position $\mathbf{x}$, velocity $\dot{\mathbf{x}}$, and stiffness $\mathbf{K}_J$ for the desired movement. The Equilibrium Model produces a trajectory of muscle lengths and activations without regard to dynamics of motion. The error between the observed and desired paths of the limb leads to formation of a Virtual Equilibrium Model that augments the equilibrium path.

control structure that adapts its model of the system with information from efferent copy and proprioceptive receptors, as shown in Figure 6. In this figure there are actually two motor maps: one is the static equilibrium map that is hypothesized to reside in the spinal cord (see Giszter et al. in this book) which maps a desired end-effector position and limb stiffness onto muscle lengths and activation: $(\mathbf{x}, \mathbf{K}_J) \rightarrow (\boldsymbol{\lambda}, \mathbf{u})$, and the other is an adaptive motor map which attempts to compensate for the limb's dynamics during the movement. This adaptive model interacts with the equilibrium model to produce a *virtual equilibrium trajectory*. Formally, the mapping that is learned by the adaptive model is: $(\mathbf{x}, \dot{\mathbf{x}}, \mathbf{K}_J) \rightarrow (\mathbf{x}', \mathbf{K}'_J)$, where the resulting variables represent error terms that add to the original trajectory in order to produce a new trajectory that compensates for limb dynamics.

For example, consider a two joint limb that we wish to move along a trajectory such as the one specified in Fig. 7A. The equilibrium trajectory specifies the end-effector position and stiffness profile during movement: the mapping to muscle activation and muscle lengths is performed in the spinal cord, e.g., in a region analogous to L4 and L5 in the cat—the area thought to be responsible for pattern generation in the scratch reflex (Berkinblit et al. 1978). For a particular stiffness profile (Shadmehr 1990), the resulting movement of the limb (referred to as the actual movement) is plotted by the dotted lines in Fig. 7A. Note that in this case, the limb lags the equilibrium trajectory and oscillates about the desired end-point before it is damped out. When a learning mechanism is in place, a virtual equilibrium trajectory can be fed to the spinal mechanism so that the actual trajectory is identical to the one that was desired. In Fig. 7B, the virtual equilibrium trajectory is the dotted set of lines. Note that in order to begin the movement, the virtual trajectory needs to accelerate the arm beyond the amount that is specified by the equilibrium trajectory. Correspondingly, to stop the motion, the virtual trajectory needs to reverse the movement to activate antagonist muscles and brake the motion. In Shadmehr (1990), we used a gross computational model of the Cerebellum, the Cerebellar Model Articulation Controller (Albus 1975), to rapidly learn this virtual equilibrium trajectory.

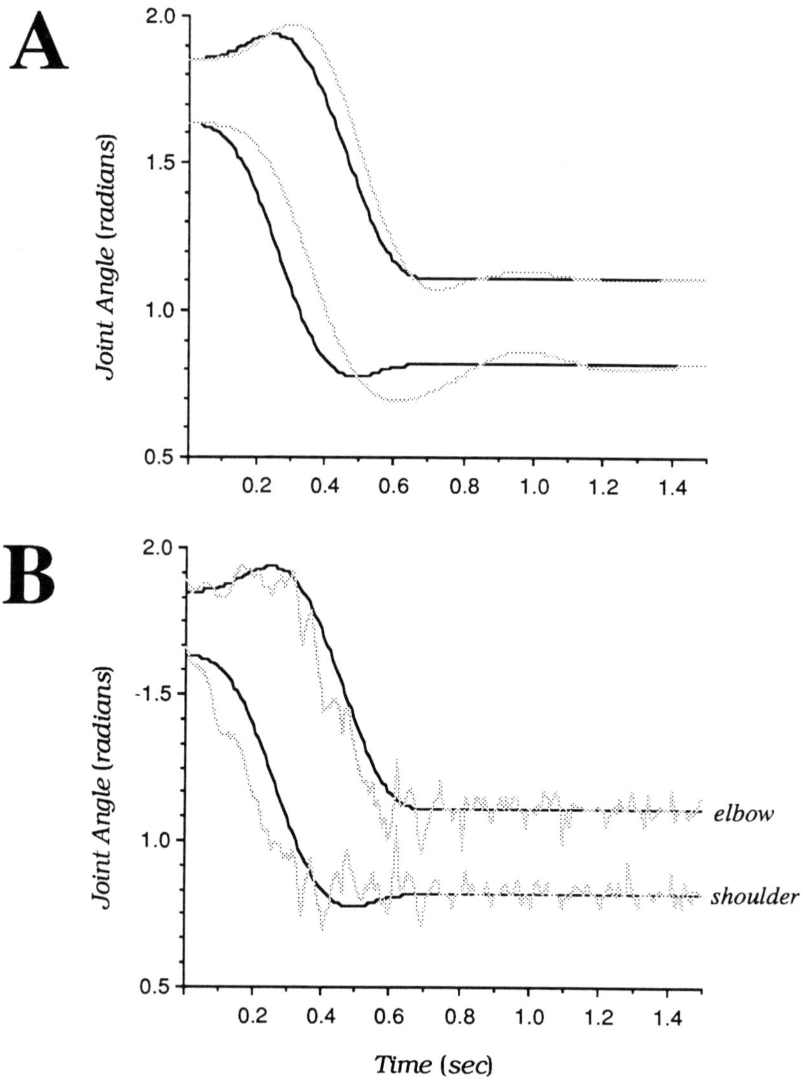

Figure 7: A learned virtual equilibrium joint trajectory that compensates for limb dynamics of a two joint arm. A) The equilibrium trajectory (solid lines) is the desired joint trajectory, while the dotted lines show the actual trajectory of the limb. B) The learned virtual equilibrium trajectory (dotted, more "noisy" lines) compensates for the limb dynamics. For this case, the actual trajectory is identical to the desired one.

6 Discussion

A fundamental question in motor control is how the CNS interacts with limbs that possess kinematic and actuator redundancies, for there are many more system parameters (e.g., α- and γ-motoneuron activation rates) than there are independent variables (e.g., position of the end-effector and limb stiffness). Our goal has been to discuss issues of actuator and kinematic redundancy within the framework of robotics in order to understand how to represent a movement so that it can uniquely specify how to perform it.

The issue of actuator redundancy arises because assignment of muscle forces to produce a given joint torque is not possible since many degrees of co-activation can lead to generation of identical joint torques. To deal with this issue, we suggested that description of a task must include not only the trajectory of the end-effector, but also, at least the trajectory of the stiffness at the end-effector. Within the framework of the *equilibrium trajectory* hypothesis (Flash 1987), this suggests that, for example, an equilibrium trajectory of the hand for reaching movements is an incomplete description of the task since the same trajectory may be performed with various degrees of joint stiffness. With reference to the λ-model (Feldman 1966), where λ is a threshold length for activation of a muscle, joint stiffness must be decided upon before the distance of the thresholds from the equilibrium joint angle can be assigned. Interestingly, for a task that requires rotating a joint from a horizontal to a vertical position, we showed that the equilibrium hypothesis predicts a *reduction* in the activity of both the agonist and the antagonist muscles, given that the limb's stiffness remains at a level just sufficient to ensure stability. This result is in sharp contrast to the muscle activity that is expected if a dynamic model of movement is used: for the same movement, this model predicts a sharp initial increase in the activity of both muscles, then a decrease and finally a braking pattern of activity by the antagonist muscle.

Mussa Ivaldi et al.'s (1988) work suggests that the kinematic redundancies of the limb can be overcome when the elastic properties of the system are considered. When an external agent displaces the end-effector, the resulting changes in joint angles can be determined if the stiffness characteristics of the limb are known. The same line of analysis can be used to relate displacements in the position of the end-effector to changes in muscle lengths (as in equation (22)). By "imitating" the effect of a foreign agent on the end-effector, the CNS can produce a trajectory in terms of joint angles and muscle lengths. The notion of an independent controller for each joint (the error vector algorithm) was shown to fit well with in this framework: the only information that is necessary for each controller is the distance of the tip of the limb to the target, and the distance from the center of the controlled joint to the end-effector. However, this mapping only describes the kinematics of the task: if muscle activations are assigned without regard to, for example, the inertial forces that act on the skeleton, then the observed trajectory of the end-effector will deviate significantly from the desired path. Learning dynamics of the muscle-load-feedback system is essential especially for execution of ballistic or precise movements.

In the adaptive control scheme that we proposed, the kinematic requirements of the task (as specified by (22)) are used by an equilibrium model, located in the spinal cord, to produce an equilibrium trajectory. This trajectory is augmented by a supra-spinal center to produce a virtual equilibrium trajectory that compensates for the dynamics of the limb. Our current work is exploring the usefulness of this approach in predicting muscle activation patterns and end-effector trajectories for control of redundant limbs.

7 References

Albus, J. S. (1975) A new approach to manipulator control: The cerebellar model articulation controller (CMAC). *Trans. ASME J. Dynamic Syst. Meas. Contr.*, 97:220–227.

Aubert, X., Roquet, M. L., and Van der Elst, J. (1951) The tension-length diagram of the frog's sartorius muscle. *Arch. Intern. de Physiol.*, 59:239–241.

Berkinblit, M. B., T. G. Deliagina, A. G. Feldman, I. M. Gelfand, and G. N. Orlovsky (1978) Generation of scratching. I. Activity of spinal interneurons during scratching. *J. Neurophys.*, 41:1040–1057.

Berkinblit, M. B., I. M. Gelfand, and A. G. Feldman (1986a) Model of the control of the movements of a multijoint limb. *Biophysics* 31(1):142–153.

Berkinblit, M. B., A. G. Feldman, and O. I. Fukson (1986b) Adaptability of innate motor patterns and motor control mechanisms. *Behav. Brain Sci.*, 9:585–599.

Feldman, A. G. (1966) Functional tuning of the nervous system with control of movement or maintenance of a steady posture—II. Controllable parameters of the muscles. *Biophysics*, 11:565–578.

Fentress, J. C. (1987) Compartments and cohesions in adaptive behavior. *J. Comp. Psychol.*, 101:254–258.

Flash, T. (1987) The control of hand equilibrium trajectories in multi-joint arm movements. *Biol. Cybern.*, 57:257–274.

Gasser, H. S., and Hill, A. V. (1924) The dynamics of muscular contraction. *Proc. Roy. Soc. B.*, 96:398–437.

Grillner, S. (1975) Locomotion in vertebrates: central mechanisms and reflex interaction. *PhysiolRev.*, 55:247–304.

Hinton, G. E. (1984) Parallel computations for controlling an arm. *J. Motor Behavior*, 16:171–194.

Hogan, N. (1984) An organizing principle for a class of voluntary movements. *J. Neuroscience*, 4:2745–2754.

Inbar, G. F., and Adam, D. (1976) Estimation of muscle active state. *Biol. Cybern.*, 23:61–72.

Mussa-Ivaldi, F. A., N. Hogan, and E. Bizzi (1985) Neural, mechanical, and geometric factors subserving arm posture in humans. *J. Neuroscience*, 5(10):2732-2743.

Mussa-Ivaldi, F. A., J. McIntyre, and E. Bizzi (1988) Theoretical and experimental perspectives on arm trajectory formation: A distributed model for motor redundancy. In *Biological and Artificial Intelligence Systems*, Eds: E. Clementi and S. Chin, Escom Press.

Sciavicco, L., and B. Siciliano (1988) A solution algorithm to the inverse kinematic problem of redundant manipulators. *IEEE J. Robotics and Automation*, 4(4):403–410.

Shadmehr, R. (1990) Learning virtual equilibrium trajectories for control of a robot arm. *Neural Computation*, 2(4):in press.

Viviani, P., and Terzuolo, C. (1982) Trajectory determines movement dynamics. *Neuroscience*, 7:431–437.

MOTOR PATTERN GENERATORS IN ANURAN PREY CAPTURE

Ananda Weerasuriya

Department of Basic Medical Sciences

Mercer University School of Medicine

Macon, GA 31207, U.S.A.

SUMMARY. Anuran prey capture, released by specific stimuli, consists of a sequence of motor synergies. This series of steps includes an approach or orientation toward the prey stimulus, a fixation of the prey in the frontal visual field and the consummatory event of snapping at the prey and swallowing it. The key stimulus that elicits prey capture is either visual, tactile or olfactory, and the outputs of their respective sensory analyzers share common access to motor pattern generators responsible for the elaboration of the appropriate motor outputs. An approach path can be altered midstream in response to movement of the prey, but an orienting turn once initiated cannot be modified. Thus these two motor components (approach and orient), while being similar in that they are produced by adjustable pattern generators, are dissimilar with respect to the degree of modifiability during execution. This distinction is largely attributable to the differences in the speed of execution and duration of the two motor patterns. The snapping stage of prey capture is a swift, stereotyped, and ballistic motor pattern with very little variability. It is the component with the smallest degree of variability in its duration and execution, is the most rapidly executed, and has several subcomponents. Snapping starts with the lunge of the head toward the prey, followed by opening of the mouth, tongue projection toward the prey, retraction of tongue with prey, closure of mouth, and swallowing of prey. The neuronal substrate responsible for this chain of events can be modeled as a linearly operating sequence of modules controlling the different subcomponents of snapping. Variations in neck muscle contractions are considered to provide the parametric control necessary to match the lunging distance with the relative location of the prey. On the other hand, jaw and tongue movements appear to be relatively invariant and stereotyped. A model with three integrator networks is proposed for the control of tongue muscles during prey capture.

INTRODUCTION

Coordinated muscle activity underlying behaviorally relevant movement is either a feedback regulated motor pattern or a ballistic stereotyped synergy. A characteristic feature of the latter group is that the relevant spatio-temporal sequence of muscle activation is orchestrated by an internuncial neuronal network - the motor pattern generator (MPG). Activation of this network by a specific pattern of impulses in predetermined afferent pathways will elicit a coordinated sequence of motoneuronal activation and inhibition leading to the expression of that particular motor synergy. These motor patterns are stereotypical, very rapidly executed and often are completed in less than a fraction of a second. On the other hand, motor patterns influenced by sensory feedback are elaborated on a slower time scale, and display an adaptive plasticity to match a given stimulus or set of environmental conditions with an appropriate behavioral response. In these muscle synergies, sensory feedback could either modify the behavior 'on-line' or adjust the movement parameters between successive trials.

Pursuit and capture of prey by toads and frogs involve both feedback regulated and stereotyped motor patterns. The sequence of events, triggered by identification and localization of an appropriate prey object, starts with an approach or orientation toward the prey, followed by fixating the prey in the binocular frontal visual field within a couple of head lengths from the tip of the snout, and is successfully terminated by a flip of the tongue, which transfers the prey into the mouth. The neural basis for this series of behaviors is a unidirectional flow of information from sense organs to sensory analyzers, motor pattern generating networks (including internal feedback loops), motoneurons, and finally to the respective muscles. This flow of information can be broadly divided into two stages, sensory processing and motor pattern generation, with an intervening sensorimotor interface. Visual prey identification and localization in toads are reviewed by Ewert ((1987) and this volume), and visual and tactual prey localization in frogs are discussed by Grobstein ((1989) and this volume). Whereas frogs and toads respond with prey catching behavior to a nearly identical range of sign stimuli, their prey acquisition modes have noteworthy differences. Frogs will sometimes orient and snap at a prey in a single smooth pursuit movement (Gans, 1961), and toads after approaching or orienting toward a prey will invariably fixate on it for a few seconds before snapping (Hinsche, 1935). The approach path and orienting turn of the anuran toward a prey is governed by a feedback regulated MPG, and the consummatory step of snapping at the prey is a stereotyped ballistic synergy. Whereas much is known about the neuronal mechanisms underlying prey identification and localization, relatively less is known about the MPGs responsible for prey acquisition, and the sensorimotor interface, which conveys sensory decisions to input portals of relevant motor pattern generators.

This essay reviews the evidence in favor of a MPG for snapping in particular and the other stages of prey capture as well, and proposes a model to account for this evidence. The model outlined below should aid in the design of further experiments to dissect the neuronal substrates of the MPG, and the results of such experiments will in turn lead to further refinements of the model. It is perhaps appropriate to point out a simplification in the categorization of motor synergies used in this discussion. Motor patterns incorporate, in varying degrees, both ballistic features and feedback principles in their neuronal organization and enactment. Thus there isn't a motor synergy that is either completely ballistic or exclusively feedback guided. Those that operate with very little feedback are considerd to be ballistic (e.g., snapping), and the ones with limited ballistic components are treated as being feedback regulated (e.g., orienting).

SENSORY PROCESSING

Visual. Prey capture by anurans is a classical example of what is referred to in the ethological literature as an innate fixed action pattern. An essential feature of such behaviors is that they are elicited by very specific key stimuli. Over the years Ewert and his group have characterized the fundamental features of the visual key stimulus for prey capture (Ewert, 1984). In addition, they have identified the retinorecipient areas of the brain that process visual information, characterized cell types found therein in terms of their responsivity to the form and movement of a range of visual stimuli that are related to prey capture, and proposed a hypothetical interactive network of these neurons to explain the processing of prey related visual information. This scheme has been modeled further by Arbib and his group to account for some quantitative aspects of the behavioral observations (Arbib, 1989), and also to predict successfully habituation patterns to prey related stimuli (Wang, Ewert and Arbib, this volume). The crucial features of a suitable prey stimulus are a rectangular shape and movement parallel to its longer axis. The configuration-dependent response of the anuran to this key stimulus is relatively invariant with respect to velocity, movement direction (x-y-z coordinates), contrast direction, structure of background, and - within limits - the size of the stimulus. The neuronal computations underlying this sensory analysis occur mainly in the tectum and the pretectum, and its outcome is conveyed to structure downstream by particular output elements of the tectum, including T5(2) neurons, with long axons descending to the hindbrain and spinal cord (Satou and Ewert, 1985). The response characteristics of T5(2) neurons to prey like stimuli correlate very well with the behavioral response of the toad to these same stimuli. This has led to the postulate that T5(2) neurons are integral elements of a command system that drives MPGs of anuran prey capture (Ewert, 1987). This is broadly analogous to the command neurons described in invertebrate motor systems (Kupfermann and Weiss, 1978). The firing patterns of T5(2) neurons during prey capture will be discussed later when the MPG for snapping will be reviewed.

Tactile. In blinded frogs, prey capture can be released by tactile stimuli (Kicliter, 1973; Comer and Grobstein, 1981a). The basic features of tactile stimuli adequate for elicitation of prey capture have not been analysed. Since similar tactile stimuli applied to different parts of the body elicit either a wiping reaction or a prey related orient/snap response (Berkinblitt et al., 1989), it is quite likely that a frog's response to a tactile stimulus as prey depends on the intrinsic properties of that stimulus as well as its location on the body surface. Chronic lesion studies have indicated that mesencephalic subtectal (toral) areas are sufficient and necessary for tactually elicited prey capture (Comer and Grobstein, 1981a;1981b). Comparison of the neural territory devoted to analysis of tactile stimuli with that for visual anaysis suggests a much more highly discriminated visual prey stimulus than a tactile prey stimulus. The ability of frogs with large toral lesions to acquire prey using visual cues, and those with complete tectal lesions to respond to tactile stimuli clearly indicate multimodal access to the MPGs responsible for prey capture. Even though visual and tactile information on prey appear to reach hindbrain structures through independent pathways, the above lesion studies also suggested that a wedge of tissue located between the ventrolateral tectum and the far lateral torus, and responding to both visual and tactile stimuli is essential for the integrity of the motor output in prey capture. It is possible that this intermediate zone between the tectum and torus contains supplementary descending pathways utilized by both the visual and somatosensory systems to increase the reliability of the motor program. Whereas the innate basis of visually guided prey capture is well established, it is yet to be demonstrated, though likely, that tactile releasers also are determined innately.

Olfactory. Frogs with bilateral enucleations will migrate toward a discrete source of mealworm odor and often respond with snapping (Shinn and Dole, 1978). It was subsequently demonstrated that odors of prey insects, the olfactory releasers of prey capture, have to be learnt by toads (*Bufo boreas*), and the rate of learning is a function of the prey species (Dole et al., 1981). Normal anurans, which migrate toward a source of mealworm odor, snap much less frequently than blind ones. It was suggested that normal animals are inhibited in their snapping by the conflicting information from olfactory and visual systems, whereas the blind animals were not subject to such a restraint. Olfactory cues are also known to facilitate visually elicited prey capture (Ewert, 1968). For example, in the presence of mealworm odor toads will orient and/or snap to a wider range of visual stimuli with a concomitant decrease in worm/antiworm discrimination. Though quite conceivable, it is not known whether olfactory cues facilitate tactually elicited prey capture. The neural structures necessary for processing olfactory information are not well understood, and the site of interaction between olfaction and vision has not been investigated. It is also not known whether an intact tectum and/or torus are necessary for prey capture to be released by olfactory cues.

In summary, prey capture by anurans appears to be released by innately determined visual and tactile sign stimuli. Prey capture released by olfactory cues is a learned response, and the range of visual stimuli capable of triggering prey capture also can be modified by experience. Visual sign stimuli appear to be more strongly discriminated than tactile ones. Olfactory cues facilitate visually elicited prey capture, though neural sites of multimodal interaction are unknown. Under natural conditions, visual stimuli are the predominant releasers of prey capture.

SENSORIMOTOR INTERFACING

In order for sign stimuli to release prey capture it is necessary for the decisions of sensory analysis to be conveyed to the MPGs. These connections, which are the output stages of the sensory analyzers as well as the afferent portals of the MPGs, are known as the sensorimotor interface. Of the three sensory modalities capable of eliciting prey capture, the electrophysiological and neuroanatomical properties of only the visual sensorimotor interface has been studied in fair detail. Earlier neuroanatomical studies, with the Golgi technique (Szekely and Lazar, 1976), suggested that the output of the tectum was mediated by large pyramidal neurons in or near layer 6 of the tectum. Subsequent studies, using horseradish peroxidase and cobaltic-lysine, enlarged this list to include ganglionic and piriform cells (Weerasuriya and Ewert, 1981; Lazar et al., 1983). The latter studies, in addition, showed that most output cells were in the lateral portion of the tectum, corresponding to the frontal visual field, and had contralaterally descendind axons. This nonuniform distribution of tectal cells has been confirmed (Kostyk and Grobstein, 1987). Three other independent lines of evidence strongly support the hypothesis that these lateral tectal output cells, with axons descending to the contralateral medulla oblongata and rostral spinal cord, form the sensorimotor interface for visually elicited prey capture by anurans. (1) Electrical stimulation of the lateral tectum evokes snapping in toads (Ewert, 1967; Matsushima et al., 1985). (2) ^{14}C-2-deoxyglucose experiments in toads during visually guided prey capture show a marked increase of metabolic activity, reflecting heightened neuronal firing, in the lateral tectum (Finkenstaedt et al., 1985). (3) Extracellular antidromic stimulation/recording studies (Satou and Ewert, 1985) and intra-cellular recording/labeling studies (Matsumoto et al., 1986) have demonstrated prey selective neurons in the lateral tectum with axons descending to the contralateral hindbrain and caudal neuraxis. Therefore, lateral tectal neurons in layer 6/7 with descending projections are a vital link in the correlation between afferent sensory messages and efferent motor commands during visually triggered prey capture.

Since tactually elicited prey capture survives ablation of the above mentioned lateral tectal regions (Comer and Grobstein, 1981b), it is fair to assume that the sensorimotor interface of tactile sensory analyzers has independent and parallel access to the relevant motor pattern generating circuitry. The principal and laminar nuclei of the torus semicircularis project to the hindbrain and spinal cord (Donkelaar et al., 1981), and could well be the substrate of the tactile sensorimotor interface. The efferent connections of the olfactory bulb are not known to reach the hindbrain directly (Northcutt and Kicliter, 1980), and since odor triggered prey capture is a learned response, it may be that the information from olfactory sensory analyzers reaches the MPGs via either the visual or tactile interface. But the nature of descending connections from olfactory bulb to tectum and/or torus are as yet unknown.

Before discussing the target of the sensorimotor interfaces - the MPGs, let us briefly review the organization and inputs of the motoneurons that are the recepients of the coordinated pattern of spatio-temporal activity arising from an appropriately triggered MPG. This would allow us to 'close in' on the MPG for snapping from both its input (sensorimotor interfaces) and output (motoneuronal pools). Thus a MPG is conceived of as a nodal point of convergence for information from sensory analyzers, and for divergence of information to the related motoneuronal pools.

MOTONEURONS INVOLVED IN SNAPPING

The motor output of snapping, the final step in prey capture, consists of the head lunging toward the prey, the tongue projecting out to strike at the prey, and returning to the mouth with the prey stuck on the dorsal surface of the tongue. We will focus on the motoneurons that control the most specific muscles used in these actions - namely, the tongue controlling motoneurons. Tracing the input to these motoneurons affords us the best opportunity to delineate the output of the MPG that controls these motoneurons during the tongue flip. Motoneurons that innervate muscles responsible for protraction (M. genioglossus) and retraction (M. hyoglossus) of the tongue are topographically distributed within the hypoglossal nucleus in the caudal medulla oblongata (Weerasuriya and Ewert, 1981). The tongue is retracted primarily by the action of the M. hyoglossus, and its eversion requires the cooordinated action of muscles that lower the tip of the mandible, stabilize the hyoid bone in the floor of the mouth, stiffen the tongue into a rod, and rotate it over a wedge of contracted muscle (Gans and Gorniak, 1982). Thus to consider only two muscles (Mm. hyoglossus and genioglossus), from a group of at least eight, is a justifiable simplification that retains a focus on the essentials components of the tongue flip.

Motoneurons innervating these two muscles do not appear to have axon collaterals, but have extensive dendritic arborizations extending far beyond the confines of the nucleus (with a few even crossing the midline) and relatively large cell bodies (Matesz and Szekely, 1977; Weerasuriya and Ewert, 1981; Stuesse et al., 1983; Satou et al., 1985). Retractor motoneurons are synaptically activated by volleys in visceral afferents coursing through the nucleus of the solitary tract (Matsushima et al., 1986), and in the nerve branch innnervating the protractor muscle (Satou et al., 1985). Dendrites of retractor, but not protractor, motoneurons extend toward the solitary tract (Satou et al., 1985). These connections are probably the neural basis for the protective reflexes that withdraw the tongue from noxious stimuli. Rejection of unpalatable objects from the mouth is probably much more complicated requiring cooordinated actions of muscles controlling the tongue and jaw. The most relevant observation for the MPG for snapping is that these motoneurons cannot be driven monosynaptically from the tectum (Satou et al., 1985). In fact, the shortest mean latency of 9 ms. suggests the intervention of at least three, and probably more, interneurons between the tectal output cells and hypoglossal motoneurons. In contrast, tectofugal impulses traverse only one or two interneurons to reach neck and forelimb motoneurons in the brachial enlargement of the spinal cord (Maeda et al., 1977). A HRP study of inputs to the hypoglossal nucleus also failed to reveal any direct connections from the tectum or any mid- and forebrain structures (Weerasuriya and Ewert, 1984). The heaviest input to the hypoglossal nucleus was from the medial reticular formation of the medulla oblongata. Therefore, all the available evidence strongly indicates that the output of the tectum reaches hypoglossal (and probably other cranial) motoneurons only through a relay of interneurons. It is these interneurons that are postulated to be part of the MPG for snapping.

MPG FOR SNAPPING

The concept of MPG for snapping was first postulated in a study on sensorimotor interfacing in prey capture (Weerasuriya and Ewert, 1981), and further elaborated subsequently (Weerasuriya, 1983, 1989). MPGs are implicated in the control of many other vertebrate motor synergies such as locomotion (Grillner, 1981), respiration (Feldman, 1986), swallowing (Doty et al., 1967; Jean, 1984), mastication (Nozaki et al., 1983), eye movement (Robinson, 1981; Fuchs and Kaneko, 1985), etc.

Since the output of the MPG would be expected to be directed to the hypoglossal nucleus, a HRP study of afferents of the hypoglossal nucleus revealed the potential site of the MPG to be the medial reticular formation (MRF) in the rostral medulla oblongata (Weerasuriya and Ewert, 1984). This area, extending from the level of the nucleus isthmi to the rostral pole of the vagal branchiomotor column, has been investigated further in terms of its inputs (Weerasuriya, 1989), intracellular (Matsushima et al., 1989; Schwippert et al., 1990) and extracellular (Ewert et al., 1990) responsivity to visual and tactile stimuli, and effects of lesions in this area of the MRF on prey catching behavior (Ewert and Weerasuriya, 1988).

Inputs of the MRF. Adjoining bilateral areas of the medial reticular formation are the major sources of input. Most often, retrogradely labeled cells were the magnocellular elements situated at the border between the grey and white matter. These cells have elaborate dendritic trees extending into the subjacent white matter. The lateral and median reticular formations also project to the MRF. It receives ascending inputs from the ipsilateral nucleus of the solitary tract and bilaterally from cells near the spinal motor column in the rostral spinal cord. The MRF receives descending bilateral inputs frm the tectum and the principal and laminar nuclei of the torus semicircularis. The tectal labeling was more extensive contralaterally, and was greater in the lateral and ventrolateral region of the tectum with most cells located in layers 6/7. There is a sparse descending input from the ipsilateral anterior and posterior ventral tegmental nuclei, and posterior nucleus of the thalamus. A descending output of the MRF to the ipsilateral hypoglossal nucleus was detected.

These connections amply support the notion that the major portion of the MPG for snapping, which needs to receive input from tactile and visual sensory analyzers and project to motoneurons of tongue muscles, is located in the MRF. The synaptic morphology of these connections, and immunohistochemical studies of MRF neurons are necessary to fill in vital details of the neural circuitry of the MPG.

Electrophysiological Properties of the MRF. The response properties of medullary neurons to visual stimuli are qualitatively similar to that of tectal neurons, with sharper discrimination properties and larger visual receptive fields. In addition, cells with large spikes and rhythmic bursting properties were a regular feature. It appears that MRF neurons integrate information from descending tectal output neurons so as to identify prey related stimuli with greater acuity in a larger visual receptive field. Does this imply that the precision of prey localization lost in these circuits is represented elsewhere in other MRF circuits devoted to accurate visual localization at the expense of identification? The bursting neurons could be activated by nonspecific visual or tactile stimuli and seemed part of a reverberating circuit since the bursting activity continued long after the original stimulus had been removed. Some MRF neurons with large visual receptive fields responded also to tactile stimuli. These neurons may be part of the circuit integrating multimodal sensorimitor interfaces. Intracellular labeling of MRF cells revealed a wide range of dendritic architectures and somata sizes that does not currently lend itself to a meaningful classification.

While the above sensory properties of MRF cells provide an essential insight of their integrative functions, it is desirable to obtain information on the synaptic physiology of the MRF-hypoglossal connection, and to record the activities of these neurons during prey capture. This would enable the elucidation of neuronal connectivity between the tectum and MRF on the one hand, and that between the MRF and hypoglossal motoneurons on the other hand. Nevertheless, the data cited above are highly consistent with the idea that during prey capture, information related to processing of visual signals is channeled through the MRF on its way to triggering a motor response.

Lesions of the MRF and Prey Capture. Bilateral and symmetrical lesions of the MRF abolish visually evoked prey catching behavior (Ewert and Weerasuriya, 1988). But extensive unilateral lesions in the MRF did not eliminate prey capture. This suggests that the MPG is bilaterally represented with each half having independent access to the relevant motoneurons on both sides. In related experiments, following bilateral transections of the hypoglossal nerve toads lunge toward mealworms with neither an accompanying tongue nor jaw movement. The lack of tongue projection and retraction is a direct consequence of the transection of axons that innervate tongue muscles. But since nerves that innervate the jaw muscles are intact, the lack of jaw opening and closing has to be attributed to an inhibition of jaw movement arising from the absence of tone in the lingual muscles detected by the glossopharnygeal (Matsushima et al., 1986) or hypoglossal (Stuesse et al., 1983) afferents. These results suggest a potential hierarchical organization of the subsynergies invoved in snapping. The components lower down in the hierarchy (tongue and jaw movements) can be uncoupled less easily than those further up (lunge vs. tongue and jaw movement) in the hierarchy. Thus there appears to be a greater interaction between tongue and jaw movement such that a background tone in the muscles involved in one subsynergy is necessary for the enactment of the other subsynergy. On the other hand, no such coupling seems to exist between the lunge and the snap.

To account for the above studies of the MRF and other structures involved in prey capture, the following functional organization (Fig. 1) of the MPGs for the lunge and snap was proposed (Weerasuriya, 1989). The degree of stereotypy in snapping, effect of the 'state of the organism' on snapping, distribution of putative MPG neurons in the medulla, interaction between premotoneuron and motoneuron pools, and spatial facilitation between the two halves of the MPG were discussed in that report.

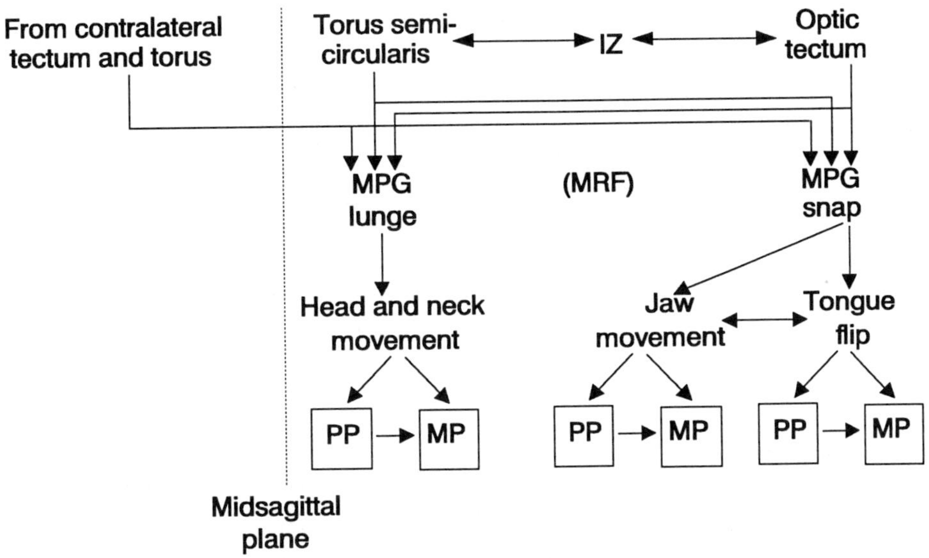

Fig. 1. Proposed functional organization of the motor pattern generator for snapping in anurans. The arrowheads indicate the direction of information flow. IZ, intermediate zone wedged between the ventrolateral tectum and far-lateral torus; MP, motoneuron pool; PP, premotoneuron pool. (Modified from Weerasuriya, 1989).

The rest of this essay will attempt to 'flesh out' the above scheme. A two stage adaptive control feedforward model (Houk, 1987) can be used to represent the lunge with its parametric control of the snapping distance (Fig.2). It is not inconceivable that a similar model with an array of adjustable pattern generators could adequately explain the orienting turn of the anuran toward the prey object. As has been pointed out (Georgopoulos and Grillner, 1989) there are no fundamental differences between the neural algorithms underlying reaching movements and locomotion, and orienting is very much a 'reaching movement' consisting of appropriate postural adjustments to locomote the animal closer to its prey. Therefore, it is quite likely that the layered cortical structures in the anuran brain, tectum and cerebellum, are essential components of the postulated array of adjustable pattern generators regulating orienting movements. Since the output of the MPG for the lunge needs to specify its parametric control in only one dimension (snout to prey distance), a mechanism qualitatively similar to that used to generate horizontal saccades would be adequate. The required transformation is from a neural code for depth perception, which anurans perform adequately with one eye, to one for the lunging distance. The importance of suitable posture for proper lunges cannot be overemphasized. At short lunging distances, the forward thrust of the head and neck is against a stable posture. At longer distances, the hindlimbs are extended rotating the upper trunk over the forelimbs. This movement reduces the distance between the snout and prey, and thus shortens the gap needed to be covered by the protruding tongue.

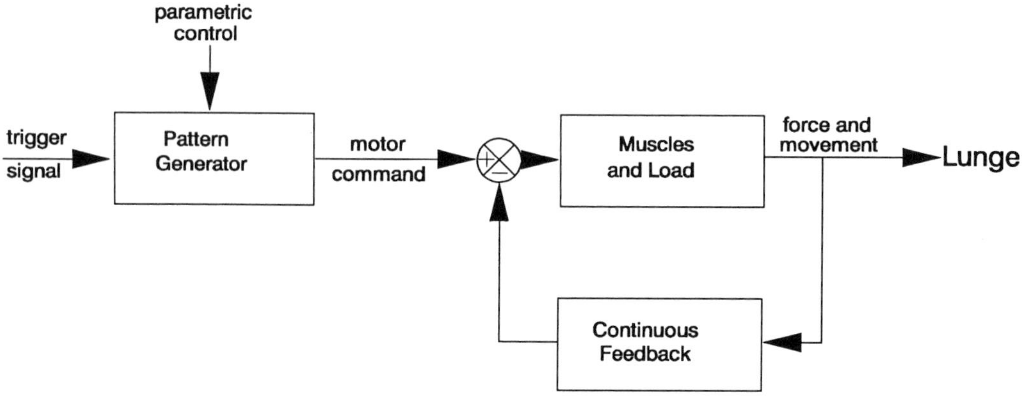

Fig. 2. A model for control of the lunge. This model is adapted from Houk (1987). Trigger signals from sensory analyzers initiate the motor output. The initial settings of the parameters are determined by hardwired codes as modified by past experience and perceived snout to prey distance. The patterned output of the first stage is fedforward as the input for the second stage. These motor commands interact with *continuous* feedback from spinal and bulbar sources to drive motoneurons resulting in movement. Feedback in the first stage is *discontinuous*,and discrepancies between outcomes and intended goals are rectified by adjusting the parameters of the adaptive control lines during intertrial intervals.

The tongue flip, consisting of jaw and tongue movements to retrieve the prey, is conceived of as a triggered ballistic movement with minimal or no parametric control. The activity of T5(2) tectal neurons during prey capture strongly supports this proposition (Schurg-Pfeiffer, 1989). These units increase their spike discharge rate before both orienting and approaching toward the

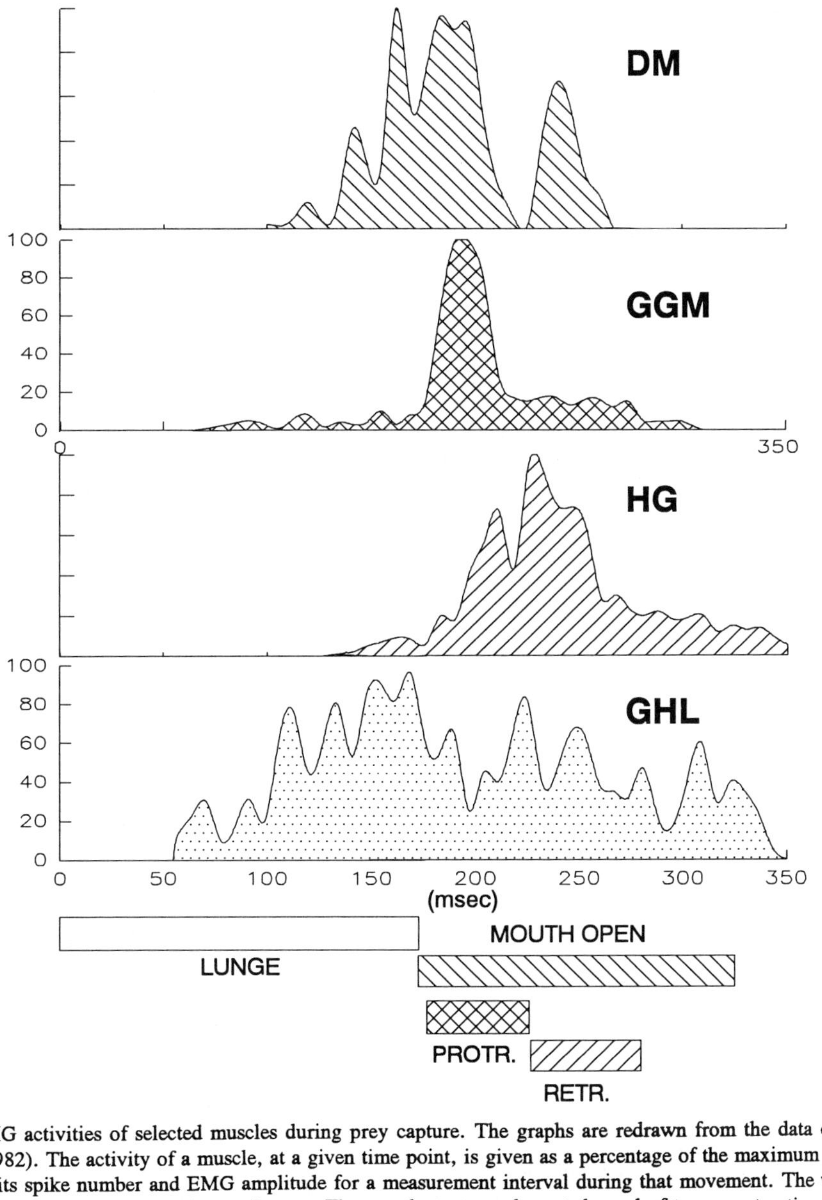

Fig. 3. EMG activities of selected muscles during prey capture. The graphs are redrawn from the data of Gans and Gorniak (1982). The activity of a muscle, at a given time point, is given as a percentage of the maximum value of the product of its spike number and EMG amplitude for a measurement interval during that movement. The width of the sampling bins in the original study was 7 msec. The mouth starts to close at the end of tongue retraction. Activity of jaw closing muscles (temporalis, pterygoideus and masseter) were not monitored in that study.

prey, and maintain the high frequency during the movements. In contrast, T5(2) units increase their firing rate before a snap, but cease firing immediately before a snap and are silent during the snap. EMG recordings of jaw and tongue muscles during snapping reflect the discharge patterns of their respective motoneurons. Fig.3 is adapted from the data of Gans and Gorniak (1982) for selected muscles involved in protruding and retracting the tongue, and lowering the jaw. The M. depressor mandibulae (DM) opens the mouth by lowering the mandible, the M. genioglossus medialis (GGM) plays a major role in tongue eversion, the M. hyoglossus (HG) retracts the tongue into the mouth with the prey, and the M. geniohyoid lateralis (GHL) stabilizes the hyoid bone in the floor of the mouth during jaw and tongue movements. DM is active during the lunge even before the jaw begins to open, and has a second burst of activity to open the mouth wider as the tongue returns with the prey. Activity of GGM is terminated more abruptly than that of HG. The forward movement of the hyoid bone, produced by the contraction of GHL, assists in lowering the mandibular symphysis, moves the soft tissue of the tongue forward to aid eversion, and shortens the stretched resting length of HG to utilize optimal length-contraction characteristics of this muscle during tongue retraction (Gans and Gorniak, 1982). This brief review of activity patterns in putative input and output channels of the MPG serves as the basis for the following discussion on a model for control of tongue muscles during prey capture.

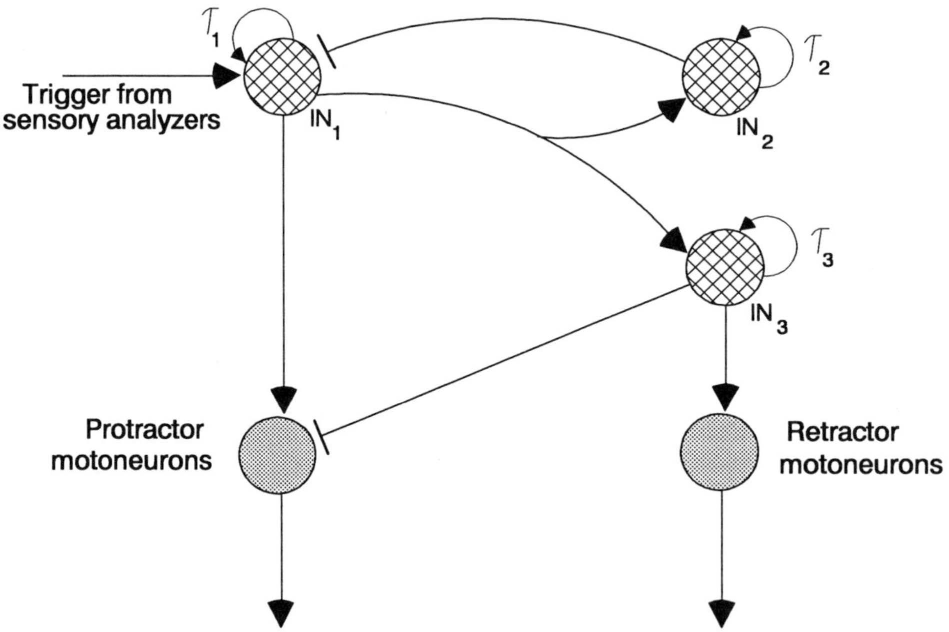

Fig. 4. A neuronal model of the circuitry controlling tongue movement during prey capture. The activity of this module has to be integrated with those governing jaw movement, and in turn they have to be coordinated with the pattern generators responsible for the head, neck and body movements involved in the lunge.

The neural circuitry subserving control of tongue muscles during the flip is modeled by the leaky integrator circuit shown in Fig. 4. The integrators in the circuit are assumed to represent clusters of interconnected interneurons organized as reverberating positive feedback loops (Robinson, 1989). The circuit is activated by a trigger signal from a sensory analyzer, and the duration of the flip is controlled by the time constants T_1 and T_2. Activity in protractor motoneurons is abruptly terminated by active inhibition from IN_3. Retractor motoneuron activity ceases more slowly, paralleling the decay of the leaky integrator IN_3. This difference in rate of termination of activity reflects well the EMG patterns in Fig. 3. A putative candidate for IN_1 is the M-10 group of neurons in the MRF (Schwippert et al., 1990). In immobilized toads, they are activated by minimal visual and tactile stimuli and fire in bursts for several seconds. The intraburst frequency is about 20 Hz, each burst has 10 - 20 spikes, and bursts repeat every second or two. Their bursting pattern is always terminated by an IPSP. This could arise from the inhibitory effect of IN_2 on IN_1. The verifiable deductions that arise from this model are (1) there are at least two, or probably three, integrators in the MRF with direct influences on hypoglossal motoneurons, (2) damage to one of them (IN_3) should prolong the burst of the protractor motoneurons and drastically reduce the activity of retractor motoneurons during snapping, and (3) lesions of IN_2 should also prolong the protractor burst but not reduce the retractor burst.

Acknowledgements. Its a pleasure to record my gratitude to Michael Arbib for seminal discussions on modeling the tongue flip module, and his warm and cordial hospitality during the workshop. To Peter Ewert a special thanks for numerous helpful discussions.

REFERENCES

Arbib, M. A. (1989) Visuomotor coordination: neural models and perceptual robotics. In:Ewert J.-P. and Arbib M. A. (eds) *Visuomotor Coordination.* Plenum Press, New York, pp 121-171.

Berkinblitt, M. B., Feldman, A. G. and Fukson, O. I. (1989) Wiping reflex in the frog: movement patterns,receptive fields, and blends. In: Ewert J.-P. and Arbib M. A. (eds) *Visuomotor Coordination.* Plenum Press, New York, pp 615-629.

Comer, C. and Grobstein, P. (1981a) Tactually elicited prey acquisition behavior in the frog, *Rana pipiens*, and a comparison with visually elicited behavior. *J. Comp. Physiol.* 142: 141-150.

Comer, C. and Grobstein, P. (1981b) Involvement of midbrain structures in tactualy and visually elicited prey acquisition behavior in the frog, *Rana pipiens. J. Comp. Physiol.* 142: 151-160.

Dole, J. W., Rose B. B. and Tachiki, K. H. (1981) Western toads *(Bufo boreas)* learn odor of prey insects. *Herpetologica* 37: 63-68.

Donkelaar, ten H.-J., de Boer-van Juizen, R., Schuoten, F. T. M. and Eggen, S. J. H. (1981) Cells of origin of descending pathways to the spinal cord in the clawed toad, *Xenopus laevis*. *Neurosci.* 16: 2297-2312.

Doty, R. W., Richmond, W. H. and Storey, A. T. (1967) Effect of medulla lesions on coordination of deglutition. *Exp. Neurol.* 17: 91-106.

Ewert, J.-P. (1967) Aktivierung der Verhaltensfolge beim Beutefang der Erdkrote (*Bufo bufo L*) durch Elektrische Mittelhirnreizung. *Z. Vergl. Physiol.* 54: 455-481.

Ewert, J.-P. (1968) Der Einfluss von Zwischenhirn defecten auf die Visuomotorik im Beute- und Fluchverhalten der Erdkrote (*Bufo bufo*). *Z. Vergl. Physiol.* 61: 41-70.

Ewert, J.-P. (1984) Tectal mechanisms that underlie prey-catching and avoidance behaviors in toads. In: Vanegas H. (ed) *Comparative Neurology of the Optic Tectum.* Plenum Press, New York, pp 247-416.

Ewert, J.-P. (1987) Neuroethology of releasing mechanisms: prey-catching in toads. *Behav. Brain Sci.* 10: 337-405.

Ewert, J.-P., Framing, E., Schurg-Pfeiffer, E. and Weerasuriya, A. (1990) Responses of medullary neurons to moving visual stimuli in the common toad. I Characterization of medial reticular neurons by extracellular recording. *J. Comp. Physiol.* in press.

Ewert, J.-P. and Weerasuriya, A. (1988) The motor pattern generator for snapping in toads. *Soc. Neurosci. Abstr.* 14: 618.

Feldman, J. L. (1986) Neurophysiology of respiration in mammals. In: Bloom F. E. (ed) *Handbook of Physiology. The Nervous System. Intrinsic Regulatory Systems of the Brain.* American Physiological Society, Bethesda, pp 463-524.

Finkenstaedt, T., Adler, N. T., Allen, T. O., Ebbeson, S. O. E. and Ewert, J.-P. (1985) Mapping of brain activity in mesencephalic and diencephalic structures of toads during presentation of visual key stimuli: a computer assisted analysis of ^{14}C-2DG autoradiographs. *J. Comp. Physiol.* 156: 433-445.

Fuchs, A. F. and Kaneko, C. R. S. (1985) A brainstem generator for saccadic eye movements. In: Evarts E. V., Wise S. P. and Bousefield D. (eds) *The Motor System in Neurobiology.* Elsevier, Amsterdam, pp 126-132.

Gans, C., (1961) The bullfrog and its prey. *Nat. Hist.* 70: 26-37.

Gans, C. and Gorniak, G. C. (1982) Functional morphology of lingual protrusion in marine toads (*Bufo marinus*). *Am. J. Anat.* 163: 195-222.

Georgopoulos A. P. and Grillner, S. (1989) Visuomotor coordination in reaching and locomotion. *Science* 245: 1209-1210.

Grillner, S. (1981) Control of locomotion in bipeds, tetrapods and fish. In: Brooks V. B. (ed) *Handbook of Physiology. The Nervous System. Motor Control.* American Physiological Society, Bethesda, pp 1179-1236.

Grobstein, P. (1989) Organization of the sensorimotor interface: a case study with increased resolution.In: Ewert J.-P. and Arbib M. A. (eds) *Visuomotor Coordination.* Plenum Press, New York, pp 537-568.

Hinsche, G. (1935) Ein Schnappreflex nach "Nichts" bei Anuren. *Zool. Anz.* 111: 113-122.

Houk, J.C. (1987) Model of the cerebellum as an array of adjustable pattern generators. In: Glickstein M., Yeo C. and Stein J. (eds) *Cerebellum and Neuronal Plastcity.* Plenum Press, New York, pp 249-260.

Jean, A. (1984) Brainstem organization of the swallowing network. *Brain Behav. Evol.* 25: 109-116.

Kicliter, E.A. (1973) Flux, wavelength and movement dicrimination in frogs: forebrain and midbrain contributions. *Brain Behav. Evol.* 8: 340-365.

Kostyk, S. K. and Grobstein, P. (1987) Neuronal organization underlying visually elicited prey orienting in the frog. II: Anatomical studies on the laterality of central projections. *Neurosci.* 21: 57-82.

Kupfermann,I. and Weiss, K. R. (1978) The command neuron concept. *Behav. Brain Sci.*1: 1-39.

Lazar, G., Toth, P., Csank, G. and Kicliter, E. (1983) Morphology and location of tectal projection neurons in frogs: a study with HRP and cobalt filling. *J. Comp. Neurol.* 215: 108-120.

Maeda, M., Magherini, P. C. and Precht, W. (1977) Functional organization of vestibular and visual inputs to neck and forelimb motoneurons in the frog. *J. Neurophysiol.* 40: 225-243.

Matesz, C. and Szekely, G. (1977) The dorsomedial nuclear group of cranial nerves in the frog. *Acta Biol. Acad. Sci. Hung.* 28: 461-474.

Matsumoto, N., Schwippert, W. and Ewert, J.-P. (1986) Intracellular activity of morphologically identified neurons of the grass frog's optic tectum in response to moving configurational visual stimuli. *J. Comp. Physiol.* 159: 721-739.

Matsushima, T., Satou, M. and Ueda, K. (1985) An electromyographic analysis of electrically-evoked prey-catching behavior by means of stimuli applied to the optic tectum in the Japanese toad. *Neurosci. Res.* 3: 154-161.

Matsushima, T., Satou, M., and Ueda, K. (1986) Glossopharyngeal and tectal influences on tongue muscle motoneurons in the Japanese toad. *Brain Res.* 265: 198-203.

Matsushima, T., Satou, M. and Ueda, K. (1989) Medullary reticular neurons in the Japanese toad: morphologies and excitatory inputs from the optic tectum. *J. Comp. Physiol.* 166: 7-22.

Northcutt, R.G. and Kicliter, E. (1980) Organization of the amphibian telencephalon. In: Ebbeson S. O. E. (ed) *Comparative Neurology of the Telencephalon.* Plenum Press, New York, pp 203-255.

Nozaki, S., Enomoto, S. and Nakamura, Y. (1983) Identification and input-output properties of bulbar reticular neurons involved in the cerebral cortical control of trigeminal motoneurons in the cat. *Exp. Brain Res.* 49: 363-372.

Robinson, D.A. (1981) Control of eye movement. In: Brooks V. B. (ed) *Handbook of Physiology. The Nervous System. Motor Control.* American Physiological Society, Bethesda, pp 1275-1320.

Robinson, D.A. (1989) Integrating with neurons. *Ann. Rev. Neurosci.* 12: 33-45.

Satou, M. and Ewert, J.-P. (1985) The antidromic activation of tectal neurons by electrical stimuli applied to the caudal medulla oblongata in the toad *Bufo bufo (L). J. Comp. Physiol.* 157: 739-748.

Satou, M., Matsushima, T. and Ueda, K. (1985) Tongue-muscle-controlling motoneurons in the Japanese toad: topography, morphology and neuronal pathways from the 'snapping-evoking area' in the optic tectum. *J. Comp. Physiol.* 157: 717-737.

Schurg - Pfeiffer, E. (1989) Behavior-correlated properties of tectal neurons in freely moving toads. In: Ewert J.-P. and Arbib M. A. (eds) *Visuomotor Coordination.* Plenum Press, New York, pp 451-480.

Schwippert, W. W., Beneke, T. W. and Ewert, J.-P. (1990) Responses of medullary neurons to moving visual stimuli in the common toad.II An intracellular recording and cobalt-lysine labeling study. *J. Comp. Physiol.* in press.

Shinn, E. A. and Dole, J. W. (1978) Evidence for a role for olfactory cues in the feeding response of leopard frogs, *Rana pipiens. Herpetologica* 34: 167-172.

Stuesse C., Cruce, W. L. R. and Powell, K.S. (1983) Afferent and efferent components of the hypoglossal nerve in the grass frog. *J. Comp. Neurol.* 217: 432-439.

Szekely, G. and Lazar, G. (1976) Cellular and synaptic architecture of the optic tectum. In: Llinas R. and Precht W. (eds) *Frog Neurobiology.* Spriger-Verlag, Berlin, pp 407-434.

Weerasuriya, A. (1983) Snapping in toads: some aspects of sensorimotor interfacing and motor pattern generation. In: Ewert J.-P., Capranica R. R. and Ingle D.J. (eds) *Advances in Vertebrate Neuroethology.* Plenum Press, New York, pp 613-627.

Weerasuriya, A. (1989) In search of the motor pattern generator for snapping in toads. In: Ewert J.-P. and Arbib M. A. (eds) *Visuomotor Coordination.* Plenum Press, New York, pp 589-613.

Weerasuriya, A. and Ewert, J.-P. (1981) Prey-selective neurons in the toads optic tectum and sensorimotor interfacing: HRP studies and recording experiments. *J. Comp. Physiol.* 144: 429-434.

Weerasuriya, A. and Ewert, J.-P. (1984) Afferents of the hypoglossal nucleus in the common European toad, *Bufo bufo. Anat. Rec.* 208: 192A.

From Tectum to Forebrain

The Striatum and Short-Term Spatial Memory: From Frog to Man

David Ingle
Northeastern University
Marine Science Center
East Point
Nahant, MA 01908

Abstract

Frogs show a robust memory for the position of recently-seen obstacles after their sudden removal, which may last for at least 60 seconds. This memory can compensate for passive rotation of the frog through at least a 45° turn: thus spatial memory is stored in real-world rather than in retinotopic coordinates. A "sliding map" network model will be presented as a speculative approach to physiological experiments. Evidence that this short-term memory depends upon the frog's striatum will be presented. Behavioral tests revealing a second mode of spatial memory (based upon landmarks) will be presented. Analogies between these two species of frog memory and contrasting classes of mammalian memory, linked to caudate and hippocampus, respectively, will be discussed. Extensions of this new approach to spatial memory to patients with Parkinson's Disease and to subjects with loss of corpus callosum will be briefly reviewed. The hypothesis will be presented that neocortex sustains short-term memory as "imagery" but that striatum maintains correlated "anticipatory response sets."

Introduction

We take for granted our ability to remember the locations of recently seen objects after turning away from them or after closing our eyes. Introspection reveals a robust mental image of the size and location of a conspicuous object, with varying imagery of its shape and color. Within 30-60 seconds these mental images fade, or may be more quickly erased as we move to another mini-environment or switch our attention to other activities. Of more interest to a theorist of visual activity is the fact that our mental image of the unseen object appears quite stationary in space as we rotate or walk about. In this review I will first show that a similar persistence of short term spatial memory occurs in the frog, and seems to be compensated during the frog's passive rotation. We shall propose a "sliding map" model of this compensation effect, and go on to argue that the striatum of the frog plays a role analogous to that of the mammalian caudate nucleus in maintaining this class of

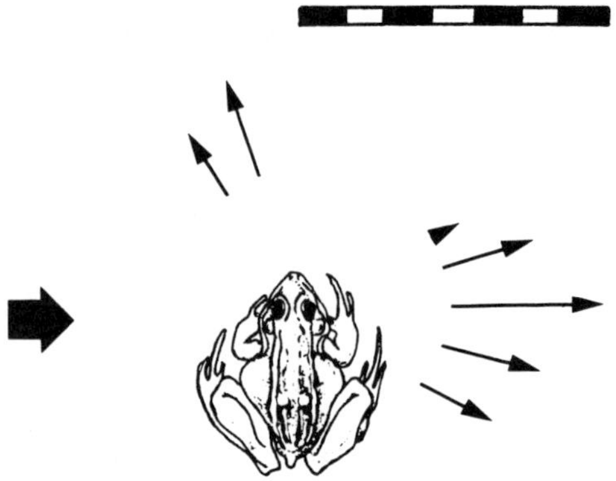

Figure 1. After viewing a barrier within the right front field for 10 sec. normal frogs will avoid that location 15 sec. after the barrier has been removed, choosing escape directions seldom taken on non-barrier trials with the same left-side threat stimulus.

short-term spatial memory. Finally, we will review our own recent studies of this phenomenon in human subjects which add new information on the dynamics of updating short-term spatial memory.

Short-Term Spatial Memory in Frogs

Although frogs are notoriously difficult to train using selective reward or classic conditioning methods, a robust kind of short-term memory is demonstrated in adult *Rana Pipiens* by a simple procedure (Ingle & Hoff, 1990). After presenting a large black-and-white striped barrier near the frog for about 5 seconds, the obstacle is suddenly removed from view (by a long wire handle) while the frog rests within an empty white-walled arena. After 10-15 seconds we move a 4 inch wide black square along the floor directly at the frog. While an unbiased frog typically jumps into the opposite forward quadrant to avoid a lateral threat (Fig. 1), a remembered barrier in that location will deflect the frog's jump direction toward one or the other side of the former barrier position. Since frogs may jump quite close to the remembered locations of the barrier edges, we infer that their memory-acuity for such edges is plus or minus 10° or less after a 10 second delay.

Using a left-vs-right memory test (Fig. 2) we find that good barrier memory is obtained even with delays of 30 and 60 seconds. In fact, jump angle histograms for these two longer intervals show that while frogs often jump into the ipsilateral rear quadrant near the remembered caudal edge of the barrier, they very seldom jump *into* the sector corresponding to its former location, as they readily do in the no-barrier control trials. Although this memory for where *not* to go lasts for at least 60 seconds in the in disturbed frog, we found that when frogs are picked up and held for five seconds, before being replaced and tested for avoidance direction 10 seconds later, barrier memory is erased. We have also noted that large spontaneous jumps of 8-10 inches away from a barrier on side will abolish any further left-right bias in response to test looms along the frontal midline.

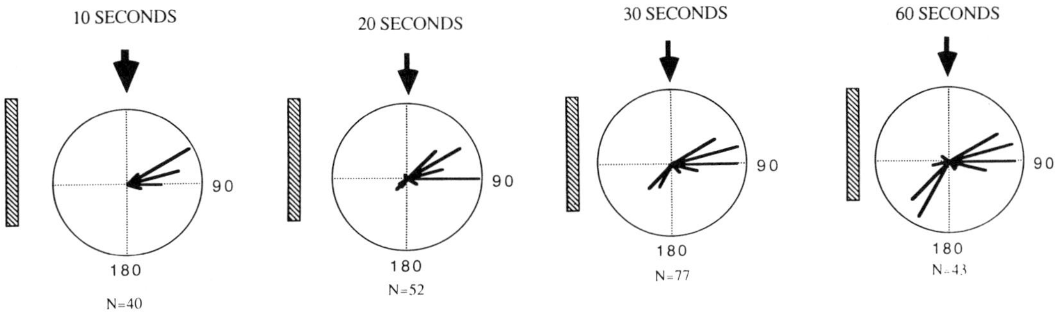

Figure 2. With a lateral barrier removed, frogs respond to frontal threat by jumping away from the remembered barrier location after 10-20 sec delays. After longer delays of 30-60 sec the same frogs sometimes jump closer to the former barrier location but they very seldom jump into that sector.

These various facts can be accommodated into a teleological explanation of why frogs ought to remember the locations of barriers which have already disappeared. Collett (1977) showed that toads are able to use information about the extent of their lens accommodation to estimate distance of prey. If a frog were thus accommodated on a nearby prey, more distant objects would be out of focus and thus of indeterminate depth. If suddenly preyed upon, the frog would have insufficient time to move his lens, focus on other objects and determine the best escape route. Such difficulty would be bypassed if the frog can remember the locations of those nearby obstacles which could block rapid escape. We should expect that such a spatial representation is *preserved* as the animal makes short translations and rotations within that area, but is *erased* and updated as it moves into a new area. According to this scheme, the intermittent erasure of memory is as important as the formation of new memory. Once the critical brain region is localized we should look for neurotransmitter systems which erase memory as well as those which fix it.

The Striatum as a Holder of Short-Term Memory: Comparative Data

Our own studies implicate the frog's striatum as necessary for the expressing of short-term "barrier memory". First we have removed either medial or lateral pallium unilaterally in a number of frogs without witnessing any loss of barrier memory (Fig. 3). In two animals bilateral ablation of medial pallium (a suggested homolog to the mammalian hippocampus) had no effect. Yet in each of seven frogs

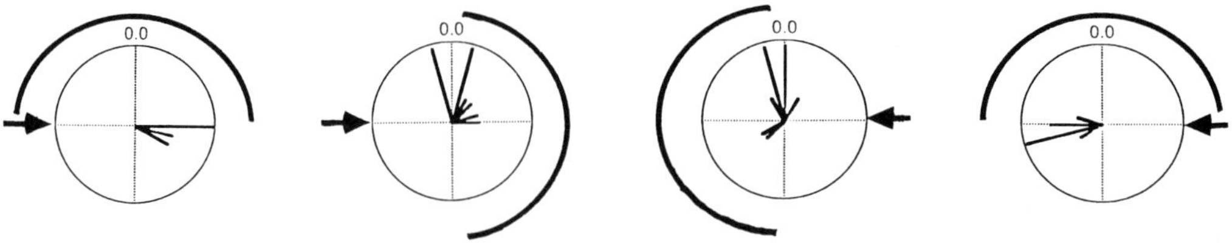

Figure 3. A frog with ablation of medial pallium normally avoids frontal and lateral barrier locations on either side. Note that the same frontal barrier memory can deflect the frog's escape route in either direction depending upon the side of looming threat.

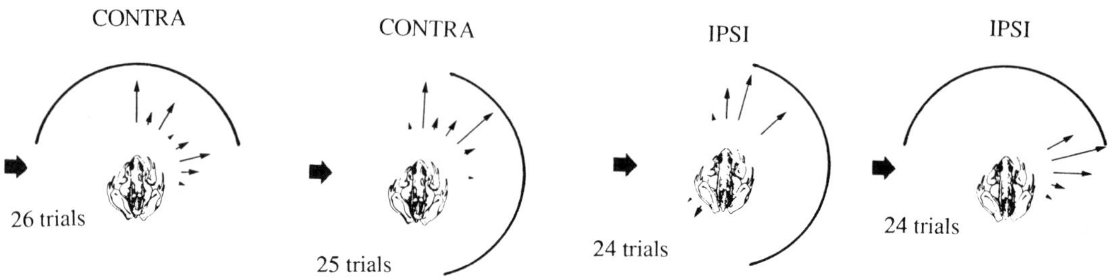

Figure 4. When the barrier edge to be cleared is contralateral to a unistriatal ablation (left side) the frog mainly jumps into the former barrier location (as if no barrier had been seen). The same frog mostly avoids the remembered barrier on the side ipsilateral to the ablation. In this figure the IPSI response directions were symmetrically reversed for comparison with the CONTRA avoidance patterns.

with unilateral ablation of striatum, we witnessed a severe or complete loss of memory for location of barrier-edges seen on the opposite side of the midline (Fig 4). Three frogs with bilateral striatum ablations lost barrier memory bilaterally. In each case, the loss of barrier memory was not accompanied by any loss of ability to avoid seen barriers. These data suggest that pretectum suffices for detection of barriers and modifications of feeding or avoidance strategies (see Ingle, 1980), but that *memory* of the same barriers depends upon integrity of the striatum.

Anatomical studies indicate that the sole major afferents to striatum emerge from both medial and lateral cell groups of the middle dorsal thalamus (auditory and visual relays). I add the suggestion that edge-sensitive cells of the pretectum

(Ewert, 1971; Ingle, unpublished) provide relevant information to the thalamo-striatal system via short axons. Such a route might explain the earlier observation of Vesselkin (1971) that electrical activity in striatum is evoked by optic nerve stimulation following ablation of tectum. Clearly single-unit recordings in striatum and in thalamus are needed to substantiate this notion, since our ablation effects do not compel us to believe that the representation of a barrier is itself stored within the striatum.

If we accept the homology between the frog's striatum and the caudate of mammals (e.g. Parent, 1986) then we find support from the fact that some monkey caudate cells discharge in correlation with the monkey's intention to look at a remembered visual target (Hikosaka & Sakamoto, 1986). We cannot tell whether the activity of these caudate cells represents a *visual* image of the target's location or a preparation to *respond* within a particular direction. We do know that the frog's barrier memory is not simply explained as a bias to move to one side or the other. In Fig. 3 we see that after removal of a wide frontal barrier the frog is prepared to clear *either* edge, depending upon the direction of threat. This memory system in the frog may impose upon lower centers a bias *not to jump* in particular directions leaving the tectal avoidance system with the final deciding vote on which direction will be taken. Perhaps in frogs the striatum participates only in this inhibitory kind of memory but in mammals also participates in the memory of goals to be approached.

Studies of caudate function in rats provide evidence for a role similar to that of frog striatum in holding an inhibitory type of spatial memory. Countless studies document the rat's spontaneous tendency to alternate turn directions in a T-maze when rewarded noncontingently with food in one or the other arm. Such behavior seems adaptive in biasing the rat to look for food in a new place, having once found it in a particular location. With specific rewards for alternating, rats can achieve 95% reliability even with imposed delays of 10 seconds between maze-runs. This competency is wiped out by lesions of the caudate, however (Divac et al., 1978).

A second approach to caudate function concerns its role in updating the animal on its spatial location during passive transport. Ingle and Hoff (1990) found that intact frogs are able to compensate for passive rotations of 45° and remember location of barrier-edges in "real world coordinates." Similarly, Abraham et al. (1983) found that rats could learn to find their way back to a once-sampled water spout (one of eight identical spouts) after being placed in a covered box and moved away through a right-angle trajectory before being released. Animals with lesions of the vestibular nuclei failed the task, and so did animals with caudate lesions. Control rats with lesions of hippocampus normally used such vestibular information to remember their passive displacement route, even though deficient in memory tests which required use of visual landmarks to locate food. This study is one of several in rats which identified the caudate as part of a system for retaining *egocentric* spatial coordinates and hippocampus are necessary for landmark-based memory. We have recently described a landmark-memory test for frogs (Ingle &

Hoff, 1990) in which location of room lights are used to remember the direction of water, toward which a frightened frog will leap. Further studies will be necessary to determine whether striatum or medial pallium (putative hippocampus) is critical to performance of our landmark-memory test.

Some Properties of Short-Term Spatial Updating in Man

Our first attempt to extrapolate from frog to man involved testing of persons with one kind of caudate dysfunction, namely those with Parkinson's Disease whose caudate levels of dopamine are typically depleted to 10% of normal values. Our test consisted of showing subjects a conspicuous target on the floor about 4-8 meters away and, after eye-closure, leading them forwards for unpredictable distances. With eyes still closed, subjects turned and pointed to the remembered location of the target. While 8 controls (the patients' spouses) average only 7° angular errors after 25 seconds delays the 8 patients averaged over 20° errors and none did as well as the worst control. A remarkable finding was that one patient with unilateral motor deficits showed normal pointing accuracy in one hemifield and made gross errors with the same hand pointing into the other hemifield. This subject resembled our frogs with unistriatum ablations.

Our study was an attempt to follow up the proposal of Potegal (1982) that the caudate nucleus plays a special role in *updating* the brain on the location of objects relative to one's own body as one moves about in space. We found that Parkinsonian subjects were fairly competent in remembering the location of a formerly seen target if they stood still with eyes closed, but were impaired after walking without vision. In a further experiment we sought to test Potegal's suggestion that an organism uses vestibular information to calibrate extent of rotations and translations while locomoting. While this is probably true for perception of passive movements (recall the displaced rats discussed above) it need not be true for monitoring active locomotion. When we tested 6 subjects with loss of vestibular function from inner ear infections, we found them just as good as normal subjects in pointing at remembered targets after closing the eyes and turning the body up to 180° in either direction. We presume that these subjects were using proprioceptive information from their own stepping movements, since they were grossly inaccurate following passive rotation in a desk-chair. It would be interesting to know whether avestibular frogs would also compensate their spatial memory following active rotations.

A final test using post-rotation localization of remembered targets was designed to test the hypothesis that during rotations past the remembered target the "movement" of the mental image from one hemifield to the other depends upon actual movement of its neural representation from one hemisphere to the other. We used 4 subjects born without a corpus callosum (that great fiber system connecting symmetrical regions of opposite cortices), plus another subject whose callosum was surgically cut in adolescence, sparing the splenium, where visual axons are known to cross. All 5 subjects were found to have "normal" transfer of

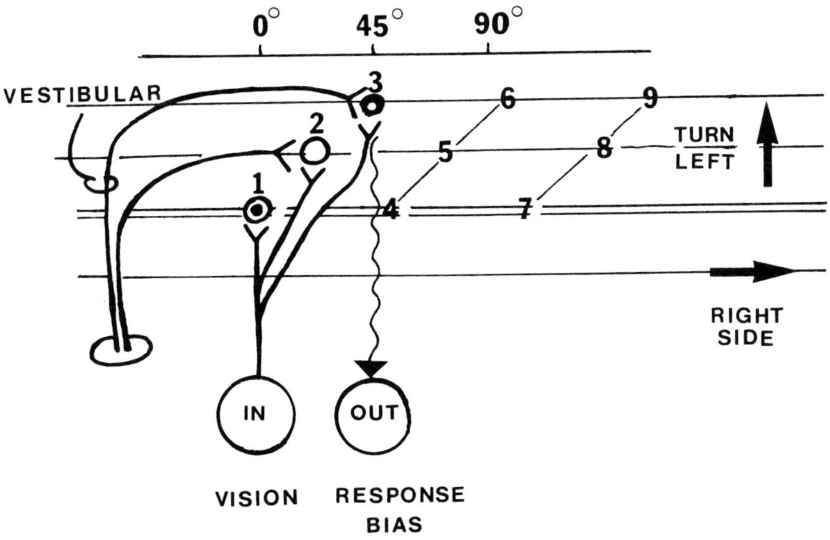

Figure 5. The author's schematic model of barrier memory storage in a maplike system which allows compensation for the frog's body rotation. Input from afferents signalling "edge-at-0°" activates an oblique row of cells at several layers. For the stationary frog cell #1 remains most active and would dominate behavior at that moment. If the frog turns by 45°, the corresponding vestibular signal optimally activates the third layer summating with ongoing activity to make cell #3 the most active. At that moment output of cell #3 dominates behavior; its activity corresponds to the real-world location of the former barrier edge.

shape and color information on standard tests, as predicted from earlier work with similar patients. Yet in our test we found no evidence for good transfer of short-term location memory as we found in 4 controls equated for age, sex and I.Q. (about 80). To make this inference, we asked subjects to point at remembered locations of targets seen initially at either 60° or 30° from their midline for about 1 second while they fixated straight ahead. When the split-brain subjects turned away from the target so that it remained within the *same* hemifield, their attempts to point at the remembered 30° target did not overlap with attempts to point at the original 60° location. This was true using either hand, and pointing into either hemifield. However, when the subjects rotated past the target locus, such that it was represented in the *opposite* hemifield, their pointing directions overlapped extensively indicating that they were largely guessing at target location.

Although these first results suggest the novel idea that a short-term memory trace (which we prefer to call "iconic memory") can actually move within the brain, it is also possible that memory is still present on the first-stimulated side, and is suppressed but not erased during the act of rotation. It becomes important to know whether a split-brain subject could retrieve his accurate location memory if re-rotated such that the representation is relocated within the original hemifield. A modification of our "sliding map" model (Fig. 5) is consistent with this possibility. On the other hand, if the iconic trace normally behaves like a tennis ball sent into the other court (i.e. hemisphere) as the body rotates, then self-rotation may erase the iconic memory in the split-brain subject, since the "ball" cannot cross to the other side via the corpus callosum and be returned again during a re-rotation of the body.

What is the relationship between the neo-cortical role in short-term spatial memory and that of the striatum? Remember that frogs display a kind of short term memory which appears to require the striatum but not a cortex. Our tentative answer is that the representation of an object as "visual image" is cortically localized but that cortex interfaces with the striatum in maintaining a "preparatory response set" (go this way, but not that way). The frog may have no precise memory of *what* the formerly-seen barrier looked like, but only *where* he should not jump. The use of imagery to guide action sequences may be a key feature of the evolution of higher brain functions in birds and mammals.

References

Abraham, L.M., Potegal, M., and Miller, S., 1983, Evidence for caudate nucleus involvement in an egocentric spatial task. *Physiol. Psychol.* 11:11-17.

Collett, T., 1977, Stereopsis in toads. *Nature*, 267: 349-351.

Divac, I., Markowitsch, H.J., and Pritzel, M., 1978, Behavioral and anatomical effects of small intrastriatal injections of kainic acid in the rat. *Brain Research* 151:523-532.

Ewert, J.-P., 1971, Single unit response of the toad caudal thalamus to visual objects. *Z. Vergl. Physiol.* 74: 81-102.

Ingle, D., 1980, Some effects of pretectum lesions on the frog's detection of stationary objects. *Behav. Brain. Res.* 1: 139-163 .

Ingle, D. and Hoff, K., 1988, Visually elicited evasive behavior in frogs. *Bioscience* 40:284-291.

Parent, A., 1986, *Comparative Neurobiology of the Basal Ganglia*, Wiley & Sons, New York, see chapter 3.

Potegal, M., 1982, *Spatial abilities: development and physiological foundations*. Academic Press, New York, see chapter 15.

Multiple Brain Regions Cooperate in Sequential Saccade Generation[1]

Peter F. Dominey and Michael A. Arbib
Center for Neural Engineering
University of Southern California
Los Angeles, CA 90089-2520, U.S.A.

Abstract: We model control of voluntary saccades to visual and remembered targets in terms of interactions between the basal ganglia (caudate and substantia nigra), superior colliculus, mediodorsal thalamus, posterior parietal cortex, frontal eye fields and the saccade generator of the brainstem. These interactions include modulation of topographic inhibitory masks that manage motor field activity, dynamic remapping of sensory maps onto the eye motor map to take account of eye movements, and sustained neural activity that embodies spatial memory. Models of these mechanisms implemented in NSL (our Neural Simulation Language) simulate behavior and neural activity described in the literature, and suggest new experiments.

1. The Saccade Paradigm

Single unit recordings during the performance of saccadic eye movement tasks provide a rich paradigm for investigation of at least three major elements of brain function involved in visual resource allocation: 1) The functional topographic relations between sensory and motor areas, including the inhibitory masks that manage motor field activities during the implementation of saccades; 2) dynamic remapping of sensory maps onto a common eye movement motor map to keep these maps aligned; and 3) the cortical and sub-cortical activity that sustains spatial memory during delay saccades. We study the functional neuroanatomy that implements these capabilities, and offer a corresponding simulation that generates neural and behavioral performance as seen in primates as they perform these tasks. Our model builds upon and extends existing models, data, and theories to arrive at a first order description of the primate saccade generation system adequate to explain sequential saccades.

Typically, in saccade experiments the primate sits in a fixed chair with its head immobilized and eyes free to move. Illuminated fixation points and saccade targets are presented on a visual screen in front of the primate. Microelectrodes record neural activity during the tasks, while eye movements are concurrently monitored and recorded. The type of data we seek to explain may be briefly summarized as follows:

---

[1]This research was supported in part by NIH under grant 7R01 NS24926 from NINCDS (Michael A. Arbib, Principal Investigator).

1. Simple saccade Task (Hikosaka and Wurtz 1983a): In the **simple saccade** task, the monkey fixates a spot which later disappears as a target appears in another location. The monkey is rewarded for making a saccade to the new target at its onset.

2. Delay saccade tasks (Hikosaka and Wurtz 1983a): In the **delay saccade** task, during the display of the fixation point, a peripheral saccade target point is briefly illuminated. The monkey is trained to saccade to the target point following the removal of the fixation point — thus showing that it has remembered the location of this target.

3. Double Saccades (Mays & Sparks 1980, Sparks & Porter 1983): In this task, following offset of the initial fixation point, targets A and B are successively presented. The total duration of their presentation is less than the time required to initiate the first saccade. Reward is contingent on successive saccades from fixation point to A to B.

The simple saccade task has been the goal of much previous modeling, and we incorporate features of such modeling in defining the saccade burst generator of Fig. 1. It is the remaining tasks that set the challenge for the present study. The delay saccade task requires us to model memory for spatial location. The double saccade task shows the need to model how the saccade system takes account of ongoing movements in updating previously coded locations.

2. The Model and its Biological Substrate

In this section, we introduce our model for the three major components of the saccadic eye movement system, namely topographical organization, dynamic remapping, and spatial memory capability. Our strategy will be to offer, for each component of the model, a brief review of the biological literature followed by a formal description of the component. As shown in Fig. 1 , our model involves the topographic relations between Superior Colliculus (SC), the substantia nigra pars reticulata (SNr) and caudate nucleus (CD) of the basal ganglia, posterior parietal cortex (PP), the frontal eye fields (FEF) of frontal cortex, the mediodorsal nuclei of the thalamus (MD), and oculomotor regions of the brain stem responsible for saccade burst generation (SG). Both SC and FEF contain topographic "maps" of saccade amplitude and direction required to attain the target with respect to its offset from the fovea, regardless of the eye's position in its orbit. The FEF interacts with the rest of the oculomotor system via multiple pathways. We will consider primarily the topographic projections to caudate, SC, thalamus and brainstem.

The model is implemented in the Neuron Simulation Language NSL (Weitzenfeld 1989) on a Sun 3/260. The major neural elements are 2-dimensional neural surfaces representing retina, PP, FEF, caudate, substantia nigra, superior colliculus, and thalamus. Functionally distinct response types are distributed in a topographic mosaic in each of PP, FEF, CD and SNr. We model this effect by using separate neuron layers for each of the functionally distinct response types. In addition, we model non-retinotopic populations of neurons in the brainstem involved in saccade generation. The eye is the interface between the world and the brain, and eye movements remap external visual information onto the retina.

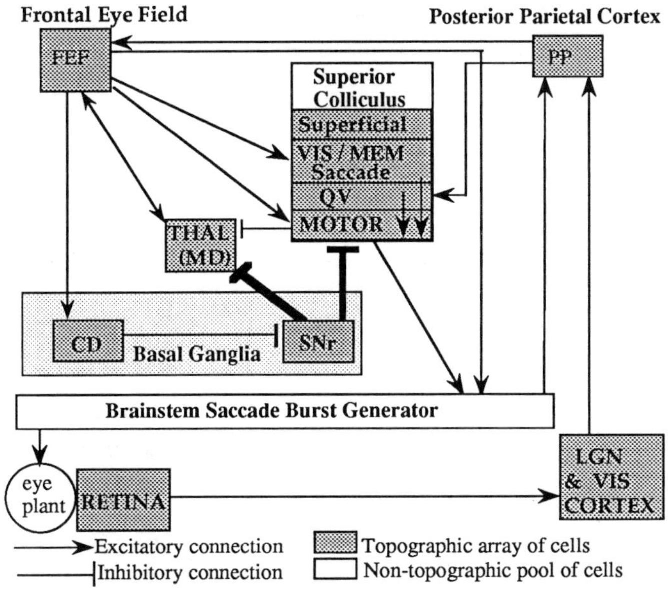

Figure 1. Saccade System Model. Connections between arrays preserve topography of saccade dimensions. The FEF to DEEP SC path includes both an excitatory topographic projection of saccade dimension, and a distributed inhibitory FOVEA-ON projection that inhibits SC and prevents saccades while a target is fixated. Output from the Brainstem Saccade Burst Generator drives a linear oculomotor plant, moving the retina with respect to the visual input projection screen (not shown), thus updating the visual image as the result of a saccade.

The cells in our model interact only via their "firing rates," while the firing rate of a cell is determined by its own membrane potential. The membrane potential at the next time step is modeled by a numerical approximation of a leaky integration of the input firing rates. The leaky integrator integrates and then dissipates its combined inputs, with the dissipation rate based on a prespecified time constant.

The "firing rate" of a cell is obtained by passing the membrane potential through a non-linear sigmoid threshold function mapping membrane potential to firing rate. We use L := R to signify that the firing rate L is obtained by summing the inputs R in a leaky integrator and passing the result through the sigmoid function. Likewise, unless otherwise specified, variables represent two-dimensional arrays of cells and operators represent componentwise operations on the arrays.

The leaky integrator acts as a very short term memory, as the sum of the inputs is gradually dissipated. We can employ coupled integrators to produce a memory that persists over a longer time period. For example, consider the dynamic relations described by equations 18 and 19:

$$\mathrm{FEFS} := \mathrm{THS} + \mathrm{FEFT} - \mathrm{FOVEA} \qquad (18)$$

$$\mathrm{THS} := \mathrm{FEFS} - \mathrm{SNRS} - \mathrm{FOVEA} - \mathrm{SC} \qquad (19)$$

We see that, in the absence of the tonic SNRS inhibition, activity in FEFS will generate a reciprocal activity in THS, producing an excitatory loop. The activity of the loop is sustained after the FEFT (sensory driven) stimulation is removed. As we shall discuss later, the memory is finally removed by a strong inhibitory burst originating in SC that overpowers the excitatory loop.

Connections between these two-dimensional arrays of neurons are defined in terms of interconnection masks which describe the synaptic weights, and are applied via a spatial convolution. The size of a connection mask represents the distance of interactions between connected neurons. The present paper does not tabulate synaptic weights or time constants.

In this simulation, the visual screen in front of the primate is modeled by a 27x27 array, INP, whose elements we turn on and off corresponding to the fixation and target points in the tasks we model. The size of INP represents the visual field available to a monkey with fixed head and free eyes, but only a 9x9 portion of this will be available on any one fixation. The topographic maps are modeled by 9x9 arrays, representing the visual field of the fixed eye. Each neuron in a given layer represents a population of real neurons.

The saccade burst generator translates the retinotopic specifications of saccades into temporal codes directing motoneurons to reposition the eye. Our model of the brainstem saccade generator is a variation of Scudder (1988) and the oculomotor plant it drives is based on Robinson's (1970) linear approximation. Further details will not be given here.

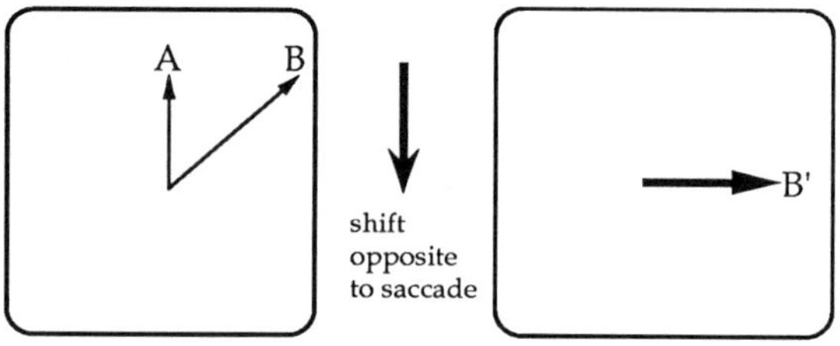

Topographic map of saccade dimension. B' denotes shift of B due to execution of movement to A.

Figure 2. Dynamic Remapping. Directionally selective modulation of the connections between neighboring neurons in a saccade-vector map results in a shift of neural activity for target B to location B', compensating for the saccade to A. This Quasi-Visual (QV) activity is seen in SC and PP.

2.1 Dynamic Remapping

In the double saccade task, the site of retinal stimulation is dissociated from the metrics of the elicited saccade — the retinal error signal for the second target is in some sense updated by the intervening saccade (see Fig. 2). Mays and Sparks (1980) detected a class of quasi-visual (QV) cells in intermediate SC that discharge with the metrics of the second saccade, even though a visual stimulus with this retinal error did not appear in the receptive field.

Gnadt and Andersen (1988) found posterior parietal (PP) neurons that have receptive fields tuned for saccades of a particular amplitude and direction, and which show a sustained activity in the delay saccade task. They show QV cell behavior in a double saccade task in which the cells that code for the second movement are not stimulated visually. Network models (Anderson & Zipser 1988, Zipser & Anderson 1989) have demonstrated how retinotopic and eye position data could be combined in PP to yield head centered spatial map. Gnadt and Andersen (1988) propose that PP may receive corollary feedback activity of saccades, suggesting that PP has access to eye position and velocity information that could be used in generating the QV shifting activity.

Berthoz and Droulez (1989) have proposed a model for this shifting. An eye velocity signal is applied to the array coding the location of a future target to continuously shift the representation of that location.[2] The temporal variation of the activity of a target representation due to eye velocity is the scalar product of the velocity with the spatial gradient of the original target activity. This results in a displacement of target activity loci in the direction opposite to the ongoing eye movement, compensating for the movement. Updating a retinotopic map by shifting activity profiles in the direction opposite to the eye movement yields the functional equivalent of a spatiotopic map embodied in retinotopic coordinates.

We suggest that an efferent signal produced upstream from the abducens nucleus is used to direct the QV shifting. Based on the detection of QV activity in area 7a (Gnadt and Andersen 1988), and the fact that other oculomotor influences are seen in the modulation of retinotopic receptive fields in that area, we assume that corollary saccade velocity information is available in this area of PP and postulate that PP is the origin of the QV activity that is seen in SC, and will likely be seen in FEF as well.

QV cells in Posterior Parietal Cortex:

Cells in our PPQV layer can communicate with all of their nearest neighbors. By selectively modulating these connections so that a cell only talks to its neighbors in a given direction, we implement the Droulez-Berthoz shifting concept. Recall that interconnections between neurons are specified by a convolution mask. The function *Shift* () generates a convolution mask from the horizontal and vertical eye velocity components coded in REBN and UEBN, respectively:

QVMASK = *Shift*(REBN,UEBN) (1)

This mask represents the local interactions between neurons in the PPQV layer that implement the QV shift of activity in direction and amplitude opposite to that of the ongoing saccade. In our simulations, the QV activity generated in PPQV provides input to intermediate SC for the correct generation of the saccades in the double saccade task. The layer PP represents the retinotopic, non-QV parietal cortex cells found by Gnadt and Andersen (1988). Application of the QV convolution mask to the parietal cortex QV cells yields

PPQV := PP + QVMASK*PPQV - FOVEAMASK & PPQV, where (2)

FOVEAMASK = 9x9 mask with central element set to one, all others zero. (3)

Here, subtracting the result of FOVEAMASK "anded" with the PPQV erases the most recent saccade target from the QV surface.

2.2 Superior Colliculus

Electrical stimulation studies have revealed that the deep SC surface maintains a topography of saccade metrics, and that activity at a given location codes for a saccade of particular amplitude and direction largely independent of initial eye position (Robinson 1972, Sparks 1986, Sparks & Mays 1990). Superficial SC receives a direct retinal projection. Sparks (1986) notes that the time delay between superficial and deep SC cellular activity indicates that the superficial SC does not play an obvious role in the normal generation of voluntary saccades.

[2] If position information were used to update the map, a single shift could update the map but would require long connections between neurons in the map. The use of a velocity signal requires only local connectivity in the map.

It has been postulated that SC employs a *winner take all* (Didday and Arbib 1975, Koch & Ullman 1985) strategy for selection of a saccade from multiple targets, like the winner-take-all competition modeled for frog tectum (Arbib 1989) by Didday (1970, 1976). Our deep SC model computes the center of mass of the strongest component of the intermediate and deep layers, as in the winner-take-all model. The intermediate and deep layers receive input from PP that represents the QV activity, and inputs on two separate channels each from FEF and SNr representing visual and memory guided saccades. Deep SC projects to the brainstem as seen in Fig. 2 above. SC also projects to the nuclei in dorsal thalamus which project to frontal eye fields (FEF) with neurons discharging in association with saccades (Ilinsky et al., 1985).

Superior colliculus intermediate layer: The visual saccade cells (SCV) are driven by FEF and inhibited by SNr (the function of this inhibition will be discussed below):

SCV := FEFV - SNRV (4)

The memory-driven movement cells (SCM) are driven by the FEF movement subpopulation (eqn. 12), and inhibited by the sustained memory subpopulation of SNr (eqn. 17):

SCM := FEFM - SNRS (5)

Superior colliculus QV activity: These intermediate layer cells (SCQV) receive the posterior parietal projection of QV activity, unless the eye is moving too fast. SACCADEMASK is a 9x9 mask whose elements are 1 when the eye velocity is less than 200°/sec, and 0 otherwise.

SCQV := SACCADEMASK & PPQV (6)

Before we can complete our model of SC, we must first mention that two types of corticotectal FEF fixation detection cells have been isolated that respond to fixation onset and offset respectively (we shall have much more to say about the frontal eye fields in the next section). These corticotectal cells are not localized to a particular location of the topographic map of FEF, and they were antidromically excited from a wide range of locations within the topographic representation of the SC (Segraves & Goldberg 1987). One role these cells may play is to inhibit the SC from producing saccades while a target is fixated. Functionally, the SC is harder to stimulate for a saccade when a target is currently fixated (Sparks, 1986). In our simulations it is this fixation related inhibition that prevents saccades from being visually triggered during the presence of a fixated target (see eqns. 7 and 8). Hikosaka et al. (1989) detected a caudate neuron whose response to the target in a delay saccade task increased when the fixation point was removed, suggesting a projection of the foveal FEF neurons to the caudate as well as the SC (see eqn. 13 below).

FEF fovea-on response:

In each of the tasks we model, the stimulus that signals the animal to initiate the saccade is the removal of the initial fixation point. The contents of the fovea, the central element of the RETINA layer, is projected to all elements of the 9x9 layer FD, that is used to provide inhibition to elements of caudate, FEF and SC.

FD := each element of this 9x9 matrix equals content of fovea (7)

Deep SC motor activity:

These are the cells in deep SC that project from superior colliculus to the saccade burst generator. Note that they are prevented from firing by the FD array while a target is fixated. The deep SC motor output is the strongest component from among the FEF visual and memory channels, and the QV activity in the intermediate layers:

SC := *winnertakeall*(SCV + SCM + SCQV - FD) (8)

2.3 Frontal Eye Fields

Bruce and Goldberg (1984) found that, as in SC, stimulation at a particular location of FEF yields saccades of a particular direction and amplitude, largely independent of stimulation parameters and the position of the eyes in their orbit when the stimulation is applied. This topography of saccade dimensions is preserved in the projection from FEF to caudate and thalamus (Stanton et al. 1988a), and from FEF to periocular regions in the midbrain pons and brainstem responsible for eye movements, including intermediate and deep SC (Stanton et al. 1988b).

Principal sulcal cortex, including FEF, participates in a precise topographically organized network of connections with posterior parietal cortex (Goldman-Rakic, 87), which receives a visuotopically organized projection from striate cortex via area PO of occipital lobe (Funahashi et al, 1989).

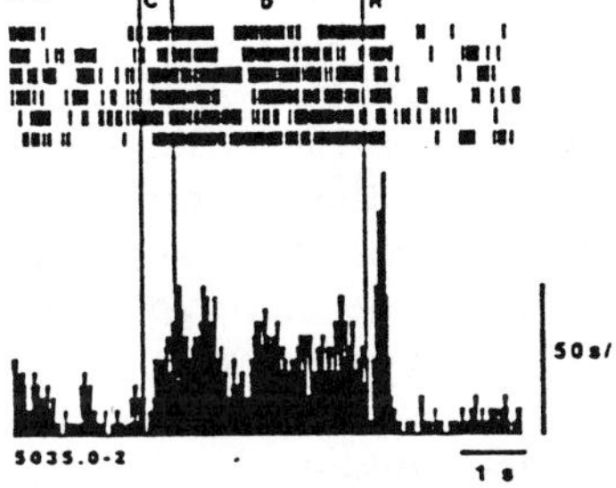

Figure 3. Spatially selective FEF neuron in delay saccade task. Two vertical lines labelled C indicate the target presentation; section labelled D is the delay, and R is the removal of the fixation point (from Funahashi et al, 1989). Note that this neuron combines a sustained memory response and a motor response (cf. eqns. 11 and 12).

We suggest that interaction between FEF and mediodorsal thalamus (MD) provides a basis for memory of target locations in the FEF. These reciprocally connect areas (Huerta et al., 1986) show sustained activity in tasks that require memory of a spatial location (Goldman-Rakic 1987). Funahashi et al. (1989) conclude that sustained activity of spatially selective FEF neurons codes spatial memory in the delay saccade task (see Fig. 3). A projection from SC to dorsal thalamus (Ilinsky et al., 1985) provides a signal that could erase this memory once the saccade has taken place (see eqn. 19).

Various FEF spatially selective neuron response types have been detected that have functional roles in the different saccade tasks. In addition to fovea-on and -off cells mentioned above, these types include visual response to both immediate and memory saccade targets, sustained response to a remembered target, and movement response for visual or remembered target (Bruce & Goldberg 1984, Segraves & Goldberg 1987, Funahashi et al. 1989).

Reward-related sensory activity in the orbitofrontal cortex may selectively gate visual information to the basal ganglia and FEF visual or memory pathways (Rolls & Williams 1987). The gating of information between these two parallel pathways is implemented with two masks - VISMASK and MEMMASK. These are 9x9 masks that allow all-or-none passage of the visual signal through FEF. The two masks are in mutual opposition, that is while one is open, the other is closed.

<u>FEF visual response to immediate saccade targets:</u>
These cells are visually driven and are active only during the visual presentation of the target for an immediate saccade.

FEFV := PP & VISMASK - FD (9)

<u>FEF transient visual response to memory saccade target:</u>
These cells are visually driven and are active only during the visual presentation of the target for the memory saccade.

FEFT := PP & MEMMASK (10)

<u>FEF sustained response to memory saccade target:</u>
These cells are initially activated by the transient cells, but continue to fire after the visual target in the memory saccade is extinguished. The sustained activity is due to reciprocal interaction with mediodorsal thalamus (THS). See Fig. 3, and eqns. 18, 19.

FEFS := THS + FEFT - FOVEA (11)

<u>FEF movement response to memory saccade target:</u>
This cell population bursts at the offset of the fixation point in the memory saccade. It is excited by the sustained cells, and codes for the saccade required to foveate the remembered target.

FEFM := FEFS - FD (12)

This system requires the time constants of populations of neurons to vary between 6 or 15 ms for the bursting SC and FEF cells respectively, up to multiple seconds for the neural ensembles that maintain the spatial memory. While the small time constants are reasonable for isolated cells or populations, the long time-constants are achieved by reciprocal excitatory connections between neural populations of FEF and the mediodorsal thalamic nucleus (see eqns. 18 and 19).

2.4 Basal Ganglia

The basal ganglia, in particular the caudate and substantia nigra, provide cortex with a mechanism to establish the "set" of the voluntary saccade system via an FEF to CD to SNr projection. Hikosaka and Wurtz (1983a) showed that the SNr provides a tonic (50 to 100 spikes/sec) inhibitory topographic projection to the SC, preventing SC from generating saccade signals, forming an *inhibitory mask* on SC (see Fig. 4B). Visually dependent and saccade-related decreases of SNr firing rate release collicular inhibition and facilitate initiation of saccades. This SNr inhibition only occurs for voluntary, not spontaneous, saccades.

Hikosaka et al. (1989) find a large number of caudate neurons that are responsive to visual saccade targets and to remembered targets. Caudate inhibits SNr via a topographic pathway, controlling SNr's inhibition of SC to selectively manage the inhibitory mask on SC (see Fig. 4A). The cortically derived dual inhibition by the basal ganglia (CD to SNr to SC) provides a mechanism of cortical control over saccades. By managing the SNr inhibitory mask on SC via caudate, the FEF can selectively control the targets for saccades, overriding collicular attempts to saccade to distracting peripheral targets.

<u>Caudate Neurons:</u> Caudate visual response to immediate saccade targets is mediated by cell layer CDV:

CDV := FEFV - FD (13)

Caudate transient visual response to memory saccade target is mediated by the array CDT. Note that this response is not inhibited by the presence of a foveated target coded by FD.

CDT := FEFT (14)

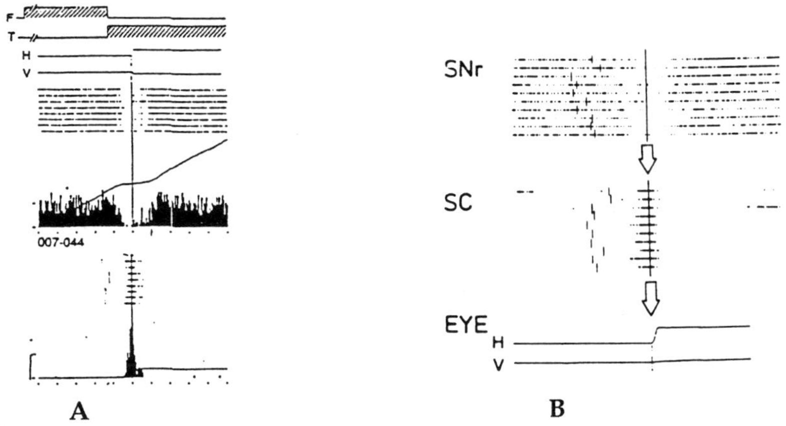

Figure 4. Striato-nigro-collicular interactions. A: Tonic activity of SNr (above) is halted by phasic burst from caudate (below.) From Hikosaka 1989 **B:** Offset of tonic SNr inhibition allows SC to generate a saccade burst. From Hikosaka & Wurtz 1983a.

Caudate sustained response to memory saccade target is mediated by array CDS:

CDS := CDT + FEFS (15)

Substantia nigra pars reticulata:

SNr visual response to immediate saccade targets is given by array SNRV. These neurons have a tonic baseline activity of 80 spikes/s. They are inhibited by the GABAergic caudate input. Their output varies from 80 to 0 as its inhibitory input varies from 0 to 50.

SNRV := 80 - CDV (16)

SNr sustained response to memory saccade targets is given by array SNRS. This neuron has a tonic baseline activity of 80 spikes/s. It is inhibited by the GABAergic caudate input.

SNRS := 80 - CDS (17)

2.5 Memory-Related Striato-thalamo-cortical interactions

There is strong evidence that sustained activity in spatially selective FEF and thalamic neurons embodies the memory required for the delay saccade (Fuster & Alexander 1973, Alexander & Fuster 1973, Goldman-Rakic 1987). FEF has excitatory reciprocal topographic projections to mediodorsal (MD) and ventral anterior (VA) nuclei of the thalamus (Ilinsky et al. 1985, Stanton et al. 1988a, Huerta et al. 1986). SNr projects a topographic inhibition to nuclei of thalamus that are involved in spatial memory circuits with the FEF (Ilinsky et al., 1985). Absence of SNr inhibition allows reciprocal connections between FEF and thalamus to generate a spatial "memory" cycle (Hikosaka 1989). The SC projects to the dorsal thalamus (Ilinsky et al. 1985), erasing the remembered target location (eqn.19).

The activity of this spatial memory is regulated by the inhibitory mask projected from SNr to MD to FEF as part of the oculomotor circuit of the basal ganglia (Ilinsky et al. 1985, Alexander et al. 1986). In the memory saccade task, while the fixation point is still present and the target briefly appears, the memory contingent sustained SNr cell reduces its inhibition on the thalamus which allows the initiation of the thalamus↔FEF cyclic excitation that embodies the memory function. The caudate participates by inhibiting SNr, setting up the conditions for the thalamocortical memory to be instantiated. Fig. 5 illustrates a caudate neuron that is preferentially tuned for sustained activity during memory guided saccades — compare this with our simulation in Fig. 6.

Thalamic interactions:

FEF sustained response to a memory saccade target is given by the array FEFS:

FEFS := THS + FEFT - FOVEA (18)

We see here half of the reciprocal connection between FEF and MD thalamus that implements the spatial memory. Note that the subtraction of the FOVEA prohibits the central region of this memory map from being activated. Mediodorsal thalamus sustained activity is represented by the array THS which satisfies

THS := FEFS - SNRS - FOVEA - SC, (19)

implementing the other part of the reciprocal thalamocortical connection. Note that SNr inhibits this path, and that once SC bursts, the memory is erased. This erasure is particularly valuable in preventing saccades to previously attained targets.

3. Simulation Results

3.1 Simulation of Cellular Activity

We simulate the saccade system starting with retinal input, through cortical and subcortical areas, the brainstem, and finally the oculomotor plant that moves the eye to generate a new retinal input. Table 1 provides the firing rate ranges that are used for some of the neurons in the simulations. We note that the neural populations we model carry information in terms of their discharge frequencies, the durations of discharge, and the latencies between stimulus and firing, and between neural events in connected regions. Our simulated results correspond well along these dimensions with the reported dynamics in the literature.

Neuron	Background Rate OBS /	SIM	Task Related Activity OBS /	SIM	Reference
a. FEF	0-20	0	100	100	Segraves & Goldberg, 1987
b. CD	0	0	50	50	Hikosaka et al., 1989
c. SNr	50-100	80	0	0	Hikosaka & Wurtz, 1983a,b
d. SC	0	0	200	200	Hikosaka & Wurtz, 1983c
e. FD	0-20	0	50	50	Segraves & Goldberg, 1987
f. Thalamus	0	0	10	10	Fuster & Alexander, 1973

Table 1. Observed (OBS) and simulated (SIM) neural activities in saccade tasks.

Delay saccade: Fig. 5 is a simulation time trace of neurons whose location in a topographic map codes for a saccade 40° up and 30° right. The duration of the trace is 1 second. Time runs from left to right. FIXATION represents the fixation point. TARGET shows the presentation of the target to be remembered. FOVEA-ON indicates whether a target is foveated (eqn. 7). FEFT indicates the transient response to the target (eqn. 10). FEFS shows the time course of the sustained activity that encodes the memory for the location (eqn. 11). FEFM is the movement cell whose

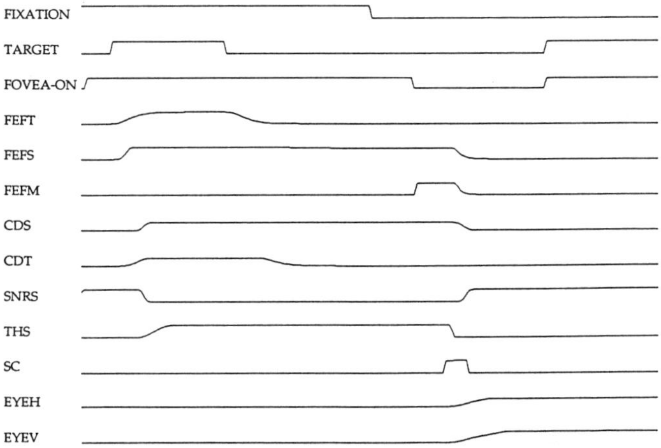

Figure 5. Memory Saccade Simulation. This output was generate by NSL for a delay saccade task in which the total experiment duration is one second. The fixation target is present for 500 ms. After the initial 50 ms of fixation point presentation, a peripheral target (40° up and 30° right) is presented for 200 ms. For the remaining 200 ms of fixation, the target is memorized. Following fixation offset a saccade is made to the remembered target. In the simulations, trials were made with the target memory retained for up to 10 simulated seconds.

burst codes for the saccade. Recall from eqn. 12 that this cell fires when the FEFT is active and FD is quiet, that is when a target is remembered and the fixation point disappears. CDS (eqn. 14) and CDT (eqn. 15) are the sustained and transient caudate cells respectively. SNRS (eqn. 17) is the sustained SNr cell whose default tonic activity inhibits SC (eqn. 5) and thalamus (eqn. 19). When this activity is blocked by caudate, the cortico-thalamic memory is enabled, and SC is only inhibited by the activity of the FOVEA-ON neurons. THS shows the sustained activity of the mediodorsal thalamus (eqn. 19). Note how this cell is co-active with FEFS, and also how it abruptly cuts off after the SC burst. This behavior is obvious when eqn. 5 is examined. SC, excited by the FEFM activity, shows the collicular burst that takes place when the fixation point is removed (eqn. 5 and 8). EYEH and EYEV indicate the horizontal and vertical eye position generated by the brainstem and oculomotor models which are not described here.

Double Saccade and QV activity: Fig. 6 illustrates the QV convolution mask and resultant shifting activity that is seen in posterior parietal cortex and in SC during the course of the double saccade task. The convolution mask is generated from the horizontal and vertical eye velocity components. When convolved with PPQV in eqn. 2, the QV mask represents local cellular interactions thought to implement the QV shifting in parietal cortex. The result is that the new locus of activity that codes for a saccade has been updated by shifting in the opposite direction and magnitude of the initial saccade. Following the completion of the first saccade, the updated locus of activity for the second target is chosen by the SC as the next saccade target. A side effect of the shifting is that the activity patterns are dispersed. This dispersion is reasonable since the areas involved, i.e., SC, FEF, PP, code in populations.

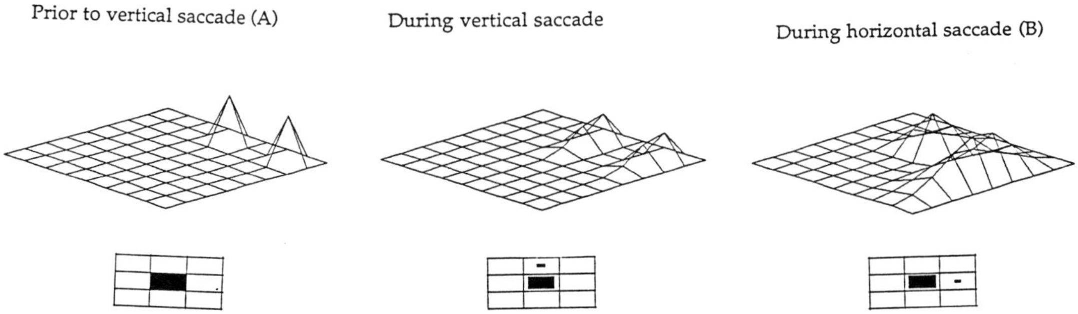

Figure 6. QV Shift Simulation Traces. In three phases of the double saccade we show the PPQV layer, and the QVMASK, a convolution mask that implements the shift opposite to the saccade. Prior to the vertical saccade, activity for the two targets are seen in PPQV. During the vertical upward saccade, the QVMASK specifies that cells get input from their upward neighbors, implementing the downward shift. Following the first saccade, the shifted activity for the second target wins the *winner-take-all* in SC, and the horizontal saccade to the right ensues.

3.2 Predictions and Experimental Design

QV activity and Sustained Memory Sequences:

The purpose of this experiment is to determine if the QV shifting mechanism can operate on prefrontal cortex spatial sequence maps. Motivated by the sequence paradigm of Barone and Joseph (1989a,b) and the double saccade of Mays and Sparks (1980) we suggest a "triple saccade" task. While the monkey is fixating a target, three peripheral targets are presented for one second each in succession in random order. One second after the third target is extinguished, the fixation point is removed. This will signal the animal to initiate a sequence of saccades to the three targets in the order they were presented. As in the double saccade, the representation of targets on the eye-movement map must be updated based on the intervening eye movements. In particular, target 2 will be shifted once (by the metrics of the first saccade), and target 3 will be shifted twice (by the metrics of saccades one and two). We predict that the animal will be able to successfully complete this task, a modification of a task previously reported by Barone and Joseph (1989). We predict that the posterior parietal cortex, upon receipt of the oculomotor efferent will initiate the QV shifting. A replica of this activity is projected to dorsomedial thalamus, and by this route initiates a change in the thalamocortical spatial memory map. An alternative route for this change is via the projection from 7a and LIP to FEF. QV shifting will likely be seen in FEF. If so, it remains to be seen if this implements the map updates, or merely reflects that it has happened.

4. Discussion

The simulation, built based on physiological data from a variety of saccade related studies, demonstrates that the following assumptions can lead to a functionally correct model: a) Managing the inhibitory mask that projects from SNr to SC allows selective cortico-striatal control of target locations for voluntary saccades to immediate and remembered targets. b) Topographic position encodings for targets of future saccades are shifted over their topographic maps to account for ongoing eye movements. This yields a spatiotopic mapping capability that is built on a retinotopic map. c) Saccades can be driven by representational memory that is hosted in reciprocal connections between mediodorsal thalamus and dorsolateral prefrontal cortex that are governed by SNr.

We must, however, continue to examine our assumptions. Though the topographic cortico-striatal projection suggest a striato-nigral topography of saccade dimensions, this has not yet been found. Similarly, we model discrete functional response types (e.g., visual, sustained, motor) in cortex and basal ganglia, whereas in the animal these responses appear to be more distributed. It is likely that the topography of saccade dimensions, and the functional response types are coded in a distributed fashion as seen in the chapter by Anastasio. In addition, the detailed relationships between the neurochemical patch/matrix organization of the caudate (Gerfen 1989) and this functional organization remain to be understood.

The QV story is not yet complete. If the FEF or SC is stimulated in mid-saccade, the location coded by the stimulated region is transformed according to the velocity and direction of the ongoing saccade (Schlag-Rey et al. 1989, Sparks & Mays 1983) to correct for the movement that would have occurred during the afferent delay from retina to the stimulated region, and the initial target is attained. Further, this phenomena is seen when one of the two regions is lesioned and the ramaining region is stimulated (Schiller & Sandell, 1983). This implies that the QV shifting is also taking place at or downstream from the FEF, possibly in the paramedian pontine reticular formation (Hepp & Henn, 1983). The shifting mechanism itself, as directional modulation of connections between neighbors, stands as a plausible model whose specific neural correlate has yet to be found.

Literature Cited

Alexander, G. E., Fuster, J. M, 1973, Effects of cooling prefrontal cortex on cell firing in the nucleus medialis dorsalis. *Brain Research*, 61:93-105

Alexander, G.E., DeLong, M. R., Strick, P.L, 1986, Parallel Organization of Functionally Segregated Circuits Linking Basal Ganglia and Cortex, *Ann. Rev.Neurosci*.9:357-381.

Arbib, M.A, 1989, *The Metaphorical Brain 2: Neural Networks and Beyond.* Wiley-Interscience.

Andersen, R.A., Zipser, D, 1988, The role of the posterior parietal cortex in coordinate transformation for visual-motor integration. *Can. J. Physiol. and Pharmacol.* 66(4):488-501.

Barone, P. and J. P. Joseph, 1989a, Prefrontal cortex and spatial sequencing in macaque monkey. *Exp Brain Res,* 78:447-464.

Barone, P. and J. P. Joseph, 1989b, Role of Dorsolateral Prefrontal Cortex in Organizing visually Guided Behavior. *Brain Behav. Evol.*, 33:132-135.

Berthoz, A. and Droulez, J., 1989, The concept of Dynamic Memory in Sensorimotor Control. Dahlem Conf. on: Principles and concepts in motor control. Berlin.

Bruce, C.J., and Goldberg. 1984. Physiology of the frontal eye fields. *TINS* November, pp. 436-441.

Didday, R.L., 1970, The Simulation and Modelling of Distributed Information Processing in the Frog Visual System. Ph.D. Thesis, Stanford University.

Didday, R. L, 1976, A model of visuomotor mechanisms in the frog optic tectum. *Math. Biosci.* 30:169-180.

Didday, R L, Arbib, M A, 1975, Eye Movements and Visual perception: A "Two Visual System" Model. *Int. J. Man-Machine Studies* .7:547-569.

Funahashi, S., C. J. Bruce, P. S. Goldman-Rakic, 1989, Mnemonic Coding of Visual Space in Monkey's Dorsolateral Prefrontal Cortex, *J. Neurophysiol.* 61(2):331-349.

Fuster, J. M. , Alexander, G. E, 1973, Firing changes in cells of the nucleus medialis dorsalis associated with delayed response behavior. *Brain Research*, 61:79-91.

Gerfen, C.R, 1989, The Neostriatal Mosaic: Striatal Patch-Matrix Organization Is Related to Cortical Lamination. Science, 246:385-388.

Gnadt, J. W., Andersen, R. A, 1988, Memory related motor planning activity in posterior parietal cortex of macaque. *Exp Brain Res* 70:216-220.

Goldman-Rakic, P. S., 1987, Circuitry of primate prefrontal cortex and regulation of behavior by representational memory, In: Handbook of physiology. Chap V. The nervous system 9:373-417.

Hepp,K., and Henn,V., 1983, Spatio-Temporal Recoding of Eye Movement Signals in the Monkey Paramedian Pontine Reticular Formation (PPRF), *Br. Research*, 52:105-120.

Hikosaka, O., and Wurtz,R, 1983a, Visual and Oculomotor functions of Monkey Substantia Nigra Pars Reticulata. I. Relation of visual and Auditory Responses to Saccades. *J. Neurophysiol.* 49(5):1230-1253.

Hikosaka, O., and Wurtz,R, 1983b, Visual and Oculomotor functions of Monkey Substantia Nigra Pars Reticulata. II. Visual Responses Related to Fixation of Gaze. *J. Neurophysiol.* 49(5):1268-1284.

Hikosaka, O., and Wurtz,R, 1983c, Visual and Oculomotor functions of Monkey Substantia Nigra Pars Reticulata. III. Memory-Contingent Visual and Saccade Responses. *J. Neurophysiol.* 49(5):1285-1301.

Hikosaka,O., Sakamoto,M, Usui,S, 1989, Functional Properties of Monkey Caudate Neurons I. Activities Related to Saccadic Eye Movements. *J Neurophysiol.* (61)4:780-798.

Hikosaka, O., 1989, Role of Basal ganglia in Initiation of Voluntary Movement. In: *Dynamic Interactions in Neural Networks: Models and Data*. Eds. Arbib, Michael, and Amari.Springer-Verlag, New York, pp 153-168.

Huerta, M.F., Krubitzer, L.H., Kass,J.H., 1986, Frontal Eye Field as Defined by Intracortical Microstimulation in Squirrel Monkeys, Owl Monkeys, and Macaque Monkeys: I. Subcortical Connections, *J. Comp. Neuro.* 253:415-439.

Ilinsky, I. A., Jouandet,M.L. and Goldman-Rakic, P, 1985, Organization of the Nigrothalamocortical system in the Rhesus Monkey. *J. Comp. Neurol.* 236:315-330.

Koch, C., Ullman, S, 1985, Shifts in Selective visual Attention: towards the underlying neural circuitry. *Human Neurobiology*. 4:219-227.

Mays, L.E. and Sparks, D.L., 1980, Dissociation of Visual and Saccade Related Responses in Superior Colliculus Neurons. *J. Neurophysiol,* 43(1):207-232

Robinson, D. A., 1970, Oculomotor unit behavior in the monkey. *J. Neurophysiol.* 33:393-404.

Robinson, D.A., 1972, Eye Movement Evoked By Collicular Stimulation In The Alert Monkey. *Vision Research* 12:1795-1808.

Rolls, E.T., Williams, G.V., 1987, Sensory and Movement-Related Neuronal Activity in Different Regions of the Primate Striatum. In:*Basal Ganglia and Motor Behavior*. Schneider & Lidsky, eds. pp. 37-59.

Schiller, P.H., Sandell,J.H., 1983, Interactions Between Visually and Electrically Elicited Saccades Before and After Superior Colliculus and Frontal Eye Field. *Exp. Br. Research*. 49:381-392.

Schlag-Rey,M., Schlag, J, and Shook, 1989, Interactions between natural and electrically evoked saccades I. Differences between sites carrying retinal error and motor error signals in monkey Superior colliculus. *Exp.Brain.Res.* 76:537-547.

Scudder, C.A., 1988, A New Local Feedback Model of the Saccadic Burst Generator. *J. Neurophysiol.* 59(5):1455-1475.

Segraves, M., and Goldberg, M.E., 1987, Functional Properties of Corticotectal Neurons in the Monkey's Frontal Eye Field. *J. Neurophysiol.* 58(6):1387-1419.

Sparks, D.L., 1986, Translation of Sensory Signals Into Commands for Control of Saccadic Eye Movements: Role of Primate Superior Colliculus. *Physiol. Rev.* 66(1):118-171.

Sparks, D. A., Mays, L. E., 1983, Spatial Localization of Saccade Targets I. Compensation for Stimulation-Induced Perturbations in Eye Position. *J. Neurophysiol.* 46:1:45-63.

Sparks, D. A., Mays, L. E., 1990, Signal Transformations Required for the Generation of Saccadic Eye Movements. *Ann. Rev. Neurosci.* 13:309-336.

Sparks, D. A., Porter, J. D., 1983, Spatial Localization of Saccade Targets II. Activity of Superior Colliculus Neurons Preceding Compensatory Saccades. *J. Neurophysiol.* 46(1):45-63.

Stanton, G. B., Goldberg, M.E., and Bruce, C.J., 1988a, Frontal Eye Field Efferents in the Macaque Monkey: I. Subcortical Pathways and Topography of Striatal and Thalamic Terminal Fields. *J. Comp. Neurol.* 271:473-492

Stanton, G. B., Goldberg, M.E., and Bruce, C.J., 1988b, Frontal Eye Field Efferents in the Macaque Monkey: II. Topography of Terminal Fields in Midbrain and Pons. *J. Comp. Neurol.* 271:493-506

Weitzenfeld, A. 1989. NSL Neural Simulation Language. Technical Report 89-02. Center for Neural Engineering, USC.

Zipser, D., Andersen, R.A, 1988, A back-propagation programmed network that simulates response properties of a subset of posterior parietal neurons. *Nature* 331:679-684.

Striato-Pretecto-Tectal Connections: A Substrate for Arousing the Toad's Response to Prey

J.-P. Ewert, N. Matsumoto), W.W. Schwippert, and T.W. Beneke*

Neurobiology, FB 19, University of Kassel
D-3500 Kassel, FR of Germany

**) Department of Biophysical Engineering*
Faculty of Engineering Science, Osaka University
Toyonaka, Osaka 560, Japan

Abstract. *Earlier studies have shown in toads that lesions to certain pretectal nuclei lead to hyperactive prey-catching in response to any moving visual object. These and other investigations suggest that in toads the discrimination of moving visual objects for prey recognition depends on inhibitory connections from pretectum to optic tectum, with both structures obtaining retinal inputs. The question thus arose how pretectum is controlled. In the present study, we examined intracellular responses of pretectal cells to electrical pulses applied to the ipsilateral ventral striatum, vSTR, or to the lateral forebrain bundle, LFB, in which striatal axons project to pretectal nuclei. The main results can be summarized in four points: (i) Pretectal neurons receive inhibitory and/or excitatory postsynaptic inputs evoked by electrical stimulation of LFB or by visual stimuli. The relatively greater percentage of purely inhibitory inputs and the occurrence of latencies between 2 and 3 ms are conspicuous. (ii) Pretectal neurons receive LFB-mediated information also via polysynaptic connections, probably involving an entopeduncular relay. (iii) Electrical stimulation of the vSTR elicits postsynaptic inhibitory or sequential excitatory/inhibitory potentials in pretectal cells; the fastest input (2 ms) was inhibitory. (iv) The fact that striatal stimulation leads to inhibitory inputs in pretectal neurons is consistent with the concept of a "disinhibitory" striato-pretecto-tectal function suitable to prime or arouse the tectal prey-catching releasing system. This is also consistent with studies showing visual "prey neglect" after ablation of vSTR.*

1. Visual Perception Through Pretecto-Tectal Inhibition

Studies applying various experimental techniques suggest that in amphibians neurons of the lateral posterodorsal (Lpd) and the very lateral part of the posterocentral (P) pretectal thalamic nuclei in response to retinal input exert on certain tectal cells inhibitory influences whose functions may be at least three-fold: (i) definition of the species-universal prey-schema expressed by the property of network integrated tectal T5.2 neurons (Ewert 1968; 1974; for details see Ewert, this volume); (ii) size constancy (Ingle 1983); and (iii) spatial limitation of the spread of tectal intrinsic lateral mutual excitation (Ewert et al. 1970; Székely & Lázár 1976). There are four major kinds of evidence for pretecto-tectal inhibition:

(1) Recordings from a T5.2 neuron in an immobilized toad or frog (Ewert 1985) or a salamander (Finkenstädt & Ewert 1983) before and after an ipsilateral pretectal Lpd/P lesion (induced electrolytically by means of direct current or chemically by injecting Kainic acid) showed dramatic changes in the neuron's visual response property post lesion: (i) loss of object discrimination, i.e., strong increase of the discharge

rate with increasing area of any moving visual stimulus, (ii) decrease of surround inhibition, i.e., responses to moving textured backgrounds, (iii) self-sustained post-stimulus activity in response to a visual input, and (iv) enlargement of the excitatory receptive field (cf. also Ewert & v. Wietersheim 1974b; for review see Ewert 1989). These results are consistent with Jung's "powder keg" analogy (Ewert, this volume).

(2) The lesion induced "dishinbition" of tectal T5.2 neurons is accompanied with hyperexcited prey-catching behavior. Studies on freely moving toads with a micro-electrode (for extracellular recording) implanted in the tectum close to a T5.2 neuron and a stimulating-electrode (for electrolytic lesion) implanted in the ipsilateral Lpd/P region showed a strict correlation between the above described lesion induced changes of the T5.2 neuron's visual response properties and corresponding changes in the toad's prey-catching behavior toward any objects traversing the neuron's receptive field (Schürg-Pfeiffer 1989). These and other results of that study provide evidence for a correlation between T5.2 neuronal activity and prey-catching activity, which is consistent with our concept of prey-selective neurons (Ewert 1974, 1989).

(3) During intracellular recording/labeling studies from a T5.2 neuron, mainly excitatory postsynaptic potentials, EPSPs, and spikes could be obtained in response to a prey-like moving stripe whose longer axis was oriented parallel to the direction of movement ("worm" configuration); however, very weak postsynaptic activities or predominantly inhibitory postsynaptic potentials, IPSPs, were recorded when the same stripe was moved nonprey-like with its longer axis oriented perpendicular to the direction of movement (Matsumoto et al. 1986) - the "antiworm" configuration known to be preferred *vs.* "worm" by pretectal thalamic TH3 neurons (Ewert & v.Wietersheim 1974a). These results demonstrate that "opposit" stimulus configurations of maximal *form contrast* and different meaning for the behavior can be distinguished in T5.2 neuron's responses by excitation and inhibition.

(4) Intracellular recordings from a prey-sensitive tectal cell shows sequential EPSP/IPSP responses to electrical stimulation of the contralateral optic tract, and the same cell displays strong IPSP responses to stimulation in the ipsilateral pretectal region (Ewert et al. 1990). Note that this tectal neuron received no spontaneous tonic IPSPs. These results provide evidence for our hypothesis that tectal visual response properties in the present context are determined by *postsynaptic inhibitory* pretecto-tectal connections.

2. How is Pretectum Controlled?

The question thus arises how pretectum is controlled. Drawing on the effects of forebrain lesions on visually released prey-catching behavior in anurans, Ewert (1967) put forward the hypothesis that the retino-tecto-bulbar/spinal pathway is influenced by connections from the forebrain to the optic tectum involving two inhibitory projections: telencephalon to pretectum and pretectum to tectum. Following bilateral ablation of frog's or toad's telencephalon, visually guided prey-catching behavior failed to occur (Diebschlag 1935; Ewert 1967); however, if in addition pretectal caudodorsal thalamic areas were destroyed, telencephalic ablated toads or frogs displayed a hyperexcited, disinhibited prey-catching toward any moving visual object, in a manner similar to pretectally lesioned toads with intact telencephalon (Ewert 1967, 1968). These results show that the telencephalon, although not involved in the primary visuomotor pathway, influences its information transfer. More specifically, Patton & Grobstein (1986) observed that frogs show a visual prey neglect if the ventral striatum (vSTR) - projecting to the pretectum within the lateral forebrain bundle (LFB) - was destroyed (confirmed for toads by Finkenstädt 1989). Since telencephalic lesions to the nucleus amygdalae or ablation of the posterior ventral medial pallium did not abolish visually guided prey-catching in toads (Finkenstädt 1989), the possibility of a striato-pretecto-tectal influence in terms of a "disinhibitory" controlling system is worth consideration (Ewert 1987, 1989).

There is neuroanatomic evidence showing that efferents from the vSTR project mainly: (i) via LFB to pretectal nuclei that are connected with the tectum, (ii) to the anterior entopeduncular nucleus that in turn projects to pretectum and tectum, (iii) to the nucleus profundus mesencephali which sends fibers to the tectum, and (iv) to the tegmentum which feeds both to tectum and to medullary reticular formations (Wilczynski & Northcutt 1983b). The vSTR receives its major afferent input via LFB from the lateral anterior thalamic nucleus (La) that is connected with the optic tectum. Furthermore, inputs from the anterior entopeduncular nucleus to vSTR exist (Northcutt & Kicliter 1980; Wilczynski & Northcutt 1983a).

In order to evaluate the hypothesis of a striato-pretecto-tectal "disinhibitory" pathway at the neuronal level, we have focussed on the study of inputs to pretectal Lpd/P nuclei by intracellular recordings of synaptic potentials to ipsilateral electrical stimulation in LFB and vSTR, respectively. Striatal input was examined in response to stimulation of LFB. For a detailed description of this subject see Matsumoto et al. (1990).

3. Methodological Comments

Common toads *Bufo bufo spinosus* (L.) were anesthetized with Halothan^R for surgery and immobilized for recording by intralymphatic injection of succinylcholine at threshold doses (5-10 mg/kg body weight). Glass microelectrodes filled with a 1:1 mixture of a Co^{3+}-lysine solution (Wako Chemicals, Japan) and a 2 M K^+-citrate solution were used for intracellular recording and iontophoretic staining at tip diameters from 0.5 to 1.2 µm and resistances between 30 and 80 MOhm. (For methodological details see Matsumoto et al. 1986, 1990). The resting potentials of pretectal and striatal neurons ranged from -35 mV to -60 mV. Depolarizing and hyperpolarizing activity in terms of excitatory (EPSPs) and inhibitory postsynaptic potentials (IPSPs) was determined by passing current through the recording electrode.

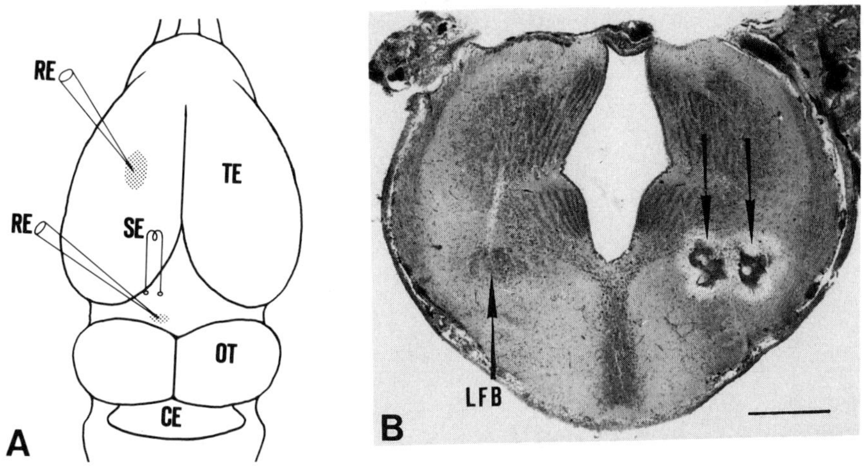

Fig. 1. A) Experimental arrangement for recording pretectal cells (RE, bottom) and striatal cells (RE, top), respectively, in response to electrical stimulation (SE) of the ipsilateral lateral forebrain bundle, LFB, in the toad's brain. CE, cerebellum; OT, optic tectum; RE, recording electrode; SE, stimulating electrode in the diencephalon; TE, telencephalon. Dotted areas indicate the region of the electrode penetrations for pretectum and striatum, respectively. B) Transverse section (Klüver-Barrera stain) of the rostral diencephalon with electrolytically applied lesions (downward arrows) marking the position of the bipolar stimulating electrodes in the area of the LFB (see upward arrow in the left half); note that the positions of the stimulating electrode tips in the centers of both lesions were close to the left and right side of LFB, so that the electric field strength covered the LFB; calibration bar: 500 µm.

For electrical stimulation (Fig. 1A) a pair of electrolytically sharpened stainless steel needles insulated except at the tip was used (distance between tips was about 500 µm). The electrical stimuli were rectangular pulses of 0.1 ms and 100-150 µA. In most experiments the stimulated sites were marked (Fig. 1B) by an electrolytic lesion (20 µA, 20 sec). For visual stimulation, three types of moving configural objects were presented, a 16°x16° black square (S), a 2°x16° black stripe whose longer axis was oriented parallel to the direction of movement (worm-configuration, W), or a 2°x16° black stripe whose longer axis was oriented perpendicular to the direction of movement (antiworm-configuration, A).

Altogether, eighty-three pretectal and striatal neurons were recorded intracellularly. Postsynaptic potentials of pretectal cells were evoked by visual stimulation or by electrical stimulation of the ipsilateral LFB or vSTR. The caudal vSTR was stimulated in an area whose cells could be activated by ipsilateral LFB stimulation.

4. Intracellular Recordings from vSTR Neurons

The striatal neurons recorded in response to ipsilateral LFB stimulation were located in a depth of 1450 to 2150 µm, mainly in the caudal vSTR, in total thirty-nine cells. Successfully labeled neurons had a somata size of 6 to 10 um in diameter. Their positions were close to the ventricle. One neuron is depicted in Fig. 2; it shows dendritic arborizations with few branches extending to the pial surface reaching the area of LFB fibers. Sequential EPSP/IPSP responses were most frequently recorded. The shortest latencies of EPSPs fell into the bin of 4-8 ms. (For further details see Matsumoto et al. 1990). About one third of the striatal cells recorded to LFB stimulation were responsive to visual stimuli. More than 50% of these did not display a clear property in response to the moving W-, A-, and S-objects tested. Two examples are shown in Fig. 3. Both cells responded to electrical LFB stimulation with an EPSP at a latency of 5 ms. In one cell the efficacy of the visual test stimuli was S = W > A (Fig. 3a), hence resembling properties of tectal class T5.1 neurons (Ewert 1989). The other cell revealed no stimulus preference (Fig. 3b), like tectal T5.0 neurons. The receptive field of both cells encompassed the whole frontal visual field, a property known from tectal T2.1 neurons.

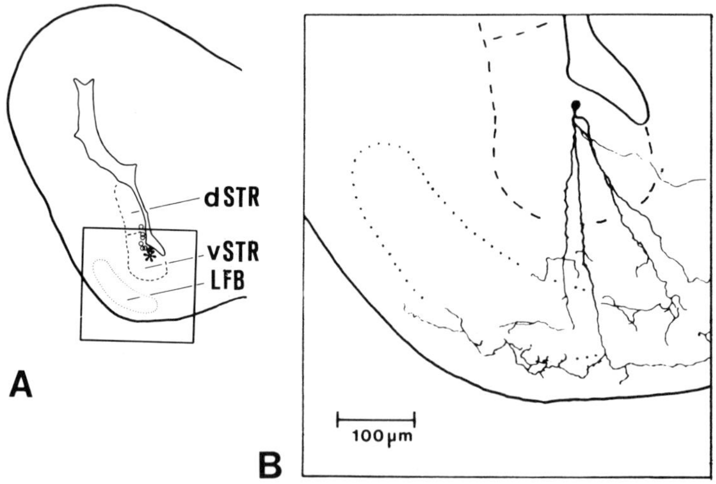

Fig. 2. Localization of neurons in the caudal ventral striatum, vSTR, responding to electrical stimulation of the ipsilateral lateral forebrain bundle, LFB. A) Schematic transverse half section of the telencephalon; circles and black asterisk indicate the positions of seven labeled somata recorded to LFB stimulation; dSTR, dorsal striatum; vSTR, ventral striatum; LFB, lateral forebrain bundle and surrounding fibers. B) Camera lucida reconstruction of a striatal neuron; for position see asterisk in (A).

To repeated presentation of a visual moving object, spike responses of many striatal neurons showed relatively fast adaptation in their discharge frequency, which is consistent with the phenomenon that to repetitive electrical LFB stimulation (1 Hz), in some cells the amplitude of the first excitatory component (e.g., in an EPSP/IPSP-type cell) was biggest to the first electric pulse. Therefore, different configural visual stimuli were presented in a random order with a pause of 15 sec.

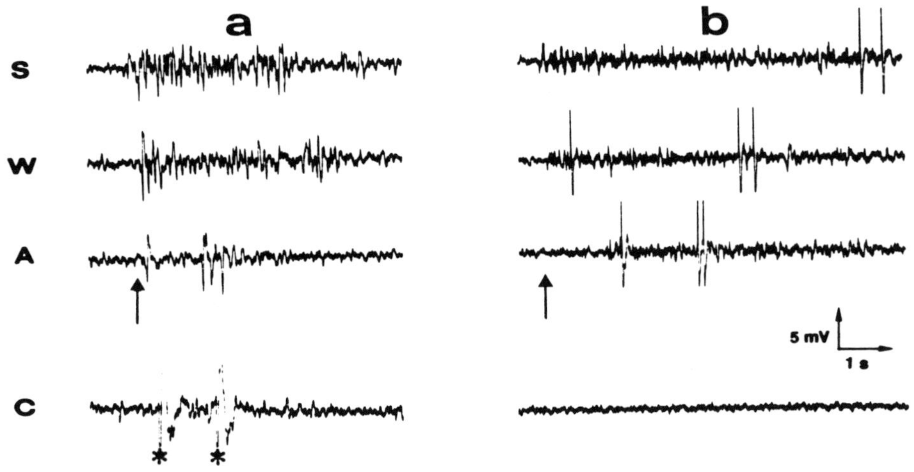

Fig. 3. Visual responses of two ventral striatal neurons (a,b). Arrows pointing upwards indicate the onset of stimulation with S (16°x16° square), W (2°x16° stripe moving in worm-configuration), or A (2°x16° stipe moving in antiworm-configuration). a) The stimulus efficacy on the neuron's postsynaptic activity was S = W › A; the control experiment without visual stimulation is indicated by "C"; an example for the neuron's responses to electrical LFB stimulation is marked by asterisks. b) This neuron responded almost equally to S, W, and A; occasionally, the cell fired. The latencies of the EPSPs of both cells fell in the bin of 4-8 ms.

5. Intracellular Recordings from Lpd/P Neurons

The twenty pretectal cells recorded in response to LFB stimulation were located in the Lpd and the very lateral P nuclei. Tests to visual stimuli showed relatively large receptive fields, responses preferentially to large moving objects, and off-activity to brisk changes of the diffuse illumination - properties known from TH3 neurons. Figure 4C depicts the reconstruction of a neuron responding with IPSP activity to LFB stimulation. This neuron displays morphological features known from pretecto-tectal projection cells which have been investigated in detail in frogs by Matsumoto (1989): from the soma situated in Lpd the dendritic shaft sent several branches with spines that spread horizontally to make finer arborizations in the pretectal neuropil, a projection field of retinal fibers; the axon projected toward the ventral tectum.

To LFB stimulation, a variety of response types of pretectal neurons could be obtained regarding sequences of EPSPs (E) and IPSPs (I). Among the twenty neurons studied, E-, EE-, EEI-, EI-, and I-types were distinguished. The pure I-type was most frequently recorded (Table 1). In order to determine the latencies of the first postsynaptic potentials (PSPs), the time delay between the stimulus artifact and the onset of the neuronal activity was measured. For sequential PSPs, latencies of responses following the first response were determined by estimation of the inflection point. Latency histograms are illustrated in Fig. 5A. Six of the nine I-type cells showed short latencies between 2 and 3 ms which indicate monosynaptic connections (e.g., see Fig. 4Ba). Taking the results together, most of the 1st EPSP and 1st IPSP responses fell into the bin of 4 ms; 2nd and 3rd responses only found in EPSPs were in the range of 8 to 16 ms.

	LFB-stimulation	vSTR-stimulation
n	20	24
E (%)	10	-
EI (%)	25	62
EE (%)	5	-
EEI (%)	15	17
EIEI (%)	-	8
I (%)	45	13

Table 1. Postsynaptic excitatory (E) and/or inhibitory (I) potentials of pretectal (Lpd/P) neurons (n=44) in response to LFB and vSTR stimulation, respectively.

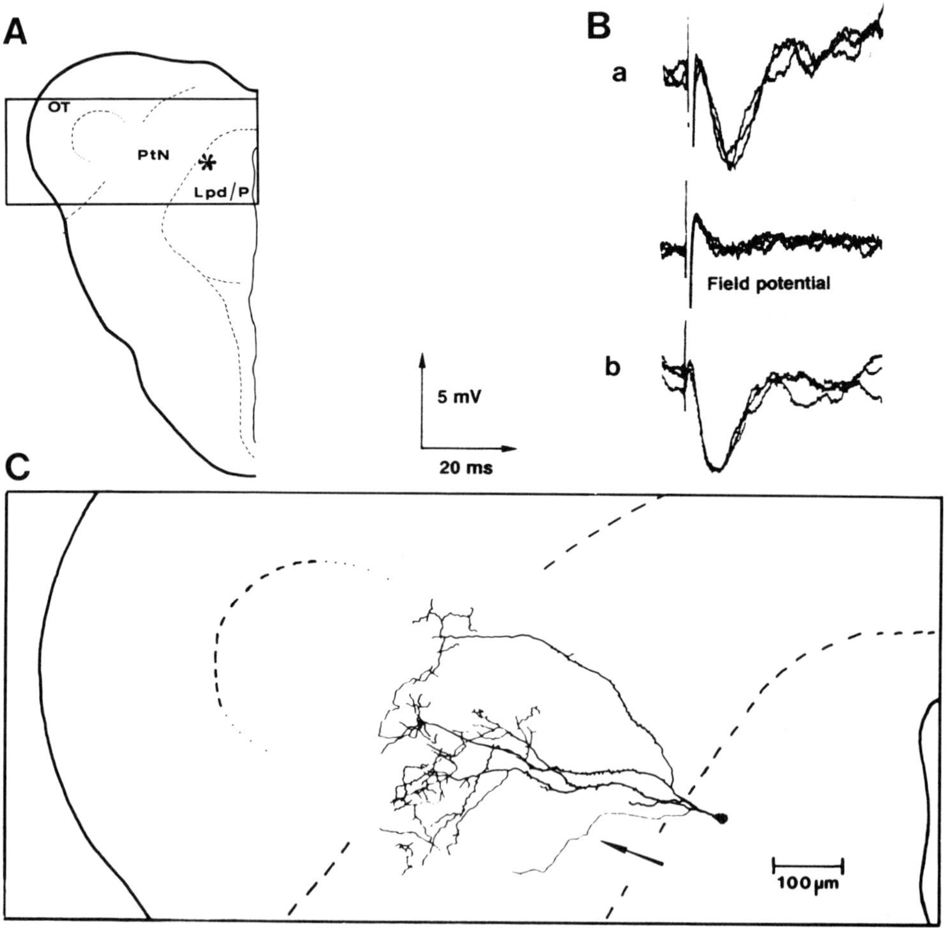

Fig. 4. Localization and postsynaptic responses of pretectal neurons. A) Schematic transverse half section of the diencephalon at the pretectal level; OT, optic tectum; PtN, pretectal neuropil; Lpd/P, pretectal lateral posterodorsal and posterior thalamic nuclei. B) IPSP response of a pretectal neuron to electrical stimulation of LFB (a) or of vSTR (b). The recording site of the cell's response shown in (a) is indicated by the asterisk in (A), and a Camera Lucida reconstruction of this pretectal neuron (arrow points to the axon) is depicted in C).

Electrical stimulation in vSTR evoked in pretectal Lpd neurons IPSPs or sequential EPSP/IPSP activity, but - in contrast to LFB stimulation - no purely depolarizing deflections (Table 1). The PSP latencies were clearly longer than those obtained for LFB stimulation (see Fig. 5B). The shortest latency - within the bin of 4 ms - was seen in pure IPSPs (Fig. 4Bb).

It is important to note that no tonic background activity (steady discharges) could be observed in the investigated pretectal neurons, i.e., the postsynaptic activity was evoked *in response to* electrical stimulation of the vSTR, the LFB, or to presentation of moving visual stimuli.

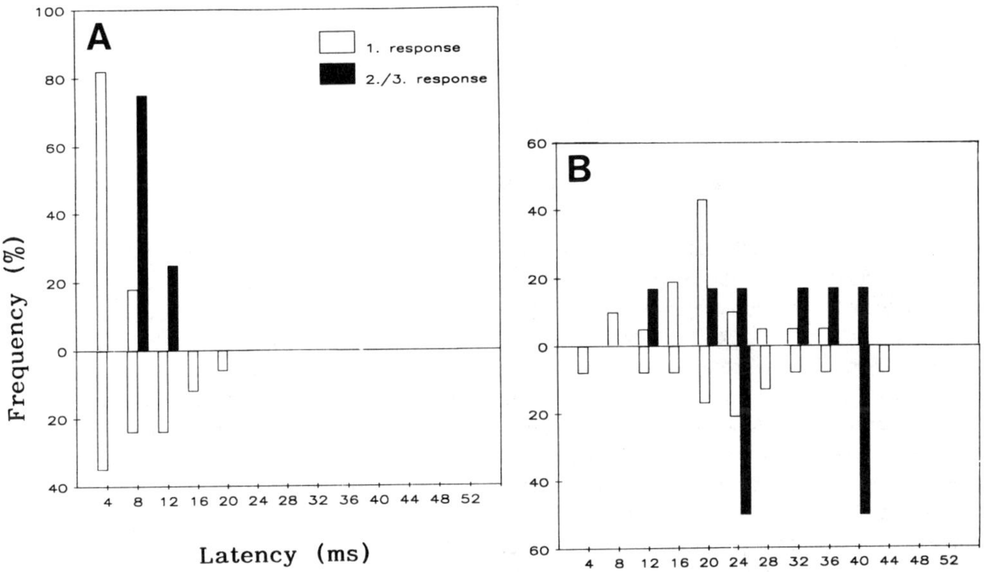

Fig. 5. Latency histograms of pretectal neurons for 1st and 2nd/3rd EPSP responses (positive scale) and IPSP responses (negative scale) to electrical stimulation of LFB (A) [n=20 cells] or vSTR (B) [n=24 cells]. The PSPs are expressed in % of their occurrence in the cells recorded. For pure and sequential PSPs see Table 1.

6. Discussion

6.1 Striato-Pretectal Inhibition

The present study provides clear evidence that visually sensitive neurons of the Lpd and lateral P region receive substantial postsynaptic pure inhibitory inputs mediated by axons traveling in the ipsilateral LFB, but there are also neurons showing excitatory or combined excitatory *and* inhibitory inputs (Table 1). Although it is difficult from the present data to calculate the exact proportions and weights of inhibitory and excitatory postsynaptic inputs to pretectal cells, the relatively great percentage of pure I-type cells (Table 1) and the fact that most of these showed latencies between 2 and 3 ms suggest massive LFB-mediated inhibition, probably monosynaptic. Other pretectal neurons obtain LFB information also via polysynaptic connections (Fig. 5A).

Electrical stimulation of the ipsilateral vSTR feeding into LFB, too, evoked pure IPSP activity in neurons of the Lpd nucleus. No cells showed pure EPSPs (Table 1). Comparing these pretectal responses with those to LFB stimulation, again the fastest input in the bin of 4 ms was inhibitory, probably monosynaptic. The purely inhibitory as well as the relatively great percentage of combined excitatory and inhibitory striatal influences to pretectal cells is consistent with the behavioral observations after striatal lesions: lack of the striatal inhibitory component leads to an increase of the activity of pretectal cells in response to visual input, so that the resulting overriding pretecto-tectal inhibition suppresses the tectal prey-catching releasing system.

The longer pretectal response latencies to vSTR- *vs.* LFB-stimulation (Fig. 5A,B) can be explained by intrinsic striatal/entopedunculus-nuclear interneuronal transmission. Regarding the higher percentage of pretectal IPSPs in response to LFB- *vs.* vSTR-stimulation (Table 1), we have to consider that *global* stimulation of the entire LFB activates the output of all neuronal populations projecting via this bundle to Lpd/P, whereas *focal* stimulation in vSTR only affects small populations of neurons descending in the bundle.

6.2 Disinhibitory Control of the Response to Prey

A brain structure can be functionally well understood only by its integration in the "dialogue" with other brain structures. Focussing on the retino-pretectal/tectal network that allows the toad for example to distinguish between prey and non prey, we realize that this system does not work in isolation, e.g. apart from certain telencephalic influences. Our hypothesis suggests that striatal influence in this context is twofold: (1) Due to the polysynaptic tecto-lateralanteriorthalamo-striato-pretecto-tectal loop, striatal visual influences on pretectal cells are delayed compared to the retino-pretectal input, so that initially pretecto-tectal inhibition overrides tectal excitation (similar to the situation after striatal lesions) This contributes to a delay which allows the toad safely to *hesitate* in the moment a putative prey object occurs. Relatively long latencies for striato-pretectal information transfer could be explained by "relay neurons" of the entopeduncular nucleus known to project via unmyelinated axons to the pretectum. (2) Upon visual and internal inputs, vSTR then directly and indirectly inhibits pretectal neurons and thus moderately "disinhibits" the tectal prey-catching releasing system. More specifically, the resulting decrease of inhibition of T5.2 tecto-bulbar/spinal projection neurons guarantees normal prey-selectivity and reduces the threshold for the release of prey-catching. Once striatal *and* visual signals concurrently arrive in the pretectal/tectal system, the readiness to respond to prey increases, a phenomenon called *priming*, *arousing*, or *warming-up* (Hinde 1954). To avoid tectal hyperexcitation (comparable to the one produced experimentally by pretectal lesions), inhibition of pretectal cells is limited by striato-pretectal excitatory influences which keep the filter properties of T5.2 neurons in behaviorally relevant ranges. The idea is that pretectal/tectal functions can be regulated by a complex network in which the striatum is integrated that in turn receives excitatory and inhibitory "adjusting" pretectal and tectal influences (feedback) via polysynaptic pathways (Matsumoto et al. 1990). Of course, this hypothesis needs further experimental proof.

The principle is to some extent familiar with the one put forward by Chevalier & Deniau (1990) and Hikosaka (1990) suggesting striato-nigro-collicular disinhibition in mammals as an arousal mechanism for the translation of perception into action: the execution of a motor response requires temporal coincidence of striatal disinhibitory influences with collicular sensorimotor commands. Regarding a functional analogy (Ewert 1987), the pretectal inhibitory action on the tectum in toads, however, arises only as a consequence of visual input. This makes the situation different from that of the striato-nigro-collicular system, where continuous inhibition is in play. In the system investigated in toads, disinhibition would arrive from the striatum in response to visual information. It is needless to emphasize that this system works more economically than the modulation of a steady stream of suppressive messages (see also Doty 1987).

Other functions of anuran striatum and pretectal nuclei - besides the one discussed here - doubtless exist (e.g., see Katte & Hoffmann 1980; Ingle 1983; Reiner et al. 1984; Hikosaka 1990). Pretecto-striato-pretectal circuitry, for example, may be involved in memory related to visual obstacles, as suggested by Ingle & Hoff (1988; see also Ingle this volume). Furthermore, the existence of fast adaptation in many visual neurons leaves the possibility open that striatum is also involved in habituation processes.

ACKNOWLEDGEMENTS. We thank Mrs. Gerda Kruk for technical assistence, Mrs. Gisela Kaschlaw for animal care, and Mrs. Ursula Reichert and Mrs. Gudrun Frühauf for text processing. During the experimental period Professor Dr. N. Matsumoto was Visiting Professor of the University of Kassel.

References

Chevalier G, Deniau JM (1990) Disinhibition as a basic process in the expression of striatal functions. TINS 13: 277-280

Diebschlag E (1935) Zur Kenntnis der Großhirnfunktionen einiger Urodelen und Anuren. Z Vergl Physiol 21: 343-394

Doty RW (1987) Has the gready toad lost its soul, and if so, what was it? A commentary. Behav Brain Sci 10: 375

Ewert J-P (1967) Untersuchungen über die Anteile zentralnervöser Aktionen an der taxisspezifischen Ermüdung der Erdkröte (*Bufo bufo* L). Z Vergl Physiol 57: 263-298

Ewert J-P (1968) Der Einfluß von Zwischenhirndefekten auf die Visuomotorik im Beute- und Fluchtverhalten der Erdkröte (*Bufo bufo* L). Z Vergl Physiol 61: 41-70

Ewert J-P (1974) The neural basis of visually guided behavior. Sci Amer 230: 34-42

Ewert J-P (1985) The Nico Tinbergen lecture 1983: concepts in vertebrate neuroethology. Anim Behav 33: 1-29.

Ewert J-P (1987) Neuroethology: toward a functional analysis of stimulus-response mediating and modulating neural circuitries. In: Ellen P, Thinus- Blanc C (eds) Cognitive processes and spatial orientation in animal and man, Vol I. Martinus Nijhoff, Dordrecht, pp 177-200

Ewert J-P (1989) The release of visual behavior in toads: stages of parallel/hierarchical information processing. In: Ewert J-P, Arbib MA (eds) Visuomotor coordination: amphibians, comparisons, models, and robots. Plenum Press, New York London, pp 39-120

Ewert J-P, Wietersheim A v (1974a) Musterauswertung durch tectale und thalamus/praetectale Nervennetze im visuellen System der Kröte (*Bufo bufo* L). J Comp Physiol 92: 131-148

Ewert J-P, Wietersheim A v (1974b) Der Einfluß von Thalamus/Praetectum-Defekten auf die Antwort von Tectum-Neuronen gegenüber bewegten visuellen Mustern bei der Kröte (*Bufo bufo* L). J Comp Physiol 92: 149-160

Ewert J-P, Speckhardt I, Amelang W (1970) Visuelle Inhibition und Exzitation im Beutefangverhalten der Erdkröte (*Bufo bufo* L). Z Vergl Physiol 68: 84-110

Ewert J-P, Schwippert WW, Beneke TW (1990) Parallel distributed processing of configural moving objects in the toad's visual system. In: Eckmiller R, Hartmann G, Hauske G (eds) Parallel processing in neural systems and computers. North-Holland, Amsterdam New York Oxford Tokyo, pp 109-112

Finkenstädt T (1989) Visual associative learning: searching for behaviorally relevant brain structures in toads. In: Ewert J-P, Arbib MA (eds) Visuomotor coordination: amphibians, comparisons, models, and robots. Plenum Press, New York London, pp 799-832

Finkenstädt T, Ewert J-P (1983) Visual pattern discrimination through interactions of neural networks: a combined electrical brain stimulation, brain lesion, and extracellular recording study in *Salamandra salamandra*. J Comp Physiol 153: 99-110

Hikosaka O (1990) Role of basal ganglia in initiation of voluntary movements. In: Arbib MA, Amari S (eds) Dynamic interactions in neural networks: models and data. Springer, New York, Berlin Heidelberg London Paris Tokyo, pp 153-167

Hinde RA (1954) Changes in responsiveness to a constant stimulus. Behaviour 2: 41-54

Ingle DJ (1983) Brain mechanisms of visual localization by frogs and toads. In: Ewert J-P, Capranica RR, Ingle DJ (eds) Advances in vertebrate neuroethology. Plenum Press, New York London, pp 177-226

Ingle DJ, Hoff S v (1988) Neural mechanisms of short-term memory in frogs. Soc Neurosci Abstr 14: 692

Katte O, Hoffmann K-P (1980) Direction specific neurons in the pretectum of the frog (*Rana esculenta*). J Comp Physiol 140: 53-57

Matsumoto N (1989) Morphological and physiological studies of tectal and pretectal neurons in the frog. In: Ewert J-P, Arbib MA (eds) Visuomotor coordination: amphibians, comparisons, models, and robots. Plenum Press, New York London, pp 201-222

Matsumoto N, Schwippert WW, Ewert J-P (1986) Intracellular activity of morphologically identified neurons of the grass frog's optic tectum in response to moving configurational visual stimuli. J Comp Physiol A 159: 721-739

Matsumoto N, Schwippert WW, Beneke TW, Ewert J-P (1990) Forebrain-mediated inhibition and disinhibition of visual prey-catching behavior in toads: intracellular study of striato-pretectal information transfer (submitted)

Northcutt RG, Kicliter E (1980) Organization of the amphibian telencephalon. In: Ebbesson SOE (ed) Comparative neurology of the telencenphalon. Plenum Press, New York London, pp 203-255

Patton P, Grobstein P (1986) Possible striatal involvement in prey orienting behaviour in the frog. Soc Neurosci Abstr 10: 61

Reiner A, Brauth SE, Karten HJ (1984) Evolution of the amniote basal ganglia. TINS 7: 320-325

Schürg-Pfeiffer E (1989) Behavior-correlated properties of tectal neurons in freely moving toads. In: Ewert J-P, Arbib MA (eds) Visuomotor coordination: amphibians, comparisons, models, and robots. Plenum Press, New York London, pp 451-480

Székely G, Lázár G (1976) Cellular and synaptic architecture of the optic tectum. In: Llinás R, Precht W (eds) Frog neurobiology. Springer, Berlin Heidelberg New York, pp 407-434

Wilczynski W, Northcutt RG (1983a) Connections of the bullfrog striatum: afferent organization. J Comp Neurol 214: 321-332

Wilczynski W, Northcutt RG (1983b) Connections of the bullfrog striatum: efferent projections. J Comp Neurol 214: 333-343

Cortical Circuitry Underlying Visual Motion Analysis in Turtles

Philip S. Ulinski, Linda J. Larson-Prior and N. Traverse Slater

Department of Organismal Biology and Anatomy, and Committee on Neurobiology, University of Chicago, Chicago, Il 60637; Department of Physiology Northwestern University, Chicago, Il 60611. USA

Abstract. *Turtles have a visual area in their cerebral cortices that lacks a topographic organization. It contains cells that respond to moving stimuli anywhere in visual space, or to the successive presentation of two stimuli at different points in space with interstimulus intervals of up to several hundred msec. This paper discusses the use of an in vitro preparation of turtle forebrain to determine how such receptive field properties are elaborated by the central visual pathways.*

The convergence of spatial information required for a cortical neuron to respond to all points in visual space occurs in a sequence of three steps involving the retinogeniculate, geniculocortical and intracortical projections. By contrast, the temporal properties of cortical neurons are determined by cortical circuitry. They can be mimicked using artificial stimulation of the geniculocortical system in an in vitro preparation of the isolated cerebral cortex. The known synaptic physiology of those cortical circuits that use excitatory amino acids and GABA can account for the major temporal properties of the cortical neurons. These results are used to make predictions about the population firing profiles of colonies of cortical neurons influencing individual striatal neurons that are likely involved in visually-guided orienting movements.

1. Visual Cortex in Turtles

The cerebral hemispheres of fresh-water turtles contain an area of cortex that is responsive to visual stimuli (Orrego, 1961; Mazurskaya, 1972; Bass et al., 1983). It has an interesting functional organization in that it lacks a topographic representation of visual space, and virtually all neurons in the visual area respond to small, moving objects located anywhere in binocular visual space (Mazurskaya, 1974). The principal neurons respond well to successive presentations of small objects at disjunct points in visual space with interstimulus intervals of up to several hundred msec, but do not respond well to two stimuli presented simultaneously. They habituate strongly to repeated presentations of an object moving in a given direction, but respond robustly when the object changes direction (Gusel'nikov and Pivovarov, 1978). These properties are consistent with a role in analyzing objects moving within the visual environment of the turtle, and imply turtle visual cortex has a role in controlling visually-guided orienting movements (Ulinski, 1988).

One goal of work in our laboratories is to understand how the receptive field properties of these cells result from the activity of the networks of neurons present in the cortex. Progress during the past few years has resulted principally from the technical advantages afforded by the unusual diving physiology of turtles, which gives turtle neural tissue a resistance to anoxia (Lutz et al., 1985). It is consequently possible to maintain the entire brain and eyes of turtles intact in an *in vitro* preparation (Kriegstein, 1987; Rosenberg and Ariel, 1990). This preparation can be used for anatomical experiments (Sjostrom and Ulinski, 1985; Heller and Ulinski, 1987; Mulligan and Ulinski, 1990; Cosans and Ulinski, 1990), for physiological experiments employing intracellular methods (Connors and Kriegstein, 1986; Kriegstein and Connors, 1986; Kriegstein, 1987; Larson-Prior et al., 1989) and for extracellular recording in conjunction with the presentation of natural visual stimuli (Rosenberg and Ariel, 1990). This chapter summarizes our current understanding of how the central visual pathways produce the spatial and temporal properties of receptive fields of turtle cortical neurons, and then speculates on how the ensemble properties of such neurons may contribute to visually-guided orienting movements.

2. Spatial Convergence in the Geniculocortical Pathway

Neurons in turtle visual cortex respond to stimuli at all loci in visual space and show no significant variation in firing rate as a function of stimulus position (Mazurskaya, 1974). This implies a convergence of spatial information from all points on the retinal surface onto each cortical neuron somewhere along the retino-geniculate-cortical pathway. Focal application of GABA to the visual cortex *in vivo* results in scotomas in the receptive fields of cortical neurons, suggesting part of the convergence occurs within the cortex itself (Mazurskaya, 1974). We (Sjostrom and Ulinski, 1985; Heller and Ulinski, 1987; Mulligan and Ulinski, 1990; Cosans and Ulinski 1990) have used the anterograde transport of horseradish peroxidase *in vitro* to determine that the convergence involves a sequence of three steps (Fig. 1).

Convergence in the Retinogeniculate Projection

Photoreceptors and retinal ganglion cells are spread inhomogeneously over the retinal surface, resulting in an increased

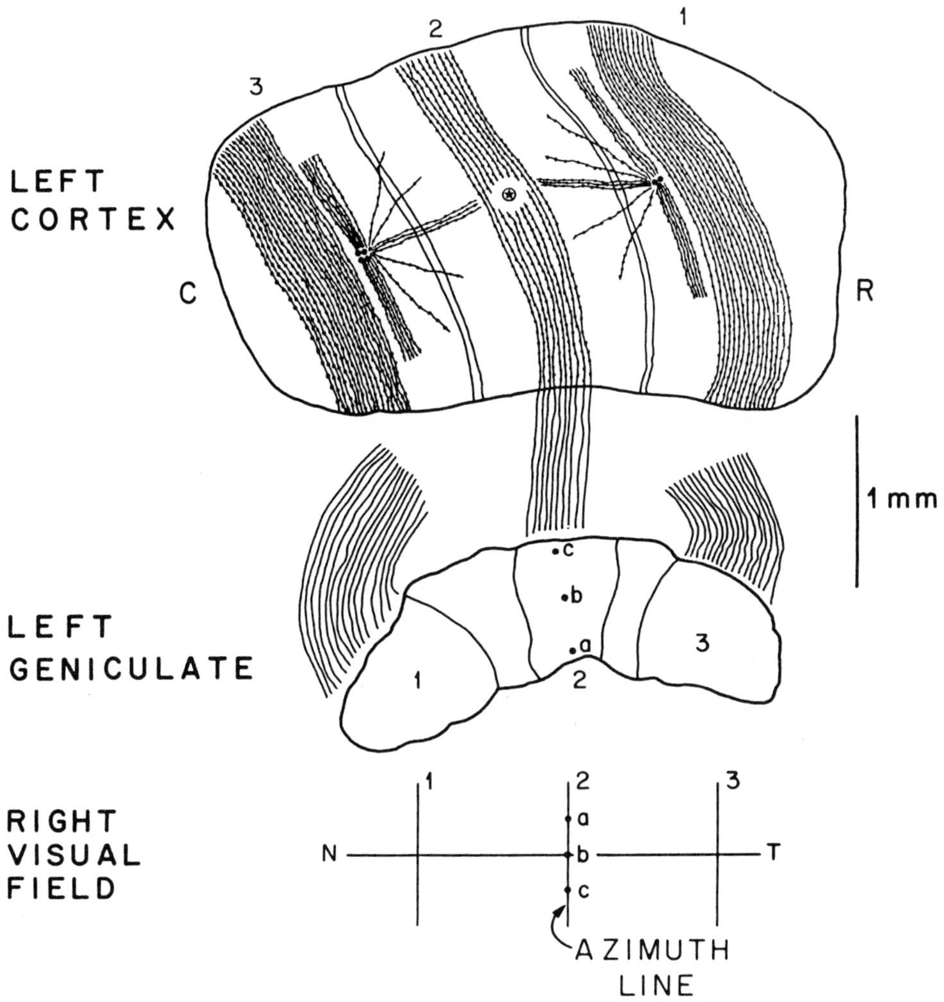

Figure 1. Spatial convergence in the retino-geniculo-cortical pathway. Organization of the projections in the geniculocortical pathway is shown using reconstructions of dorsal lateral geniculate complex and visual cortex. The nasal-temporal axis (NT) of the right visual field and three azimuth lines (1,2,3) are indicated at the bottom of the figure. Three elevations (a,b,c) are indicated on azimuth line 2. Each of the azimuth lines is represented on a band of cells (1,2,3) in the geniculate complex. Representations of the three elevation points (a,b,c) are shown on azimuth band 2. Axons of geniculate neurons run into the visual cortex in fascicles that cross over each other as they near the lateral edge of cortex. Thus, the nasal-temporal axis of the visual field is represented along the rostral (R) - caudal (C) axis of the cortex. Each azimuth line is represented as an isoazimuth lamella in the cortex. Cortico-cortical connections between lamellae result in a complete convergence of all points in visual space upon any one cortical cell. Such as the one indicated by a star.

density of cells along a visual streak that extends along the nasotemporal axis of the retina (Brown, 1969; Granda and Haden, 1970; Peterson and Ulinski, 1982). There is a convergence ratio of between three and six photoreceptors to each ganglion cell (Peterson and Ulinski, 1982), and ganglion cells have oval excitatory centers about 5 degrees of visual arc in diameter (Bowling, 1980; Granda and Fulbrook, 1989). They form several morphological (Kolb, 1982) and physiological (Bowling, 1980; Fulbrook, 1982; Marchiafava et al., and 1983; Jensen and Devoe, 1983; Granda Fulbrook, 1989) groups, and are tuned to moving stimuli with relatively specific directions (Bowling, 1980; Jensen and DeVoe, 1983), velocities (Fulbrook, 1982) or both.

The majority of ganglion cells project to both the optic tectum and the dorsal lateral geniculate complex via axons that bifurcate in the optic tract (Sjostrom and Ulinski, 1985). Retinogeniculate arbors are spatially restricted, between 20 and 50 μm in diameter, and comprise two major groups that differ in the number and size of varicosities per arbor. They also vary in the diameter of their parent axons, which suggests they are derived from ganglion cell populations with different conduction velocities (Woodbury and Ulinski, 1986) and functional properties (Marchiafava et al., 1983).

The contralateral retinogeniculate projection is the most extensive (Bass and Northcutt, 1981; Kunzle and Schnyder, 1984; Ulinski and Nautiyal, 1988). It is retinotopically organized with the nasotemporal axis of the retina represented along the rostrocaudal axis of the geniculate complex and the dorsoventral axis of the retina represented along the ventrodorsal axis of the geniculate (Ulinski and Nautiyal, 1988). There is also a projection from the ventral and temporal rim of the ipsilateral retina to the geniculate complex. The majority of geniculate neurons have dendritic fields that are conical in shape with major diameters of between 200 to 500 μm (Rainey and Ulinski, 1986). The relative sizes of the retinogeniculate arbors and geniculate dendritic fields suggest a substantial spatial convergence occurs in the retinogeniculate projection, resulting in geniculate receptive fields about 30 degrees in diameter (Boiko, 1980). The populations of retinogeniculate arbors show some laminar organization within the geniculate complex (Sjostrom and Ulinski, 1985) so that each geniculate cell probably receives inputs from functionally distinct ganglion cell populations along different segments of its dendritic arbor.

Convergence in the Geniculocortical Projection

Axons of individual geniculate neurons course from the thalamus to the visual cortex via the lateral forebrain bundle (Heller and Ulinski, 1987). They maintain a specific order within the forebrain bundle as it approaches the visual cortex: axons from the rostral pole of the geniculate course in the caudal edge of the forebrain bundle while those from the caudal pole of the geniculate course in the rostral edge of bundle (Mulligan and Ulinski, 1990). Axons cross over each other within the cortex so that fibers originating in the rostral pole of the geniculate run into the caudal part of visual cortex and vice versa. Individual axons run in relatively straight arcs along the full lateral to medial extent of the cortex, bearing about 7 varicosities on each 100 μm segment of axon.

The visual cortex will thus contain a band of cells occupying about a third of its rostrocaudal length that receives inputs from geniculate neurons located along a particular dorsoventral transect through the geniculate. Since this transect of geniculate neurons receives inputs from a vertical (or azimuth) line of visual space, the band of cortical cells is termed an isoazimuth lamella (Mulligan and Ulinski, 1990). Visual cortex can therefore be conceptualized as a series of overlapping isoazimuth lamellae, each receiving inputs from points of all elevations along a given azimuth line of visual space. Cells situated within an isoazimuth lamella will respond to stimuli moving through the corresponding azimuth line with varying directions (Gusel'nikov and Pivovarov, 1978) and velocities (Gusel'nikov et al., 1972).

Convergence within Visual Cortex

The spatial convergence that occurs within the retinogeniculate and geniculocortical projections is sufficient to explain how all cortical neurons can respond to points of differing elevation along a given azimuth line, but does not explain how the same neuron responds to stimuli situated at different azimuths. Convergence of information about different eccentricities along the horizontal meridian of visual space occurs through projections effected by collaterals of cortical neurons (Cosans and Ulinski, 1990). Each restricted patch of cortical cells gives rise to a wreath of axon collaterals that extend for up to 500 μm within visual cortex. They show a distinct preference to run along isoazimuth lamellae, but do extend in all directions. Information from any region of visual cortex can reach all other regions through a sequence of about three relays through the cortex.

3. Temporal Properties of Cortical Neurons

Firing Probability Curves in Vivo

In addition to their wide receptive field properties, neurons in turtle visual cortex are distinguished by their responses to visual stimuli appearing at disjunct points in visual space with some interstimulus interval (Mazurskaya, 1974). Such stimuli are useful because they mimic moving objects that would appear at two distinct loci with an interstimulus interval determined by the object's velocity. Neurons throughout the depth of visual cortex respond to such stimuli, but those situated superficially in the cortex have different temporal tuning properties than those situated deep in the cortex (Fig. 2).

Turtle visual cortex is a trilaminar cortex in which the outer layer 1 contains a few scattered neurons (Davydova and Goncharova, 1979; Desan, 1984, 1988; Ulinski, 1990). Their responses to successive visual stimuli can be depicted by plotting their average firing rate as a function of the time interval that separates the two stimuli (Mazurskaya, 1974). The curves in Figure 2 show fractional firing rates because they are derived from an average of ten stimulus presentations. They give some index of the probability that a cell will fire at a given interstimulus interval and what its firing rate will be if it does fire. Layer 1 neurons (Fig. 2A) show a strong response to a simultaneous presentation of the two stimuli (i.e an interstimulus interval of zero). This is followed by a response at an interval of about 50 msec, and then by a broad period of several hundred msec during which there is a significant probability that the cell will fire.

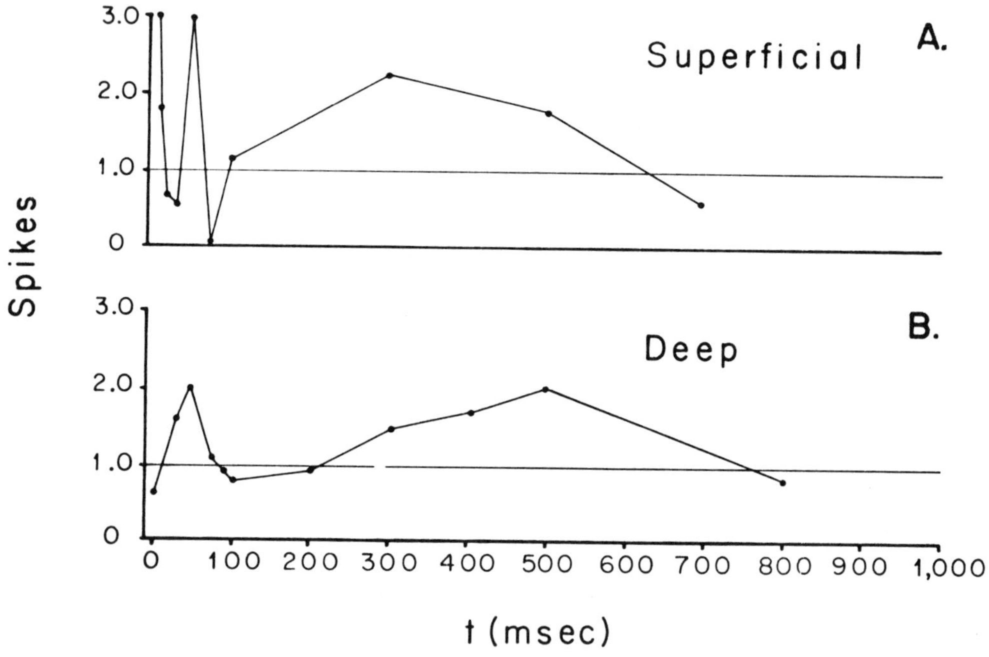

Fig. 2. Firing curves of superficial and deep cells *in vivo*. The response of cells in the superficial (A) and deep (B) layers of visual cortex are shown. The average number of action potentials produce over ten stimulus presentations are plotted on the vertical axis. The horizontal axis is the time interval between two stimuli presented in succession at two different points in visual space. Replotted from Mazurskaya (1974).

The second and third layers of the cortex contain many neurons. Those with somata positioned in layer 2 are mostly double pyramidal cells

(Ulinski, 1990) whose dendrites extend into both layers 1 and 3. Layer 3 contains a high proportion of neurons whose dendrites are oriented horizontally, concentric with the ventricular surface. Neurons in both layers have collaterals that form a substrate for interactions within visual cortex (Desan, 1984; Connors and Kriegstein, 1986), but some are the source of efferent projections to the striatum and thalamus (Ulinski, 1986). The temporal tuning properties of neurons situated in layers 2 and 3 differ from those of neurons situated in layer 1 in that they do not respond strongly to two spatially disjunct stimuli presented simultaneously (Fig. 2B). Like layer 1 neurons, though, they show two phases of facilitation -- one occuring with an interstimulus interval around 50 msec and the second occuring after several hundred msec. The timing of both facilitatory peaks varies as a function of stimulus position in a way that is not yet understood.

Firing Probability Curves In Vitro

The receptive field properties shown in Figure 2 were recorded in awake, paralyzed animals using presentations of visual stimuli on a tangent screen. They presumably are an accurate reflection of the normal

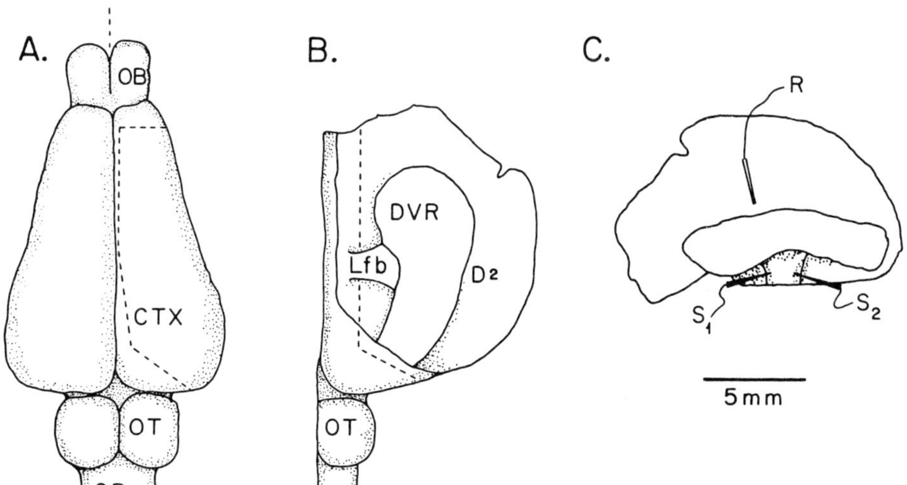

Figure 3. *In vitro* preparation of turtle forebrain. Steps in the preparation of an isolated cortex and lateral forebrain bundle (Lfb) are shown. A. Dorsal view of intact brain showing olfactory bulb (OB), cortex (CTX), optic tectum (OT) and cereabellum (CB). Dotted lines indicate lines along which cuts are made. B. Cortex is folded back to expose the Lfb, dorsal ventricular ridge (DVR) and visual cortex (D2). C. Preparation as it appears in the recording chamber. S1 and S2 are stimulating electrodes placed in the Lfb. R is a recording electrode in the cortex.

firing properties of cortical neurons, but potentially result from the activity of many different visual pathways and circuits. It is not evident *a priori*, then, what role cortical circuits play in determining the temporal tuning properties of cortical neurons. We have developed an *in vitro* preparation that consists of the isolated visual cortex and lateral forebrain bundle in order to more clearly delimit how neural circuits within visual cortex contribute to the receptive field properties of cortical neurons (Fig. 3). Since geniculocortical fibers within the lateral forebrain bundle retain a topographic order as they approach the lateral edge of the cortex (Mulligan and Ulinski, 1990), electrical stimulation at two separate loci within the bundle should activate fibers carrying information from different regions of visual space. Thus, successive activation of the two electrodes should provide at least an approximation of the pattern of activity that flows from the geniculate to the visual cortex as a stimulus moves across the retina.

Figure 4A shows responses recorded intracellularly from a neuron deep in visual cortex following activation by two electrodes placed at the rostral and caudal edges, respectively, of the lateral forebrain bundle. Its firing properties and position in the cortex identify it as a double

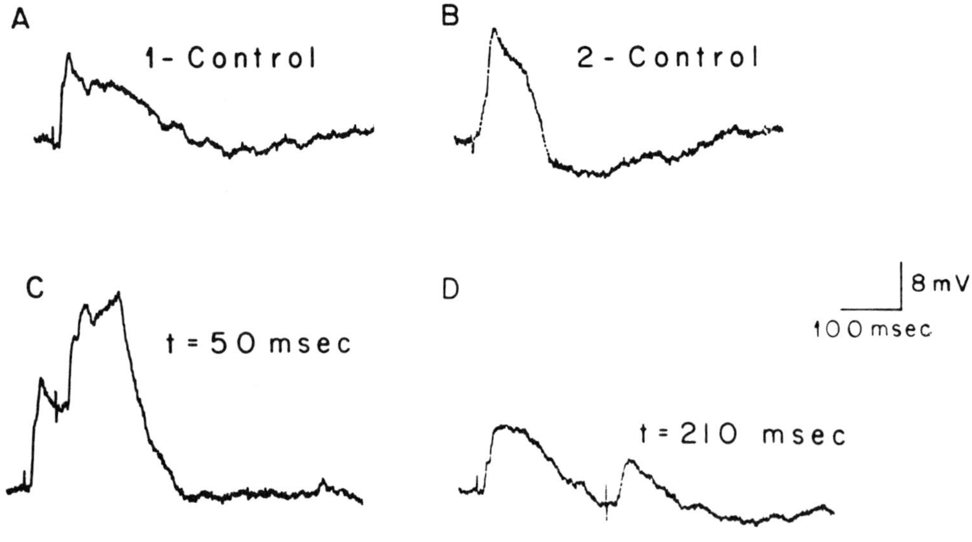

Figure 4. Responses of cortical neurons to sequential stimulation of the LFB. A, B, postsynaptic responses in cortical neurons resulting from activation of electrode S1 or S2, respectively, alone. C, D, responses in the same cell to sequential activation of the two electrodes with interstimulus intervals of 50 and 210 msec, respectively. Note that each response shows two stimulus artifacts.

pyramidal cell (Connors and Kriegstein, 1986). Its membrane potential was held about 10 mV negative to its resting potential of - 50 mV during this set of observations, and activation of either electrode produced subthreshold depolarizing potentials. They differ in latency, reflecting the additional intracortical conduction time required by one of the activation pathways. Figure 4 shows examples from a series of responses in which the two electrodes were activated in succession with intersimulus intervals varying from 25 to 1,000 msec. Traces 4A and 4B

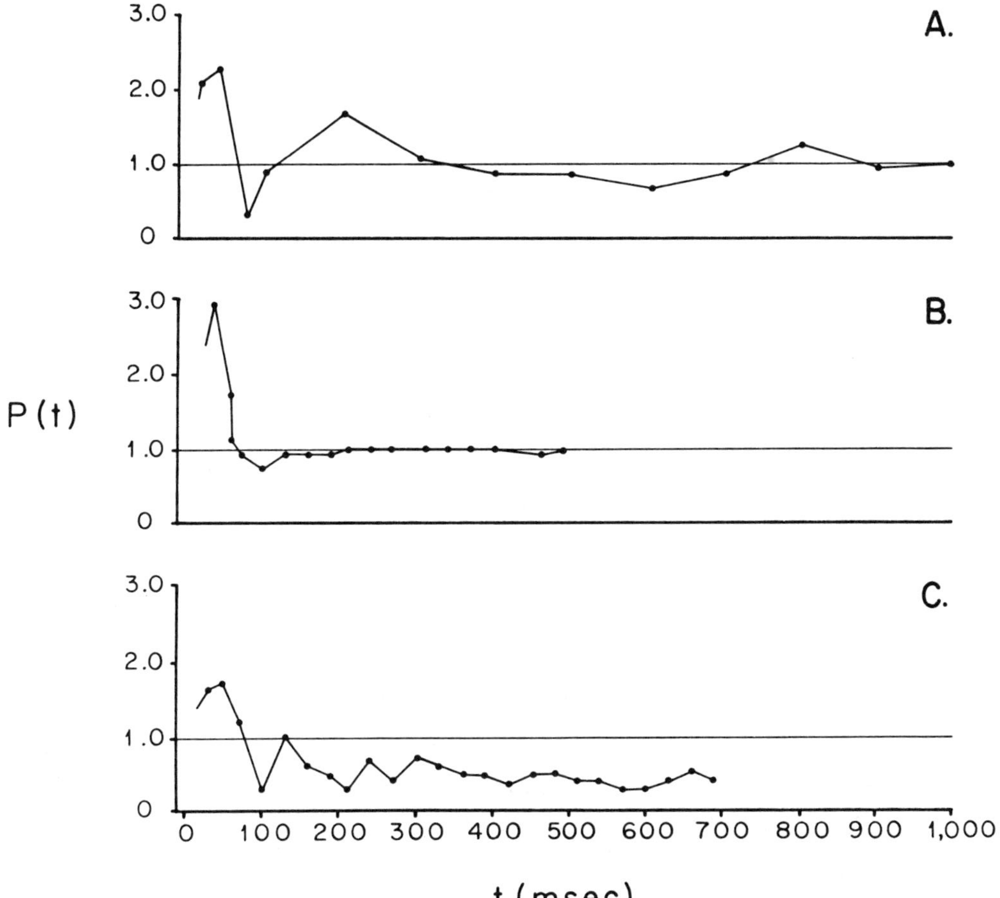

Figure 5. P(t) curves of cortical cells *in vitro*. Data such as those shown in Figure 4C, D are shown for three cortical neurons. Each curve shows the ratio of the amplitude of the second depolarization in a sequential presentation to the amplitude of the depolarization produced by electrode S2 in isolation, plotted as a function of the interstimulus interval.

show responses to activation of either electrode 1 (Fig. 4A) or Electrode 2 (Fig. 4B) in isolation. Figures 4C and 4D show responses to successive activation of first electrode 1 and then electrodes 2 with delays of 25 and

1990; Watkins et al., 1990). Thus, activation of cortical circuitry by geniculocortical fibers involves a fast EPSP mediated by quisqualate/ kainate receptors.

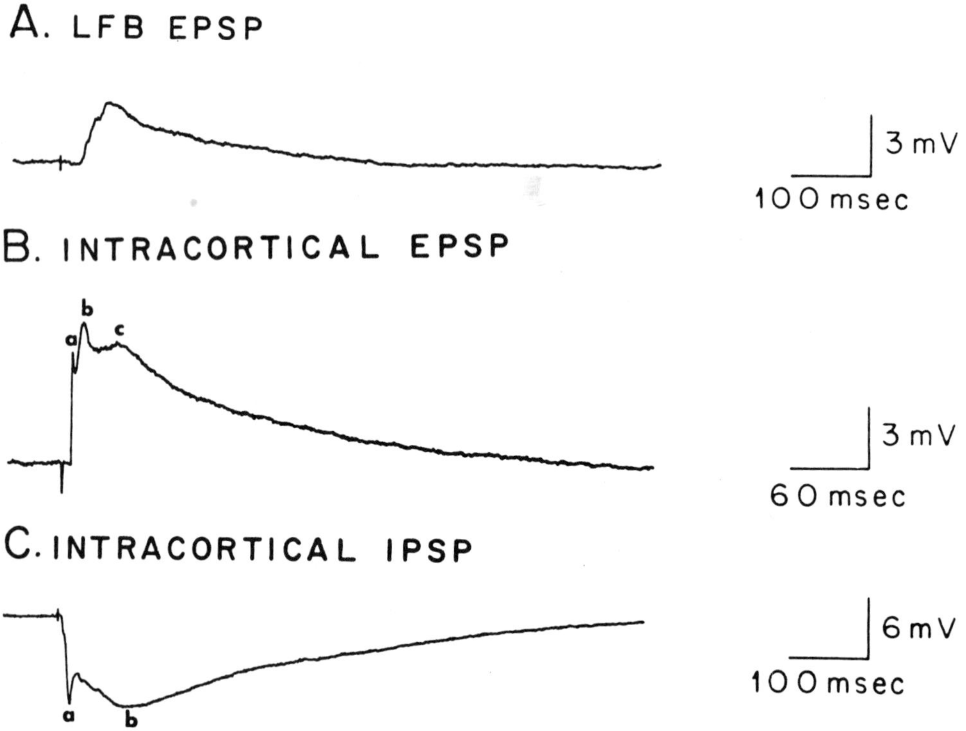

Figure 6. Postsynaptic responses of cortical neurons. Postsynaptic responses recorded *in vitro* from a cortical neuron. A. EPSP produced by electrical stimulation of the lateral forebrain bundle. B. EPSP produced by electrical stimulation of intracortical fibers. The EPSP is a compound event with fast (a), intermediate (b) and slow (c) components. C. IPSP produced by electrical stimulation of intracortical fibers. The IPSP is a compound event with a fast (a) and slow (b) component.

Since the neurons present in layer 1 are apparently all GABAergic (Blanton et al., 1987) interneurons that synapse upon layer 2 cells, it is not surprising that activation of the geniculocortical or corticocortical fibers often produces IPSPs (Fig. 6c) in layer 2 or 3 neurons (Pivovarov and Trepakov, 1972; Kriegstein and Connors, 1986). These IPSPs can occur either in isolation, or in combination with EPSPs. In many cases they

210 msec, respectively. The responses in these experiments can be simple EPSPs or IPSPs, or relatively complex waveforms. Most are subthreshold, but some cause the cell to fire action potentials.

The temporal tuning properties of this and two other cells are depicted as P(t) curves in Figure 5. These curves show the amplitude of the second depolarization relative to its control value, plotted as a function of interstimulus interval. Curves of this sort universally show a strong facilitation over the range of interstimulus intervals falling between 25 and 75 msec. The shape of the curve for longer interstimulus intervals is variable. Some cells (Fig. 5 C) show a strong, long-lasting inhibition. Others show a shorter, less pronounced inhibition (Fig. 5B). Still others show a strong facilitation that develops after 100 msec and can last for several hundreds of msec.

These results are striking in that the P(t) curves depicting subthreshold events in the isolated cortex closely resemble in shape the firing rate curves obtained from cortical neurons obtained *in vivo* with natural stimulation. Our preliminary studies of the biophysical properties of cortical neurons indicate that it will be possible to account for the shape of the firing rate curves as a function of the P(t) curves and the firing rate-current relations and subthreshold conductance properties of the individual cortical cells. Thus, at least the general features of the temporal tuning properties of cortical neurons seem to be determined largely by the intracortical circuitry.

Neural Circuitry Underlying P(t) Curves

Our knowledge of the anatomy and physiology of turtle visual cortex is too preliminary to offer a complete analysis of which cortical neurons contribute to the elaboration of these receptive fields, but we have used the *in vitro* preparation of visual cortex to determine some of the basic features of geniculocortical and intracortical circuitry (Fig. 6). Our findings provide the basis for some initial ideas.

The geniculocortical fibers synapse upon layer 1 interneurons and the ascending dendrites of layer 2 double pyramidal cells (Ebner and Colonnier, 1978; Smith et al., 1980). The synapses are predominantly upon dendritic spines and involve clear, round synaptic vesicles and asymmetric active zones. Ionophoretic application of the exitatory amino acid L-glutamate to cortical neurons produces a depolarization (Kriegstein and Connors, 1986), and bath application of the broad-spectrum excitatory amino acid antagonist kynurenic acid (Watkins et al., 1990) reversibly abolishes the monosynaptic EPSP (Fig. 6A) evoked in cortical neurons by electrical stimulation of geniculocortical fibers (Larson-Prior et al., 1989). The geniculocortical EPSP is also abolished by quinoxalinediones (Larson-Prior et al, 1989) such as DNQX and CNQX (Honore et al., 1988) which are believed to act specifically upon the quisqualate/kainate subtypes of excitatory amino acid receptors (Collingridge and Lester, consist of two distinct events. The fast IPSP (Fig. 6c,a) has a reversal potential around - 70 mv, is sensitive to bicuculline and picrotoxin and can be attributed to activation of $GABA_A$ receptors and chloride channels (Kriegstein and Connors, 1986). The slow IPSP (Fig. 6c, b) has a reversal potential about -90 mv, is sensitive to phaclofen and can be attributed to activation of $GABA_B$ receptors. The layer 1 neurons show a strong and persistent firing profile upon activation (Connors and Kriegstein, 1986), and have axons that branch extensively in layer 1 (Desan, 1984). Thus, neurons in layers 2 and 3 frequently demonstrate IPSPs that are sufficient to completely shunt depolarizing responses.

Intracortical collaterals of double pyramidal cells ascend obliquely through layers 2 and 3 for up to 500 μm (Cosans and Ulinski, 1990) and probably synapse upon both other double pyramidal cells and layer 1 interneurons (Kriegstein and Connors, 1986). They can be activated by focal electrical stimulation within the cortex (Larson-Prior et al., 1989). Intracortical fibers produce compound EPSPs in double pyramidal cells. An initial fast component is insensitive to kynurenic acid and presumably reflects the activity of a neurotransmitter that is not an excitatory amino acid, perhaps substance P (Reiner et al., 1984). The intermediate latency component is sensitive to DNQX and reflects activation of quisqualate/kainate receptors. The late component is sensitive to the antagonist APV, which is selective for the NMDA subtype of excitatory amino acid receptor (Collingridge and Lester, 1990; Watkins et al., 1990). The late component demonstrates a dramatic potentiation upon repetitive stimulation.

These features of the anatomy and physiology of visual cortex are sufficient to account for the general aspects of the P(t) curves of neurons located deep in the cortex. The relatively strong excitatory drive from the geniculocortical fibers to the layer 1 interneurons provides a rapid and strong activation of the layer 1 cells when stimuli occur simultaneously at two or more loci in visual space. This results in robust and persistent firing of the stellate cells, which in turn elicits a strong inhibition of the double pyramidal cells due to the activation of chloride channels via $GABA_A$ receptors. The resulting conductance shunt of EPSPs appears as the initial low values of the P(t) curve at t = 0.

Successive presentation of two stimuli with non-zero t-values has a different effect. Presentation of the first stimulus activates cortical cells within a given isoazimuth lamella due to the activity of a fascicle of geniculocortical fibers, but it is unlikely that this activity will be sufficient to trigger action potentials in more than a few cortical cells. However, presentation of the second stimulus activates a sequence of intracortical axons that reactivates the cortical cells after several tens of msec. This produces a potentiation of the fast and intermediate components of the intracortical EPSPs that will be sufficient to bring many cells to threshold. The synaptic mechanisms underlying this potentiation are not yet known, but the latency of the potentiation accounts for the first peak in the P(t) curve, which occurs in the range of 25 to 75 msec.

Presentation of the second stimulus with t-values in excess of about 100 msec is unlikely to potentiate the fast and intermediate components of intracortical EPSPs, but will lead to a repetitive activation of NMDA receptors on double pyramidal cells via the intracortical collaterals of other double pyramidal cells, and a consequent potentiation of the late component of the intracortical EPSPs that results in the second facilatatory peak in the P(t) curves. Again, the synaptic mechanisms underlying the potentiation remain to be analyzed. We also do not yet understand how the variability observed in the second peaks occurs, but it is likely that very complex interactions between the IPSPs and voltage-dependent block of the NMDA receptors by magnesium ions plays a major role.

It may be possible, then, to eventually understand in some detail how the circuitry within visual cortex determines the P(t) curves of the double pyramidal cells. It should be emphasized here that the *in vitro* preparation of visual cortex facilitates such analyses. We have begun, for example, to look at the effects of antagonists to specific receptor types on the P(t) curves to determine the roles that transmitter-specific circuits play in generating the P(t) curves.

4. Ensemble Properties of Colonies of Cortical Neurons

The previous sections have concentrated on understanding the firing properties of individual cortical neurons. However, it is likely -- of course -- that visual guidance behaviors require the activity of thousands of neurons, so a future challenge is to understand how large ensembles of cortical neurons act in unison to help control visually-guided behaviors. The principal route whereby cortical neurons gain access to the premotor circuitry in the brainstem reticular formation is via projections from the visual cortex to the striatum.

The striatum is the component of the telencephalon that forms the ventrolateral wall of the cerebral hemisphere. It receives inputs from the visual cortex and a second telencephalic visual structure, the dorsal ventricular ridge (Ulinski, 1983). Striatal neurons can influence the optic tectum via structures in the pretectum and the midbrain tegmentum (Reiner et al., 1980; Parent, 1986). Neurons in the optic tectum give rise to extensive projections to the brainstem reticular formation (Sereno,1985; Sereno and Ulinski, 1985).

Axons of cortical neurons diverge as they pass through the striatum and bear many varicosities *en passant*. Individual striatal neurons will, thus, receive inputs from many cortical neurons and the relevant unit structure in the corticostriatal projection is the set of cortical neurons that synapse upon an individual striatal neuron. Borrowing a term from the literature on the corticospinal tract, we have termed this set of cells a colony (Landgren et al., 1962). Imagine, then, a colony of double pyramidal cells spread throughout the visual cortex (Fig. 7A). Each neuron in the colony will have a P(t) function that determines its firing rate as a stimulus moves past the turtle with some specific velocity and direction, and for some period of time. Figure 7B shows idealized examples of such P(t) functions. As a first approximation towards characterizing the population P(t) curve, we can construct the simple sum of individual P(t) functions to gain an impression of what the time course of the synaptic drive to the striatal cell might look like. Figure 7C shows such estimates of the population firing profile in two cases, one when the stimulus moves through visual space for a total of 200 msec and one for a total movement duration of 500 msec. Because of the convergences that have occured *en route* to the cortex, the shape of the firing profile is independent of the specific velocity or direction of the stimulus. In both cases, the movement elicits a brisk discharge from the colony of cortical neurons between 25 and 75 msec after the movement has begun, followed by a plateau of discharge that will persist as long as the stimulus continues to move.

The moving stimulus will, therefore, result in two very different types of inputs reaching the optic tectum over two very different time intervals. Those collaterals of ganglion cell axons that project directly to the tectum will carry information about the stimulus as it moves through a local domain of visual space. They code information about stimulus velocity and direction, but not about the total duration of the movement. Those ganglion cell collaterals that carry information through the geniculate to the cortex will result ultimately in information reaching the optic tectum carrying information about the total duration of the movement, throughout visual space.

5. Summary

We have used *in vitro* preparations of the turtle visual system to examine the manner in which the spatial and temporal properties of neurons in the visual cortex are elaborated by the various components of the visual pathways. The spatial convergence needed to produce the wide receptive fields of cortical neurons occurs in three steps within the retinogeniculate, geniculocortical and intracortical projections. By contrast, the major temporal features of the cortical receptive fields can be produced in an *in vitro* preparation of an isolated cortex, suggesting they are elaborated largely by the cortical circuitry. Some hypotheses about how this is accomplished and the functional significance of the receptive fields are explored.

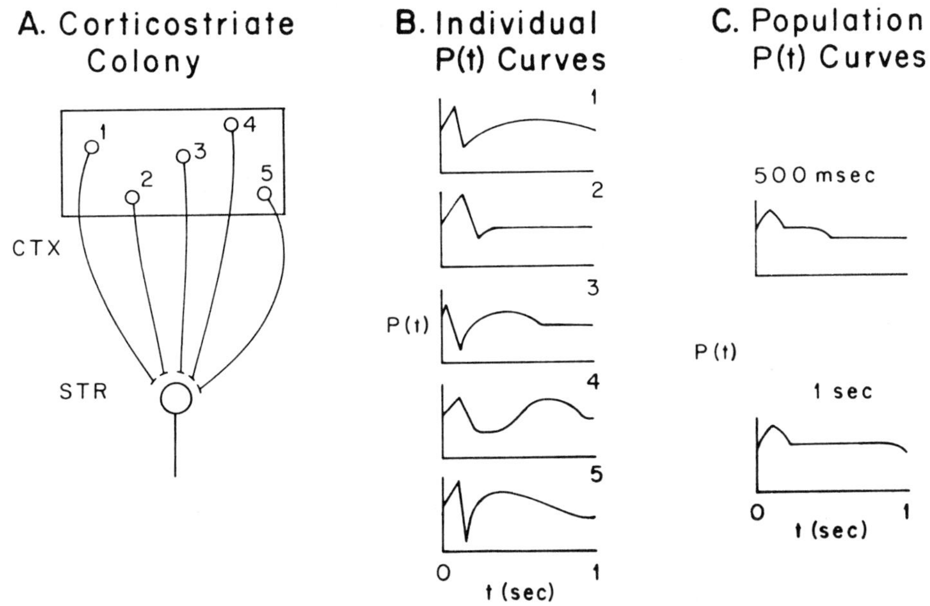

Figure 7. Synaptic organization of a cortico-striatal colony. A. Spatial organization of a hypothetical corticostriate colony. Four cortical cells (1,2,3,4) are shown projecting to a single striatal neuron. B. Hypothetical P(t) curves of each of the four neurons shown in A. C. Predictions of the P(t) curves show the population response for the four neurons in the cases of a stimulus moving through the visual field for 500 msec and 1 sec.

Acknowledgements

Supported by PHS Grants EY08352 to P.S.U. and NS 17489 and NS 25682 to N.T.S. Roberta Mims typed the manuscript.

6. Bibliography

Bass, AH and Northcutt, RG, 1981, Retinal recipient nuclei in the painted turtle, *Chrysemys picta*: An autoradiographic and HRP study. J. Comp. Neurol. 199: 97 - 112.

Bass, AH, Andry, ML and Northcutt RG, 1983, Visual activity in the telencephalon of the painted turtle, *Chrysemys picta*. Brain Res. 26: 201-210.

Blanton, MG, Shen, JM and Kriegstein, AR, 1987, Evidence for the inhibitory neurotransmitter γ-aminobutyric acid in aspiny and sparsely spiny nonpyramidal neurons of the turtle dorsal cortex. J. Comp. Neurol. 259: 277-297.

Boiko, VP, I980, Response to visual stimuli in thalamic neurons of the turtle *Emys orbicularis*. Neurosci. Behav. Physiol. 10: 183-188.

Bowling, DM, 1980, Light responses of ganglion cells in the retina of the turtle. J. Physiol. (Lond.) 299: 173-196.

Brown, KT, 1969, A linear area centralis extending across the turtle retina and stabilized to the horzion by non-visual cues. Vision Res. 9: 1053-1062.

Collingridge, GL and Lester, RJ, 1990, Excitatory amino acids in the vertebrate central nervous system. Pharm. Rev. 40: 143-210.

Connors, BW and Kriegstein AR, 1986, Cellular physiology of the turtle visual cortex: Distinctive properties of pyramidal and stellate neurons. J. Neurosci. 6: 164-177.

Cosans, CE and Ulinski, PS, 1990, Spatial organization of axons in turtle visual cortex: Intralamellar and interlamellar projections. J. Comp. Neurol. 296: 548-558.

Davydova, TV and Goncharova, NV, 1979, Comparative characterization of the basic forebrain cortical zones in *Emys orbicularis* (Linnaeus) and *Testudo horsefieldi* (Gray). J. Hirnforsch. 20: 245-262.

Desan, PH, 1984, The organization of the cerebral cortex of the pond turtle, Pseudemys scripta elegans. Ph.D. Dissertation, Harvard University.

Desan, PH, 1988, Organization of the cerebral cortex in turtle. In: Schwerdtfeger, WK and Smeets WJAJ (eds) The Forebrain in Reptiles. Karger, Basel, pp 1 - 11.

Ebner, FF and Colonnier, M, 1978, Synaptic patterns in the visual cortex of turtles: An electron microscopic study. J. Comp. Neurol. 160: 51-80.

Fulbrook, JE, 1982, Motion sensitivity of optic nerve axons in turtle, *Pseudemys scripta elegans*. Ph.D. dissertation, University of Delaware.

Granda, AM and Fulbrook, JE, 1989, Classification of turtle retinal ganglion cells. J. Neurophysiol., 62: 723-737.

Granda, AM and Haden, KW, 1970, Retinal oil globule counts and distributions in two species of turtles: *Pseudemys scripta elegans* (Wied) and *Chelonia mydas* (Linnaeus). Vision Res. 10: 79-84.

Gusel'nikov, VI, Morenkov, ED and Pivovarov, AS, 1972, Unit responses of the turtle forebrain to visual stimuli. Neurosci. Behav. Physiol. 5: 235-242.

Gusel'nikov, VI and Pivovarov, AS, 1978, Postsynaptic mechanism of habituation of turtle cortical neurons to moving stimuli. Neurosci. Behav. Physiol. 9: 1-7.

Heller, SB and Ulinski PS, 1987, Morphology of geniculocortical axons in turtles of the genera *Pseudemys* and *Chrysemys*. Anat. Embryol. 175: 505-515.

Honore, TS, Davies, N, Drejer, J, Fletcher, EJ, Jacobsen, P, Lodge, D and Nielsen, FE, 1988, Quinoxalinediones: Potent competitive non-NMDA glutamate receptor antagonists. Science 241: 701-703.

Jensen, RJ and DeVoe, RD, 1983, Comparisons of directionally selective with other ganglion cells of the turtle retina: Intracellular recording and staining. J. Comp. Neurol. 217: 271-287.

Kolb, H., 1982, The morphology of the bipolar cells, amacrine cells and ganglion cells in the retina of the turtle *Pseudemys scripta elegans*. Phil. Trans. R. Soc. Lond. B 298: 355-393.

Kriegstein, AR, 1987, Synaptic responses of cortical pyramidal neurons to light stimulation in the isolated visual system. J. Neurosci. 7: 2488-2492.

Kriegstein, AR and Connors, BW, 1986, Cellular physiology of the turtle visual cortex: Synaptic properties and intrinsic circuitry. J. Neurosci. 6: 178-191.

Kunzle, H and Schnyder, H, 1984, Do retinal and spinal projections overlap within the turtle thalamus? Neurosci. 10: 161-168.

Landgren, S., Phillips, CG and Porter, R, 1962, Cortical fields of origin of the monosynaptic pyramidal pathways to some alpha motoneurones of the baboon's hand and forearm. J. Physiol. (Lond.) 161: 112-125.

Larson-Prior, LJ, Ulinski, PS, and Slater, NT, 1989, Excitatory amino acid receptor mediated transmission in geniculocortical and intracortical pathways in visual cortex. Soc. Neurosci. Abstr. 15: 947.

Lutz, PL, Rosenthal, M and Sick, TJ, 1985, Living without oxygen: turtle brain as a model of anaerobic metabolism. Mol. Physiol. 8: 411-425.

Marchiafava, PL, Weiler, R and Strettroi, E, 1983, Intracellular recording with horseradish peroxidase reveals distinct functional roles of retinal plexiform layers. In: Grinnel, AD and Moody, WJ, JR. (eds) The Physiology of Excitable Cells. A. Liss, New York, pp. 549-556.

Mazurskaya, PZ, 1972, Study of projection of the retina to the forebrain of the tortoise *Emys orbicularis*. J. Evol. Biochem. Physiol. 7: 532-536.

Mazurskaya, PZ, 1974, Organization of receptive fields in the forebrain of *Emys orbicularis*. Neurosci. Behav. Physiol. 7: 311-318.

Mulligan, KA and Ulinski, PS, 1990, Organization of geniculocortical projections in turtles: Isoazimuth lamellae in visual cortex. J. Comp. Neurol. 296: 531- 547.

Orrego, F, 1961, The reptilian forebrain. I. The olfactory pathways and cortical areas in the turtle. Arch. ital. Biol. 99: 425-445.

Parent, A, 1986, Comparative Neurobiology of the Basal Ganglia. John Wiley, New York.

Peterson, EH and Ulinski, PS, 1982, Quantitative studies of retinal ganglion cells in a turtle, *Pseudemys scripta elegans*. I. Number and distribution of ganglion cells. J. Comp. Neurol. 186: 17-42.

Pivovarov, AS and Trepakov, VV, 1972, Intracellular analysis of unit responses to afferent stimulation in the general and hippocampal cortex of turtles. Neurosci. Behav. Physiol. 6: 144-150.

Rainey, WT and Ulinski, PS, 1986, Morphology of neurons in the dorsal lateral geniculate complex in turtles of the genera *Pseudemys* and *Chrysemys*. J. Comp. Neurol. 253: 440-465.

Reiner, A., Brauth, SE, Kitt, CA and Karten, HJ, 1980, Basal ganglionic pathways to the tectum: Studies in reptiles. J. Comp. Neurol. 193: 565-589.

Reiner, A., Krause, JE, Eldred, WD and JF McKelvy, 1984, The distribution of substance P in the turtle nervous system: A radioimmunoassay and immunohistochemical study. J. Comp. Neurol. 226: 50 -75.

Rosenberg, AF and Ariel, M, 1990, Visual-response properties of neurons in turtle basal optic nucleus in vitro. J. Neurophysiol. 63: 1033-1045.

Sereno, MI, 1985, Tectoreticular pathways in the turtle, *Pseudemys scripta l.* I. Morphology of tectoreticular axons. J. Comp. Neurol. 233: 48-90.

Sereno, MI and Ulinski, PS, 1985, Tectoreticular pathways in the turtle, *Pseudemys scripta*. II. Morphology of tectoreticular cells. J. Comp. Neurol. 233: 91-114.

Sjostrom, AM and Ulinski, PS, 1985, Morphology of retinogeniculate terminals in the turtle, *Pseudemys scripta elegans*. J. Comp. Neurol. 238: 107-120.

Smith, LM, Ebner, FF and Colonnier, M, 1980, The thalamocortical projection in *Pseudemys* turtles: A quantitative electron microscopic study. J. Comp. Neurol. 190: 445-462.

Ulinski, PS, 1986, Organization of corticogeniculate projections in the turtle, *Pseudemys scripta*. J. Comp. Neurol. 254: 529-542.

Ulinski, PS, 1988, Functional architecture of turtle visual cortex. In: Schwerdtfeger, WK and Smeets, WJAJ (eds) The Forebrain of Reptiles. Karger, Basel, pp 151-161.

Ulinski, PS, 1990, Cerebral cortex in reptiles. In: Jones, EG and Peters, A (eds) Cerebral Cortex. Plenum, New York, Vol. 8, in press.

Ulinski, PS and Nautiyal, J, 1988, Organization of retinogeniculate projections in turtles of the genera *Pseudemys* and *Chrysemys*. J. Comp. Neurol. 276: 92-112.

Watkins, JC, Krogsgaard-Larsen, P and Honore, T, 1990, Structure-activity relationships in the development of excitatory amino acid receptor agonists and competitive antagonists. Trends Pharm. Sci. 11: 25-33.

Woodbury, PB and Ulinski, PS, 1986, Conduction velocity, size and distribution of optic nerve axons in the turtle, *Pseudemys scripta elegans*. Anat. Embryol. 174: 253-263.

Computational Significance of Lamination of the Telencephalon

Toru Shimizu and Harvey J. Karten

Department of Neurosciences, School of Medicine
University of California, San Diego
La Jolla, California 92093-0608, USA

Abstract. *How is sensory information processed in nonmammalian vertebrates? Is the mechanism of information processing similar to that in mammalian brains? In the nonmammalian telencephalon (birds and reptiles in particular), there are neuronal populations corresponding to cell groups in the neocortex of mammals in terms of connections, single unit-responses, and functions. Some areas of these populations in sauropsids (reptiles and birds), however, are arranged in a nonlaminar, rather than laminar fashion. The major nonlaminar multinucleate area, the dorsal ventricular ridge (DVR), is an intraventricular protrusion of cells within the core of the cerebral hemispheres. Comparative studies of visual representation with the DVR of birds and mammalian neocortex has emphasized the need for a reassessment of the functional roles of lamination of the mammalian neocortex. Thus, for example, complex visual and cognitive performance can be accomplished by neural circuits in the DVR of sauropsids equivalent to those of the neocortex but without laminar organization. If the same computational strategies in the DVR are employed in the mammalian forebrain, what is the functional significance of lamination in the neocortex? The role of neural circuits and laminar organization can be and should be differentiated in order to understand computational strategies in the neocortex.*

1. Introduction

Are there general rules of information processing shared by all vertebrates? Can ccmputational models constructed for the nonmammalian telencephalon be applied to the human neocortex, or vice versa?

We hope that this chapter will provoke the reader to reconsider his/her views regarding the functional nature of lamination for constructing computational models of neural circuits, particularly for the cerebral cortex. Lamination is a frequent feature of neural circuits in many central nervous system structures among vertebrates (and also invertebrates). Laminar configuration itself, however, is not necessarily the definitive feature of a given structure when the structure is compared in different classes of vertebrates. Homologous neuronal populations arranged in laminae in one species may be arranged in a nonlaminar configuration in another species. The degree of lamination of equivalent neurons may vary, from a nonlaminar configuration to a highly developed laminar configuration, depending on species. Such a differentiation of the degree of lamination might be related in part to phyletic differences of uncertain nature and, in part, to the degree to which the animal in question has utilized various structures in its interactions with its environment. For example, the vagal gustatory lobe of fishes, the optic tectum of nonmammals, or the olfactory bulbs of highly macrosomatic animals are much more developed than those of primates.

In the central nervous system of vertebrates, the mammalian isocortex is characterized by a laminar organization which has been considered unique to the mammalian telencephalon (*neo*-cortex). The greater degree of lamination of the neocortex has been associated with the *purportedly* higher cognitive abilities of mammals. This interpretation was justified as no apparent laminar structures corresponding to the neocortex had been described in the nonmammalian telencephalon. The common assumption regarding the nonmammalian brain used to be that nonmammals lacked any structures which could be comparable to the neocortex of mammalian brains, and that the *neo*cortex was *new* both in regard to its constituent neuronal populations and its laminar configuraton. Nonmammals had been considered to be "lower" vertebrates which had little capacity for the higher cognitive functions associated with the neocortex. Recent behavioral and anatomical studies, however, have shown that these assumptions regarding nonmammalian brains are untenable. The current view of the nonmammalian telencephalon is that the basic pattern of organization of the forebrain is common to all vertebrates (Northcutt, 1981). In nonmammals, birds and reptiles in particular, there are distinct neural groups equivalent to the mammalian neocortex. Sophisticated techniques to study the behaviors of animals have demonstrated that nonmammalian vertebrates have the significantly greater "higher cognitive functions" than was previously realized.

As a consequence of this realization - that lamination of the neocortex and other structures is a variable way of configuring similar (homologous) neurons - we may then ask what are the computational properties of circuits assembled in laminar versus nonlaminar arrangements. We have to differentiate those aspects of computational strategies which are unique to lamination (i.e., possible only found to occur within the neocortex) from those associated with nonlaminated homologous neural circuits operational in both neocortical *equivalent* systems as well as the neocortex itself.

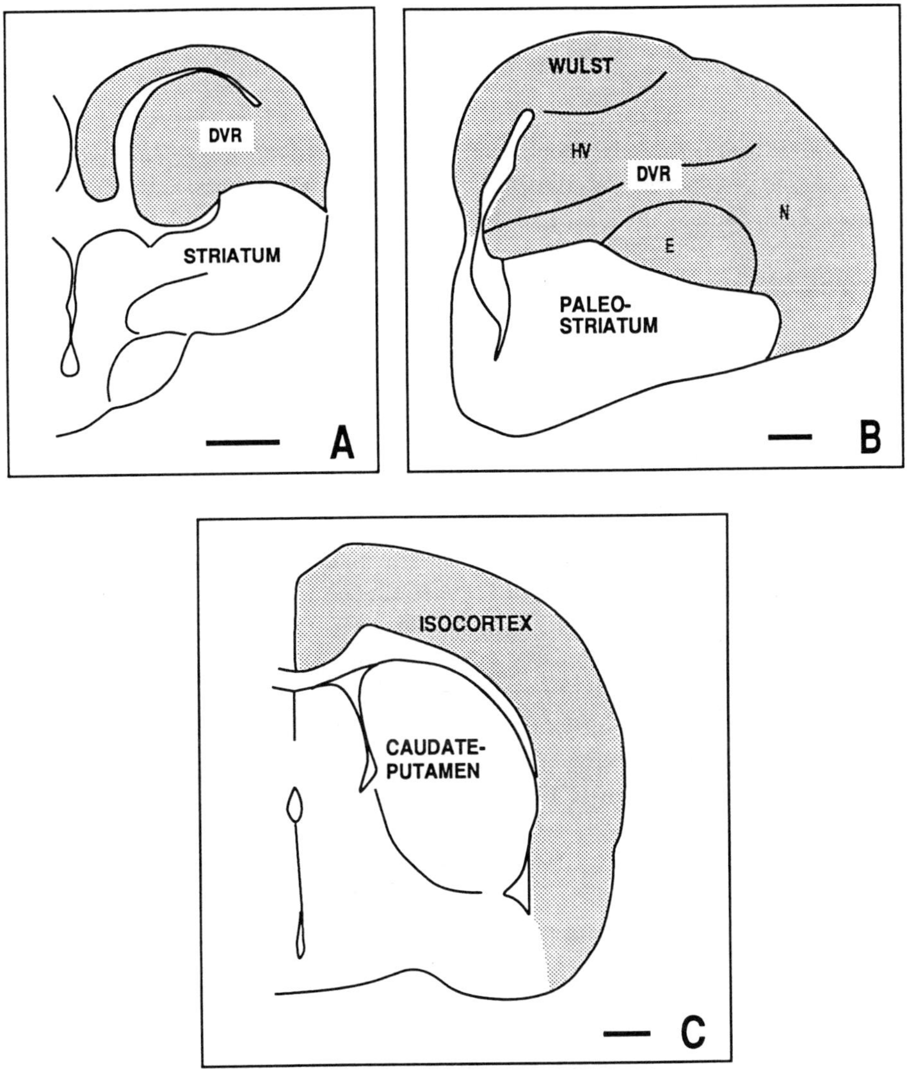

Figure 1. Organization of the telencephalon of a reptile, a bird, and a mammal. Shaded areas indicate the region of the telencephalon derived from the pallium. **A**, a transverse section of the brain of an adult tegu lizard. **B**, a transverse section of the brain of an adult pigeon. **C**, a transverse section of the brain of an adult rat. Each bar indicates 1 mm.

2. Neocortical Equivalents in the Nonmammalian Brain

In the nonmammalian telencephalon (birds and reptiles in particular), there are at least two major regions of neuronal populations corresponding to cell groups in the neocortex of mammals in terms of connections, single unit-responses, and functions: 1) the dorsal portion of the telencephalon; and 2) the lateral portion of the telencephalon. The dorsal portion of the telencephalon is very limited in size, though organized in a laminar fashion, and receives ascending visual input from thalamic nuclei via the thalamofugal pathway (i.e., not relayed by the tectum). This area is mainly comparable to the visual striate cortex of mammals, and is called the visual wulst in birds and the dorsal cortex in reptiles (see Ulinski and Margoliash, in press). The visual wulst has a number of characteristics in common with the mammalian striate visual cortex: 1) a well defined cytoarchitectonic laminar organization (Pettigrew, 1979), 2) afferent and efferent connections with subcortical visual structures (Karten et al., 1973; Bagnoli and Burkhalter, 1983; Miceli et al., 1987), and 3) retinotopic organization of the receptive fields (see Miceli et al., 1979). Although there are many similarities between the wulst and the mammalian striate cortex, significant differences in microcircuitry and chemistry also exist (Reiner and Karten, 1983; Shimizu and Karten, 1990).

The lateral portion of the telencephalon, though comparable to the neocortex in a number of major characteristics, is arranged in a nonlaminar, rather than laminar fashion. The major nonlaminar multinucleate area, the dorsal ventricular ridge (DVR), is an intraventricular protrusion of cells within the core of the cerebral hemispheres (see Figs. 1 and 2). The DVR had previously been misinterpreted as a unique hypertrophy of the "basal ganglia" because of its superficial apparent resemblance to the caudate putamen of mammals (Ariëns Kappers et al., 1936). Modern neuroanatomical studies, however, clearly indicate that the only a limited portion of the sauropsid telencephalon corresponds to the basal ganglia of mammals, and the DVR is most directly comparable to the pallium of mammals. Despite the lack of obvious lamination and pyramidal cells, the DVR has some features in common with the neocortex in mammals. Individual regions of the DVR of sauropsids receive sensory projections from modality-specific thalamic nuclei, conveying visual, auditory, and somatic sensory inputs to discrete domains within the DVR (see Ulinski, 1983 for a review). The pattern of the thalmotelencephalic connections are organized as follows (see Figs. 2 and 3): the visual center which receives input via the thalamic nuclei is located laterally in the DVR (Karten, 1979; Pritz, 1980), the somatosensory center is located centrally (Wild, 1987; Pritz, 1980), and the auditory center is located medially (Karten, 1967; Pritz, 1980, Wild et al., 1990). The visual center in the DVR receives a tectofugal visual pathway. This input is conveyed from retina to optic tectum, thence to the nucleus rotundus of the thalamus, and finally into a core nucleus of the DVR. While the visual wulst in birds and the visual cortex in reptiles are retinotopically organized, no obvious retinotopic organization has been found in the visual centers in the DVR or the nucleus rotundus. Neurons in the visual centers in the DVR

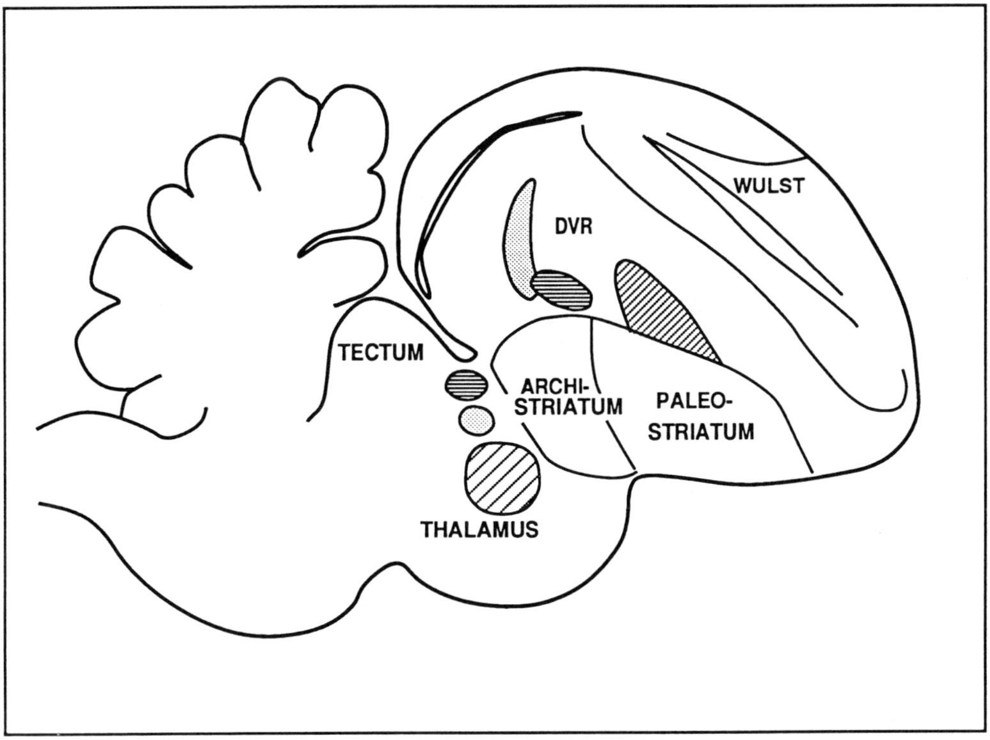

Figure 2. A schematic drawing of a sagittal section of a pigeon brain. Areas with dots, horizontal stripes and diagonal stripes indicate the regions of auditory, somatosensory, and visual systems, respectively.

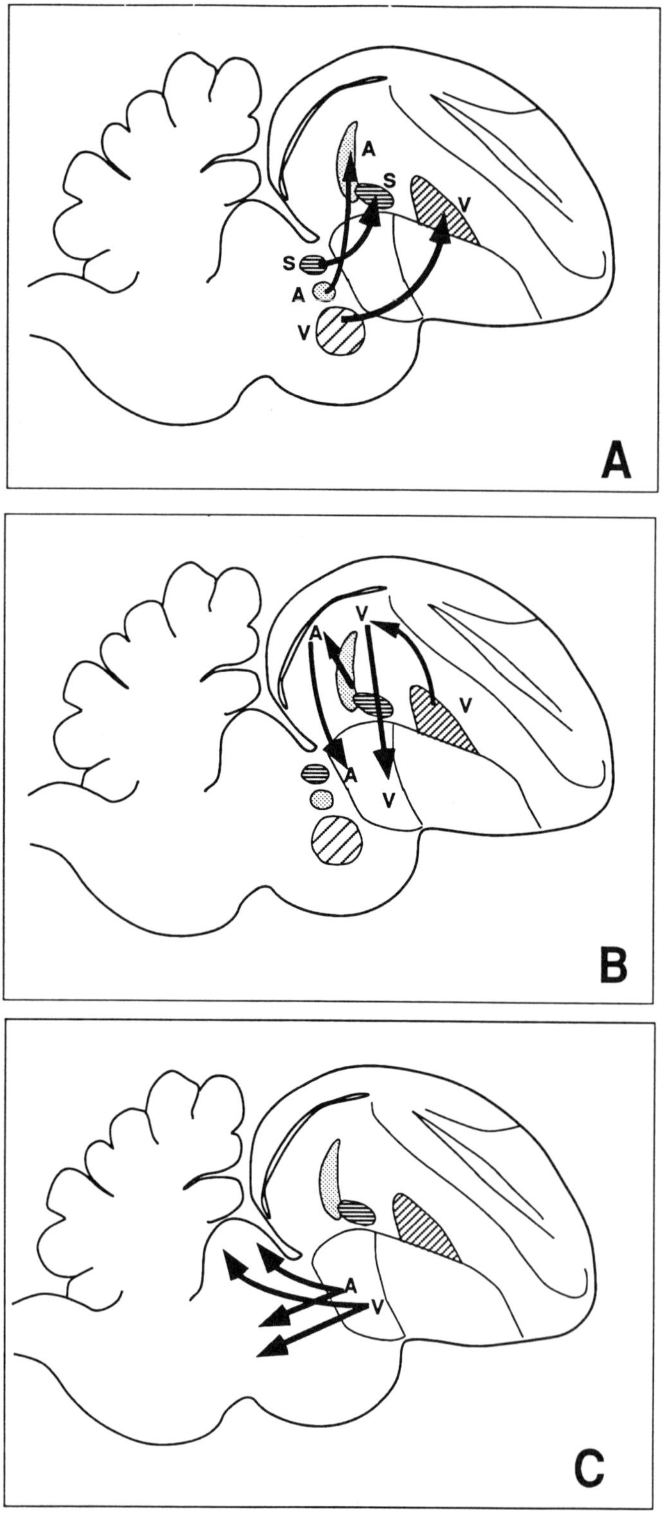

Figure 3 Schematic drawings of the flow of sensory information through the DVR. Fig. **A**, auditory (A), somatosensory (S), and visual (V) information from the sensory specific nuclei in the thalamus travel to the sensory-specific regions in the DVR of the telencephalon. Fig. **B**, sensory information travels within the DVR and reaches the archistriatum. Fig. **C**, information is sent back to various areas of the brainstem, including the thalamus, tectum and inferior colliculus.

have very large receptive fields that are sensitive to moving stimuli, often with clear directional preferences. The basic connectivity of the sauropsid tectofugal visual pathway, i.e., retinal ganglion cells project to the optic tectum, which projects to the nucleus rotundus, which projects to the DVR, is similar to the connection of the mammalian retino-collicular pathway to the extrastriate visual cortex. Thus, retinal input terminates in the superior colliculus, which projects to the pulvinar and/or lateral posterior nucleus of the thalamus, which in turn sends efferents to mainly layer IV of the extrastriatal cortices.

3. Intratelencephalic Connections in the Nonmammalian Brain

In birds, the tectal visual efferents project upon the nucleus rotundus, and then upon the core region of the DVR designated as the ectostriatum (Ec). The Ec appears to correspond to thalamic recipient neurons (layer IV) of the temporal visual cortex. Ritchie and Cohen (1979) and Shimizu et al. (1989) showed that the Ec projects on its surrounding region (including the neostriatal intermedium laterale; NIL) in the DVR, and this region in turn projects on another restricted area in the telencephalon, the archistriatum (Ai). (The readers should keep in mind that term neostriatum as used in birds and reptiles does not refer to the similarly named neostriatum of mammals). Brecha, Hunt and Karten (1976) demonstrated that Ai projects on laminae 11, 12, and 13 of the ipsilateral optic tectum. These sequential projections through the individual "nuclear" entities in the DVR may correspond to the interlaminar connections of the neocortex; i.e., from layer IV (Ec) to layers II and III (NIL), then to layers V and VI (Ai), then to the optic tectum (Karten, 1969; Nauta and Karten, 1970; Karten and Shimizu, 1989). Indeed, the pattern of projections is reminiscent of the neocortex, such as areas 18, 19, and 22 and the inferotemporal cortex of mammals.

Similar studies of the ascending auditory pathways demonstrated the existence of a distinct lemniscal pathway from the cochlear nuclei to the central nucleus of the inferior colliculus, to the nucleus ovoidalis thalami, and then upon Field L in the DVR of the telencephalon. The nucleus ovoidalis is comparable to the principal division of the medial geniculate nucleus of mammals, the main auditory thalamic cell group transmitting auditory input to the auditory cortex. Field L appears comparable to the specific thalamic-recipient neurons of the auditory cortex (Layer IV). The general pattern of subsequent connections of Field L within the DVR are similar in organization to those of the visual pathway (Bonke, Bonke and Scheich, 1979). Neurons in Field L project upon adjacent populations of interneurons (L1 and L3) then to another area in the DVR, the neostriatum caudale, and from the neostriatum to a distinct subset of the archistriatum (Aivm) clearly separated from the visual area (Aivl) of the archistriatum. The auditory Aivm projects upon a region surrounding the central nucleus of the inferior colliculus (Wild et al., 1990), an area concerned with spatial localization of auditory stimuli

(Knudsen et al., 1987). Analogous connections have been reported regarding the somatosensory system (Wild, 1987). Figure 3. summarizes the general pattern of organization of the visual, auditory and somatosensory systems of the DVR.

4. Cognitive Ability in Birds

For years, researchers have known that birds have well developed "cognitive" ability (Macphail, 1982, 1987). This ability used to be considered unique to mammals and/or primates and had been assumed to be functionally associated with the presence of neocortex. This cognitive capacity in birds includes the following examples:

1) Several species of birds, such as chikadees, store food (seeds and invertebrate prey) in widely dispersed cache sites (Sherry, 1985). An individual bird can store as many as several hundred food items in separate cache sites per day and several thousand through a year. Each cache site is usually in a novel location. Birds can accurately retrieve cached food after a few days or after many months.

2) A series of experiments by Pepperberg (e.g., 1987) has demonstrated that birds (e.g., African gray parrot) have highly cognitive abilities equivalent to those found in nonhuman primates. Such capacities include categorization of objects, labeling of particular numerical quantities, comprehension of the abstract concepts of same-different, and understanding of object permanence.

3) Vaughan and Greene (1984) used as stimuli more than 300 diverse photographic slides of outdoor scenes including trees, water, houses, flowers, and so on. Pigeons were required to learn to categorize each photograph according to the two groups the experimenters had created. The animals were not only capable of memorizing the individual stimulus according to this categorization, but also remembered the categorization accurately after two years without further exposure to the stimuli.

4) Blough (1982) demonstrated that pigeons can learn to differentiate alphabetic letters. The author analyzed the errors made during learning by pigeons and scaled letters of similarity in a multidimensional space. The results turned out to be very similar to perception by human subjects.

What is the neural substrate of such cognitive abilities in birds? In the visual system, the tectofugal pathway and its telencephalic targets in the nonlaminated DVR seem to play an important role in many types of visual information processing. In birds, the tectofugal visual system is generally highly developed. Structures related to the tectofugal pathway, such as the optic tectum, the nucleus rotundus, and the visual center in the DVR, are all well-developed entities. In some birds, the thalamofugal pathway to the wulst is also well developed. The increase of the wulst size is often associated with the higher ability of binocular vision and stereopsis. Thus the wulst seems to be well developed in the birds that have large binocular fields including owls (Pettigrew, 1986).

Lesions of the tectofugal pathway, including the optic tectum, the nucleus rotundus, or the ectostriatum, cause severe deficits in standard visual discrimination tasks such as pattern, color and brightness (Hodos, 1976). Thus we can assume that the ectostriatum and/or DVR are closely related to the visual processing of incoming information. This is indirectly supported by the fact that lesions in the laminated wulst do not have devastating effects on visual information processing, but rather more indirect and often subtle effects (Shimizu and Hodos, 1989).

5. Consequences of Lamination

Complex visual and cognitive performance appear to be accomplished by neural circuits in the DVR of sauropsids. These circuits are anatomically and functionally equivalent to those of the neocortex, but without laminar organization. These recent comparative studies of visual and auditory representation in the DVR of birds and the mammalian neocortex have emphasized the need for a reassessment of the functional roles of lamination. What are the benefits of lamination in the neocortical system? The discussion above does not argue against the fact that lamination is of significance to many sensory, motor, and cognitive functions. Well developed laminar organization is found in various brain structures other than the telencephalon, including structures as diverse as the retina, cerebellum, hippocampus, optic tectum (e.g., birds), vagal lobes (e.g., fishes), and the olfactory bulbs (e.g., rodents). There is apparently some correlation between the degree of development of lamination and specialized functions. Our suggestion, however, is that two distinct attributes of cortical structures, i.e. neural circuitry and lamination, may contribute to the functional capacities of the cortex. Each may provide unique contributions to cortical function. The equivalent neural circuits, whether laminated or not, may be responsible for those most general operations concerned with cognition and behavior that are observed equally in mammals and nonmammals. While the laminar organization, more common among mammals, may confer upon mammals their presumed distinctive functional capacity, the actual operational benefits of lamination have not been adequately identified. Previous studies have failed to distinguish the role of each of these separate aspects of cortical organization, and have generally assumed that the two properties of connections and lamination were inseparable.

In general, lamination refers to the presence of horizontal bands of a uniform nature juxtaposed against a horizontal band of a different character. Most commonly, one layer consists of neurons, and the second layer may consist of a cell-free synaptic region (e.g., retina) or the second layer may consist of cells of different character (e.g., the neocortex).

In the simplest condition, all cells in a single layer possess similar morphology, biochemical characteristics (e.g., transmitters, modulators, receptors, and second messengers), and have similar afferent and efferent connections. An example of such uniformity is found in cerebellar Purkinje cells.

There are at least two major qualities that must be considered in the discussion of benefits associated with lamination: 1) the extent of interactions between layers; and 2) the extent of interactions between adjacent vertically associated groupings, or columns.

1) Interaction between layers

The degree of lamination might be correlated to the efficiency of the local circuits. The interconnections between the clustered entities of nonlaminated organization (such as the DVR) are presumably restricted to axo-somatic, -dendritic, and -axonic synapses. The lack of laminar apposition of these nucleate populations limits the interlaminar dendro-dendritic and other short connections. On the other hand, short connections appear to be a fundamental characteristic of the neocortex. The laminar organization may be advantageous, particularly when information must be most efficiently transferred to operationally critical components across layers perpendicular to the surface (i.e., columnar organization). In fact, in addition to the many incoming and outgoing fibers in the neocortex, intracortical processes of neurons are vertically oriented.

These interlaminar interactions may also facilitate the efficiency of control of the subjacent neural circuits. In the neocortex, a layered neuronal array may contain a representation of a topographical map. Thus, the close apposition and connection between arrayed layers may provide a basis for potentially fast point-to-point operational circuitry. Such congruent mapping provides a substrate for a more highly ordered control of output neurons by discrete populations of sensory neurons. For instance, in the neocortex, this permits more direct control of the output neurons of layer V and VI in relationship to the sensory inputs to layer IV. The thalamo-cortical fibers often send collaterals to layer V and/or VI as well as layer IV. Such circuitry may be useful in the synchrony of input and output of the column.

2) Interactions between columns

The cortical areas are not uniformly organized. For instance, the thalamic input from the two eyes constitute the nonoverlapping stripe patterns in layer IV (occular dominant columns). Similar pattern of stripes can be found in the barrel field of the somatosensory areas of rodents. Based on such characteristics of the arrangement, one may speculate that the cortex is organized in a vertical "modules" (Mountcastle, 1978). It is yet unknown, however, how interactions between these information processing columns occur. Recent findings by Singer and his co-workers (Gray et al., 1989) indicated that such interactions may be correlated to synchronous firings of neurons in widely separated columns. According to the authors, the cross-columnar synchronization provides a mechanism for the formation of neuronal assemblies representing objects in the visual field. If such neural oscillation turns out to be the substrata for these events, it would imply that temporal coding is used between columns in the cortex.

6. Conclusion

The dichotomy of laminar organization and neural circuitry for possible cortical processing has provided the basis for a reassessment of sensory processing in the telencephalon of nonmammalian amniotes. This dichotomy may also be important for understanding the evolutionary origins of the mammalian neocortex. Two distinctive events might have occurred in the evolution of the cortex (Karten and Shimizu, 1989). One event consists of a defined microcircuitry. The second major event - presumably independent of the development of specific populations - is assemblage of these populations into a laminar structure as in the cortex. Thus, these neuronal populations may be found in "clusters" in nonmammalia, or as individual laminae in the neocortex of mammals. Further discussion, however, will be explored when more information about anatomical, biochemical, and functional aspects of the nonmammalian and mammalian telencephalon become available.

Acknowledgements

Supported by ONR Contract N00014-88-0504, NINDS Grant NS24560, and NEI Grant EY06890-06. The authors are grateful to Ms. Ellanora B. Watelet for secretarial assistance.

Bibliography

Ariëns Kappers, C. U., Huber, G. C. and Crosby, E. C., 1936, The comparative Anatomy of the Nervous System of Vertebrates, Including Man. Republished in 1960, New York: Hafner.

Bagnoli, P. and Burkhalter, A., 1983, Organization of the afferent projections to the Wulst in the pigeon. Journal of Comparative Neurology, 214: 103-113.

Blough, D., 1982, Pigeon perception of letters of the alphabet. Science, 218: 397-398.

Bonke, B. A., Bonke, D., and Scheich, H., 1979, Connectivity of the auditory forebrain nuclei in the guinea fowl *(Numida meleagris)*. Cell Tissue Research, 200: 101-121.

Brecha, N., Hunt, S. P. and Karten, H. J., 1976, Relations between the optic tectum and basal ganglia in the pigeon. Society for Neuroscience Abstract, 1, 95.

Gray, C. M., König, P. Engel, A. K. and Singer, W., 1989, Oscillatory responses in cat visual cortex exhibit inter-columnar synchronization which reflects global stimulus properties. Nature. 338: 334-337.

Hodos, W., 1976, Vision and the visual system: A bird's eye-view. In J. M. Sprague & A. M. Epstein (Eds.), Progress in Psychobiology and Physiological Psychology (pp.29-62). New York: Academic Press.

Karten, H. J., 1969, The organization of the avian telencephalon and some speculations on the phylogeny of the amniote telencephalon. In C. Noback & J. Petras (Eds.), Comparative and evolutionary aspects of the vertebrate central nervous system. Annals New York Academy of Sciences, 167: 146-179.

Karten, H. J., 1967, Telencephalic projections of the nucleus ovoidalis in the pigeon (Columba livia). Anatomical Record, 157: 268.

Karten, H. J. and Shimizu, T., 1989, The origins of neocortex: Connections and lamination as distinct events in evolution. Journal of Cognitive Neuroscience, 1: 291-301.

Karten, H. J., 1979, Visual lemniscal pathways in birds. In A. M. Granda & J. H. Maxwell (Eds.), Neural mechanisms of behavior in the pigeon (pp. 409-430). New York: Plenum Press.

Karten, H. J., Hodos, W., Nauta, W. J. H., and Revzin, A. M. (1973). Neural connections of the "visual wulst" of the avian telencephalon. Experimental studies in the pigeon (*Columba livia*) and owl (*Speotyto cunicularia*). Journal of Comparative Neurology, 150: 253-277.

Knudsen, E. I., du Lac, S., and Esterly, S. D., 1987, Computational maps in the brain. Annual Review of Neuroscience, 10: 41-65.

Macphail, E. M., 1982, Brain and intelligence in vertebrates. New York: Oxford University Press.

Macphail, E. M., 1987, The comparative psychology of intelligence. Behavioral and Brain Sciences, 10: 645-695.

Miceli, D., Gioanni, H., Repérant, J. and Peyrichoux, J., 1979, The avian visual wulst: I. An anatomical study of afferent and efferent pathways. II An electrophysiological study of the functional properties of single neurons. In A. M. Granda & J. H. Maxwell (Eds.), Neural mechanisms of behavior in the pigeon (pp. 223-254). New York: Plenum press.

Miceli, D., Repérant, J., Villalobos, J. and Dionne, L., 1987, Extratelencephalic projections of the avian Wulst. A quantitative autoradiographic study in the pigeon *Columba livia*. Journal für Hirnforschung, 28: 45-57.

Mountcastle, V. B., 1978, An organizing principle for cerebral function: The unit module and the distributed system. In G. M. Edelman & V. B. Mountcastle (Eds.), The mindful brain (pp. 7-50). Cambridge, MA: MIT press.

Nauta, W. J. H. and Karten, H. J., 1970, A general profile of the vertebrate brain with sidelights on the ancestry of the cerebral cortex. In F. O. Schmitt (Ed.), The Neurosciences: Second Study Program. New York: Rockefeller Press.

Northcutt, R. G., 1981, Evolution of the telencephalon in nonmammals. Annual Review of Neuroscience, 4: 301-350.

Pepperberg, I, M., 1987, Interspecies communication: A tool for assessing conceptual abilities in the African Grey parrot *(Psittacus erithacus)*. In G. Greenberg and E. Tobach (Eds.), Language, cognition and consciousness: Integrative levels (pp.31-56). Hillsdale, NJ: Erlbaum.

Pettigrew, J. D., 1979, Binocular visual processing in the owl's telencephalon. Proceedings of the Royal Society (London), Series B, 204: 435-454.

Pritz, M. B., 1980, Parallels in the organization of auditory and visual systems in crocodiles. In S. O. E. Ebbesson (Ed.), Comparative neurology of the telencephalon (pp. 331-342). New York: Plenum Press.

Reiner, A. and Karten, H. J., 1983, The laminar source of efferent projections from the avian Wulst. Brain Research, 275: 349-354.

Ritchie, T. C. and Cohen, D. H., 1979, The avian tectofugal visual pathway: Projections of its telencephalon target ectostriatal complex. Society for Neuroscience Abstract, 2: 119.

Sherry, D. F., 1985, Food storage by birds and mammals. Advances in the Study of Behavior, 15: 153-188.

Shimizu, T. and Hodos, W., 1989, Reversal learning in pigeons: effects of selective lesions of the wulst. Behavioral Neuroscience, 103:262-272.

Shimizu, T. and Karten, H. J., 1990, Immunohistochemical analysis of the visual wulst of the pigeon *(Columba livia)*. Journal of Comparative Neurology, 300: 2-25.

Shimizu, T., Woodson, W., Karten, H. J., and Schimke, J. B., 1989, Intratelencephalic connections of the visual areas in birds *(Columba livia)*. Society for Neuroscience Abstract, 15: 1398.

Ulinski, P. S., 1983, Dorsal ventricular ridge, New York: John Wiley & Sons.

Ulinski, P.S. and D. Margoliash, in press, Neurobiology of the reptile-bird transition In E. G. Jones and A. Peters (Eds.)., Cerebral Cortex, vol. 8., Evolution and Comparative Anatomy of Cerebral Cortex. New York: Plenum Press.

Vaughan, W., Jr and Greene, S. L., 1984, Pigeon visual memory capacity. Journal of Experimental Psychology: Animal Behavior Processes, 10: 256-271.

Wild, J. M., 1987, The avian somatosensory system: Connections of regions of body representation in the forebrain of the pigeon. Brain Research, 412: 205-223.

Wild, J. M., Frost, B. J., and Karten, H. J., 1990, Some aspects of the organization of the auditory forebrain and midbrain in the pigeon. Society for Neuroscience Abstract. (In press).

Information Processing In The Temporal Lobe Visual Cortical Areas Of Macaques

Edmund T. Rolls
University of Oxford, Department of Experimental Psychology
Oxford OX1 3UD, England

Visual pathways project by a number of cortico-cortical stages from the primary visual cortex, the striate cortex, until they reach the inferior temporal visual cortex, area TE (Cowey, 1979; Gross, 1973; Desimone and Gross, 1979; Seltzer and Pandya, 1978; Rolls, 1990a,d; see Fig. 1). The inferior temporal cortex is among the highest unimodal visual cortical areas known. In addition, there are a set of cortical areas in the anterior part of the superior temporal sulcus which also receive visual inputs (Seltzer and Pandya, 1978; Baylis, Rolls and Leonard, 1987) (see Fig. 1). Of these areas, TPO receives inputs from temporal, parietal and occipital cortex; PGa and IPa from parietal and temporal cortex; and TS and TAa primarily from auditory areas (Seltzer and Pandya, 1978).

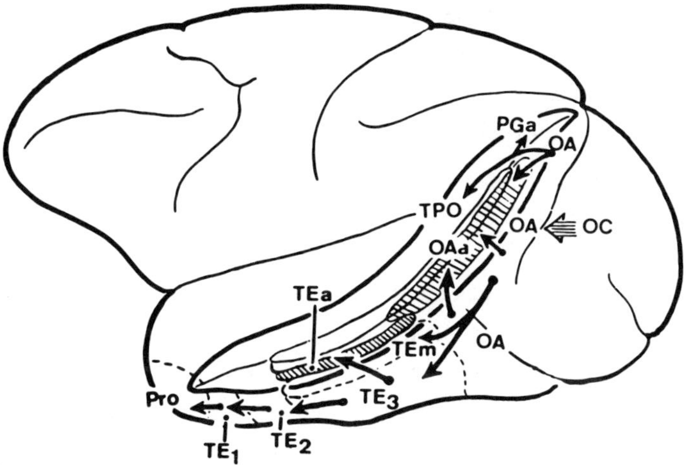

Fig. 1. Forward projections in the temporal lobe from visual areas of the macaque monkey. OC is the primary (striate) cortex, the TE areas are the inferior temporal visual cortex, and the superior temporal sulcus has been opened. (From Seltzer and Pandya, 1978).

In order to investigate the information processing being performed by these parts of the temporal lobe, the activity of single neurons has been analysed in each of these areas in the rhesus monkey (Baylis, Rolls and Leonard, 1987). There is considerable specialization of function in these different temporal cortical areas concerned with vision. For example, areas TPO, PGa and IPa are multimodal, with neurons which respond to visual, auditory and/or somatosensory inputs; the inferior temporal gyrus and adjacent areas (TE3,TE2,TE1,TEa and TEm) are primarily unimodal visual areas; areas in the cortex in the anterior and dorsal part of the superior temporal sulcus (e.g. TPO, IPa and IPg) have neurons specialized for the analysis of moving visual stimuli; and neurons responsive primarily to faces are found predominantly in areas TPO, TEa and TEm (Baylis, Rolls and Leonard, 1987).

The analysis of visual information taking place in some of these areas will be considered in this paper, to highlight the ways in which information is represented in these areas, and to consider ways in which some of the computations may be being performed. One of the populations of neurons considered to illustrate these points is that with responses selective for faces (see Rolls, 1984,1990a,b,1991). The responses of these neurons to faces are selective in that they are 2-10 times as large to faces as to gratings, simple geometrical stimuli, or complex 3-D objects (Perrett, Rolls and Caan, 1982; Rolls, 1984; Baylis, Rolls and Leonard, 1985, 1987).

Ensemble encoding of object identity

In experiments to investigate how a particular object in the environment might be represented by the activity of neurons in temporal lobe visual cortical areas, the responses of the face selective neurons to the faces of different individuals have been examined. It has been shown that in many cases (77% of one sample), these neurons are sensitive to differences between faces (as shown by an analysis of variance (Baylis, Rolls and Leonard, 1985)). However, each neuron does not respond only to one face. Instead, each neuron has a different relative response to each of the members of a set of faces. To quantify how finely these neurons were tuned to the faces of particular individuals, a measure derived from information theory, the breadth of tuning metric developed by Smith and Travers (1979), was calculated. This is a coefficient of entropy (H) for each cell which ranges from 0.0, representing total specificity to one stimulus, to 1.0 which indicates an equal response to the different stimuli. The breadth of tuning of the majority of the neurons analyzed was in the range 0.8 - 1.0. It was thus clear from this and other quantitative measures of the tuning of these face-responsive neurons that they did not respond only to the face of one individual, but that instead typically each neuron responded to a number of faces in the stimulus set (which included 5 different faces) (Baylis, Rolls and Leonard, 1985). Such evidence shows that only across a population or ensemble of such cells is information conveyed which would be useful in making different behavioral responses to different faces.

It may also be noted that it is unlikely that there are further processing areas beyond those described where ensemble coding changes into grandmother cell encoding, in that anatomically there do not appear to be a whole further set of visual processing areas present in the brain; and from the temporal lobe visual areas such as those described, outputs are taken to limbic and related regions such as the amygdala and via the entorhinal cortex the hippocampus. Indeed, tracing this pathway onwards, Leonard, Rolls, Baylis and Wilson (1985) have found a population of neurons with face-selective responses in the amygdala, and in the majority of these neurons, different responses occur to different faces, with ensemble not unique coding still being present.

Given this empirical evidence for ensemble encoding, that is for distributed representation, in these parts of the visual system, we can ask what the advantages are of this type of coding. It has been suggested (Rolls, 1987, 1989a, 1990a,c,d) that the advantages arise from the fact that the outputs of these cortical areas reach structures such as the amygdala which are involved with making associations between visual representations of objects and with primary reinforcing stimuli such as the taste of food using pattern association neuronal networks; and such as the hippocampus which are involved in recognition and episodic memory using auto-association neuronal networks. The advantages of distributed representations for the inputs to such association neuronal networks include pattern completion, generalization, and graceful degradation (or graceful performance with an imperfect specification of the connectivity during development) (see Rolls, 1987, 1989a). The type of tuning found in at least some temporal lobe cortical visual areas is thus seen as a delicate compromise between very fine tuning, which has the advantage of increasing the number of patterns which can be stored with low interference in some neuronal networks of the type which may operate in the brain but the disadvantage of losing the emergent properties of storage in neuronal networks, and broad tuning, which has the advantage of allowing the emergent properties of neuronal networks to be realised but of leading to interference between the different memories stored in the network (see Rolls, 1987, 1989a). Fine tuning implies a sparse representation of patterns in input vectors to neuronal networks, that is a representation in which relatively few neurons are active for any one input pattern. Taking the case of associative and autoassociative memories, we have formalized the above arguments, and shown that sparse encoding increases the number of patterns which can be stored in many variations of such networks, for example in the case of networks with neurons with graded but non-linear responses and with covariance learning rules or with learning rules with non-linearity in the post-synaptic contribution such as might be implemented by NMDA receptors (Treves and Rolls, 1990; Treves, 1990a,b; Treves and Rolls, in preparation).

The development of specificity of the neuronal responses

Given the fundamental importance of a computation which results in relatively finely tuned neurons which across ensembles but not individually specify objects in the environment, we have investigated whether experience plays a role in determining the selectivity of single neurons which respond to faces. The hypothesis being tested was that visual experience might guide the formation of the responsiveness of neurons so that they provide an economical and ensemble-encoded representation of items actually present in the environment. To test this, we investigated whether the responses of the face-selective neurons were at all altered by the presentation of new faces which the monkey had never seen before. It might be for example that the population would make small adjustments in the responsiveness of its individual neurons, so that neurons would be provided with filter properties which would enable the population as a whole to discriminate between the faces actually seen. We thus asked whether when a set of totally novel faces was introduced, the responses of these neurons were fixed and stable from the first presentation, or instead whether there was some adjustment of responsiveness over repeated presentations of the new faces. First, it was shown for each neuron tested that its responses were stable over 5-15 iterations of a set of familiar faces. Then a set of new faces was shown in random order, and the set was repeated with a new random order over many iterations. It was found that some of the neurons studied in this way altered the relative degree to which they responded to the different members of the set of novel faces over the first few (1-2) presentations of the set (Rolls, Baylis, Hasselmo and Nalwa, 1989). Thus there is now some evidence from these experiments that the response properties of neurons in the temporal lobe visual cortex are modified by experience, and that the modification is such that when novel faces are shown, the relative responses of individual neurons to the new faces alter. It is hypothesized that alteration of the tuning of individual neurons in this way results in a good discrimination over the population as a whole of the faces known to the monkey.

In searching for models to help understand better the representation of information across ensembles of neurons, and how optimal categorisation of the inputs actually presented is developed by the population, we have found neuronal net models with competition to be helpful. These, and the way in which their operation may be guided by backprojections in the cerebral cortex, are considered below and elsewhere (Rolls, 1989a,1989b).

A Neuronal Representation Showing Invariance

One of the major problems which must be solved by a visual system is the building of a representation of visual information which allows recognition to occur relatively independently of size, contrast, spatial frequency, angle of view, etc. To investigate whether these neurons in the temporal lobe visual cortex are at a stage of processing where such invariance is being represented in the responses of neurons, the effect of such transforms of the visual image on the responses of the neurons has been investigated.

To investigate whether the responses of these neurons show some of the perceptual properties of recognition including tolerance to isomorphic transforms (i.e. in which the shape is constant), the effects of alteration of the size and contrast of an effective face stimulus on the responses of these neurons were analysed quantitatively in macaque monkeys (Rolls and Baylis, 1986). It was shown that the majority of these neurons had responses which were relatively invariant with respect to the size of the stimulus. The median size change tolerated with a response of greater than half the maximal response was 12 times. Also, the neurons typically responded to a face when the information in it had been reduced from 3D to a 2D representation in gray on a monitor, with a response which was on average 0.5 that to a real face. Another transform over which recognition is relatively invariant is spatial frequency. For example, a face can be identified as a line drawing (in which only high spatial frequencies are present), and when it is blurred (when it contains only low spatial frequencies). It has been shown that if the face images to which these neurons respond are low-pass filtered in the spatial frequency domain (so that they are blurred), then many of the neurons still respond when the images contain frequencies only up to 8 cycles per face. Similarly, the neurons still respond to high-pass filtered images (with only high spatial frequency edge information) when frequencies down to only 8 cycles per face are included (Rolls, Baylis and Leonard, 1985; Rolls, Baylis and Hasselmo, 1987). Face recognition shows similar invariance with respect to spatial frequency. These neurons also show considerable invariance with respect to retinal translation (Azzopardi and Rolls, 1989).

An Object-Centered Representation of Visual Information

The findings that the responses of these neurons show considerable invariance over changes in the size, contrast and retinal translation of face stimuli is evidence that the encoding is not in retinal coordinates. Alternatives are that it is in viewer-centered or object-centered coordinates.

Some neurons with responses selective for faces only respond if the face is moving (Perrett et. al., 1985). It is noted that encoding of gestures is likely to be important in the social behavior of primates. We took advantage of the fact that these neurons respond to moving faces to investigate whether the encoding of faces by these neurons is in viewer-centered or object-centered coordinates (Hasselmo, Rolls, Baylis and Nalwa, 1989). For 10 neurons it has been shown that the neuron responds to particular movements which can only be described in object-centered coordinates. For example, four neurons responded vigorously to a head undergoing ventral flexion, irrespective of whether the view of the head was full face, of either profile, or even of the back of the head. These different views could only be specified as equivalent in object-centered coordinates. Further, for all of the neurons that were tested in this way, the movement specificity was maintained across inversion, responding for example to ventral flexion of the head irrespective of whether the head was upright or inverted. In this procedure, retinally encoded or viewer-centered movement vectors are reversed, but the object-centered description remains the same. It was of interest that the neurons tested generalized across different heads performing the same movements.

Further evidence supporting the hypothesis that some of the neurons in this region use object-centered descriptions is that their selectivity between the faces of different individuals is maintained across anisomorphic transforms of the stimulus. For example, some neurons reliably responded differently to the faces of two different individuals independently of viewing angle. However, in most cases (16/18 neurons) although the identity of the face was reflected in the neuronal response, the response was not perfectly view-independent, and viewing angle also influenced the response. It is possible that these latter neurons represent an intermediate stage in the computation of object-centered descriptions (Hasselmo, Rolls, Baylis and Nalwa, 1989).

The finding that some neurons in this region of temporal visual cortex have object-based rather than viewer-centered responses provides a form of invariance which is very useful in providing an input to associative memories in for example the hippocampus and amygdala (see Rolls, 1987, 1989a), for it means that the associative memories can store a single rather than multiple representations of the object, and moreover can respond appropriately when a different view of an object is seen.

The functions of backprojections in neuronal networks in the neocortex in the storage and recall of visual memories

In order to form new (declarative) memories, including memories of visual objects such as faces, limbic structures such as the hippocampal system and perhaps the amygdala are necessary (Mishkin, 1982; Squire, 1986; Squire and Zola-Morgan, 1988; Rolls, 1990a-c). (Declarative memories can be described by humans, in contrast to procedural memories for motor and related skills which are not dependent on limbic strucures.) Although these limbic structures are necessary for the formation of these memories, the memories are not stored in the limbic strucures in the long term but instead are stored in the neocortical areas which perform for example visual processing, in that old memories are at least partially spared in patients with damage to limbic structures. High order neocortical areas such as the inferior temporal visual cortex send projections to these limbic strucures, and in turn receive backprojections from these limbic structures (see Rolls, 1989a, 1990c). Because of the importance of these backprojections in the storage of visual information in the neocortex, their connections to the visual association cortex, and their functions in the operation of the visual association cortex, are considered next.

The forward and backward projections which will be considered are shown in Fig. 2 (for further anatomical information see Peters and Jones, 1984).

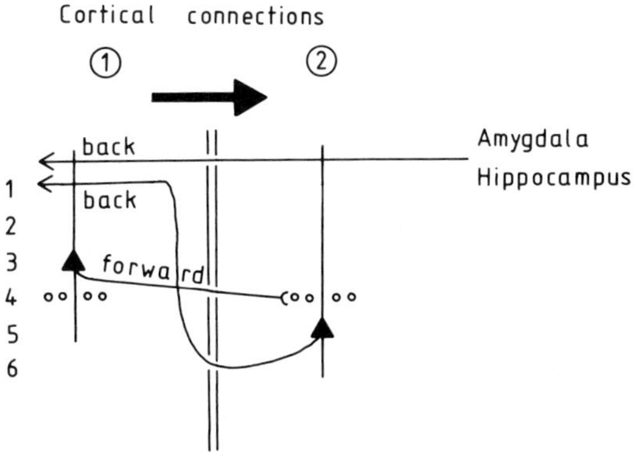

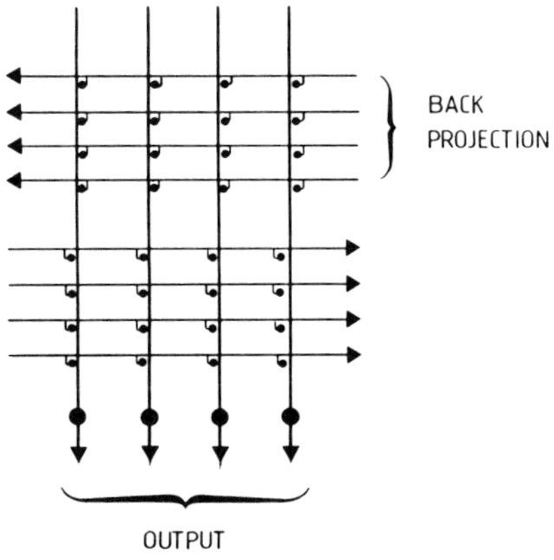

Fig. 2 (left). Schematic diagram of forward and backward projections in the neocortex. The arrow from left to right connecting the cortical area 1 to 2 shows the forward direction. The superficial pyramidal cells (triangles) in layers 2 and 3 project forward to terminate in layer 4 of the next cortical area. The deep pyramidal cells in the next area project back to layer 1 of the preceding cortical area, in which there are apical dendrites of pyramidal cells. The hippocampus and amygdala also are the source of backprojections which end in layer 1. Spiny stellate cells are represented by small circles in layer 4. See text for further details. Fig. 3 (right). The architecture used to simulate the properties of backprojections. The forward ('input stimulus') and backprojected axons make Hebb modifiable synapses onto the same set of vertical dendrites, which represent cortical pyramidal cells.

In primary sensory cortical areas, the main extrinsic 'forward' input is from the thalamus, and ends in layer 4, where synapses are formed onto spiny stellate cells. These in turn project heavily onto pyramidal cells in layers 3 and 2, which in turn send projections forward to terminate strongly in layer 4 of the next cortical layer (on small pyramidal cells in layer 4 or on the basal dendrites of the layer 2 and 3 (superficial) pyramidal cells). (Although the forward afferents end strongly in layer 4, the forward afferents have some synapses also onto the basal dendrites of the layer 2 pyramidal cells, as well as onto layer 6 pyramidal cells and inhibitory interneurons). Inputs reach the layer 5 (deep) pyramidal cells from the pyramidal cells in layers 2 and 3 (Martin, 1984), and it is the deep pyramidal cells which send backprojections to end in layer 1 (predominantly) of the preceding cortical area (see Fig. 2), where there are apical dendrites of pyramidal cells. In addition to these cortoco-cortical backprojections, there are also axons and terminals in layer 1 in later areas in the cortical hierarchy including the inferior temporal visual cortex from the amygdala and (via the subiculum, entorhinal cortex and parahippocampal cortex) from the hippocampal formation (see Fig. 2) (Van Hoesen, 1981; Turner, 1981; Amaral and Price, 1984; Amaral, 1986, 1987).

Now the amygdala and hippocampus are stages of information processing at which the different sensory modalities (such as vision, hearing, touch, and taste for the amygdala) are brought together, so that correlations between inputs in different modalities can be detected in these regions, but not at prior cortical processing stages in each modality, as these cortical processing stages are unimodal. Now, as a result of bringing together the two modalities, significant correspondences between the two modalities can be detected. One example might be that a particular visual stimulus is associated with the taste of food. Another example might be that another visual stimulus is associated with painful touch. Thus at these limbic stages of processing, but not before, the significance of for example visual and auditory stimuli can be detected and signalled, and sending this information back to the neocortex thus can provide a signal which indicates to the cortex that information should be stored, but even more than this, provides a signal which could help the neocortex to store the information efficiently. It is proposed below that one function of backprojections from these limbic strucures is to guide the formation of long-term memories in structures such as the inferior temporal visual cortex so that neuronal networks in these cortical areas need not operate in an unsupervised way, but can benefit from feedback from further on in processing.

The way in which backprojections could assist learning in the cortex can be considered using the architecture shown in Fig. 3. The (forward) input stimulus occurs as a vector applied to (layer 3) cortical pyramidal cells through modifiable synapses in the standard way for a competitive net (see Rolls, 1989a). The lower set of synapses on the pyramidal cells would then start by competitive learning to set up representations on the lower parts of these neurons which would represent correlations in the input information space, and could be said to correspond to features in the input information space, where a feature is defined simply as the representation of a correlation in the input information space.

Consider now the application of one of the (forward) input stimulus vectors with the conjunctive application of a pattern vector via the backprojection axons with terminals in layer 1. Given that all the synapses in the matrix start with random weights, some of the pyramidal cells will by chance be strongly influenced both by the (forward) input stimulus and by the backprojecting vector. These strongly activated neurons will then compete with each other as in a standard competitive net, to produce contrast enhancement of their firing patterns. Next, Hebbian learning takes place as in a competitive net, with the addition that not only the synapses between forward projecting axons and active postsynaptic neurons are modified, but also the synapses in layer 1 between the backward projecting axons and the (same) active postsynaptic neurons are modified.

This functional architecture has the following properties. First, the backprojections, which are assumed to carry signals which reflect information derived by convergence of different cortical areas in the next cortical processing stage or in the hippocampus and amygdala, help the neurons to learn to respond differently to (and thus separate) input stimuli (on the forward projection lines) even when the forward input stimuli are very similar. (For neuronal network simulations of this see Rolls, 1989a). Another way in which the learning of pyramidal cells can be influenced is that if two relatively different input stimuli are being received, but the same backprojecting signal occurs with these two somewhat different forward inputs, then the pyramidal cells will be guided by the tutor to build the representation of the forward input stimuli on the same pyramidal cells. If the representations of the forward inputs built by modification of the synapses on the pyramidal cells are considered as vectors in a multidimensional space, then the representations in this case will be forced towards each other as a result of the operation of the common backprojecting tutor.

In the cerebral cortex, the backprojecting tutors can be of two types. One originates from the amygdala and hippocampus, and by benefiting from cross-modal comparison, can provide a useful backprojected vector. This backprojection moreover may only be activated if the multimodal areas detect that the visual stimulus is significant, because for example it is associated with a pleasant taste. This provides one way in which guidance can be provided for a competitive learning system as to what it should learn, so that it does not attempt to lay down representations of all incoming sensory information, but only those shown to be significant as a result of a comparison made with inputs originating in a different modality. Another way for this important function to be achieved is by activation of neurons which 'strobe' the cortex when new or significant stimuli are shown. The cholinergic system originating in the basal forebrain (which itself receives information from the amygdala), and the noradrenergic input to layer 1 of the cortex from the locus coereleus may also contribute to this function (see Rolls, 1987; Bear and Singer, 1986). However, in that there are relatively few neurons in these basal forebrain and noradrenergic systems, it is suggested that these projections only provide a simple 'strobe' (see Rolls, 1987). In contrast there are sufficient axons in the backprojecting systems from the hippocampus and amygdala to carry pattern-specific information, so that the hypothesis presented here is therefore that these backprojections guide how information is consolidated, by influencing how information is stored on pyramidal cells in the way described above. The situations in which the amygdala and hippocampal backprojections guide learning (and also recall and attention as discussed below) are probably different in the following way. It is suggested that the amygdala is particularly involved when associations are made between a neutral stimulus (such as a visual stimulus) and a reinforcing stimulus (arising from for example the taste or touch inputs it receives) (see Rolls, 1985, 1990b). In contrast, it is suggested that the hippocampus is particularly involved when conjunctions must be detected between spatial inputs (received by the major inputs of the hippocampus which originate from the parietal cortex) and its other inputs (from for example the inferior temporal visual cortex), in for example conditional spatial and object-place learning tasks (see above and Rolls, 1989a, 1990c).

The second type of backprojection is that from the next cortical area in the hierarchy. This next cortical area operates in the same manner, but because it receives convergent inputs from different cortical regions in the preceding cortical processing stage, is able to build representations which reflect categories which can only be formed as a result of the convergence. This next cortical stage then projects back these representations as tutors to the preceding stage, to effectively build at the preceding stage better filters for the diagnosis of the categories being found at the next stage.

A second property of this architecture is that it provides for recall. Recall of a previously stored representation by another stimulus or event occurs in the following way. If only the tutor is presented, then the neurons originally activated by the forward projecting input stimuli are activated. This occurs because the synapses from the backprojecting axons onto the pyramidal cells have been modified only where there was conjunctive forward and backprojected activity during learning. Recall could operate as a result of signals backprojected directly from the amygdala or parahippocampal gyrus, or indirectly through series of back-connected cortical stages.

A third property of this architecture is that attention could operate from higher to lower levels to selectively facilitate only certain pyramidal cells by using the backprojections. Indeed, the backprojections described could produce many of the "top-down" influences which are common in perception. A fourth property is that semantic priming could operate by using the backprojecting neurons to provide a small activation of just those neurons in earlier stages which are appropriate for responding to that semantic category of input stimulus. A fifth property of such a return mechanism would be a positive feedback effect which would result in better storage in the cerebral cortex in the long term. This would operate so that when a useful tutor had been built at a higher level this would influence the earlier stage(s) so that a better filter would be built at the earlier level. This in turn would lead to formation of an even better tutor at the higher stage, and so on repeatedly. This might provide a neurophysiological and computational basis for any gradient of retrograde amnesia which may occur for the period just before disruption of temporal lobe function (Squire, 1986; Squire and Zola-Morgan, 1988). A sixth property of the backprojections is that they might assist the stability of the preceding competitive networks by providing a relatively constant guiding signal as a result of associations made at a higher stage, to for example an unconditioned taste or somatosensory input. A seventh property of such a return mechanism would be a dynamic (short term) positive feedback effect by which higher levels might affect the tuning of lower levels to help the system settle into the best fit to multiple constraints. This process might also, perhaps by temporal correlations, help to bind together representations at an earlier stage which are sub-parts of a representation being activated at a higher stage (see von der malsburg, 1985). An eighth property of the backprojection system is that it can speed the rate of learning above that possible in a simple competitive learning system.

This hypothesis of the functions of backprojections in the cerebral cortex requires a large number of backprojecting axons, as pattern-specific information (used to guide learning by providing a set of mutually orthogonal guidance signals, or to produce recall) must be provided by the backprojections. It also helps to solve the deaddressing problem, in that higher stages (such as the hippocampus) do not need to know exactly where in the cortex information should be stored. Instead, the backprojection signal spreads widely in layer 1, and the storage site is simply on those neurons which happen to receive strong (and precise) forward activity as well as backprojected activity, so that Hebbian learning by conjunction occurs there. This scheme is consistent with neocortical anatomy, in that it requires the same pyramidal cell to receive both forward and (more diffuse) backprojected activity, which the arrangement of pyramidal cells with apical dendrites which extend all the way up into layer 1 achieves (see Peters and Jones, 1984).

REFERENCES

Amaral,D.G. (1986) Amygdalohippocampal and amygdalocortical projections in the primate brain. Pp. 3-17 in Excitatory Amino Acids and Epilepsy, eds. R.Schwarz and Y.Ben-Ari. Plenum: New York.

Amaral,D.G. (1987) Memory: anatomical organization of candidate brain regions. Pp. 211-294 in Handbook of Neurophysiology. Section 1 : The Nervous System. Vol. V. Part 1. American Physiological Society: Washington, D.C.

Amaral,D. and Price,J.L. (1984) Amygdalo-cortical projections in the monkey (Macaca fascicularis). J. Comp. Neurol. 230: 465-496.

Azzopardi,P. and Rolls,E.T. (1989) Translation invariance in the responses of neurons in the inferior temporal visual cortex of the macaque. Society for Neuroscience Abstracts 15: 120.

Barlow, H.B. (1972) Single units and sensation: a neuron doctrine for perceptual psychology ? Perception 1: 371-394.

Baylis,G.C., Rolls,E.T. and Leonard,C.M. (1985) Selectivity between faces in the responses of a population of neurons in the cortex in the superior temporal sulcus of the monkey. Brain Research 342: 91-102.

Baylis,G.C., Rolls,E.T. and Leonard,C.M. (1987) Functional subdivisions of temporal lobe neocortex. Journal of Neuroscience 7: 330-342.

Cowey,A. (1979) Cortical maps and visual perception. Quart. J. Exp. Psychol. 31: 1-17.

Desimone,R. and Gross,C.G. (1979) Visual areas in the temporal lobe of the macaque. Brain Res. 178: 363-380.

Gross,C.G. (1973) Inferotemporal cortex and vision. Progr. Physiol. Psychol. 5: 77-123.

Hasselmo,M.E., Rolls,E.T. and Baylis,G.C. (1989) The role of expression and identity in the face-selective responses of neurons in the temporal visual cortex of the monkey. Behav. Brain Res. 32: 203-218.

Hasselmo,M.E., Rolls,E.T., Baylis,G.C. and Nalwa,V. (1989) Object-centered encoding by face-selective neurons in the cortex in the superior temporal sulcus of the monkey. Experimental Brain Research 75: 417-429.

Leonard,C.M., Rolls,E.T., Wilson,F.A.W. and Baylis,G.C. (1985) Neurons in the amygdala of the monkey with responses selective for faces. Behavioural Brain Research 15: 159-176.

Martin,K.A.C. (1984) Neuronal circuits in cat striate cortex. Chapter 9, pp. 241-285 in The Cerebral Cortex, Vol. 2, Functional properties of cortical cells. Eds. E.G.Jones and A.Peters. Plenum: New York.

Mishkin,M. (1982) A memory system in the monkey. Phil. Trans. Roy. Soc. Lond. B 298: 85-95.

Perrett,D.I., Rolls,E.T. and Caan,W. (1982) Visual neurons responsive to faces in the monkey temporal cortex. Exptl. Brain Res. 47: 329-342.

Peters,A. and Jones,E.G. (1984) The Cerebral Cortex, Vol. 1, Cellular Components of the Cerebral Cortex. Plenum: New York.

Rolls,E.T. (1984) Neurons in the cortex of the temporal lobe and in the amygdala of the monkey with responses selective for faces. Human Neurobiology 3: 209-222.

Rolls,E.T. (1985) Connections, functions and dysfunctions of limbic structures, the prefrontal cortex, and hypothalamus. Pp. 201-213 in The Scientific Basis of Clinical Neurology, eds. M.Swash and C.Kennard. London: Churchill Livingstone.

Rolls,E.T. (1987) Information representation, processing and storage in the brain: analysis at the single neuron level. Pp. 503-540 in The Neural and Molecular Bases of Learning, eds. J.-P. Changeux and M.Konishi. Wiley: Chichester.

Rolls,E.T. (1989a) The representation and storage of information in neuronal networks in the primate cerebral cortex and hippocampus. Ch. 8, pp. 125-159 in The Computing Neuron, eds. R.Durbin, C.Miall and G.Mitchison. Addison-Wesley: Wokingham, England.

Rolls,E.T. (1989b) Functions of neuronal networks in the hippocampus and cerebral cortex in memory. Pp. 15-33 in Models of Brain Function, ed. R.M.J.Cotterill. Cambridge University Press: Cambridge.

Rolls,E.T. (1990a) The processing of face information in the primate temporal lobe. In Processing Images of Faces, eds. V.Bruce and M.Burton. Ablex: Norwood, New Jersey.

Rolls,E.T. (1990b) A theory of emotion, and its application to understanding the neural basis of emotion. Cognition and Emotion 4, in press.

Rolls,E.T. (1990c) Functions of the primate hippocampus in spatial processing and memory. Ch. 12, pp. 339-362 in Neurobiology of Comparative Cognition, eds. D.S.Olton and R.P.Kesner. L.Erlbaum: Hillsdale,N.J.

Rolls,E.T. (1990d) Visual information processing in the primate temporal lobe. In Models of Visual Perception: from Natural to Artificial, ed. M.Imbert. Oxford University Press: Oxford.

Rolls,E.T., Baylis,G.C., and Leonard,C.M. (1985) Role of low and high spatial frequencies in the face-selective responses of neurons in the cortex in the superior temporal sulcus. Vision Research 25: 1021-1035.

Rolls,E.T. and Baylis,G.C. (1986) Size and contrast have only small effects on the responses to faces of neurons in the cortex of the superior temporal sulcus of the monkey. Exp. Brain Res. 65: 38-48.

Rolls,E.T., Baylis,G.C. and Hasselmo,M.E. (1987) The responses of neurons in the cortex in the superior temporal sulcus of the monkey to band-pass spatial frequency filtered faces. Vision Research 27: 311-326.

Rolls,E.T., Baylis,G.C., Hasselmo,M.E. and Nalwa,V. (1989) The effect of learning on the face-selective responses of neurons in the cortex in the superior temporal sulcus of the monkey. Experimental Brain Research 76: 153-164.

Seltzer,B. and Pandya,D.N. (1978) Afferent cortical connections and architectonics of the superior temporal sulcus and surrounding cortex in the rhesus monkey. Brain Res. 149: 1-24.

Smith,D.V. and Travers,J.B. (1979) A metric for the breadth of tuning of gustatory neurons. Chemical Senses 4: 215-229.

Squire,L. (1986) Mechanisms of memory. Science 232: 1612-1619.

Squire,L.R. and Zola-Morgan,S. (1988) Memory: brain systems and behavior. Trends in Neurosciences 11: 170-175.

Treves,A. (1990a) Threshold-linear formal neurons in auto-associative nets. J. Physics A 23: 2631-2650.

Treves,A. (1990b) Graded-response neurons and information encodings in autoassociative memories. Physical Review A, in press.

Treves,A. and Rolls,E.T. (1990) Neuronal networks in the hippocampus involved in memory. Neural Networks: Proceedings of the XI Sitges Conference. Springer: Berlin.

Turner,B.H. (1981) The cortical sequence and terminal distribution of sensory related afferents to the amygdaloid complex of the rat and monkey. Pp. 51-62 in The Amygdaloid Complex, ed. Ben-Ari,Y. Elsevier: Amsterdam.

Van Hoesen,G.W. (1981) The differential distribution, diversity and sprouting of cortical projections to the amygdala in the rhesus monkey. Pp. 79-90 in The Amygdaloid Complex, ed. Y.Ben-Ari. Elsevier: Amsterdam.

Van Hoesen,G.W. (1982) The parahippocampal gyrus. New observations regarding its cortical connections in the monkey. Trends in Neurosciences 5: 345-350.

von der Malsburg,C. (1985) Nervous structures with dynamical links. Ber. Bunsenges. Phys. Chem. 89: 703-710.

Acknowledgements. The author has worked on some of the experiments and neuronal network modelling described here with A.Bennett, G.C.Baylis, P.Cahusac, J.D.Feigenbaum, M.Hasselmo, R.P.Kesner, G.Littlewort, Y.Miyashita, H.Niki, R.Payne, T.R.Scott, and A.Treves and their collaboration is sincerely acknowledged. Discussions with David G. Amaral of the Salk Institute, La Jolla, were also much appreciated. This research was supported by the Medical Research Council.

Development, Modulation, Learning, and Habituation

NEURAL MECHANISMS OF SONG LEARNING IN A PASSERINE BIRD

Sarah W. Bottjer
University of Southern California

Abstract. The neural control of vocal learning and behavior in songbirds is vested in a highly localized system of interconnected brain nuclei. In zebra finches (Poephila guttata), these neural circuits undergo profound developmental changes during the time when song behavior is being learned, and both song learning and its neural substrate are sensitive to gonadal hormones such as testosterone and estrogen. Work from our lab has suggested that there may be separate neural circuits that are associated with early versus late stages of vocal learning in zebra finches. Thus, lesioning the song-control circuit that includes the forebrain nucleus IMAN (lateral magnocellular nucleus of the anterior neostriatum) disrupts song behavior only during early stages of vocal learning, whereas making the same lesion during later stages of song development has little or no effect on song behavior. In contrast, lesioning the song-control nucleus HVc (caudal nucleus if the ventral hyperstriatum) produces substantial disruption of song behavior in older juvenile and adult birds, but seems to exert less of a disruptive effect early in the song learning process. Both IMAN and HVc are only two synapses away from the motor neurons that control the vocal muscles, but the circuit containing IMAN regresses during vocal development while the circuit containing HVc grows. Interestingly, neurons are lost from IMAN during early stages of song learning whereas many newly generated neurons are added to HVc throughout the time when song is being learned. This pattern may correlate with the finding that the circuit containing HVc appears to assume increasing control over song behavior as the circuit containing IMAN is relinquishing its role.

Introduction and Background: A Brain-Behavior System

Songbirds learn the sounds used for vocal communication during development. That is, they can learn about acoustic signals in their early environment, and they can also learn to produce vocalizations that match those acoustic signals. Zebra finches (Poephila guttata) are a species of passerine bird that learn a specific song pattern during a restricted period of development, and then maintain production of that song pattern throughout their adult life. Vocal learning in all passerine birds is thought to consist of two main phases (see Bottjer & Arnold, 1986, for review). During the first phase juvenile males hear a specific song pattern (usually their father's). Based on this auditory experience, they form a relatively permanent memory of the song pattern, referred to as a "template" (Marler, 1970). The second phase of vocal learning begins as the bird starts to produce his first song-like vocalizations, which are initially quite variable and bear little resemblance to the bird's final song pattern. This second phase appears to involve a process of auditory-motor integration, during which the bird uses auditory feedback of self-produced vocalizations to refine his vocal production until it matches the sounds stored in his auditory template (Konishi, 1965; Price, 1979). By the time zebra finches are fully adult (100 days of age), the song pattern they produce is a faithful copy of the song heard earlier in life, and this song pattern is maintained in a highly stereotyped fashion throughout adulthood.

There is much evidence that the initial phase of auditory learning is age-limited in many species: birds that are raised in the absence of exposure to song never develop normal vocalizations, even if they receive extensive "tutoring" later in life (e.g., Marler, 1970). However, it is not known whether the auditory-motor phase of vocal learning is also restricted to an early period of development. It **is** known that deafening juvenile zebra finches during the period of auditory-motor integration produces profound disruption of normal song development, whereas deafening adult birds that have learned to produce a highly stereotyped vocal pattern yields little or no disruption (Price, 1979). Thus, the results of deafening studies indicate that auditory feedback is necessary in order to learn to produce a good copy of the song model, but is no longer required following the successful development of stereotyped song patterns. However, it is not clear whether auditory feedback can be used by birds only during a "sensitive" (restricted) period of development, or whether it could be used by birds of any age prior to the emergence of stereotyped song. That is, the ability to use feedback of self-produced vocalizations could be age-limited (as in the case of initial auditory exposure to an external song model), or it could be open-ended, and perhaps dependent on experiential factors. In any case, once zebra finches have learned to produce a stereotyped song pattern, they maintain production of that song without changes throughout their adult life, and are apparently incapable of learning new song patterns.

The neural control of vocal learning and behavior is vested in a highly localized system of interconnected brain regions (see Figure 1). Furthermore, both vocal behavior and its neural substrate are regulated by circulating levels of sex hormones such as testosterone and estradiol. Thus, songbirds provide a "model" brain-behavior system for studies of the neural and hormonal bases of learned behaviors. Early studies of the neural substrate for song control demonstrated that lesions of HVc (caudal nucleus of the ventral hyperstriatum) disrupted production of already-learned song patterns by adult birds (Nottebohm et al., 1976). In addition, multi-unit electrophysiological recordings from HVc of awake singing birds reveal a pattern of activity that correlates precisely with the production of individual song syllables (McCasland & Konishi, 1981; McCasland, 1987). However, HVc does not serve a purely motor function, as it also contains auditory neurons that are especially responsive to conspecific song (Margoliash, 1983; Margoliash & Konishi, 1985). This pattern of findings suggests that HVc is importantly involved with encoding sensorimotor aspects of learned vocal behavior.

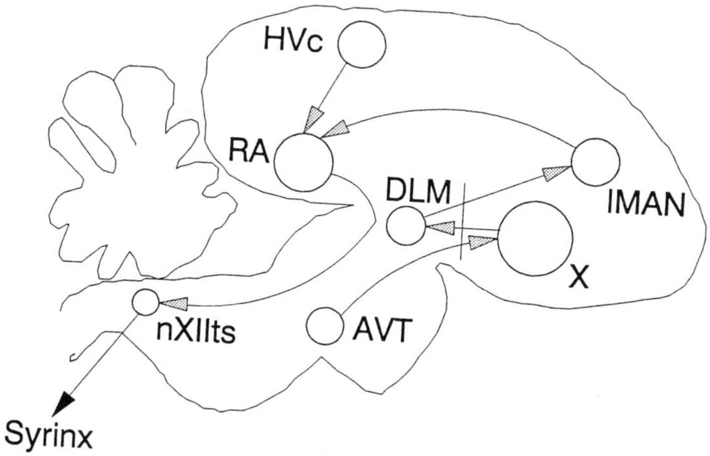

Figure 1. Schematic diagram of some of the major brain nuclei involved in song control in passerine birds (sagittal view); some connections are omitted for the sake of clarity. HVc and IMAN both project to RA, which in turn projects to the hypoglossal motor neurons that innervate the vocal organ (syrinx). Abbreviations: HVc, caudal nucleus of the ventral hyperstriatum; RA, robust nucleus of the archistriatum; nXIIts, tracheosyringeal portion of the hypoglossal nucleus; IMAN, lateral magnocellular nucleus of the anterior neostriatum; DLM, medial dorsolateral nucleus of the anterior thalamus; X, area X; AVT, vental area of Tsai.

Functional Changes in Neural Circuits during Vocal Learning

Because the ability to learn song is limited to a sensitive period of development in zebra finches, it was tempting to speculate that we might be able to isolate discrete neural circuits that would underlie the initial capacity for vocal learning. With this idea in mind, we made bilateral electrolytic lesions of the forebrain nucleus lMAN (lateral magnocellular nucleus of the anterior neostriatum) at different stages of vocal development (Bottjer et al., 1984). lMAN seemed a likely place to begin our search, since lMAN shares two characteristics with HVc: lMAN neurons contribute to an efferent pathway to the vocal muscles (Fig. 1), and a high proportion of cells in lMAN concentrate androgenic hormones in adult males. We found that lesioning lMAN in juvenile birds that were just beginning to produce song sounds produced profound disruption of vocal behavior. However, making the very same lesion in adult birds (>90 days of age) had no effect on song performance. Thus, a neural circuit that includes the efferent and/or afferent connections of lMAN seems to be critically involved in vocal learning during only a restricted period of development. This result may indicate that functions important for song learning are being carried out by lMAN neurons during a particular period of development, but that control of vocal behavior is taken over by other vocal-control areas of the brain after this period. In relation to our general understanding of neural mechanisms of learned behaviors, this result suggests that even if a specific brain region is intimately involved in the initial representation of some behavior, the permanent neural representation of that behavior may reside in a quite separate location in the brain.

We do not know what the specific function of lMAN may be. However, the time course of the effectiveness of lesions in MAN may provide some clues to this question. Lesioning lMAN in birds ranging from 30 to 50 days of age produced the same result: immediate disruption of song behavior with no improvement over time. However, lMAN lesions made between 50 and 60 days disrupted behavior in proportion to song development for individual birds: birds with reasonably stereotyped song patterns suffered little disruption of behavior, whereas birds with highly variable song patterns showed the same degree of disruption as younger birds. Interestingly, the stereotypy in the song patterns derived not so much from the consistent production of individual song notes, as from the systematic repetition of a particular note sequence. That is, as soon as the temporal order of the notes stabilized, lMAN lesions became much less effective, even though the morphology of individual notes was still somewhat variable. Thus, the time course of effectiveness of lMAN lesions seems to parallel the development of song as a motor sequence or pattern. This outcome may indicate that lMAN neurons are involved with some aspect of motor learning for song, or with the transcription of auditory coding of vocalizations to motor coding. Whatever the exact function of lMAN may be, it seems likely that lMAN neurons may be directly involved in controlling song behavior early in development, even though they are clearly not on the main motor pathway in adulthood.

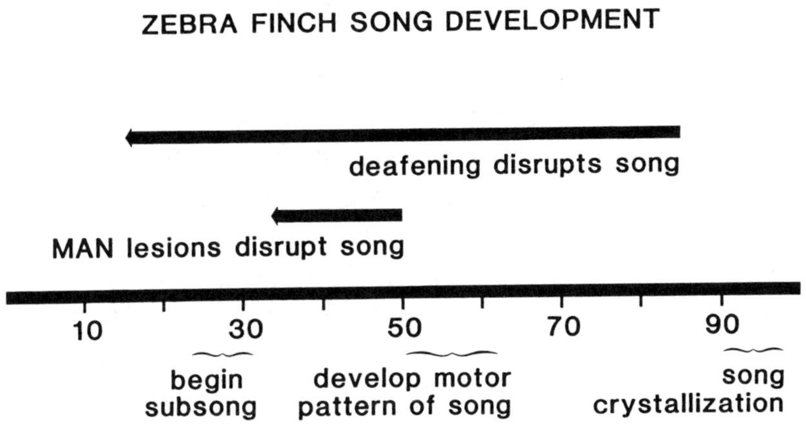

Figure 2. Time line of zebra finch song development. Zebra finches are sexually mature by 90-100 days of age. They begin to produce their first song-related vocalizations around 25 days of age, and develop a reasonable degree of stereotypy in their song pattern between approximately 50 and 60 days, which is when lesions of lMAN become ineffective. In contrast, deafening continues to disrupt song behavior until 80-85 days of age.

An interesting question pertains to whether the time course for lMAN lesions relates to critical periods for song learning. Behavioral studies of song learning have suggested that there are at least two periods involved in vocal learning: one involving formation of an auditory memory of a specific song pattern, and a second involving a phase of auditory-motor integration during which auditory feedback is used to produce correct efferent commands (see above). The exact timing of the first of these two sensitive periods (when exposure to conspecific vocalizations is necessary) is not known in zebra finches. However, it is known that the duration of the second period (when auditory feedback is necessary) extends almost to adulthood: deafening zebra finches as old as 80 days of age results in severe deterioration of the song pattern (Price, 1979). Lesions of lMAN have become completely ineffective at this age, even though auditory feedback is evidently still necessary. At the present time, we do not know whether the decreased effect of lMAN lesions correlates with the end of the first sensitive period, or with some other, as yet undiscovered aspect of song learning.

As one approach to investigating the function of lMAN, we have been identifying the pathways that are afferent to lMAN (Bottjer et al., 1989). Figure 1 describes our initial results, showing that the song-control nucleus Area X projects to IMAN via the thalamic region DLM (medial dorsolateral nucleus of the anterior thalamus). IMAN, in turn, makes a single efferent projection to RA (robust nucleus of the archistriatum), and Area X receives its input from the midbrain region AVT (area ventralis of Tsai). Knowing something about the axonal connectivity of this circuit enabled us to make lesions at different levels of this pathway and compare the resultant behavioral deficits to those produced by lesions of IMAN. As a first step, we lesioned a forebrain region that contains axons that are afferent to X, DLM, and IMAN (see Figure 1). To our surprise, such lesions produce disruption of vocal learning rather late in development - almost up until the time when birds reach adulthood (Halsema & Bottjer, 1990). However, lesions of these fibers in birds that are fully adult have no effect on maintenance of stable song patterns (cf. Scharff & Nottebohm, 1989; Sohrabji et al., 1990). This result seems paradoxical: why should lesioning the input to IMAN continue to disrupt behavior **after** lesions of IMAN lose effectiveness? One possible reason is that afferents to IMAN somehow serve to hold it "at bay" in order to prevent IMAN neurons from continuing their influence over song behavior. Holding IMAN in abeyance, as it were, might enable other circuits, including the HVc-to-RA pathway, to assume greater control over vocal behavior.

The disruptive effect of these axonal lesions seems exciting in another respect, which is that the time course of effectiveness of such lesions may parallel the time course for reliance on auditory feedback. As mentioned above, it is known that deafening disrupts song development in juvenile male zebra finches, but has little or no disruptive effect in adult birds. Furthermore, deafening continues to be effective in birds as old as 80 to 85 days, even though they are already producing a stereotyped song pattern (Price, 1979). Thus, the time course of reliance on auditory feedback does not correlate directly with the emergence of a stable song pattern. Rather it seems possible that some period of "consolidation" is required following the development of stereotyped song behavior before auditory feedback is no longer required. Could this period correspond to the interval between approximately 60 and 80 days of age when lesions of IMAN no longer disrupt behavior, but lesions involving the afferent pathway to IMAN are effective? What is particularly exciting to us is the idea that lesions of brain regions that project (either directly or indirectly) to IMAN may produce the same pattern and timing of results as those obtained by deafening. Because Area X is thought to receive auditory information from the song-control nucleus HVc (Katz & Gurney, 1981), it seems possible that X might receive input based on auditory feedback. Perhaps such input is necessary during this period of "consolidation" in part to help transfer control of song behavior away from IMAN. Following this period, it seems that lesions anywhere in this circuit are ineffective in disrupting vocal behavior.

Development of Brain Regions that Control Vocal Behavior

Given the striking change in the effectiveness of IMAN lesions during vocal development, we became curious about the normal development of IMAN in juvenile males, as well as of other song-control regions. In fact, the apparent change in the function of IMAN occurs during a time of substantive changes in the normal development of the song-control system (Figure 3; Bottjer et al., 1985; 1986; Hermann & Bischof, 1986; Nordeen & Nordeen, 1988a; b). The number of IMAN neurons is high at 25 days when birds are first learning to produce their song pattern, but decreases sharply by 35 days (Bottjer & Sengelaub, 1989). Thus, at least half of the original contingent of neurons contained in IMAN is lost during the time when the bird begins to produce its incipient song vocalizations.

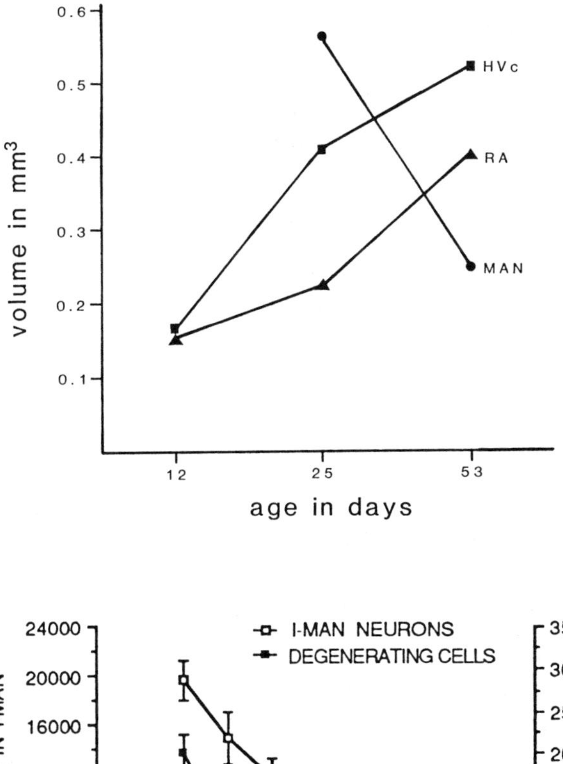

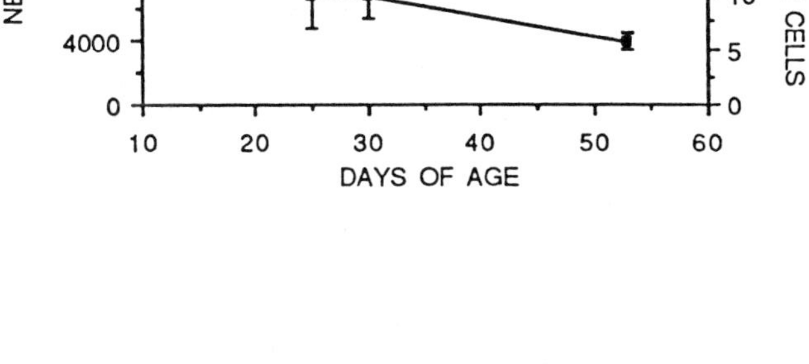

Figure 3. Top figure shows the overall volume changes in HVc, RA, and IMAN. HVc and RA grow dramatically in size, as does area X (data not shown). The growth of HVc and X is due to the incorporation of a large number of newly generated neurons. In contrast, IMAN shrinks in size and undergoes a massive loss of neurons. Bottom figure shows total number of live neurons as well as total number of degenerating cells in IMAN at 5-day intervals from 20 to 53 days of age; the number of pyknotic cells is greatest at the time when the loss of live neurons is high. Neuronal loss is complete by 30-35 days of age, well before lesions of IMAN lose the ability to disrupt song behavior.

In contrast, the number of neurons in HVc increases sharply during this time: there is a fifty percent increase in neuronal number some time between 25 and 55 days due to ongoing genesis and migration of neurons into HVc. Interestingly, the rate of incorporation of new neurons into HVc is much greater in juvenile males during song learning than in adult birds. Furthermore, a large proportion of these newly generated HVc neurons form efferent connections to RA during the period of song learning, thereby contributing to a massive ingrowth of HVc fibers into RA that occurs between 30 and 35 days in juvenile male zebra finches (Nordeen & Nordeen, 1988b; Konishi & Akutagawa, 1985). The volume of RA also increases during this time, but this growth is attributable primarily to a decrease in neuronal density with no net change in neuron number. These changes occur at a time when other brain regions that are not known to be involved with song control show little or no change in size. Thus, the development of the song-control system seems to occur relatively late, suggesting that these neural changes may be somehow timed to occur during vocal learning. The timing of these changes may in fact be necessary for song learning to proceed - it is particularly striking that part of the major limb subserving song control in adult birds (i.e., the HVc-to-RA pathway) is actually created as song learning progresses.

The pattern of results seen from lesion studies as well as from description of the normal ontogeny of the song-control system suggest that lMAN exerts decreasing control over song behavior, while HVc may exert an increasing degree of control. This hypothesis is consistent with the fact that HVc is known to be directly involved in the control of adult song behavior, whereas lMAN appears to play a role only in early song development and is clearly not directly involved with song control in adulthood. Within HVc, the increased number of neurons may be somehow necessary to encode the adult song pattern. The loss of neurons from lMAN seems more difficult to interpret. Cell death appears to be the mechanism underlying the loss of lMAN neurons, although re-differentiation of neurons and/or a secondary migration out of lMAN cannot be ruled out (Bottjer & Sengelaub, 1989). Whatever the mechanism, the loss of lMAN neurons is difficult to understand in a functional sense. The number of neurons in lMAN seems to be maximal at the time that birds are producing their first song sounds, suggesting that this larger number of neurons may be necessary when birds are initially learning to produce song. However, the timing of neuronal loss does not correlate with the time course for lMAN lesions: neuronal loss is complete by around 35 days of age, well before lesions of lMAN become ineffective (Bottjer & Sengelaub, 1989; Fig. 3). Thus, there is a fixed population of neurons in lMAN following the period of cell loss that apparently undergo a substantial change in function at the time lesions lose effectiveness (around 55-60 days). A prepossessing question remains: what is the switch that triggers this dramatic change in the function of lMAN neurons? We have thus far detected only one change within lMAN that occurs around the time when lesions are losing the ability to disrupt behavior, which is that a large proportion of androgen target cells that do not project to RA lose the ability to concetrate androgens some time between 45 and 60 days (Korsia & Bottjer, 1989). Interestingly, androgen target cells that do project to RA do not lose the ability to concentrate androgens, suggesting that maintaining an afferent projection reduces the probability of losing hormone sensitivity. In any case, it is possible that the nonprojecting androgen target cells that lose the ability to concentrate hormone are initially involved with the maintenance of a high degree of plasticity in song behavior. Loss of the capacity to accumulate androgens in these cells might help to signal the termination of highly plastic vocal production or the onset of stereotypy.

Study of the overall pattern of the time of development of various song-control circuits has yielded some interesting results. As indicated above, the genesis of HVc neurons occurs at a high rate during early stages of song development. Although many HVc-to-RA projecting neurons are generated post-hatch, a separate population of HVc-to-X projecting neurons is born prior to hatching (Nordeen & Nordeen, 1988b; Alvarez-Buylla et al., 1988). Interestingly, the target of HVc-to-X axons - i.e., Area X neurons themselves - appear to be born fairly late during song development. In fact, Area X is not visible in 10-day old juvenile males, but grows to be the largest of the song-control nuclei by approximately 55 days of age (Bottjer et al., 1985; Johnson & Bottjer, 1990; Kirn & DeVoogd, 1989; Nordeen & Nordeen, 1988a). In contrast, all of the neurons in both RA and lMAN appear to be generated in ovo (Nordeen & Nordeen, 1988a; Kirn & DeVoogd, 1989; Konishi & Akutagawa, 1990). Thereafter, RA grows in overall size, but does not increase in neuron number, whereas lMAN loses neurons (see above). Recent data from our lab has shown that DLM, like lMAN, regresses in size during song learning; we do not yet know whether this regression entails a loss of neurons (Johnson & Bottjer, 1990). Thus, the overall pattern suggests that X, HVc, and RA grow during song development, whereas DLM and lMAN regress.

In terms of axonal connectivity, HVc axons congregate around the borders of RA as early as 15 days, before invading it around 30 days of age (Konishi & Akutagawa, 1985). Unpublished data from our lab indicate that lMAN fibers ramify throughout RA as early as 20 days of age, a finding consistent with the notion that lMAN neurons may have some direct role in song production during early development. In addition, afferents from DLM have arrived in lMAN by at least 20 days of age, suggesting that the DLM-lMAN-RA circuit develops early. However, the fact that Area X neurons appear to develop relatively late may indicate that development of the X-to-DLM pathway is correspondingly delayed. In accord with this notion, we have seen growth cones heading from X to DLM in birds as old as 30 days of age (Johnson & Bottjer, unpublished data).

One basic question raised by these descriptive studies of development regards the factors that control regulation of neuron number in song-control nuclei. We have examined this question only in regard to the loss of neurons from IMAN. Somewhat to our surprise, neither the property of projecting to an efferent target (RA), nor the ability to concentrate androgenic hormones are able to spare neurons from dying (Korsia & Bottjer, 1989). Another factor that has been shown to regulate cell survival is afferent input (e.g., Clarke & Egloff, 1988; Furber et al., 1987). In this regard, the later development of X and its efferent projection to DLM is intriguing. It seems possible that the development of inputs to DLM could act to curtail its regression, which could in turn work to stabilize neuron number in IMAN.

Rampant Speculation and a Possible Model for Song Learning?

It is interesting to speculate in an a priori way about the factors that are important for vocal learning as a first step towards thinking about a possible model for mechanisms of vocal learning. We know that auditory feedback is important as the bird is learning to produce its vocal pattern. It seems likely that auditory feedback of self-produced sounds must be compared with the auditory "template" - i.e., with the representation of the song that the bird has heard earlier. One might also wonder if input from the vocal muscles might play a role (Figure 4). It is known that the tracheosyringeal branch of the hypoglossal nerve (which innervates the syrinx) contains afferent fibers (Bottjer & Arnold, 1982). These fibers might represent a pathway for proprioceptive feedback. Unfortunately, the central termination of these syringeal afferents is not known, although it has been established that afferents from the main hypoglossal trunk terminate throughout the trigeminal tract. Section of these afferent fibers has no effect on production of already-learned song patterns in adult birds, even if they are deafened (Bottjer & Arnold, 1984). This finding suggests that normal development of a stereotyped song pattern in zebra finches may entail the establishment of a central motor program that no longer requires peripheral input for its maintenance. Unfortunately, it has not been possible to section these hypoglossal afferents in juvenile birds, so we do not know whether they are important during the period of vocal learning.

In order to elucidate structure-function relationships in the song-control system, it would be useful to know the circuitry in greater detail. We have attempted to determine the complete trans-synaptic pathway that is afferent to the X-DLM-IMAN pathway. A midbrain region known as AVT (ventral area of Tsai) projects to X, and this projection is known to be heavily catecholaminergic (Bottjer et al., 1989; Lewis et al., 1981). Preliminary evidence from our lab indicates that AVT receives trigeminal input from the principal sensory nucleus of V (PrV; Tsutsui & Bottjer, 1989). This result is potentially very exciting, given that hypoglossal afferents terminate throughout the trigeminal tract (see above). It seems irresistible to

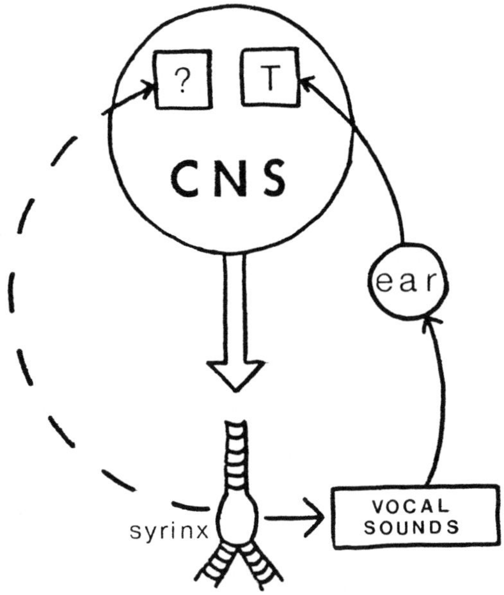

Figure 4. Diagram indicating factors that are likely to be involved with song learning. When a sound is produced, auditory feedback of that sound is relayed into the CNS, where it is presumably compared with the template (T). The anatomical location of the template is not known. It is also not known whether the syringeal muscles themselves send input back to the CNS. If so, then it might be suspected that auditory and proprioceptive feedback are directly compared in order to adjust efferent commands to produce sounds that match those stored in the template.

speculate that afferent fibers from the syrinx may terminate in PrV, and provide a substrate for proprioceptive feedback from vocal muscles. Such a result would be extremely exciting in two major respects. First, it would establish evidence for a trans-forebrain loop, originating in the syrinx, relaying through a pathway that seems to be uniquely associated with vocal learning, and conveying an efferent message back to the syrinx and motor neurons through the lMAN-RA pathway. Such a circuit could certainly represent a means by which the bird could compare the sounds it produces with the auditory template of the song pattern the bird has heard earlier in development, and modify the efferent commands it produces in order to adjust those sounds to match those stored in the template. Secondly, evidence that input from the vocal organ is feeding into pathways involved with song learning would certainly be consistent with the notion that proprioceptive feedback is important during song development, and would enable this hypothesis to be tested directly. We know that neither auditory feedback nor input from hypoglossal afferent fibers appear to be important for maintenance of already-learned song patterns in adult birds (see above). However, both types of input may be important during vocal learning. For example, it might be the case that self-produced sounds could generate feedback of both an auditory and proprioceptive nature, and that the simultaneous occurrence of both types of feedback in the brain could be an important determinant of how the neural substrate for song control develops. Once the neural circuits that encode the bird's song pattern have stabilized, they may not require any peripheral input in order to be maintained.

BIBLIOGRAPHY

Alvarez-Buylla, A., Theelen, M., and Nottebohm, F. (1988) Birth of projection neurons in the higher vocal center of the canary forebrain before, during, and after song-learning. Proc. Natl. Acad. Sci. USA **85:** 8722-8726.

Bottjer, S.W., and Arnold, A.P. (1982) Afferent neurons in the hypoglossal nerve of the zebra finch (Poephila guttata): Localization with horseradish peroxidase. J. Comp. Neurol. **210:** 190-197.

Bottjer, S.W., and Arnold, A.P. (1984) The role of feedback from the vocal organ: Maintenance of stereotypical vocalizations by adult zebra finches. J. Neurosci. **4:** 2387-2396.

Bottjer, S.W., and Arnold, A.P. (1986) The ontogeny of vocal learning in songbirds. In: Handbook of Behavioral Neurobiology, Vol. 8, E.M. Blass (Ed.), Plenum, New York, pp. 129-161.

Bottjer, S.W., and Foster, E. (1990) Blocking steroid hormones during vocal learning extends susceptibility to deafening in zebra finches. Soc. Neurosci. Abstr.

Bottjer, S.W., Glaessner, S.L., and Arnold, A.P. (1985) Ontogeny of brain nuclei controlling song learning and behavior in zebra finches. J. Neurosci. **5:** 1556-1562.

Bottjer, S.W., Halsema, K.A., Brown, S.A., and Miesner, E.A. (1989) Axonal connections of a forebrain nucleus involved with song learning in zebra finches. J. Comp. Neurol. **279**: 312-326.

Bottjer, S.W., Miesner, E.A., and Arnold, A.P. (1984) Forebrain lesions disrupt development but not maintenance of song in passerine birds. Science **224:** 901-903.

Bottjer, S.W., Miesner, E.A., and Arnold, A.P. (1986) Changes in neuronal number, density and size account for increases in volume of song-control nuclei during song development in zebra finches. Neurosci. Lett. **67:** 263-268.

Bottjer, S.W., and Sengelaub, D.R. (1989) Cell death during development of a forebrain nucleus involved with vocal learning in zebra finches. J. Neurobiol. **20:** 609-618.

Clarke, P.G.H., and Egloff, M. (1988) Combined effects of de-afferentation and de-efferentation on isthmo-optic neurons during the period of their naturally occurring cell death. Anat. & Embryol. **179:** 103-108.

Furber, S., Oppenheim, R.W., and Prevette, D. (1987) Naturally-occurring neuron death in the ciliary ganglion of the chick embryo following removal of preganglionic input: Evidence for the role of afferents in ganglion cell survival. J. Neurosci. **7**: 1816-1832.

Halsema, K.A., and Bottjer, S.W. (1990) Lesioning axonal connections of a thalamic nucleus disrupts song development in zebra finches. Soc. Neurosci. Abstr.

Hermann, K., and Bischof, H-J. (1986) Delayed development of song control nuclei in the zebra finch is related to behavioral development. J. Comp. Neurol. **245:** 167-175.

Johnson, F., and Bottjer, S.W. (1990) Developmental changes in a thalamic nucleus involved with vocal learning in zebra finches. Soc. Neurosci. Abstr.

Katz, L.C., and Gurney, M.E. (1981) Auditory responses in the zebra finch's motor system for song. Brain Res. **211:** 192-197.

Kirn, J.R., and DeVoogd, T.J. (1989) Genesis and death of vocal control neurons during sexual differentiation in the zebra finch. J. Neurosci. **9**: 3176-3187.

Konishi, M. (1965) The role of auditory feedback in the control of vocalization in the white-crowned sparrow. Z. Tierpsychol. **22**: 770-783.

Konishi, M., and Akutagawa, E. (1985) Neuronal growth, atophy and death in a sexually dimorphic song nucleus in the zebra finch brain. Nature **315:** 145-147.

Konishi, M., and Akutagawa, E. (1990) Growth and atrophy of neurons labeled at their birth in a song nucleus of the zebra finch. Proc. Natl. Acad. Sci. USA **87:** 3538-3541.

Korsia, S., and Bottjer, S.W. (1989) Developmental changes in the cellular composition of a brain nucleus involved with song learning in zebra finches. Neuron **3**: 451-460.

Lewis, J.W., Ryan, S.M., Arnold, A.P., and Butcher, L.L. (1981) Evidence for a catecholaminergic projection to Area X in zebra finch. J. Comp. Neurology **196**: 347-354.

Margolish, D. (1983) Acoustic parameters underlying the responses of song-specific neurons in the white-crowned sparrow. J. Neurosci. **3:** 1039-1057.

Margolish, D., and Konishi, M. (1985) Auditory representation of autogenous song in the song system of white-crowned sparrows. Proc. Natl. Acad. Sci. **82:** 5997-6000.

Marler, P. (1970) A comparative approach to vocal learning: Song development in white-crowned sparrows. J. Comp. Physiol. Psychol. **71(2):** 1-25.

McCasland, J.S., and Konishi, M. (1981) Interaction between auditory and motor activities in an avian song control nucleus. Proc. Natl. Acad. Sci. **78:** 7815-7819.

McCasland, J.S. (1987) Neuronal Control of Bird Song Production. J. Neuroscience **7**: 23-39.

Nordeen, E.J., and Nordeen, K.W. (1988a) Sex and regional differences in the incorporation of neurons born during song learning in zebra finches. J. Neurosci. **8:** 2869-2874.

Nordeen, K.W., and Nordeen, E.J. (1988b) Projection neurons within a vocal motor pathway are born during song learning in zebra finches. Nature **334:** 149-151.

Nottebohm, F., Stokes, T.M., and Leonard, C.M. (1976) Central control of song in the canary, Serinus canarius. J. Comp. Neurol. **225:** 1046-1048.

Price, P.H. (1979) Developmental determinants of structure in zebra finch song. J. Comp. Physiol. Psychol. **93:** 260-277.

Scharff, C. & Nottebohm, F. (1989) Lesions in Area X affect song in juvenile but not adult male zebra finches. Soc. Neurosci. Abstr. **15:** 618.

Sohrabji, F., Nordeen, E.J., Nordeen, K.W. (1990) Selective impairment of song learning following lesions of a forebrain nucleus in the juvenile zebra finch. Beh. Neural Biol. **53:** 51-63.

Tsutsui, K., and Bottjer, S.W. (1989) Afferent input from the vocal organ projects to forebrain nuclei that are involved with song learning in zebra finches. International Congress for Neuroethology **2**.

Hormonal Regulation of Motor Systems: How Androgens Control Amplexus (Clasping) in Male Frogs

Albert A. Herrera and Michael Regnier
Neurobiology Section, Department of Biological Sciences
University of Southern California
Los Angeles, California, 90089-2520, U.S.A.
aherrera%alhena.usc.edu.@usc.edu

Abstract

In this article we review the literature concerning the hormonal, neural, and muscular basis of amplexus (clasping) in male frogs and toads. We also present our recent results on the effects of testosterone on the clasping musculature of Xenopus laevis. Amplexus, the prolonged embrace of females by males during mating, has been characterized as a spinal reflex that is subject to modification by higher neural centers. Many of the cells involved in this behavior have high levels of androgen receptors and cellular structure and function can be regulated by androgens. We find that testosterone modifies the flexor carpi radialis muscle (FCR) such that it has an enhanced ability to maintain low levels of tension while retaining the ability to contract rapidly. These properties are well matched to the functional demands of clasping. Histology reveals that there are regional differences in testosterone sensitivity within the FCR. Tension recordings show that the more androgen-sensitive region of the muscle is composed of fibers that contract more slowly than other regions. These results reveal some of the ways in which the FCR is seasonally remodelled for the specialized task of clasping.

Acknowledgments: This work was supported by a grant (NS27209) and Research Career Development Award (NS00951) from NIH. We thank Dr. Richard Dunia, Naomi Nagaya-Stevens, and Peter Miller for commenting on the manuscript and Barbara Tower for technical assistance. Our special thanks go to Naomi Nagaya-Stevens for help in reviewing the literature.

1. Reproductive Behavior in Male Frogs and Toads

In males of most groups of anuran amphibians, two reproductive behaviors commonly occur, mate calling and amplexus. Males use mate call vocalizations to attract females for mating. In amplexus, males use their forelimb flexor muscles to clasp onto the dorsal side of females so that their cloacae are brought into proximity. This positioning facilitates external fertilization at the time of egg laying. In many species, there is intense competition between males for a limited number of females. This competition constitutes a strong selective pressure for the evolution of mechanisms to ensure reproductive success (Wells, 1977). Of particular importance is the ability of males to seize females rapidly and clasp for hours, days, or even weeks without interruption. The long term goal of our research is to understand the neurological and physiological basis of this remarkable motor performance. In the course of this work, we hope to learn much about neuronal and muscular plasticity that will be generally applicable.

Amplexus has been particularly well described for *Xenopus laevis* (Russell, 1954, 1960; Hutchison & Poynton, 1963; Kelley & Pfaff, 1976). Hutchison & Poynton (1963) pointed out that the forelimbs are used in two different ways during the behavior. When the female is at rest, the male hangs on loosely, generating a constant low level of tension with forelimb flexors. If the female moves suddenly, as she does when going to the surface for air or in an escape response, the male rapidly tightens his grip to avoid being dislodged. During clasping, therefore, forelimb motoneurons and muscles must generate two different patterns of motor activity that differ markedly in their speed and duration. These twin patterns of forelimb motion may be contrasted with a single motion seen most often outside of breeding season. This single motion, which is very rapid, has been referred to as "flicking" (Russell, 1954; Hutchison, 1965). Flicking consists of repeating cycles of forelimb flexion and extension and is used to gather and stuff food into the mouth. The forelimbs are of course used for other purposes, such as warding off obstacles during slow locomotion and wiping the mouth, but flicking is a predominant motion. Since *Xenopus* is entirely aquatic, the forelimbs do not support the body as in other anurans nor do the forelimbs function directly in swimming.

2. Hormonal and Neural Regulation of Clasping

Steroid sex hormones have profound effects on neurons that control mating and other reproductive behaviors (Kelley & Pfaff, 1976; McEwen, Davis, Parsons & Pfaff, 1979; Kelley, 1988). The location of such hormone-sensitive neurons can be identified by treating living animals with radiolabeled hormones then using autoradiography to localize hormone binding sites in tissue sections. Such techniques were used by Kelley and her coworkers to map the principal androgen- concentrating neurons in the central nervous system of adult and developing *Xenopus* males (Kelley, Morrell & Pfaff, 1975; Kelley, 1980, 1981; Wetzel, Haerter & Kelley, 1985; Gorlick & Kelley, 1986, 1987). Androgen-concentrating neurons were found in the anterior preoptic area, certain thalamic nuclei, the ventral infundibular nucleus, anterior pituitary, the laminar nucleus of the torus semicircularis, the pretrigeminal nucleus of the dorsal tegmental area of the medulla, and inferior reticular nuclei. In addition, neurons that send projections to the medullary motoneurons that innervate vocal muscles have been identified (Wetzel et al., 1985). In male *Rana pipiens*, electrical stimulation of neurons in the preoptic area generated neural correlates of mate calling in laryngeal motor nerves (Schmidt, 1983). Since hormonal manipulations cause parallel changes in mate calling and clasping (Kelley & Pfaff, 1976), it is likely that many of these androgen concentrating regions are also involved in clasping behavior. Unfortunately, the specific neural circuitry involved in clasping has not been mapped.

High levels of androgen receptors are also found in the motoneurons and muscle fibers used for mating behavior. In the mate calling system, for example, laryngeal motoneurons located in cranial nerve nucleus IX-X and laryngeal muscle fibers (Segil, Silverman & Kelley, 1987) are androgen targets. In the clasping system, androgen- concentrating motoneurons are located in spinal segments 2 and 3, within the boundaries of motor pools for forelimb flexors. These spinal segments also contain high testosterone reductase activity (Kelley, 1980; Erulkar, Kelley, Jurman, Zemlan, Schneider & Krieger, 1981). Since androgens have potent and selective effects on clasping muscles in vivo (Muller, Galavazi & Szirmai, 1969; Thibert, 1986; see below), it is likely that they too have high levels of androgen receptors, but this has not been directly demonstrated.

The motor behavior associated with amplexus is so stereotyped that it has been described as a "clasp reflex." In intact males, the reflex is well developed at the peak of breeding season (Smith, 1938). Alternatively, clasping can be induced by treatment with the androgens testosterone or dihydrotestosterone or by treatment with human chorionic gonadotropin, a mixture of hormones that stimulates androgen production by the testes. The reflex is triggered by contact with a sexually receptive female and consists of a rapid seizing of the female, orientation into the pelvic clasp position, and maintenance of the clasp for many hours. As described above, the reflex includes brief periods of rapid tightening to maintain the grasp while the female moves. Attempts to forcibly remove a clasping male are also met with sudden tightening, as well as clawing at the intruder by means of forward kicking of the hindlimbs (Hutchison & Poynton, 1963).

Perhaps the most dramatic demonstration of the reflexive nature of the clasp was shown in preparations of male frogs in which the entire body anterior and posterior to the thoracic region was removed. These preparations clasped objects avidly (Smith, 1938). Later experiments on *Rana pipiens* (Aronson & Noble, 1945) and *Xenopus laevis* (Hutchison & Poynton, 1963; Hutchison, 1965) involved selective lesions of certain brain regions. In *Xenopus*, if the brain was transected in the anterior medulla between cranial nerves V and VII, operated frogs would clasp strongly onto an experimenter's fingers or other unnatural objects. Intact males do not normally clasp such objects. When tested in an artificial clasping apparatus, it was clear that the frogs were responding to tactile stimulation of the throat skin and the inner surface of the forelimbs, as well as stretching of the forelimbs. The behavior showed most of the attributes of a normal clasp, including tightening and kicking in response to attempts to dislodge the frog. The interpretation of these results was that transections in the anterior medulla left intact all of the circuitry necessary for reflexive clasping, but removed higher centers responsible for discriminating appropriate objects. If the brain was transected just anterior to the medulla, frogs would not grasp artificial objects, consistent with the notion that such a lesion spares the higher discriminatory centers.

3. Effects of androgens on clasping muscles

Most of the research on the clasping system has focused on the effector organs, the flexor muscles of the male forearm. A variety of species of frogs and toads have been employed in these studies. In all cases, the two principal muscles involved are the coracoradialis (sternoradialis), a chest muscle that flexes the shoulder and elbow, and the flexor carpi radialis, a forearm muscle that flexes the elbow and wrist. These muscles exhibit a striking sexual dimorphism, with males muscles being larger and containing more fibers than in females (Oka, Ohtani, Satou & Ueda, 1984; Rubinstein, Erulkar & Schneider, 1983). Male muscles also have slower contractile kinetics (Melichna, Gutmann, Herbrychova & Stichova, 1972; Hayatsu, Kosaka, Tsutsuura & Nagai, 1981; Kirby, 1983). These differences vary seasonally, being most pronounced during breeding (Muller et al., 1969). Androgens have been shown to control certain properties of the clasping muscles. For example, castration causes a decrease in the cross sectional area of muscle fibers (Muller et al., 1969; Thibert, 1986; Regnier & Herrera, 1989, 1990) and in the estimated percentage of tonic fibers (Rubinstein et al., 1983). When breeding males (high androgen levels) are compared with non-breeding males (low androgen levels), it can be inferred that hormones have positive effects on the strength of contraction and twitch duration (Melichna et al., 1972). These androgen-sensitive properties are specific to the forelimb flexors, with most forelimb extensors and other muscles showing no sexual dimorphism (Oka et al., 1984) or androgen sensitivity (Thibert, 1986).

To characterize these effects in greater detail and to begin our study of the cellular mechanisms of clasping, we have examined the effects of androgens on the flexor carpi radialis muscle (FCR) of male *Xenopus laevis* (39-46 g body weight; supplied by Nasco, Ft. Atkinson, WI). Frogs were anesthetized by immersion in 0.1% aqueous tricaine methanesulfonate, chilled on ice, and castrated by removing both testes through small incisions on either side of the abdomen. In half the operated frogs, Silastic tubes (O.D. 0.65 in, I.D. 0.3 in, Dow Corning) filled with testosterone crystals (Steraloids) and sealed at the ends were implanted subcutaneously in the dorsal lymph sac. These animals are referred to as **C + T**, signifying that they were castrated and received testosterone. The other half of the operated frogs were implanted with empty tubes, and are referred to as the **C**, or castrated group. Some studies were also done on unoperated frogs (35-68 g) that received no treatment.

Figure 1 shows the plasma concentrations of testosterone measured by radioimmunoassay (Diagnostic Products) in each group of animals. Here and throughout this paper, data are presented as means $\pm$ standard errors of the mean. The C group had very low levels of the hormone, confirming that the gonads are the source of most endogenous testosterone. The C + T group had levels that were 67 times higher. Levels in the C + T group were not unreasonably high, however, being only about 2 1/2 times that measured in unoperated males (U in Fig. 1). Although it is not shown by the standard error bars in Fig. 1, unoperated frogs showed the greatest range in levels of circulating testosterone. This probably reflects seasonal differences in hormone levels, but the data were insufficient to reliably analyze seasonal patterns.

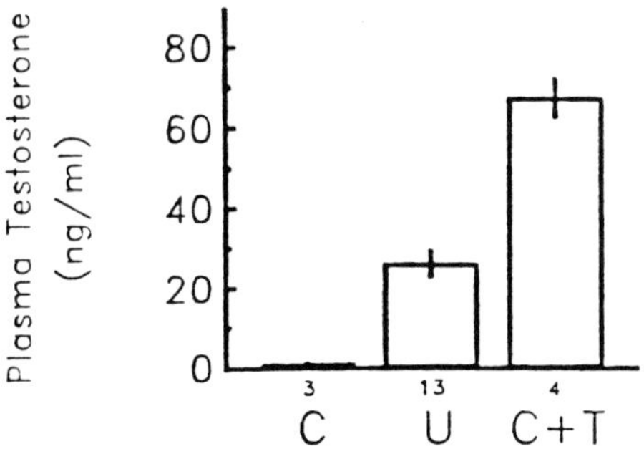

FIGURE 1: Concentration of testosterone in the plasma of frogs that were castrated (C), unoperated (U), and castrated then given implants containing testosterone (C+T). Number of frogs shown below bars. Abbreviations C and C+T also used in Figures 2 and 3.

Preliminary studies showed that differences between the C and C + T groups were quite pronounced by 8 weeks after surgery, so acute experiments on operated frogs were done at that time. For these experiments, frogs were anesthetized, decapitated, and pithed, and the FCR was dissected with its origin onto the humerus and its innervation intact. Excised muscles were superfused with Ringer solution (in mM: NaCl 116, KCl 2, $CaCl_2$ 1.8, TES buffer 5, pH 7.2) at 22-24 °C. The humerus was held in a fixed position and the wrist tendon was attached to an isometric force transducer (Grass FT03C or Statham UC2). The angle and length of the muscle was adjusted to a position that elicited maximum twitch tension. Tension signals were digitized and stored for later analysis. Contractions were elicited by stimulating the nerve with a suction electrode.

It was immediately apparent that twitches produced by nerve stimulation were more prolonged in muscles from testosterone-treated frogs. The prolonged contraction was not due to a difference in twitch rise time, which was identical in C + T frogs (24 $\pm$ 1.2 ms, N= 4) and C frogs (24 $\pm$ 0.7 ms, N=3). Instead, the difference in twitch duration was entirely due to a difference in relaxation time. The consequences of slower twitch relaxation in testosterone-treated frogs were clearly seen when the nerve was stimulated at physiological frequencies. Figure 2A shows examples of the tension generated when nerves to C + T and C muscles were stimulated at 10 Hz for one second. Because of the slower relaxation of each twitch, tension summed to a greater degree in the testosterone-treated muscles. For the records shown, the ratio of the amplitudes of the tenth to the first twitch was 45% higher for the C + T muscle. Figure 2B shows the time courses of the fall in tension following the tenth twitch. Tension was normalized to the peak of the tenth twitch so that data from 3 C + T and 3 C records could be averaged. It can be seen that relaxation was much more prolonged in C + T muscles. The time required to relax to 10% of the peak tension, for example, was 69% longer in C + T muscles (172 $\pm$ 18 ms) compared to C muscles (102 $\pm$ 31 ms).

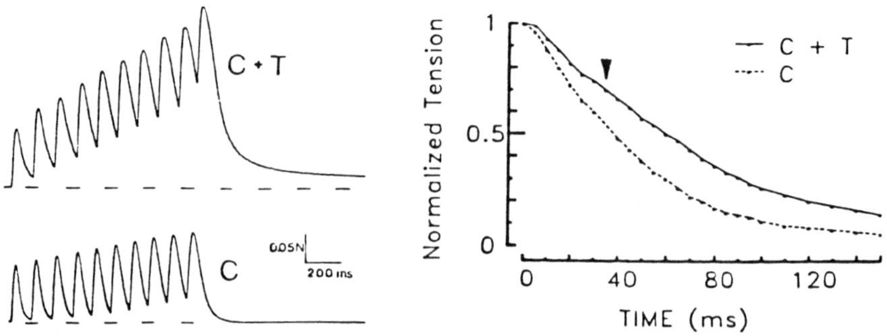

FIGURE 2: Left, isometric tension generated by 10 Hz nerve stimulation in androgen treated and control frogs. Right, time course of tension relaxation after the last twitch in a 10 Hz, 1 s train, normalized and averaged for 3 muscles in each treatment group. Arrow marks when the difference in decay is first significant. Symbols as in Figure 1.

We concluded from these measurements that testosterone-treated muscles were much more efficient in maintaining a constant low level of tension. Moreover, they achieved this efficiency without sacrificing the ability to contract rapidly. These abilities appear to be well matched to the functional demands of clasping.

As described above, previous work has shown that androgens affect muscle fiber size. In order to measure changes in fiber size, whole muscles were pinned at maximum twitch tension length and frozen in isopentane cooled with liquid N_2. Muscles were transferred to a cryostat and cross sections were cut at 10 um thickness. Sections were dried onto coverslips and reacted for myofibrillar ATPase activity using a modification of the method of Guth & Samaha (1970). Video microscope images of the stained sections were captured with a digital image processor (Image-1, Universal Imaging). The cross- sectional areas of every fiber in the section were measured with an image analysis program (IM-5000, Analytical Imaging).

Testosterone was found to have strong effects on muscle fiber size. Figure 3A shows that mean fiber cross-sectional area for a C + T muscle (5,836 $\pm$ 90 um^2, N=933 fibers) was over twice as great as that in a matched C muscle (2,670 $\pm$ 45 um^2, N=816) 8 weeks after surgery. Closer examination revealed a differential sensitivity to the androgen within muscles. Figure 3B is a schematic drawing of the FCR muscle showing its origin along the humerus and insertion at the wrist. Differences in cross-sectional area between C + T and C fibers were much more pronounced for fibers originating near the shoulder (S in Fig. 3B) than for fibers originating near the elbow (E in Fig. 3B). These differences are illustrated in Figure 3C for random samples of 100 fibers taken from each of these compartments. In the shoulder region, fiber size was over three times greater in the C + T muscle (7,985 $\pm$ 248 um^2) compared to the C muscle (2,553 $\pm$ 50 um^2). Testosterone effects were less striking in the elbow region, where fiber size in the C + T muscle (5,266 $\pm$ 275 um^2) was only one and a half times as large as in the C muscle (3,564 $\pm$ 152 um^2). These findings, together with the contraction measurements described above, suggested that there might be functional differences between fibers in different muscle compartments. Histochemical staining, which revealed a fairly homogeneous population of large fibers with high ATPase activity in the shoulder region, supported this suggestion (Regnier & Herrera, unpublished observations).

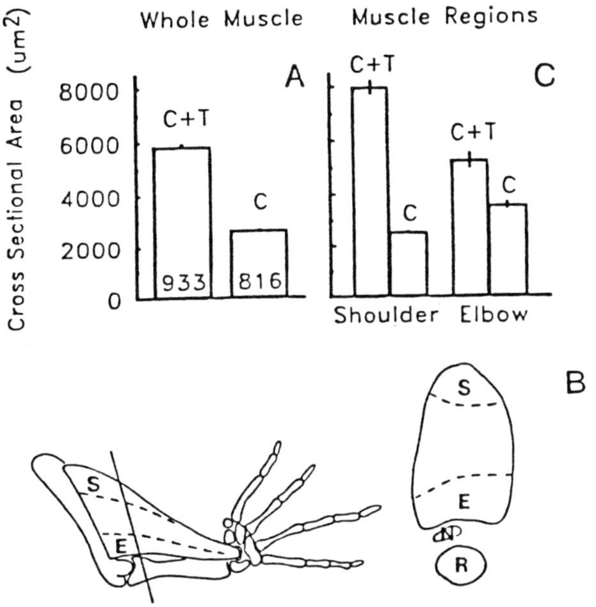

FIGURE 3: **A:** Mean cross sectional areas for all fibers (number of fibers shown) in one androgen-treated and one control muscle. **B:** Left, a schematic drawing of the FCR muscle in relation to forelimb bones. Right, a cross section taken at the level shown by the slanted line in left drawing. Abbreviations: S, shoulder region; E, elbow region; N, nerve bundle; R, radioulna bone. **C:** Mean cross sectional areas in shoulder and elbow regions of the same muscles used in A.

The compartmentalization of testosterone effects in muscles from operated frogs prompted us to examine the contractile properties of these same regions in unoperated frogs. Muscles were cut into longitudinal strips approximately along the dashed lines shown in Figure 3B. The strips, which each contained 100-200 fibers, were attached to an isometric force transducer and stimulated directly with Pt/Ir wires. The recordings of twitches shown in Figure 4 illustrate our observation that shoulder fibers both contracted and relaxed more slowly than elbow fibers. Table 1 shows averaged data from 6 shoulder and 6 elbow strips. The rise time of the twitch was 73% longer in the shoulder compartment compared to the elbow compartment. The time required to fall to 50% or 90% of peak tension was about one and a half times as long in shoulder fibers compared to elbow fibers, although the 90% relaxation time was not significantly different.

FIGURE 4: Isometric twitches of FCR muscle strips from shoulder (dashed line) and elbow (solid line) regions (see Fig. 3B). Amplitude scale is arbitrary.

Table 1: The time course (in ms) of single twitches from shoulder and elbow strips of the FCR muscle.

	Time to Peak	50% Relaxation	90% Relaxation
Shoulder	30 ± 1.7	26 ± 3.0	78 ± 15.3
Elbow	17 ± 1.5	18 ± 1.2	51 ± 6.0
P	<0.005	<0.05	>0.05

These results serve as a reminder that whole-muscle tension measurements, whether elicited by nerve or direct stimulation, give at best only a rough approximation of the performance of muscles in vivo. The smallest functional unit of motor activity is the motor unit, and motor units are known to be selectively recruited during particular movements and behaviors. The compartmentalization of testosterone effects within the FCR muscle suggests that the motor units that are most active in clasping are concentrated in the shoulder region, since fibers there are more sensitive to the hormone and have slower contractile kinetics.

4. Effects of androgens on motoneurons

We will first consider evidence that androgens affect the central end of the motoneuron, namely the soma, dendrites, and associated spinal circuitry. Later, the subject of androgen-dependent changes at neuromuscular junctions will be considered. In general, male motoneurons involved in sexual behavior are different than in females and have high levels of androgen receptors. Particularly well studied examples are motoneurons in the spinal nucleus of the bulbocavernosus in rats, which innervates muscles associated with the penis (Breedlove, 1986), syringeal motoneurons in songbirds (Luine, Nottebohm, Harding & McEwen, 1980), laryngeal motoneurons in frogs (Simpson, Tobias & Kelley, 1986), electromotoneurons in weakly electric fish (Bass, 1989), and sonic motoneurons in fish (Fine, Economos, Radtke & McClung, 1984; Bass & Marchaterre, 1989). In the case of SNB motoneurons, it has been shown that soma size and dendritic extent is sensitive to androgens even in adults (Kurz, Sengelaub & Arnold, 1986; Matsumoto, Micevych & Arnold, 1988). In *Xenopus*, there is good evidence that motoneurons supplying the FCR and coracoradialis (sternoradialis) muscles have high levels of androgen receptors (Erulkar, et al., 1981). Interestingly, androgen- concentrating motoneurons are restricted to the anterior third of the motor nuclei of both muscles. We are currently testing whether, in the case of the FCR, this compartmentalization of the motor nucleus corresponds to the compartmentalization found in the muscle.

Finally, there is intriguing evidence that androgens can cause functional changes in the spinal cord of adult *Xenopus*. Within a few hours of application to an isolated male spinal cord in vitro, dihydrotestosterone caused an increase in the excitation of coracoradialis (sternoradialis) motoneurons and an increase in the amount of ventral root activity produced by stimulation of dorsal roots (Erulkar et al., 1981). If clasping is fundamentally a spinal reflex, these physiological changes would likely enhance reflex activity and clasping in breeding males. Moreover, the changes may specifically enhance the excitability and recruitment of those motor units most involved in clasping. When results from various experimental systems are combined, it can be seen that androgens could have both short term effects on excitability and synaptic function as well as long term effects on synaptic structure.

Much less work has been done on androgen-dependent changes at neuromuscular junctions. During the development of the rat levator ani muscle, testosterone treatment postpones synapse elimination, the normal postnatal decrease in the number of motor axons contacting each muscle fiber (Jordan, Letinsky & Arnold, 1989a,b). In adult rats, castration causes a decrease in neuromuscular junction size in the bulbocavernosus (Breedlove, Balice-Gordon & Lichtman, 1988) and levator ani (Bleisch & Harrelson, 1989). In the syrinx of male birds, androgens have positive effects on the content of cholinergic enzymes and acetylcholine receptors (Luine et al., 1980; Bleisch, Luine & Nottebohm, 1984). In *Xenopus*, sexual dimorphisms in synaptic function have been described (Tobias & Kelley, 1988) as well as androgen effects on opening of acetylcholine receptor channels (Erulkar & Wetzel, 1989). Our laboratory and others have shown that frog neuromuscular junctions are highly dynamic synapses in which structure is continually being remodelled (Wernig & Herrera, 1986; Herrera, Banner & Nagaya, 1990). Some of these morphological and physiological changes can be correlated with season, suggesting that at least a portion of the plasticity seen at adult frog neuromuscular junctions may be the result of hormonal regulation. Consistent with this view, histology has revealed an extraordinary degree of remodelling in neuromuscular junctions in the FCR muscle (Nagaya & Herrera, 1989). We are currently using intracellular recording and histology to study neuromuscular junctions in C + T and C frogs to test whether synaptic function or structure is regulated by androgens.

5. Modeling of Clasping Behavior

The sterotyped nature and robustness of the clasping reflex suggest that it may be a good system for computational models of the generation and control of motor behavior. In the 1950s, attempts along these lines were made (Russell, 1954, 1960) but little theoretical work has been done since. It may in fact be premature to attempt to model the complete behavior without more detailed knowledge of the neural centers involved, their interconnections, and the electrical activity of component neurons. As a first step, however, it might soon be possible to model the spinal and peripheral components of the clasping reflex. The neurons involved are large and well studied, and the frog spinal cord is a particularly hardy electrophysiological preparation, surviving for many hours even when isolated in vitro. It would be particularly interesting to model motor control at a detailed level by considering the selective recruitment of motor units. As knowledge of the role of higher centers becomes available, models of clasping behavior could evolve by adding these modulatory influences to the basic model of the reflex.

The androgen sensitivity of clasping circuitry presents particular challenges and opportunities for modeling studies. The onset of the breeding season brings with it several striking changes in behavior. Forelimbs acquire a new motor capability, the generation of tonic tension, but do not lose their ability to tighten rapidly. The pattern of motor unit recruitment probably changes, so that muscle fibers that are best suited for the behavior are selectively activated. From such models may emerge general principles concerning how hormones orchestrate neuronal and synaptic plasticity to change the function and performance of neural circuits.

6. References

Aronson, L.R. & Noble, G.K. (1945) The sexual behavior of Anura. 2. Neural mechanisms controlling mating in the male leopard frog, *Rana pipiens*. Bull. Am. Mus. Nat Hist. 86:83-140.

Bass, A.H. (1989) Comparative neurobiology: Morphological and neurophysiological correlates of hormonal influences on the electromotor system of fishes. In: ***Perspectives in Neural Systems and Behavior***, pp. 137-150, Alan Liss, N.Y.

Bass, A.H. & Marchaterre, M.A. (1989) Sound-generating (sonic) motor system in a teleost fish (*Porichthys notatus*): Sexual polymorphisms and general synaptology of sonic motor nucleus. J. Comp. Neurol. 286:154-169.

Bleisch, W.V. & Harrelson, A. (1989) Androgens modulate endplate size and ACh receptor density at synapses in the rat levator ani muscle. J. Neurobiol. 20:189-202.

Bleisch, W., Luine, V.N. & Notttebohm, F. (1984) Modification of synapses in androgen-sensitive muscles. I. Hormonal regulation of acetylcholine receptor number in the songbird syrinx. J. Neurosci. 4:786-792.

Breedlove, S.M. (1986) Cellular analyses of hormone influence on motoneuronal development and function. J. Neurobiol. 17:157-176.

Breedlove, S.M., Balice-Gordon, R.J. & Lichtman, J.W. (1988) Neuromuscular junctions expand and shrink as muscle fibers change size: studies in an androgen sensitive muscle. Soc. Neurosci. Abstr. 14:1209

Erulkar, S.D., Kelley, D.B., Jurman, M.E., Zemlan, F.P., Schnieder, G.T. & Krieger, N.R. (1981) Modulation of the neural control of the clasp reflex in male *Xenopus laevis* by androgens: A multidisciplinary study. Proc. Nat. Acad. Sci. 78:5876-5880.

Erulkar, S.D. & Wetzel, D.M. (1989) 5-alpha-dihydrotestosterone has nonspecific effects on membrane channels and possible genomic effects on ACh-activated channels. J. Neurophysiol. 61:1036-1052.

Fine, M., Economos, D., Radtke, R. & McClung, J. (1984) Ontogeny and sexual dimorphism of motor nucleus in the oyster toadfish. J. Comp. Neurol. 225:105-110.

Gorlick, D.L. & Kelley, D.B. (1986) The ontogeny of androgen receptors in the CNS of *Xenopus laevis* frogs. Dev. Brain Res. 26:193-200.

Gorlick, D.L. & Kelley, D.B. (1987) Neurogenesis in the vocalization pathway of *Xenopus laevis*. J. Comp. Neurol. 257:614-627.

Guth, L. & Samaha, F.J. (1970) Procedure for the histochemical demonstration of actomyosin ATPase. Exp. Neurol. 28:365-367.

Hayatsu, Y., Kosaka, I., Tsutsuura, M. & Nagai, T. (1981) Potassium contracture in the tonic bundle isolated from the enlarged flexor carpi radialis muscle of the frog. Jap. J. Physiol. 31:403-415.

Herrera, A.A., Banner, L.R. & Nagaya, N. (1990) Repeated, in vivo observation of frog neuromuscular junctions: remodelling involves concurrent growth and retraction. J. Neurocytol. 19:85-99.

Hutchison, J.B. (1965) Investigations on the neural control of clasping and feeding in *Xenopus laevis* (Daudin). Behaviour 24:47-66.

Hutchison, J.B. & Poynton, J.C. (1963) A neurological study of the clasp reflex in *Xenopus laevis* (Daudin). Behaviour 22:41-63.

Jordan, C.L., Letinsky, M.S. & Arnold, A.P. (1989a) The role of gonadal hormones in neuromuscular synapse elimination in rats. I. Androgen delays the loss of multiple innervation in the levator ani muscle. J. Neurosci. 9:229-238.

Jordan, C.L., Letinsky, M.S. & Arnold, A.P. (1989b) The role of gonadal hormones in neuromuscular synapse elimination in rats. II. Multiple innervation persists in the adult levator ani muscle after juvenile androgen treatment. J. Neurosci. 9:239-247.

Kelley, D.B. (1980) Auditory and vocal nuclei in the frog brain concentrate sex hormones. Science 207:553-555.

Kelley, D.B. (1981) Locations of androgen-concentrating cells in the brain of *Xenopus laevis*: autoradiography with ^{3}H-dihydrotestosterone. J. Comp. Neurol. 199:222-231.

Kelley, D.B., Morrell, J.I. & Pfaff, D.W. (1975) Autoradiographic localization of hormone-concentrating cells in the brain of an amphibian (*Xenopus laevis*). I. Testosterone. J. Comp. Neurol. 164:47-61.

Kelley, D.B. and Pfaff, D.W. (1976) Hormone effects on male sex behavior in adult South Africa clawed frogs (*Xenopus laevis*). Horm. Behav. 7:159-182.

Kirby, A.C. (1983) Physiology of the sternoradialis muscle: sexual dimorphism and role of amplexus in the leopard frog (*Rana pipiens*). Comp. Biochem. Physiol. 74A:705-709.

Kurz, E.M., Sengelaub, D.R. & Arnold, A.P. (1986) Androgens regulate the dendritic length of mammalian motoneurons in adulthood. Science 232:395-398.

Luine, V., Nottebohm, F., Harding, C. & McEwen, B.S. (1980) Androgen affects cholinergic enzymes in syringeal motor neurons and muscle. Brain Res. 192:89-107.

Matsumoto, A., Micevych, P.E. & Arnold, A.P. (1988) Androgen regulates synaptic input to motoneurons of the adult rat spinal cord. J. Neurosci. 8:4168-4176.

McEwen, B.S., Davis, P.G., Parsons, B. & Pfaff, D.W. (1979) The brain as a target for steroid hormone action. Ann. Rev. Neurosci. 2:65-112.

Melichna, J., Gutmann, E., Herbrychova, A & Stichova, J. (1972) Sexual dimorphism in contraction properties and fibre pattern of the flexor carpi radialis muscle of the frog (*Rana temporaria* L.). Experientia 28:89-91.

Muller, E.R.A., Galavazi, G. and Szirmai, J.A. (1969) Effect of castration and testosterone treatment on fiber width of the flexor carpi radialis muscle in the male frog (*Rana temporaria* L.). Gen. Comp. Endocrinol. 13:275-284.

Nagaya, N. & Herrera, A.A. (1989) Matching of pre- and postsynatpic size in neuromuscular junctions of androgen-sensitive muscles. Soc. Neurosci. Abstr. 15:578.

Oka, Y., Ohtani, R., Satou, M. & Ueda, K. (1984) Sexually dimorphic muscles in the forelimb of the Japanese toad, *Bufo japonicus*. J. Morphol. 180:297-308.

Regnier, M. & Herrera, A.A. (1989) Androgenic regulation of motor units in sexually dimorphic muscle of the frog. Soc. Neurosci. Abstr. 15, 579.

Regnier, M. & Herrera, A.A. (1990) Differential sensitivity to androgens within sexually dimorphic muscles of male frogs (*Xenopus*). Soc. Neurosci. Abstr. 16, in press.

Rubinstein, N.A., Erulkar, S.D. & Schneider, G.T. (1983) Sexual dimorphism in the fibers of a "clasp" muscle of *Xenopus laevis*. Exp. Neurol. 82:424-431.

Russell, W.M.S. (1954) Experimental studies of the reproductive behavior of *Xenopus laevis*. I. The control mechanisms for clasping and unclasping and the specificity of hormone action. Behaviour 7:113-188.

Russell, W.M.S. (1960) Experimental studies of the reproductive behavior of *Xenopus laevis*. II. The clasp positions and mechanisms of orientation. Behaviour 15:253-283.

Schmidt, R.S. (1983) Neural correlates of frog calling. Masculinization by androgens. Horm. Behav. 17:94-102.

Segil, N., Silverman, L. & Kelley, D.B. (1987) Androgen-binding levels in a sexually dimorphic muscle of *Xenopus laevis*. Gen. Comp. Endocrinol. 66:95-101.

Simpson, H.B., Tobias, M.L. & Kelley, D.B. (1986) Origin and identification of fibers in the cranial nerve IX-X complex of *Xenopus laevis*: Lucifer yellow backfills in vitro. J. Comp. Neurol. 244:430-444.

Smith, C.L. (1938) The clasping reflex in frogs and toads and the seasonal variation in the development of the brachial musculature. J. Exp. Biol. 15:1- 9.

Thibert, P. (1986) Androgen sensitivity of skeletal muscle: Nondependence on the motor nerve in the frog forearm. Exp. Neurol. 91:559-570.

Tobias, M.L. & Kelley, D.B. (1988) Electrophysiology and dye-coupling are sexually dimorphic characteristics of individual laryngeal muscle fibers in *Xenopus laevis*. J. Neurosci. 8:2422-2429.

Wells, K.D. (1977) The social behaviour of anuran amphibians. Anim. Behav. 25:666-693.

Wernig, A. & Herrera, A.A. (1986) Sprouting and remodelling at the nerve-muscle junction. Prog. Neurobiol. 27:251-291.

Wetzel, D., Haerter, U. & Kelley, D.B. (1985) A proposed efferent pathway for mate calling in South African clawed frogs, *Xenopus laevis*. J. Comp. Physiol. A 157:749-761.

Sensorimotor Learning and the Cerebellum

Gábor T. Bartha and Richard F. Thompson
University of Southern California

Mark A. Gluck
Stanford University

Abstract. *This paper describes our current work on integrating experimental and theoretical studies of a simple form of sensorimotor learning: the classically conditioned rabbit eyelid closure response. We first review experimental efforts to determine the neural basis of the conditioned eyelid closure response and these support the role of the cerebellum as the site of the memory trace. Then our current work to bring the modeling in closer contact with the biology is described. In particular, we extend our earlier model of response topography to be more physiological in the circuit connectivity, the learning algorithm, and the conditioned stimulus representation. The results of these extensions include a more realistic conditioned response topography and reinforcement learning which accounts for an experimentally established negative feedback loop.*

1 Introduction

Classical conditioning is a valuable paradigm for the study of sensorimotor learning because of the high degree of experimental control possible. In a typical experiment, a neutral stimulus is presented in close succession with a response evoking stimulus called the unconditioned stimulus (US). After repeated pairings, the previously neutral stimulus, called the conditioned stimulus (CS), evokes a conditioned response (CR) which is similar to the US evoked response termed the unconditioned response (UR). The time between the CS onset and the US onset is defined as the inter-stimulus-interval (ISI). By varying the ISI and the intensity, duration, number, and type of stimuli, a tremendous variety of conditioning phenomena can be explored.

The rabbit eyelid closure response preparation has proved to be particularly good for classical conditioning studies because of low background response rates. In this preparation, the rabbit is restrained in a plexiglass box during the presentation of stimuli while the movement of the nictitating membrane (third eyelid) is measured. Most commonly, the CS is a light or tone and the US is a corneal air-puff or paraorbital shock. The movement of the third eyelid across the cornea corresponds to both the CR and the UR.

The rabbit eye-blink has been well characterized by nearly three decades of studies in many laboratories (see Gormezano *et al.* 1983 for review). We can only mention a few of the behavioral phenomena here. Eye-blink conditioning can occur with ISIs from 100 msec to well over 1 second. The rate of learning and the asymptotic response level are optimal at an ISI of about 250 msec. A standard training paradigm in our lab consists of a 350 msec tone, a 100 msec air-puff, and an ISI of 250 msec so that the CS and US terminate together. After about 200 trials, rabbits will give CRs on over 90% of the trials (Figure 1). If a trained rabbit is repeatedly presented with unreinforced CSs (CS alone trials), the CRs will decrease and eventually extinguish altogether.

Reacquisition following extinction is fast, usually occurring in less than 20 trials (Scavio and Thompson 1976). The CR onset initially develops near the US onset and gradually moves earlier in the trial over the course of training. In a well trained animal the CR generally peaks near the onset of the US.

More complex conditioning phenomena can be observed when more than one CS is used. Compound conditioning involves the simultaneous presentation of two or more CSs. If a rabbit is trained with a compound light and tone stimulus, CRs to a light test-stimulus are typically much smaller than in control animals trained with light alone. This phenomenon is called overshadowing (Pavlov 1929). A related phenomenon called blocking (Kamin 1969) involves an initial pretraining phase with one CS followed by compound training. With sufficient pretraining, CRs to the other CS presented alone may be absent or blocked (Kehoe 1982).

There has been a large theoretical interest in classical conditioning motivated by the desire to understand basic laws of learning. The bulk of the modeling work has been directed at the reproduction of behavioral conditioning phenomena. The Rescorla-Wagner model (1972) is one of the most influential models of classical conditioning and forms the core of many recent models. The major goal of this model is to account for stimulus context phenomena such as overshadowing, blocking, and conditioned inhibition. The Rescorla-Wagner model is closely related to the LMS rule (Widrow and Hoff 1960) which is well known in the adaptive network literature (cf. Sutton and Barto 1981).

Many models have been specifically applied to the rabbit eye-blink preparation (Moore *et al.* 1986; Klopf 1988; Sutton and Barto 1989) and these have been successful at accounting for numerous behavioral phenomena. Although the number of behavioral phenomena a model can account for is a good measure of its success, there are likely to be several alternative behavioral level models which can account for a similar set of phenomena. A more complete measure of a model's success would include how well the anatomical and physiological basis of the behavior is accounted for. The focus on learning rules, however, has resulted in models which are difficult to relate to the neural circuits subserving the behavior. In addition, the learning rules try to account for multiple phenomena which may involve different but overlapping brain structures. All of this points to a need for models which incorporate some of the voluminous data collected on the neural basis of the rabbit eye-blink.

2 Empirical studies of the neural basis of the rabbit eyelid closure response

The long-term goal of our experimental work on the rabbit eye-blink preparation is to localize the site(s) of the essential memory trace so that the mechanisms of its formation can be analyzed (see Thompson 1986, 1989 for reviews). Early on, it became clear that the memory trace cannot be localized until the entire neural network critically involved in the learned behavior has been identified and its projection neurons characterized. In this section, we briefly summarize the major empirical findings.

Decerebrate animals can retain the conditioned response (CR) if the red nucleus is not damaged, indicating that the essential memory trace must be below the level of the thalamus (Mauk and Thompson 1987). In a multiple unit recording brainstem mapping study, neural responses correlated with the behavioral responses were localized to the cerebellar cortex, the pontine nuclei, the tegmental reticular nucleus, the red nucleus, the superior colliculus, the periaqueductal gray, and various reticular regions (McCormick *et al.* 1983). Additionally, multiple neural unit (McCormick and Thompson 1984) and single unit (Foy *et al.* 1984) recordings from the lateral

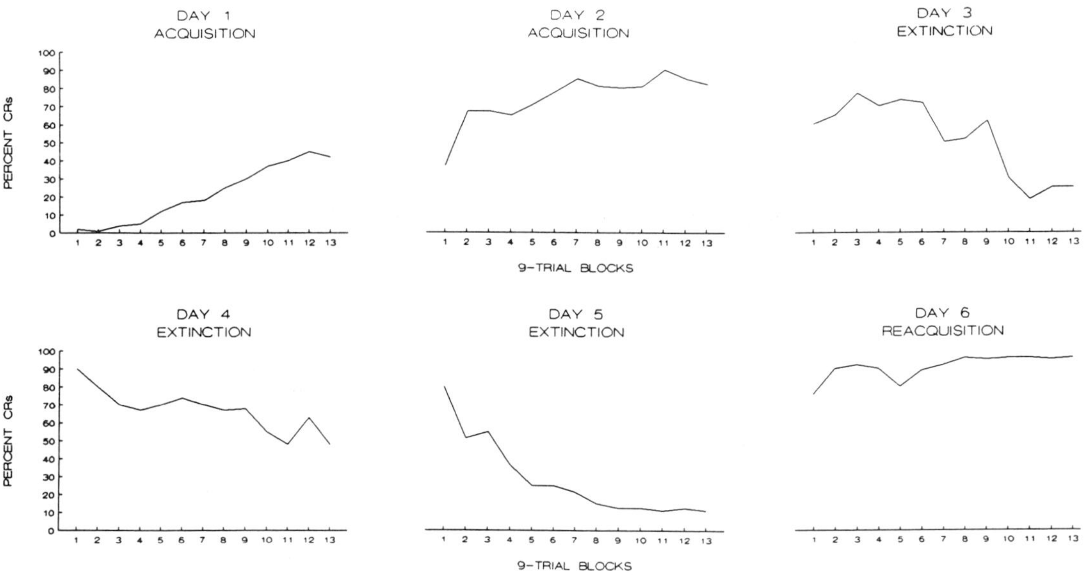

Figure 1: Behavioral phenomena of acquisition, extinction, and reacquisition of the classically conditioned eye-blink response in the rabbit. Note that extinction is a slow process but reacquisition is extremely rapid compared to initial acquisition. Composite of data from Scavio and Thompson (1979) and Berger and Thompson (1978, 1982).

interpositus nucleus show changes which precede and are correlated with the form of the CR over the course of training (Figure 2). A very small kaini[illegible]id lesion of the ipsilateral anterior interpositus nucleus abolished the CR without affecting [illegible]e UR and without degeneration in the inferior olive (Lavond *et al.* 1985). Normal learning was attained on the contralateral eye. These data suggest that the memory trace is formed in the cerebellum or systems for which the cerebellum is a mandatory efferent.

The essential CR pathway appears to consist of ipsilateral interpositus fibers projecting through the superior cerebellar peduncle, crossing to the contralateral magnocellular red nucleus, and then crossing back via the rubrobulbar tract to act on motoneurons (Chapman *et al.* 1988, Haley *et al.* 1988, Rosenfield and Moore 1983). Anatomical studies show that the motoneurons innervating the muscles responsible for the CR reside the abducens, accessory abducens, and facial nuclei (Gray *et al.* 1981; Disterhoft *et al.* 1985).

Lesions of the middle cerebellar peduncle prevent acquisition and immediately abolish retention of the CR to all CS modalities (Solomon *et al.* 1986) while lesions of the pontine nuclei can selectively abolish the CR to an acoustic CS (Steinmetz *et al.* 1987) indicating that the mossy fiber projections to the cerebellum are part of the essential CS pathway. Further support for this comes from anatomical studies (Thompson *et al.* 1986), electrophysiological studies (Logan *et al.* 1986), and electrical stimulation studies (Steinmetz *et al.* 1986). The visual CS pathway has been less studied but appears to consist of a projection to the superior colliculus which then projects to a portion of the lateral pons before making its way to the cerebellum as mossy fibers.

Small electrolytic lesions of the dorsal accessory olive (DAO) result in extinction of the CR similar to controls given CS only (McCormick *et al.* 1985). This supports the hypothesis that the

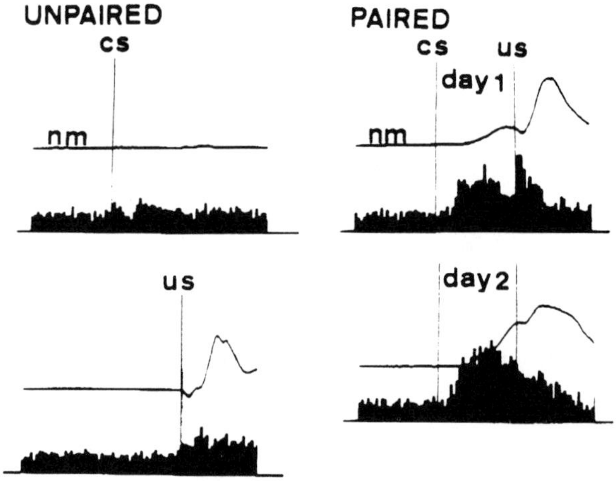

Figure 2: Nictitating membrane movement (top trace) and histograms of multiple unit recordings from the interpositus nucleus. Note the correlation between these two measurements for the paired training and that the neural responses precede the behavior and the US in time. CS-US onset interval is 250 msec. From McCormick and Thompson (1984).

memory is not formed in the DAO and that the climbing fibers make up an essential reinforcing pathway. Electrical microstimulation of mossy fibers and climbing fibers can substitute for a peripheral CS and US, producing behavioral learning (Steinmetz *et al.* 1989). This makes a strong case for the cerebellum as the site of the trace but the possible involvement of other brainstem structures has not been definitely ruled out.

Removal of cerebellar cortical lobule HVI in well trained rabbits resulted in an initial impairment or abolition of CRs but these animals were able to relearn (Lavond *et al.* 1987). Acquisition of CRs is possible in rabbits in which the cerebellar cortex is removed before training (Lavond and Steinmetz 1989) but the learning is prolonged and the CRs are of poor quality. Very large lesions of cerebellar cortex including all of HVI, flocculus, and paraflocculus has yielded mixed results on acquisition of the CR (Logan *et al.* 1989). However, if electrical stimulation of the dorso-lateral pontine nuclear region is used as a CS then a cortical lesion limited to HVI and Crus I and II can permanently abolish the CR (Knowlton *et al.* 1986). These results suggest that the cerebellar cortex is normally involved in the learning of CRs but it may not be essential. The data also support, by elimination, the interpositus as a critical site of plasticity.

Prior to training, Purkinje neurons that respond to the tone CS predominantly show increases in simple spike frequency (Donegan *et al.* 1985, Foy and Thompson 1986). After training, however, the preponderant response is a decrease in simple spike frequency that appears to be closely correlated with the occurrence of the behavioral CR (Figure 3). In this context it is worth noting the phenomenon of Long-Term-Depression (LTD) in cerebellar cortex. Repeated conjoint activation of parallel fibers and climbing fibers results in a prolonged decrease in excitability of the parallel fibers synapses on Purkinje cell dendrites (Ito *et al.* 1989). Interestingly, repeated activation of parallel fibers alone results in increased excitability of their synapses on Purkinje neurons (Sakurai 1987).

Early in training the onset of the corneal air-puff US typically evokes complex (climbing fiber) spikes in activated Purkinje neurons. However, late in training the US rarely evokes complex spikes. Evidence for direct and indirect inhibitory pathways from the interpositus to the DAO may account for this (Weiss *et al.* 1985, Nelson and Mugnoini 1989). Further

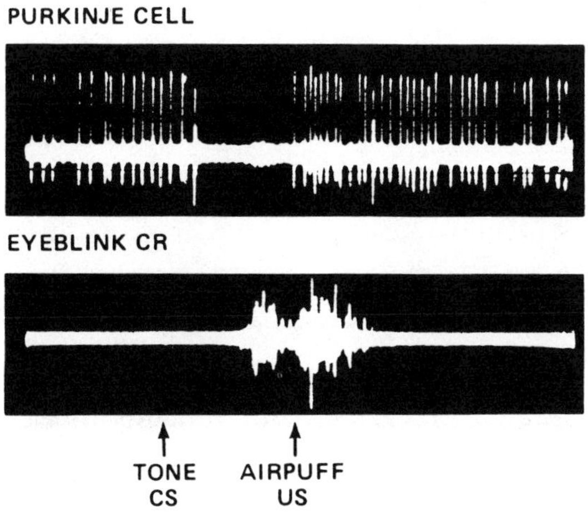

Figure 3: A Purkinje cell record from a well trained rabbit showing a decrease in simple spike firing that precedes the EMG record of the conditioned eye-blink response. CS-US onset interval is 250 msec. Unpublished data from Foy and Thompson (1986).

support for this inhibitory pathway comes from multiple unit recordings in the DAO showing a substantial decrease in US related activity after training but no decrease when the US is presented alone (Figure 4).

In a collaborative effort with Greenough, (Anderson *et al.* 1989), rabbits were conditioned using paired stimulation of mossy fibers and corneal air-puff, after which the cerebella were stained using the Golgi method. The number of bifurcating and terminating branches per distil spiny branchlet on Purkinje cells showed a significant decrease in the trained cerebellar hemisphere (HVI) versus the untrained hemisphere. Preliminary studies of rabbits receiving unpaired stimulation have shown a numerical but not statistically significant increase in number of branches on the stimulated side versus the unstimulated side (unpublished observations). These results suggest that plasticity in Purkinje cells involves morphological changes and that the direction of the change is a decrease in synaptic strength for paired stimuli and (perhaps) an increase for unpaired stimuli. Note that the phenomenon of LTD parallels these morphological changes.

When a cold probe is placed in the cerebellar white matter near the interpositus so as to selectively and reversibly inactivate the cell bodies in its vicinity, one finds that no learning occurs at all when the animal is given standard training (Lavond *et al.* 1990 and personal communication). This suggests that learning is not occurring in an afferent structure. Together with the lesion data cited above which indicates that learning takes place in the cerebellum or an afferent structure, this strongly supports the view that the memory traces are in fact formed in the cerebellum.

The majority of the experimental work on the neural basis of eye-blink conditioning has focused on finding the site of the essential memory trace and these data point with increasing strength to the cerebellum as the best candidate. In the process, considerable progress has been made in identifying the neural circuit subserving the rabbit eye-blink response and this is summarized in Figure 5. Perhaps the largest gap in the experimental data is the characterization of single unit responses in the circuit, particularly in the interpositus. This is important for the understanding of the dynamics of the neural network for the performance of the CR. One cannot fully understand the learning process without a detailed understanding of what is being learned in terms of the neural dynamics.

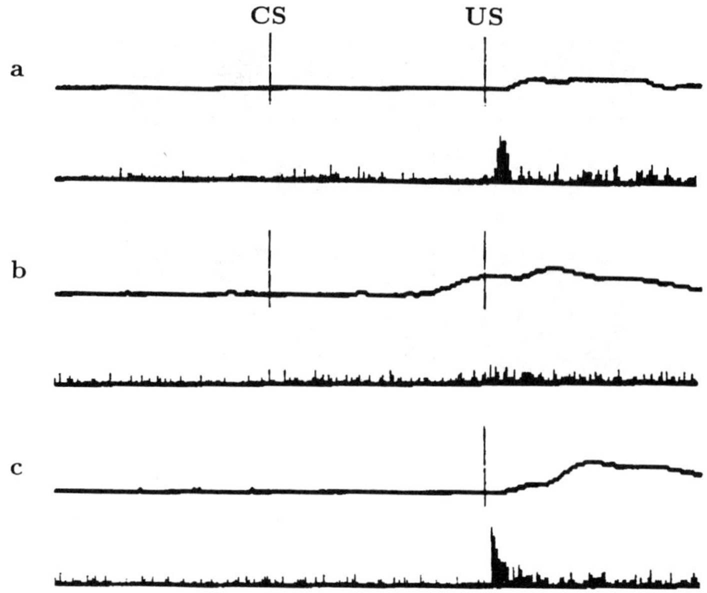

Figure 4: Nictitating membrane movement (top trace) and multiple neural unit recordings from the dorsal accessory olive. a) Early in training. b) Well trained rabbit showing substantial reduction in olivary activity. c) US alone response indicating that acquired CS related activity is responsible for the decrease. ISI is 250 msec. From Sears and Steinmetz (in press).

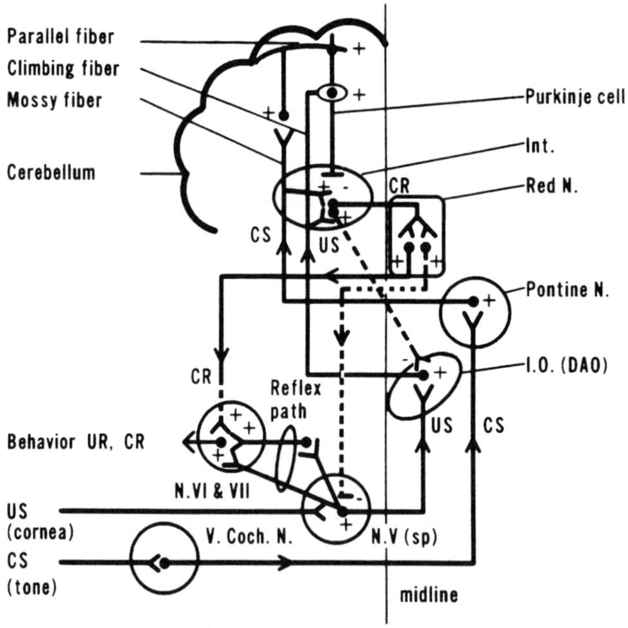

Figure 5: Simplified schematic of the neural circuits identified as subserving the conditioned eye-blink response. Pluses indicate excitatory and minuses inhibitory synaptic action. Abbreviations: Int, interpositus; IO, inferior olive; N V (sp), N VI, and N VII, spinal fifth, sixth and seventh cranial nuclei; V Coch N, ventral cochlear nucleus. Adapted from Thompson (1986).

3 Relating Models to the neural basis of the rabbit eyelid closure response

An important challenge is to show how model computations are distributed over different brain regions for the performance and learning of the eyelid closure response. One approach is to begin with a well defined and tested behavioral model and then try to map the model's computational subprocesses onto identified neural circuits (e.g., Donegan *et al.* 1989; Moore and Blazis 1989; Moore *et al.* 1989). Our approach, described here, is to start with an oversimplified model of the real-time learned behavior and incrementally incorporate biological knowledge. We then evaluate its ability to account for physiological and behavioral phenomena.

We take the LMS spectrum analyzer model of response topography as our starting point (Gluck *et al.* 1990). The main assumption of this model is that the CS is internally represented as a collection of sine waves of specific frequencies and phases. The essential function of the LMS algorithm is to approximate a sample of the Fourier coefficients of the US representation (Widrow *et al.* 1987). These coefficients correspond to the weights of the CS lines and determine the form of the CR. The better the approximation, the more the CR will resemble the internal US representation. The model accounts for the following aspects of CR topography: 1) the amplitude of the CR peaks near the locus of US onset, 2) the CR peak shifts discretely with a change in the ISI, and 3) double peaked CRs occur with mixed-ISI training. A serious limitation of the model stems from its success in approximating the US. So only when then CR is well described by the shape and timing of the US will the model appropriately characterize the CR. For the rabbit eye-blink paradigm, the LMS spectrum analyzer model fails to predict the anticipatory nature of the CR whereby the CR onset moves earlier in time over the course of training.

3.1 Model Description

We have extended our work on the LMS spectrum analyzer model of response topography by making the following changes drawn from studies of the neural basis of the eyelid closure response: 1) anatomical data on the connectivity of the cerebellar circuit, 2) a learning rule based on Purkinje cell physiology, 3) two sites of plasticity, and 4) more plausible assumptions for the CS representation.

The building blocks of our circuit are frequency coded leaky integrator elements. Neural elements of this type have been used as simple models of single neurons and populations of neurons (Arbib and Lara 1982; Baird 1990). The activation of element i is given by:

$$\Delta a_i = (\sum_j w_{ij} f_j - a)\Delta t / \tau_i$$

where the w_{ij} are the weights of the incoming frequency coded signals, f_j, and τ_i is the time constant. The output of the element is 0 until the activation level exceeds a threshold, θ, at which point the output function is $f_i = r_i + s_i(a_i - b_i)$. If the activation exceeds a saturation value, σ, the output is constrained from further increases. For all simulations in this paper, $\theta = 15, \sigma = 200, s_i = .5$, and $r_i = b_i = 0$.

The basic circuit for our model incorporates anatomical data on the connectivity between the cerebellar cortex, the interpositus nucleus, and the inferior olive (see Figure 5). A single neural element is used to represent populations of cells in each of these structures. Information about the US passes through the inferior olive and has collaterals connecting to both the interpositus element and the cortex element. CS information also has collaterals to both the interpositus and the cerebellar cortex. The output of the cortex element, which represents the activity of Purkinje cells, has an inhibitory connection to the interpositus element. The output of the interpositus has an inhibitory connection to the inferior olive element. Based on multiple unit recordings in the interpositus nucleus (see Figure 2), we take the activity in the interpositus element as a measure of the CR.

Like the LMS spectrum analyzer model of response topography, this model represents the CS internally as a set of sinusoidal oscillations of firing rates lasting for the duration of the CS. We assume that precerebellar as well as cerebellar structures are involved in generating these oscillations (see Discussion). The LMS spectrum analyzer model uses CS signals which correspond to a large number of terms in a Fourier series. An important change in the current model is that the CS representation consists of only a few low frequency components. In addition, these components do not have such specifically arranged frequencies which is more biologically plausible. The simulations shown in this paper use 12 CS signals with periods of 1750, 1400, 1050, 700, 350, and 200 ms and phases of π and $3\pi/2$ radians. The firing rate of all signals oscillate between 0 and 200 Hz. The internal representation of the 100 msec air-puff US to the inferior olive neural element is a short (10 ms) high intensity (432 Hz) burst. This is qualitatively consistent with recordings of complex spikes in the cerebellar cortex and the physiological properties of the inferior olive (see Figure 4).

The empirical evidence reviewed in Section 2 has led us to incorporate plasticity (learning) in the model at the CS connections to both the cerebellar cortex and the interpositus. We base our learning rule at the CS-cortex synapses on a rule by Houk *et al.* (1989) which was derived from considerations of Purkinje cell LTD. Our learning rule is:

$$\Delta w_{Pj} = [\alpha - \beta(\lambda f_c + f_P)] f_j \Delta t$$

where α, β, and λ control learning rates, f_c is the climbing fiber activity, f_P is the Cortex/Purkinje cell output, and the f_j are the CS signals which can be identified with mossy and parallel fiber activities. Note that f_c does not contribute to the element activation in our model. Although our rule appears quite distinct from that presented by Houk *et al.*, the two rules reduce to the same form if $\alpha = \beta$ (in both rules) and $\lambda = 1$. The remaining difference between the learning rules is that we use graded frequency coded outputs which we feel are more realistic than Houk *et al.*'s assumption of two-state neurons (maximum firing rate or quiescent) for both the Purkinje cell and inferior olivary cell.

Houk *et al.* (p. 26) suggested that their rule does not fit into the usual categories associated with network learning algorithms. We can see otherwise by representing the Houk rule as a contingency table (Table 1). In this form its relationship to Rosenblatt's (1962) perceptron rule (Table 2) is evident. In order to make a full correspondence between the learning rules we must

f_c \ f_P	0	1
0	$+\alpha f_j$	0
1	0	$-\alpha f_j$

Table 1: Houk rule contingency table.

	$A \cdot \phi \leq 0$	$A \cdot \phi > 0$
F^+	$+\phi(X)$	0
F^-	0	$-\phi(X)$

Table 2: Perceptron rule contingency table.

note that: 1) The f_j can vary continuously in time whereas the activity patterns, $\phi(X)$, in the perceptron rule are fixed. If one applies the Houk learning rule at small discrete intervals then the patterns can be considered fixed during these intervals. 2) Patterns which occur during the US period are in the F^- class which means the Purkinje cell output, f_P, should be low for these patterns. 3) Purkinje cells in the Houk rule must be nodes which linearly sum their inputs, have a threshold of 0, and have binary output states. Houk *et al.* actually used two thresholds which yields a hysteresis effect.

Based on cortical lesion data and the series arrangement of the cortex and interpositus, we assume that learning in the cortex and interpositus elements works toward a common goal in terms of the cerebellar output. So we propose an interpositus learning rule that is the excitatory cell analog of the cortex/Purkinje cell rule. The analog is given by inverting the rows in Table 1 and allowing for graded frequency coded outputs and separate learning rates for weight increases and decreases. More formally, the learning rule for the CS-interpositus synapses is:

$$\Delta w_{Ij} = \alpha_I(\lambda_I f_c - \beta_I f_I) f_j \Delta t$$

where α_I, β_I, and λ_I control interpositus learning rates, f_I is the interpositus neural activity, and the other parameters are as before. This rule is equivalent to the Widrow-Hoff LMS rule, upon which the LMS spectrum analyzer model of response topography (Gluck *et al.* 1990) was based.

3.2 Simulation Results

We simulated CR acquisition using the standard delay paradigm described in Section 1. We assume that a high level of background activity on the parallel fibers exists during the non-CS periods and this is responsible for the baseline activity of the cortex/Purkinje cell element (Figure 6a). The CS-cortex weights are initialized so that the CS is a neutral stimulus. No baseline activity is assumed for the interpositus element. The cortex/Purkinje cell element's activity completely depresses around the time of the US onset (Figure 6bcd) which compares favorably with the neurophysiology (Figure 3). Activity in the interpositus, which we have taken to represent the CR, initially begins after the US onset (Figure 6b). As training progresses the CR gradually moves forward in time and becomes larger in amplitude (Figure 6cd). The interpositus element activity (Figure 6d) is reasonably isomorphic with both multiple unit recordings and the eye-blink behavior (Figure 2). So this model accounts for important features of response topography that the LMS spectrum analyzer model could not account for; namely 1) the CR advances in time during the course of training to precede the US onset and 2) the CR is isomorphic with behavior using a plausible US representation. These advances are partly the result of the inability of the learning rule to reconstruct the US representation using low frequency oscillations. The topography of the CR is now more a function of the CS representation than a function of the US. The low frequencies effectively force the CR to be extended in time both before and after the US period. This accounts for both the proper form of the CR and the CR onset preceding the US.

In addition to the development of CR topography, the inhibitory projection of the interpositus element reduces the activity of the inferior olive element to less than one fifth its original value by the end of training (Figure 6d). The residual level of inferior olive activity is largely a function of the interpositus learning rate parameters. Cerebellar cortex learning also contributes but in a less direct way. If λ_I, which controls weight increases, is very large relative to β_I, which controls the rate of decreases, then the reinforcement signal from the inferior olive can be reduced to a relatively small value and still maintain the CR. To be consistent with experiment (Figure 4), the activity in the inferior olive must be reduced substantially but not completely. If the US signal from the inferior olive were completely shut off the CR would extinguish in this model. Larger learning rates for weight increases relative to learning rates for decreases are also consistent with faster behavioral acquisition relative to extinction (Figure 1).

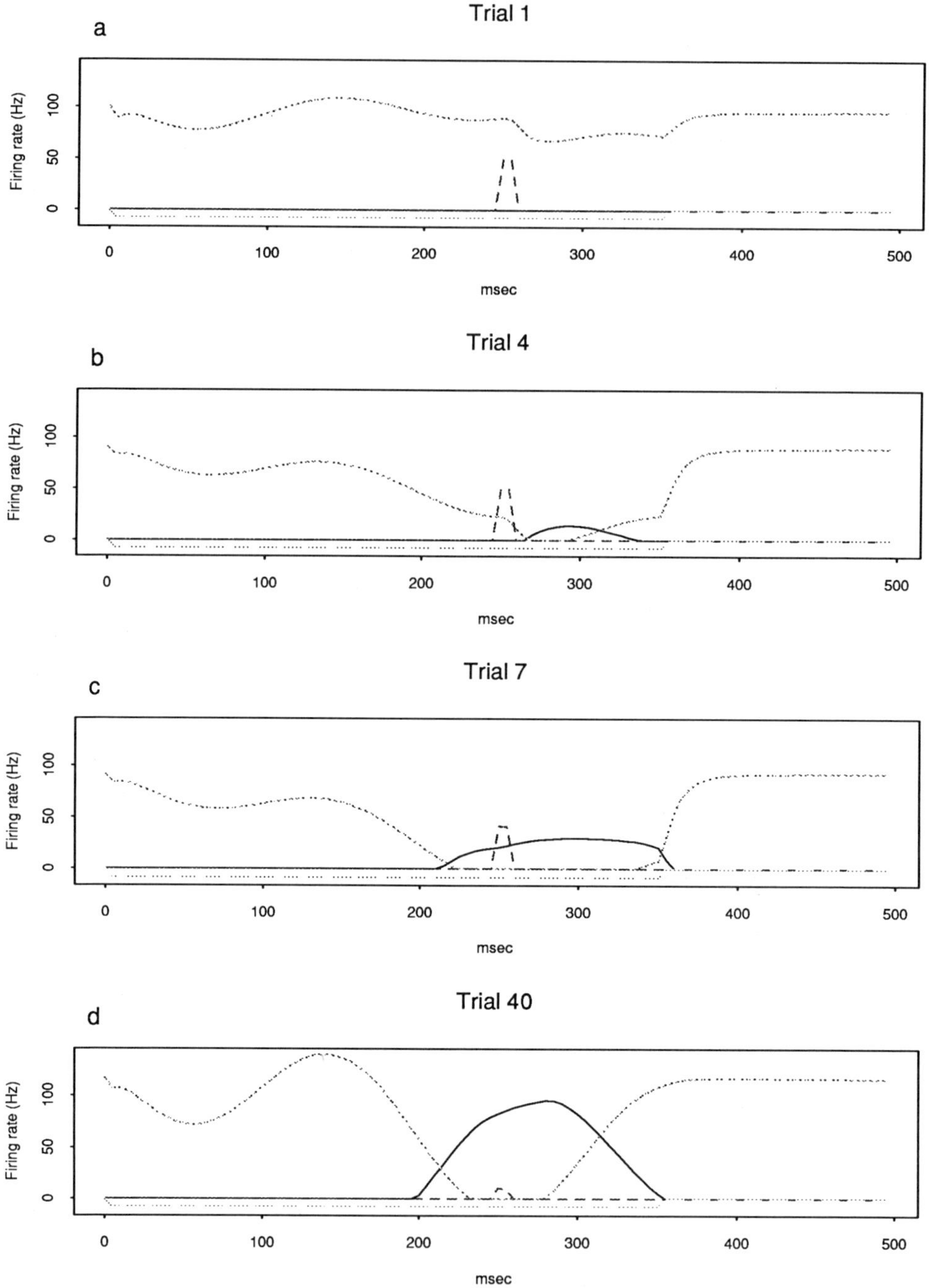

Figure 6: Simulation of interpositus (solid line), Purkinje cell (dotted line), and inferior olive (dashed line) activity during conditioned response acquisition. Note that the interpositus activity advances in time to precede the US (inferior olive) onset, the Purkinje cell completely inactivates around the time of the US, and the inferior olive activity is substantially reduced late in training. The negative going trace represents the CS period. To facilitate comparison to Figure 4, full scale for inferior olive activity is 400 Hz. Parameters: $\alpha = 2. \times 10^{-6}, \beta = 2. \times 10^{-8}, \lambda = 15.5, \alpha_I = 5. \times 10^{-8}, \beta_I = .01, \lambda_I = 1, w_{IP} = -1.5, w_{OI} = -4., \tau_P = \tau_I = 10, \tau_O = 1., \Delta t = 1$ ms.

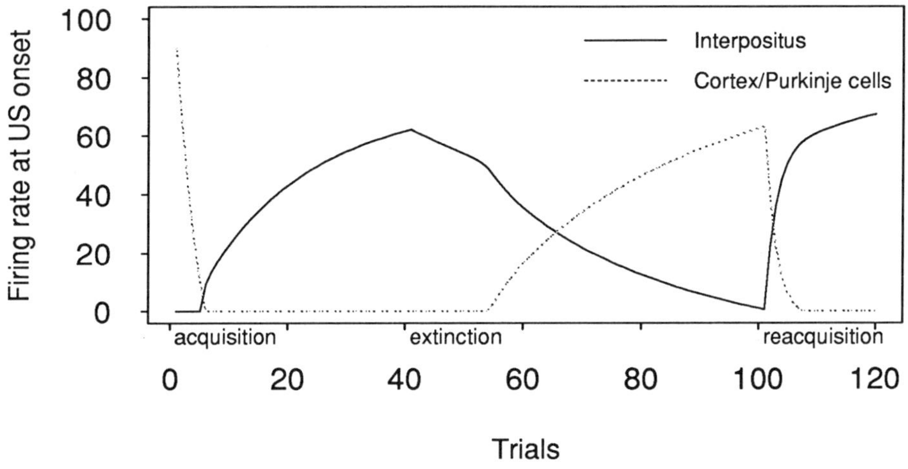

Figure 7: Simulation of acquisition, extinction, and reacquisition learning curves for interpositus and cortex. Note that reacquisition is faster than initial acquisition. Since the model produces a deterministic response, we have taken the interpositus firing rate at the onset of the US as a measure of the CR learning which can be compared with percentage CRs reported experimentally. Note that the learning rates have been increased for practical reasons but the qualitative features of the learning curves are preserved. Parameters identical with Figure 6 except: $\beta_I = .05, \lambda_I = .75$.

In simulations of acquisition, extinction, and reacquisition, we find that reacquisition is more rapid than initial acquisition (Figure 7). Faster reacquisition is primarily the result of the two-layer learning architecture as suggested by the theoretical work of Kehoe (1988). Kehoe's model uses a two-layer feedforward network in which the US projects to both layers. His model depends on high threshold values for reacquisition savings. The high thresholds allow behavioral extinction to occur well before the weights decrease to their initial values. In addition, once the first layer stops activating the second layer no more unlearning can occur in the second layer even if extinction training is continued after behavioral extinction. Kehoe's scheme does not apply directly to the cerebellar architecture because of the inhibitory action of the cortex and more importantly, because of the CS collaterals to the interpositus. In our model, fast reacquisition depends strongly on different learning rates for synaptic weight increases relative to decreases. Unlearning in the interpositus is very slow so that the cortex/Purkinje element unlearning controls the rate of behavioral extinction. When the activity of the interpositus is eliminated by cortical inhibition, no further interpositus weight decrements occur. During reacquisition, the cortex relearns quickly releasing inhibition of the interpositus (which has mostly intact weights) and the net result is fast behavioral reacquisition. Another factor which contributes to faster reacquisition in the model is that the cortex weights following extinction can be closer to the weights acquired for the CR than the initial weights.

We are currently exploring a variety of alternative schemes for mapping multiple CSs onto this circuit structure. One possibility is to conceive of multiple CSs as distinct elements in the cerebellar cortex and the interpositus sharing the same inferior olive signals. In a preliminary exploration of this idea, we have assumed that different CS signals (CS1 and CS2) are segregated in the cortex but combined in the coarser representation of the interpositus. The combined output of the interpositus provides negative feedback to the inferior olive element. This provides a possible biological mechanism to account for blocking (Kamin 1969). If we train this model with CS1, it will eventually come to inhibit inferior olive activity. Continued training with a compound CS1+CS2 will promote little new learning because the reinforcement signal is too

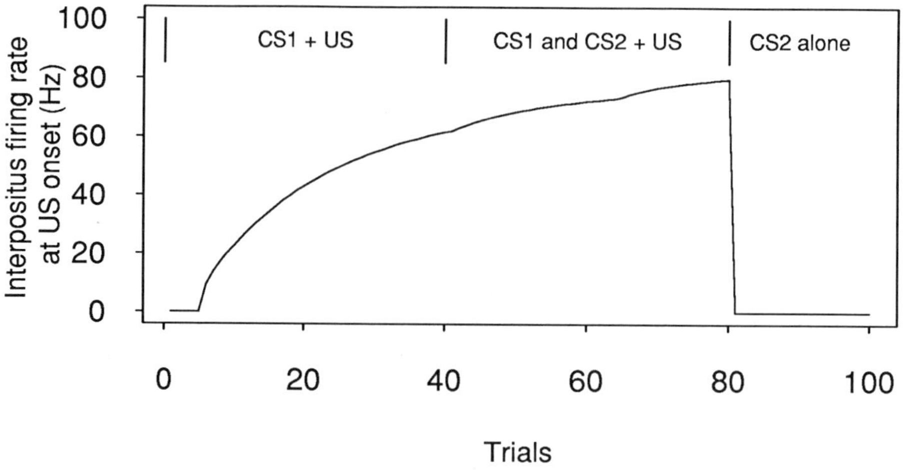

Figure 8: Simulation of blocking. Note that there is no response to the CS2 alone after 40 trials of compound training with CS1. Parameters are the same as for Figure 7 for each of the two pairs of cortex and interpositus elements.

weak. Subsequent tests with CS2-alone result in no CR (Figure 8) because the learning has been blocked by the presence of the first CS. If the negative feedback pathway is "lesioned" in the model then the second CS will not be blocked although it will still be reduced relative to training alone due to the properties of the synaptic learning rules and their relationship to the Rescorla-Wagner account of blocking. Thus, blocking arises from two processes: 1) the learning rule at the neural element level, and 2) the negative-feedback loop to the inferior olive inhibiting the ability of the US to promote learning. This two-part explanation of blocking is functionally related to the SOP model of conditioning (Wagner 1981) as described more fully in Donegan *et al.* (1989).

4 Discussion

The negative feedback from the interpositus to the inferior olive poses a paradox for models which assume that climbing fibers make up the essential reinforcing pathway. Reinforcement is required for the acquisition and maintenance of the CR but the CR is correlated with the negative feedback that acts to eliminate the reinforcement. Strong inhibition of the inferior olive by CR related activity in the interpositus would result in a rabbit for which paired CS/US trials would be equivalent to extinction training (see also Gluck and Thompson 1987 for a discussion of a related problem in models of *Aplysia* conditioning). However, we have shown in our model that, with an appropriate choice of learning rates for weight increases and decreases, the US signal can be reduced to levels consistent with experimental observation while still providing sufficient reinforcement to maintain the CR. The choice of learning rates that account for this physiological data is also consistent with the behavioral data on faster acquisition relative to extinction.

An alternate account of how CRs can be maintained in the presence of the negative feedback comes from the behavioral observation that even a well-trained animal sometimes fails to make a CR. When this happens there will be no inhibition of the US evoked activation of climbing fibers and so reinforcement will occur. This account also depends on faster acquisition rates relative to extinction rates.

What is the role of the negative feedback loop? One possible answer, suggested by our simulations of multiple CS phenomena, is that it could function to implement blocking. However, as noted in Section 3.2, the synaptic learning rules in our model are sufficient themselves to implement blocking without the negative feedback loop. If it were demonstrated that blocking depends on the negative feedback loop, this would argue against an account of blocking at the neural element level. In addition to its role in classical conditioning, the negative feedback may also subserve some motor control function that our model is not able to address.

Gormezano *et al.* (1983) suggested that the following experimental data is evidence that extinction is a separate process from the unlearning of the association formed during acquisition: 1) extinction is slow, 2) reacquisition is much faster than initial acquisition, and 3) animals show spontaneous recovery of CRs between extinction training sessions. Our model accounts for the first two of these without the need for a separate process. Prolonged extinction is accounted for simply by having different rates for synaptic weight increments and decrements. Faster reacquisition is the result of the circuit connectivity and relative learning rates. The current model does not account for spontaneous recovery.

The VET model (Desmond 1988, Desmond and Moore 1988, Moore *et al.* 1989) and the spectral timing model (Grossberg and Schmajuk 1989) are two major alternative accounts of rabbit eye-blink topography. These models are are related to our model in the sense that the CS is internally represented as an spectrum of signals. In the VET model, the CS produces a spectrum of pure time delays using a tapped delay-line. In the spectral timing model, the CS drives a population of neurons that have a spectrum of time constants. Both these models are successful at reproducing many of the important features of response topography described here.

Our major critique of the VET model is that it assumes as input to the model a "CR image" signal which is essentially the desired CR topography except for timing. Moore *et al.* (1989) suggest that the CR image is formed in the supratrigeminal reticular formation, the sensory trigeminal system, or other brain stem structures. How the CR image is formed is unspecified. The learning rule combined with the tapped delay-line architecture effectively allows the model to shift the CR image to occur at the proper ISI. Another problem with the VET model is that the delay-lines must be on the order of hundreds of msec which is physiologically unrealistic.

One critique of the spectral timing model is that no proposal is made as to how its computations are distributed over different brain structures. In addition, Grossberg and Schmajuk (1989) propose single leaky integrator neuron time constants as long as 400 msec which is implausible. Perhaps some other mechanism could account for these long time constants but no such mechanism is suggested in their work.

An important critique of our model is that the CS representation by oscillations currently has no experimental support. In defense of our proposal we note that low frequency oscillations do exist in the brain; the hippocampal theta rhythm is a good example. In addition, there are rich recurrent connections in the precerebellar structures such as those between the tegmental reticular nucleus and the interpositus which could support oscillations. In light of the current alternative ideas and empirical data, the hypothesis of neural oscillations is viable and will continue to influence some of our future experimental and theoretical work.

In our modeling we have broken away from consideration of purely behavioral phenomena by trying to answer the question of how computations are distributed over different brain regions. The current model has made several advances over our earlier LMS spectrum analyzer model of response topography. These can be summarized as follows: the model 1) incorporates important anatomical and physiological data, 2) accounts for more features of response topography, 3) accounts for additional behavioral phenomena including reacquisition and blocking, 4) accounts for the reduction of inferior olive activity in a way consistent with the maintenance of the CR, 5) provides an account of the role of the negative feedback loop, and 6) shows how extinction does not require a separate process in the context of the cerebellar network.

We recognize that our integration of experimental and theoretical work on the classical conditioning of the rabbit eyelid closure response is still at an early stage. Nevertheless, important biological data has been incorporated into our model and the predictions of the model are beginning to influence our experimental work. We are confident that future efforts will tighten the cycle of interaction between theory and experiment. The elucidation of the processes of classical conditioning may eventually lead to a better understanding of more complex learning phenomena. In addition, an understanding of the cerebellar substrates and dynamics for a simple sensorimotor behavior may lead to a broader understanding of cerebellar function and motor control.

Acknowledgements

This research was supported by an Office of Naval Research grant (N00014-88-K-0112) to R.F. Thompson and M.A. Gluck.

References

Anderson, B.J., Lee, S., Thompson, J., Steinmetz, J.E., Logan, C.G., Knowlton, B.J., Thompson, R.F., Greenough, W.T. 1989. Decreased branching of spiny dendrites of rabbit cerebellar Purkinje neurons following associative eyeblink conditioning. *Soc. Neurosci. Abstr.* **15**, 640.

Arbib, M.A., Lara, R. 1982. A neural model of the role of the tectum in prey-catching behavior. *Biological Cybernetics* **44**, 185-196.

Baird, B. 1990. Associative memory in a simple model of oscillating cortex. In *Advances in neural information processing systems 2*, D. Touretzky, ed., Morgan Kaufmann Publishers, San Mateo, CA.

Berger, T.W., Thompson, R.F. 1982. Hippocampal cellular plasticity during extinction of classically conditioned nictitating membrane behavior. *Behavioral Brain Research* **4**, 63-76.

Berger, T.W., Thompson, R.F. 1982. Neuronal plasticity in the limbic system during classically conditioning of the rabbit nictitating membrane response. *Brain Research* **145**(2), 323-346.

Chapman, P.F., Steinmetz, J.E., Thompson, R.F. 1988. Classical conditioning does not occur when direct stimulation of the red nucleus or cerebellar nuclei is the unconditioned stimulus. *Brain Research* **442**, 97-104.

Desmond, J.E. 1988. Temporally adaptive conditioned responses: Representation of the stimulus trace in neural-network models. COINS Technical Report 88-80.

Desmond, J.E., Moore, J.W. 1988. Adaptive timing in neural networks: The conditioned response. *Biological Cybernetics* **58**, 405-415.

Disterhoft, J.F., Quinn, K.J., Weiss, C., Shipley, M.T. 1985. Accessory abducens nucleus and conditioned eye retraction/nictitating membrane extension in rabbit. *J. Neuroscience* **5**, 941-950.

Donegan, N.H., Gluck, M.A., Thompson, R.F. 1989. Integrating behavioral and biological models of classical conditioning. In *Computational models of learning in simple neural systems*, R. Hawkins, G. Bower, eds., Academic Press, New York, NY.

Donegan, N.H., Wagner, A.R. 1987. Conditioned diminution and facilitation of the UCR: A sometimes-oponent-process interpretation. In *Classical conditioning II: Behavioral, neurophysiological, and neurochemical studies in the rabbit*, I. Gormezano, W. Prokasy, R. Thompson, eds., Lawrence Erlbaum Associates, Hillsdale, NJ.

Donegan, N.H., Foy, M.R., Thompson, R.F. 1985. Neuronal responses of the rabbit cerebellar cortex during the performance of the classically conditioned eyelid response. *Soc. Neurosci. Abstr.* **11**, 245.8.

Foy, M.R., Thompson, R.F. 1986. Single unit analysis of Purkinje cell discharge in classically conditioned and untrained rabbits. *Soc. Neurosci. Abstr.* **12**, 518.

Foy, M.R., Steinmetz, J.E., Thompson, R.F. 1984. Single unit analysis of cerebellum during classically conditioned eyelid response. *Soc. Neurosci. Abstr.* **10**, 122.

Gluck, M.A., Reifsnider, E.S., Thompson, R.F. 1990. Adaptive signal processing and the cerebellum: Models of classical conditioning and VOR adaptation. In *Neuroscience and Connectionist Theory*, M. Gluck, D. Rumelhart, eds., Lawrence Erlbaum Associates, Hillsdale, NJ.

Gluck, M.A., Thompson, R.F. 1987. Modeling the neural substrates of associative learning and memory: A computational approach. *Psychological Review* **94**, 176-191.

Gormezano, I., Kehoe, E.K., Marshal, B.S. 1983. Twenty years of classical conditioning with the rabbit. *Progress in psychobiological and physiological psychology* **10**, 197-275.

Gray, T.S., McMaster, S.E., Harvey, J.A., Gormezano, I. 1981. Localization of retractor bulbi motoneurons in the rabbit. *Brain Research* **226**, 93-106.

Grossberg, S., Schmajuk, N.A. 1989. Neural dynamics of adaptive timing and temporal discrimination during associative learning. *Neural Networks* **2**, 79-102.

Haley, D.A., Thompson, R.F., Madden IV, J. 1988. Pharmacological analysis of the magnocellular red nucleus during classical conditioning of the rabbit nictitating membrane response. *Brain Research* **454**, 131-139.

Houk, J.C., Singh, S.P., Fisher, C., Barto, A.G. 1989. An adaptive sensorimotor network inspired by the anatomy and physiology of the cerebellum. COINS Technical Report 89-108.

Ito, M. 1989. Long term depression. *Annual Review of Neuroscience* **12**, 85-102.

Kamin, L.J. 1969. Predictability, surprise, attention and conditioning. In *Punishment and aversive behavior*, B. Campbell, R. Church, eds., Appleton-Century-Crofts, New York, NY, 279-296.

Kehoe, E.J. 1988. A layered network model of associative learning: Learning to learn and configuration. *Psychological Review* bf 95(4), 411-433.

Kehoe, E.J. 1982. Overshadowing and summation in compound stimulus conditioning of the rabbit's nictitating membrane response. *J. experimental psychology: Animal behavior processes* **8**, 313-328.

Klopf, H.A. 1988. A neuronal model of classical conditioning. *Psychobiology* **16**(2), 85-125.

Knowlton, B.J., Lavond, D.G., Thompson, R.F. 1986. The effect of lesions of cerebellar cortex on retention of the classically conditioned eyeblink response when stimulation of the lateral reticular nucleus is used as the conditioned stimulus. *Behavioral and Neural Biology* **49**, 293-301.

Lavond, D.G., Kanzawa, A.S., Esquenazi, V., Clark, R.E., Zhang, A.A. 1990. Effects of cooling interpositus during acquisition of classical conditioning. *Soc. Neurosci. Abstr.* **16**

Lavond, D.G., Steinmetz, J.E. 1989. Acquisition of classical conditioning without cerebellar cortex. *Behavioral Brain Research* **33**, 113-164.

Lavond, D.G., Steinmetz, J.E., Yokaitis, M.H., Thompson, R.F. 1987. Reacquisition of classical conditioning after removal of cerebellar cortex. *Exp. Brain Research* **67**, 569-593.

Lavond, D.G., Hembree, T.L., Thompson, R.F. 1985. Effect of kainic acid lesions of the cerebellar interpositus nucleus on eyelid conditioning in the rabbit. *Brain Research* **326**, 179-185.

Logan, C.G., Lavond, D.G., Thompson, R.F. 1989. The effects of combined lesions of cerebellar cortical areas on the acquisition of the rabbit conditioned nictitating membrane response. *Soc. Neurosci. Abstr.* **15**, 640.

Logan, C.G., Steinmetz, J.G., Thompson, R.F. 1986. Acoustic related responses recorded from the region of the pontine nuclei. *Soc. Neurosci. Abstr.* **12**, 754.

Mauk, M.D., Thompson, R.F. 1987. Retention of classically conditioned eyelid responses following acute decerebration. *Brain Research* **403**, 89-95.

McCormick, D.A., Steinmetz, J.E., Thompson, R.F. 1985. Lesions of the inferior olivary complex cause extinction of the classically conditioned eyeblink response. *Brain Research* **359**, 120-130.

McCormick, D.A., Thompson, R.F. 1984. Neuronal responses of the rabbit cerebellum during acquisition and performance of a classically conditioned nictitating membrane-eyelid response. *J. Neuroscience* **4**, 2811-2822.

McCormick, D.A., Lavond, D.G., Thompson, R.F. 1983. Neuronal responses of the rabbit brainstem during performance of the classically conditioned nictitating membrane (NM)/eyelid response. *Brain Research* **271**, 73-88.

Moore, J.W., Desmond, J.E., Berthier, N.E. 1989. Adaptively timed conditioned responses and the

cerebellum: a neural network approach. *Biological Cybernetics* **62**, 17-28.
Moore, J.W., Blazis, D.E.J. 1989. Simulation of a classically conditioned response: A cerebellar network implementation of the Sutton-Barto-Desmond model. In *Neural models of plasticity: Experimental and theoretical approaches*, J. Byrnes, W. Berry, eds., Academic Press, New York, NY, 187-207.
Moore, J.W., Desmond, J.E., Berthier, N.E., Blazis, D.E.J., Sutton, R.S., Barto, A.G. 1986. Simulation of the classically conditioned nictitating membrane response by a neuron-like adaptive element: Response topography, neuronal firing, and interstimulus intervals. *Behavioral Brain Research* 21, 143-154.
Nelson, B.J., Mugnaini, E. 1989. Origins of GABAergic inputs to the inferior olive. *Exp. Brain Research* Series 17. Springer-Verlag, Berlin.
Pavlov, I. 1927. *Conditioned reflexes.* Oxford University Press, London.
Rescorla, R.A., Wagner, A.R. 1972. A theory of Pavlovian conditioning: Variations in the effectiveness of reinforcement and nonreinforcement. In *Classical Conditioning II*, A. Black, W. Prokasy, eds., Appleton-Century-Crofts, New York, NY, 64-99.
Rosenblatt, F. 1962. *Principles of Neurodynamics.* Spartan, New York.
Rosenfield, M.E., Moore, J.W. 1983. Red nucleus lesions disrupt the classically conditioned nictitating membrane response in rabbits. *Behavioral Brain Research* **10**, 393-398.
Sakurai, M. 1987. Synaptic modification of parallel fibre-Purkinje cell transmission in *in vitro* guinea-pig cerebellar slices. *J. Physiology* **394**, 463-480.
Scavio M.J., Thompson, R.F. 1979. Extinction and reacquisition performance alternations of the conditioned nictitating membrane response. *Bulletin of the Psychonomic Society* **13**, 57-60.
Sears, L.L. and Steinmetz, J.E. (in press). Dorsal accessory inferior olive activity diminishes during acquisitioı of the rabbit classically conditioned eyelid response. *Brain Research.*
Solomon, P.R., Lewis,J.L., LoTurco, J., Steinmetz, J.E., Thompson, R.F. 1986. The role of the middle cerebellar peduncle in acquisition and retention of the rabbit classically conditioned nictitating membrane response. *Bulletin of the Psychonomic Society* **24**(1), 75-78.
Steinmetz, J.E., Lavond, D.G., Thompson, R.F. 1989. Classical conditioning in rabbits using pontine nucleus stimulation as a conditioned stimulus and inferior olive stimulation as an unconditioned stimulus. *Synapse* **3**, 225-233.
Steinmetz, J.E., Logan, C.G., Rosen, D.J., Thompson, J.K., Lavond, D.J., Thompson, R.F. 1987. Initial localization of the acoustic conditioned stimulus projection system to the cerebellum essential for classical eyelid conditioning. *Proc. Nat. Acad. Sci.* **84**, 3531-3535.
Steinmetz, J.E., Rosen, D.J., Chapman, P.F., Lavond, D.G., Thompson, R.F. 1986. Classical conditioning of the rabbit eyelid response with a mossy-fiber stimulation CS: I. Pontine nuclei and middle cerebellar peduncle stimulation. *Behavioral Neuroscience* **100**, 878-887.
Sutton, R.S., Barto, A.G. 1981. Toward a modern theory of adaptive networks: Expectation and prediction. *Psychological Review* **88**, 135-170.
Sutton, R.S., Barto, A.G. 1989. Time-Derivative models of Pavlovian reinforcement. In *Learning and Computational Neuroscience*, J. Moore, M. Gabriel, eds., MIT press, Boston, MA.
Thompson, R.F. 1989. Neural circuit for classical conditioning of the eyelid closure response. In *Neural models of plasticity: Experimental and theoretical approaches*, J. Byrnes, W. Berry, eds., Academic Press, New York, NY, 160-177.
Thompson, R.F. 1986. The neurobiology of learning and memory. *Science* **233**, 941-947.
Thompson, J.K., Lavond, D.G., Thompson, R.F. 1986. Preliminary evidence for a projection from the cochlear nucleus to the pontine nuclear region. *Soc. Neurosci. Abstr.* **12**, 754.
Weiss, C., McCurdy, M.L., Houk, J.C., Gibson, A.R. 1985. Anatomy and physiology of dorsal column afferents to forelimb dorsal accessory olive. *Soc. Neurosci. Abstr.* **11**, 182.
Wagner, A.R. 1981. SOP: A model of automatic memory processing in animal behavior. In *Information processing in animals: Memory mechanisms*, N. Spear, G. Miller, eds., Lawrence Erlbaum Associates, Hillsdale, NJ.
Widrow, B. 1987. A fundamental relationship between the LMS algorithm and the discrete Fourier transform. *Proceedings of the IEEE International Symposium on Circuits and Systems*, May 4-7, Philadelphia, PA.
Widrow, B., Hoff, 1960. Adaptive switching circuits. *Institute for Radio Engineers, Western show and convention, Convention Record* 4, 96-194.

Modulation of Prey-Catching Behavior in Toads: Data and Modeling

*Francisco Cervantes-Pérez**
*Angel D. Guevara-Pozas*** *and Alberto A. Herrera-Becerra**
* *Instituto de Fisiología Celular, Universidad Nacional Autónoma de México, Apdo. Postal 70-600, México D.F., C.P. 04510*
** *Area de Ingeniería Biomédica, Universidad Autónoma Metropolitana, Iztapalapa, México 13 D.F., C.P. 03300*

Abstract

In the first part of this paper we present a study of toads' learning capabilities during prey-catching behavior. Under our experimental paradigm, that we defined as ***Motor Response Inhibition (MRI)****, animals are repeatedly stimulated with worm-like dummies, and are allowed to display their whole prey-catching behavioral repertoire when interacting with the stimulus. We analyze how the intensity and duration of toads' response to potential preys are modulated by learning. Our results show that: a) toads' motor response towards moving "worm-like" stimuli gradually decreases until complete inhibition; b) the process of MRI is stimulus-specific, there is almost no analogical generalization; c) MRI dynamics (i.e., response frequency and inhibition-time) depends on the stimulus attractivity; and d) the animal's motor response slowly returns during rest, that is, MRI shows spontaneous recovery. In the second part of the chapter, we present a conceptual model for Long-Term Memory in amphibia, and, in addition, we use a simplified model of the retino-tectal-pretectal interactions in order to study how the computational properties of this neural machinery might underlie the MRI learning process. We analyze the mathematical properties of the neural net model, using techniques of the qualitative theory of differential equations, to define a set of parameter values that produce the proper dynamic behavior.*

1 Introduction

In neuroethological studies, amphibia have been a very good biological model to explore possible explanations for sensorimotor coordination in terms of anatomical and physiological data. It has been shown that amphibian prey-catching responses to a potential prey are determined by different factors: a) stimulus characteristics, e.g., form, size, and their relationship with respect to the direction of motion, velocity, contrast, etc., (Grüsser and Grüsser-Cornehls, 1976; Ewert, 1984; Ingle, 1982);

b) changes in motivation, e.g., season of the year, hour of the day, presence of odors, food deprivation, etc., (Ewert and Siefert, 1974; Ingle, 1983; Shinn and Dole, 1979); c) previous experiences with the stimulus, e.g., learning and conditioning, (Ewert and Ingle, 1971; Ewert and Kehl, 1978; Finkenstädt, 1989a, 1989b); and d) brain lesions (e.g., pretectal and tectal lesions) (Ewert, 1980, 1984; Ingle, 1977).

Finkenstädt (1989) showed that conditioning modulates the way toads interact with moving stimuli. He conditioned naive animals to elicit prey-catching behaviors towards a non-prey dummy (black moving squares of 8 x 8 cm), by pairing it with a "worm-like" stimulus. He found associative generalization along the size parameter and that the configural worm/antiworm discrimination, as defined by Ewert (1980, 1984), is largely reduced.

Another learning process that has been studied in amphibia is habituation. Ewert and Ingle (1971) showed that habituation of prey-catching activity in frogs and toads is specific to the visual field region being stimulated, and that there is an excitatory after-effect over other regions of the visual field at some distance from that part where the habituation process was carried out. Based on the results of pretectal lesions and physiological recordings from neural elements of the optic tectum, these authors explained their behavioral observations in terms of tectal-pretectal interactions. In toads, Ewert and Kehl (1978) showed that habituation of the orienting response towards a prey-like object is stimulus-specific, and that it is related to pattern recognition; that is, if the animal is stimulated with a different stimulus from that used during the habituation process, the orienting response reappears.

The possible neural mechanisms underlying behavior and learning in amphibia have also been studied through theoretical approaches. Cervantes-Pérez *et al.* (1985) analyzed a family of neural net models to show how the retino-tectal-pretectal interactions might be the substrate of prey-predator discrimination and size preference. The model allowed us to test the postulates of direction invariance of prey-predator recognition being a consequence of the tectal architecture, and of size preference and response latency depending on the motivational state of the animal.

Lara and Arbib (1985) extended this model to include the analysis of neural mechanisms possibly responsible for habituation, testing the hypothesis of stimulus-specific habituation being the result of a build up of inhibition from pretectum upon tectum. They also considered, following Sokolov (1975), that this process requires in some stage the construction of a model of the activity produced by the current stimulus. This model is to be compared with the activity produced by novel stimuli, presented afterwards, and when there is a match between these activities then the build up of inhibition starts. Lara and Arbib's neural net model includes specific hypotheses in relation to those neural elements forming an habituation column. These elements are postulated to be external to the visual pathway, as well as being responsible for the decrease of tectal activity, which yields a different response of the animal towards the same stimulus. In addition, they also assumed that the level of pretectal inhibition over tectum depends on the association of visual stimuli with the expected consequences of the animal's past motor responses.

Lara and Arbib considered only prey-orienting behavior, whose consequence is to bring the stimulus to the frontal part of the animal's visual field (Ewert, 1980, 1984). If we consider Lara and Arbib's hypothesis of the role played by the expected consequence of the motor responses executed by the animal during habituation, there are other behaviors in the toad's prey-catching repertoire (e.g., approaching and snapping) that may have stronger effects on the dynamics of its learning capabilities because of the importance associated to their expected consequences, which even includes the ingestion of the prey in the case of snapping.

The emphasis in the first part of this chapter is in analyzing the toad's learning capabilities under an experimental paradigm that we have defined as **Motor Response Inhibition (MRI)**. This learning process occurs during a period of time where toads are stimulated repetitively with a visual prey-like dummy. The animals are allowed to display their whole prey-catching repertoire while interacting with the stimulus, and the MRI process ends when the animal stops trying to catch the stimulus, even though it remains in the environment. We analyze how the intensity and duration of these animals' response to prey-like stimuli might be modulated by learning; that is, we propose the MRI process as a means to analyze the nature of aspects of the toad's prey-recognition and learning capabilities.

In the second part, we present a conceptual model to explain the MRI learning process in terms of changes in the activity of those structures subserving short-term (STM) and long-term (LTM) memory processes. This new model is based on Lara and Arbib's (1985) habituation model, and it includes an interaction between STM and LTM mechanisms. Additionally, we analyze the mathematical properties of a simplified model of the retino-tectal-pretectal interactions, in order to study how the computational characteristics of this neural machinery might subserve the MRI. We follow the same approach as in Cervantes-Pérez and Arbib (1990), where we analyzed possible correlates between behavioral, anatomical and physiological data. In that paper, we showed that, by using techniques from the qualitative theory of differential equations, we can establish the critical parameter values under which the model of the retino-tectal-pretectal interactions may explain the prey-catching facilitation phenomenon.

2 Motor Response Inhibition: The Experimental Paradigm

Adult toads *Bufo marinus horribilis* were used as experimental subjects. During the experiments, the toads were taken to a room with twilight lighting conditions, and placed in front of a TV screen, turned off at the time, for over 15 minutes before the experiment. The animals were stimulated with visual worm-like dummies, which were played back from a VCR (Fig. 1). Stimuli were previously generated by using a microcomputer with color graphic capabilities, in order to have complete

Figure 1. Experimental set-up. Toads within an acrylic cage are placed in front of a black and white TV set. The animal is stimulated by playing back the image of a "worm-like" object. The stimuli were previously generated with a microcomputer and recorded using a TV closed circuit.

control over the stimulus characteristics (e.g., form, size, velocity, etc.), and recorded on video cassette with a TV closed circuit.

The general conditions of the stimulation procedure were: a) all dummies were black rectangles; b) moving on a white background; c) as worm-like stimuli (i.e., their longest axis parallel to the direction of motion); d) with a constant velocity of 4.1 cms per second; e) at a height of 2 cms from the substrate where the animal was placed; and f) during the experiment, stimuli traversed the TV screen in the horizontal direction moving back and forth from one extreme to the other. During the experiments an observer recorded those motor responses directed by the toad towards the stimulus (i.e., orienting, approaching, following and snapping). It should be pointed out that the ethogram displayed by toads *Bufo marinus horribilis* is similar to that described by Ewert (1980) for the common toad *Bufo Bufo*; it only differs in that our toads do not present the wiping-mouth response. A complete description of the experimental set is given by Cervantes-Pérez and Guevara-Pozas (1990).

The learning process of interest to us occurs during the continuous stimulation paradigm and it ends when the toad stops making motor responses towards the visual stimulus, even though the stimulus still remains in the visual field. To make sure that the animal's motor responses to the dummy have been completely inhibited, the observer kept recording for up to thirty minutes after the animals' interactions had stopped. If the animal did not interact with the stimulus then we considered that the MRI learning process had been established.

We defined our learning paradigm as **Motor Response Inhibition (MRI)** to try to describe what happens. During repeated stimulation with a worm-like dummy the intensity of toads' motor response decreases until its complete inhibition (see below). We do not refer to it as *habituation*, as Ewert and coworkers did (Ewert and Ingle, 1971; Ewert and Kehl, 1978), because the definition of habituation includes the possibility of reverting the process not only through spontaneous recovery but also by a *dishabituation* process (Harris, 1943; Thompson and Spencer, 1966; Kandel *et al.*, 1976). Our animals show spontaneous recovery under this paradigm but, so far, we have not been able to produce dishabituation. Ewert and coworkers (Ewert and Ingle, 1971; Ewert and Kehl, 1978) considered the reappearance of toads' motor responses when the animals are confronted with a different prey-like stimulus from that used during the habituation process, as a kind of dishabituation. Since MRI and their habituation process are stimulus-specific, we interpret this response reappearance not as meaning that the learning process has been reverted but as showing that there is no analogical generalization between those two stimuli. We do not consider this as dishabituation, because dishabituation has to be related to the motor response recovery towards the same stimulus used during the habituation process (Thompson and Spencer, 1966; Leaton and Tighe, 1976).

In our analysis, we considered the toad's **Response Frequency** (RF) — number of prey-catching motor actions in a time block of five minutes — and the **Inhibition-Time** (IT) — the time required by the animals to "learn" not to respond towards a specific stimulus. When presented in graphs, these data were normalized to a percentage of activity, taking the highest frequency in the experiment as the reference point (100%) and grouping the frequency responses in time blocks of five minutes.

3 Temporal Characteristics of MRI

The results obtained by training nine animals with a wormlike stimulus (black rectangle of 1.5 x 0.4 cms) are shown in Fig. 2. Even though there is great variability among individual toads with respect to the temporal dynamics of the normalized value of RF and of the IT, it is clear that there is a tendency to gradually decrease the percentage of motor actions presented by the animals. However, we must point out that this response decrement is not continuous; furthermore, in some animals the maximum number of interactions occurs at intermediate times during the stimulation period. The dynamics of the activity goes up and down, usually with the valleys getting deeper as the experiment progresses towards the complete MRI. When the animals kept ignoring the visual dummy for a period of 30 minutes, we assumed they had "learned" that the stimulus was a non-catchable prey. We postulate that this decrement in toad's current motor response is due, in part, to the failure of the expected consequence of previous actions elicited by the animal to catch the prey, especially to the snap response whose expected consequence is the ingestion of the prey. However, an alternative explanation might be that the toads had "learned" that there is a barrier stopping them from catching the prey. This is analyzed below.

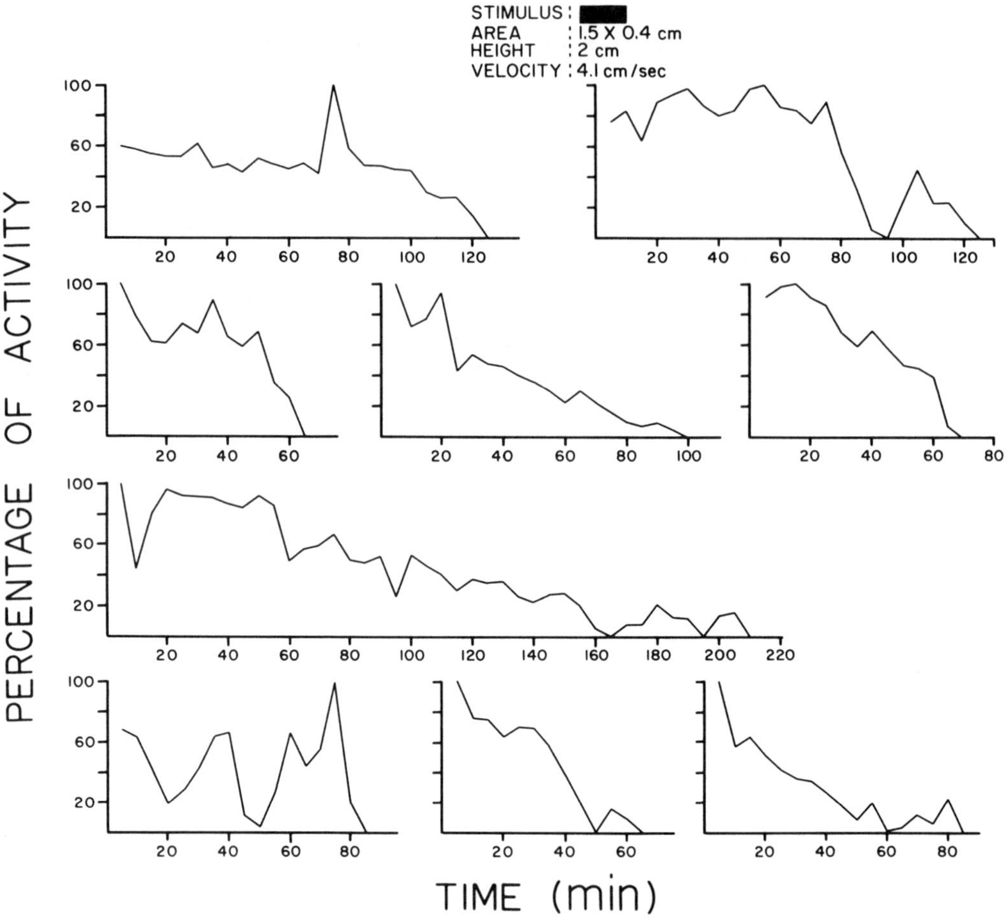

Figure 2. Individual toads **MRI** profiles. Nine animals were chosen to show the great variability in the response intensity (i.e., **RF** and **IT**), displayed by toads when trained under the **MRI** learning paradigm.

4 Stimulus-Specificity of MRI

In order to explore the possibility that toads had "learned" there was a barrier in between them and the prey, we conducted a set of experiments where five animals were trained first with stimulus A ("worm-like" dummy of 1.5 x 0.4 cms) until the MRI was reached; and then, after 10 minutes of rest, they were stimulated, again until the completion of the MRI, with stimulus B ("worm-like" dummy of 5.5 x 0.4 cms, moving at 4.2 cms per second). The results are shown in Table 1. When the stimulus is changed, toads' prey-catching motor responses reappear supporting the hypothesis that the animals "learn" that stimulus A was a non-catchable prey, eliminating the alternative hypothesis related to the presence of a barrier.

Table 1. Effects of changes in size of a "worm-like" stimulus on the Frequency Response (FR) of toad's Prey-Catching behaviour. The * indicates that toads were stimulated for 30 more minutes without eliciting a motor response. Stimulus A (1.5 x 0.4 cms) and B (5.5 x 0.4 cms).

STIMULI ORDER: A		B		STIMULI ORDER: B		A		(CONT): A	
TIME	FR	TIME	FR	TIME	FR	TIME	RF	TIME	RF
5	122	155	91	5	86	150	78	270	9
10	107	160	86	10	80	155	68	275	7
15	96	165	84	15	88	160	60	280	1
20	89	170	45	20	74	165	49	285	0
25	91	175	38	25	62	170	42	290	3
30	83	180	39	30	54	175	39	295	0
35	87	185	32	35	42	180	24	300	11
40	85	190	24	40	40	185	22	305	11
45	66	195	24	45	27	190	19	310	3
50	60	200	16	50	18	195	13	315	0
55	53	205	2	55	16	200	14	320	5
60	58	210	13	60	19	205	13	325	7
65	46	215	10	65	16	210	6	330	4
70	47	220	2	70	14	215	11	335	3
75	44	225	2	75	13	220	8	340	4
80	21	230	5	80	11	225	18	345	5
85	1	235	6	85	9	230	17		*
90	0	240	2	90	7	235	22		
95	6	245	3	95	7	240	22		
100	12	250	4	100	4	245	21		
105	6		*	105	4	250	13		
110	6			110	1	255	15		
115	3				*	260	14		
	*					265	8		

These results also suggest that the MRI process is stimulus-specific. To complete this part, we trained another six toads, stimulating them in the inverse order; that is, we presented stimulus B first and, after the MRI was completed, the stimulus A. The results are also shown in Table 1. Again, toads' motor response reappears when the stimulus is changed, that is, the second stimulus is treated by the toad as a different prey. However, it is clear that: a) the intensity of their interactions (RF) is not as strong as in the case where animals have not been previously stimulated with another stimulus; and b) in relation to the IT, there is no effect on the results obtained for stimulus A, in both cases, but with stimulus B it decreases when the an-

Table 2. Spontaneous recovery after an MRI training. Frequency of Response = FR; Inhibition Time = IT. See text for explanation.

STIMULUS (cm)	INTER-STIMULATION PERIOD (days)	EFFECT	FR	IT (min)
1.5 x 0.4	11	No-Response		
"	13	No-Response		
"	20	No-Response		
"	45	No-Response		
"	49	Response	22	28
"	56	No-Response		
"	58	No-Response		
"	64	Response	1	10
"	95	No-Response		
"	95	Response	1.8	45
"	133	Response	7	115
"	141	Response	4	30
"	169	Response	5	106
"	172	Response	12	75
"	188	Response	13.3	35
"	197	Response	8.4	100
"	301	Response	7.9	125

imals have been previously trained with another stimulus. The latter result might be due to interindividual variability, rather than to the stimulation process. Thus, these results confirm that the MRI process is stimulus-specific, presenting only a small analogical generalization effect for changes in the stimulus dimensions.

In addition, comparing how the animals' RF and IT (see Table 1) vary over time, it can be observed that: a) the RF towards stimulus A is stronger than to stimulus B; and b) the IT required by the toads to accomplish MRI is longer when confronted with stimulus A. These results suggest a certain dependency of the MRI learning process with respect to the stimulus "attractivity" of a potential prey; that is, there are visual stimuli that represent a better prey than others.

5 MRI is a Reversible Process

To analyze if the MRI is a reversible process, we trained seventeen toads using a stimulus of 1.5 x 0.4 cms. After a period of rest, different for each animal (minimum

11 days), we conducted a second trial. The results are shown in Table 2. It is clear that toad's prey-catching response towards a specific stimulus, in animals trained under the MRI paradigm, reappears spontaneously. However, we observed, again, great variability among individual toads with respect to the RF, ranging from 1 to 22 responses per minute, and to the IT, from 10 to 125 minutes. For example, on day 49 after the first trial, one toad responded for 28 minutes, eliciting 22 interactions per minute; whereas other animal did not respond at all even after 95 days.

6 Towards a Conceptual Model for Prey-catching Modulation

Because of the long duration of MRI, with spontaneous recovery taking over 30 days, our results suggest that this learning process is associated with long-term memory (LTM). A previous study (Ewert and Kehl, 1978) analyzed a similar learning process but with very short duration. It took toads five minutes to stop responding to the stimulus and about half an hour to recover the response. Therefore, we consider it as a learning process related to short-term memory (STM). In addition, following Ewert and Kehl's study, Lara and Arbib (1985) developed and analyzed a neural net model that explained the blocking of the prey-orienting response in terms of neural mechanisms extrinsic to the primary visual pathway. They proposed a neural circuit that includes a model generator MO that creates a model, within a certain time course, of the neural activity produced by the presence of the stimulus. The output of MO is compared to the activity produced by the current stimulus by a comparator element CO. When these two signals match, CO activates a habituation element H. At this moment H starts a build up of inhibition upon those brain regions in the visuomotor system that control the orienting behavior, with a timing that depends on the stimulation time period. This inhibition was proposed to modulate the activity of tectal (T5.2) and pretectal (TH3) elements (Ewert, 1984, 1989), by increasing the level of pretectal inhibition upon tectal neural elements. According to the assumed timing and the style to retain the information from the training period, Lara and Arbib's model accounts very well for this STM process, but it could not explain our data. If we stimulate toads with other stimuli, after being trained with a specific stimulus A until the MRI completion, Lara and Arbib's model predicts that if we confront the animals with stimulus A, again, they should respond with prey-catching behaviors. This occurs because the output of the MO has changed trying to match the activity produced by the novel stimuli, which implies that it would not match the activity produced by stimulus A during the second presentation.

Our results clearly show that once the MRI to a specific "worm-like" stimulus has been established, it lasts even after stimulating the animals with different stimuli, including natural prey. Toads eat very well after MRI training. Thus, our paradigm offers great advantages to study, both theoretically and experimentally, the possible processes in the CNS that underlie LTM. If we combine our results with those obtained by Ewert and Kehl (1978) and Lara and Arbib (1985), we may study the relationships among the mechanisms controlling STM and LTM.

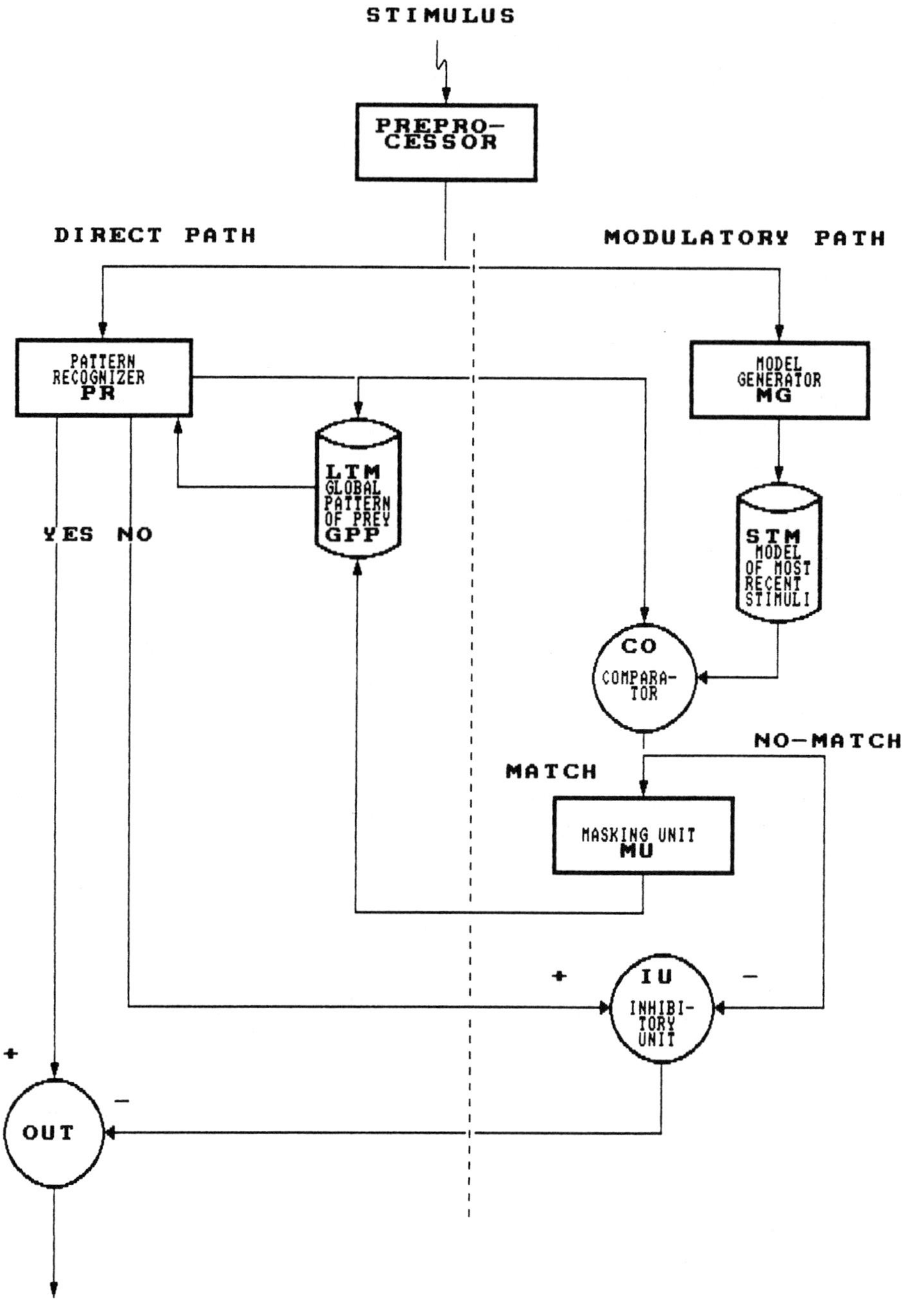

Figure 3. A Conceptual Model for the Motor Response Inhibition (MRI) Learning Paradigm. See text for explanation.

The dynamics followed by the MRI characteristics (RF and IT) allows us to establish the requirement for a more complex memory system than that proposed by Lara and Arbib (1985). Fig. 3, show a conceptual model for the processes that might be carried out by the toad's brain during the MRI. The presence of the stimulus is detected by a preprocessor that activates simultaneously two different pathways:

a) The direct path starts with the pattern recognizer (PR), which, in combination with the long-term memory unit (LTM), represents the neural machinery responsible for defining whether or not the current stimulus fits the prey category. PR accesses the information stored in the LTM, and its output is affirmative (**yes**) if

the stimulus belongs to the Global Pattern of Prey (GPP), or negative (**not**) otherwise. If the output is affirmative, PR activates the efferent element (OUT) with the level that corresponds to the effectiveness of the stimulus to elicit prey-catching behaviors. In case the PR response is negative, then the inhibition unit (IU) is activated to decrease OUT activity.

b) The modulatory path starts with the model generator (MG). It has the role of building up a short-term memory (STM) with information that might modulate the OUT level of activity produced by the interactions between PR and the LTM mechanisms. The modulatory effect of this pathway over *the direct path* must account for the qualitative and quantitative changes observed in the toad's behavior during the course of MRI.

The modulation process starts when the comparator (CO) detects a matching between the model (STM) generated by MG and the activity produced by the current stimulus in the PR module. This activates the masking unit (MU) that modifies the effectiveness associated to that specific visual stimulus. The stimulus effectiveness decreases gradually as long as the stimulus remains in the animal's visual field. The temporal effect of this decrement will depend on the IT and on the time period we still stimulate the animal after the MRI was completed. While the stimulus effectiveness yields a level of activity in the OUT element above a threshold value, what we should observe is a decrement in the intensity of the animal's motor response. This could explain our data during the first stage of the MRI process. If the experiment continues, the level of OUT activity goes below that threshold value; this would stop any motor response towards the dummy, meaning that the stimulus cannot be considered as a potential prey anymore. This would explain the second stage of MRI where the motor response is completely inhibited. When the stimulus is retired from the animal's visual field, then a slow recovery of the stimulus effectiveness to yield prey-catching motor responses takes place, which would account for the spontaneous recovery observed after the MRI.

7 Simplified Neural Model (SNM) of a Facilitation Tectal Column

In order to investigate if the interactions among retina, optic tectum and pretectum might subserve changes in neural activity — such as those postulated for the OUT element during the course of MRI — in this section we analyze a simplified neural model (SNM) of a Facilitation Tectal Column (FTC).

Figure 4. Simplified Neural Model (SNM) of a Facilitation Tectal Column (FTC) Model. O means excitation, and - inhibition. See text for explanation.

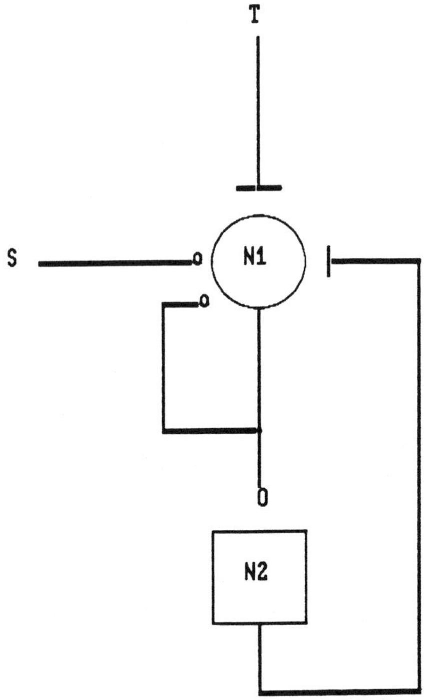

The SNM includes the computational properties that have been identified in neurophysiological and neuroanatomical studies (Schürg-Pfeiffer, 1989; Lázár, 1984; Matsumoto, 1989), and analyzed in theoretical models (Ewert and von Seelen, 1974; Lara *et al.*, 1982; Cervantes-Pérez *et al.*, 1985; Cervantes-Pérez, 1985; Cervantes-Pérez and Arbib, 1990). These properties are: a) a direct projection from retinal ganglion cells to tectum and pretectum; b) cooperation/ competition between positive and negative feedback loops within tectum; and, c) an overall inhibitory effect from pretectum upon tectal activity.

Cervantes-Pérez and Arbib (1990) found that most of the qualitative behaviors that a FTC might present involve an oscillatory activity in that part of the tectal circuit formed by the Large-Pear (LP) cell and the Stellate neuron (SN). Thus, the neural machinery taken from the FTC model to define the SNM is formed by those elements driving the oscillatory activity of the LP-SN network (see Fig. 4). It is comprised by three elements: a constant **S**, representing the retinal input; a constant **T**, representing the level of pretectal activity; and, a two neuron circuit (**N1 and N2**) representing the intratectal processing of information. In the circuit: a) **N1** receives excitation from **S** (the retina) and inhibition from **T** (the pretectum); b) it is excitatory and has recurrent axons, forming a positive feedback loop; c) neuron N2 is activated by **N1**; and, d) the synapse from **N2** to **N1** is inhibitory, forming a negative feedback loop.

In our analysis, each neuron was represented by two quantities: its membrane potential, and its firing rate. For the dynamic changes in the membrane potential we used the *leaky integrator model* (Lara, Arbib, and Cromarty 1982), where the properties of spatial and temporal summation are described as follows,

$$\tau_j \frac{dm_j(t)}{dt} = - m\, I_j(t)(t) + I_j(t) + M_{oj}$$

where τ_j is the time constant characteristic of neuron j, $m_j(t)$ is the membrane potential of neuron j at time t, M_o is its the resting potential, and $I_j(t)$ simulates the spatial summation property as

$$I_j(t) = \sum_{i=1}^{h} w_{ij}F_i(m_i(t); par) - \sum_{k=h+1}^{n} w'_{kj}F_k(m_k(t); par)$$

representing the weighted effects of the h excitatory (first sum) and (n-h) inhibitory (second sum) synapses impinging upon neuron j. Functions F(m(t);par) are nonlinear functions, and they simulate the generation of action potentials and their transmission to the axon terminals.

Thus, the dynamic behavior of the SNM circuit is defined by the following system of ordinary differential equations,

$$\dot{m}_1(t) = \sigma_1 [- m_1(t) + S - T + w_{11}F_1(m_1(t); \theta_1) - w_{21}F_2(m_2(t);[k, \theta_2])]$$

$$\dot{m}_2(t) = \sigma_2 [- m_2(t) + w_{12}F_1(m_1(t); \theta_1)]$$

where $\sigma_1 = 1/\tau_1$; $\sigma_2 = 1/\tau_2$, all the weighting values are positive, and the value of the combined effect of **S** and **T (S-T)** over neuron **N1** can be negative, 0, or positive.

To model the output function of **N1**, $F_1(m_1(t); \theta_1)$, we use a simple piecewise linear model, a step function,

$$F(m(t); \theta) = \begin{cases} 1 & \text{if } m(t) \geq \theta \\ 0 & \text{if } m(t) < \theta \end{cases}$$

assuming, from Ingle's data (Ingle, 1973), that it shows a constant firing rate. The activity of **N2**, $F_2(m_2(t);[k, \theta_2])$, is modeled by a function of the form,

$$F_2(m_2(t);[k, \theta_2]) = \begin{cases} k m_2(t) & \text{if } m_2(t) \geq \theta_2 \\ 0 & \text{if } m_2(t) < \theta_2 \end{cases}$$

so that the **N2** level of activity becomes stronger as **N2** membrane potential increases, in order to control the level of excitation within the circuit.

The Phase Plane associated to the SNM can be divided into four different regions (see Fig. 5), where the network dynamics can be described by a set of linear differential equations:

A) Region $R_1 = \{(m_1, m_2) \mid m_1 < \theta_1, m_2 < \theta_2\}$, whose dynamic behavior is described by the linear system ($\mathbf{L_1}$),

$$\dot{m}_1(t) = \sigma_1 [- m_1(t) + S - T]$$

$$\dot{m}_2(t) = - \sigma_2\, m_2(t)$$

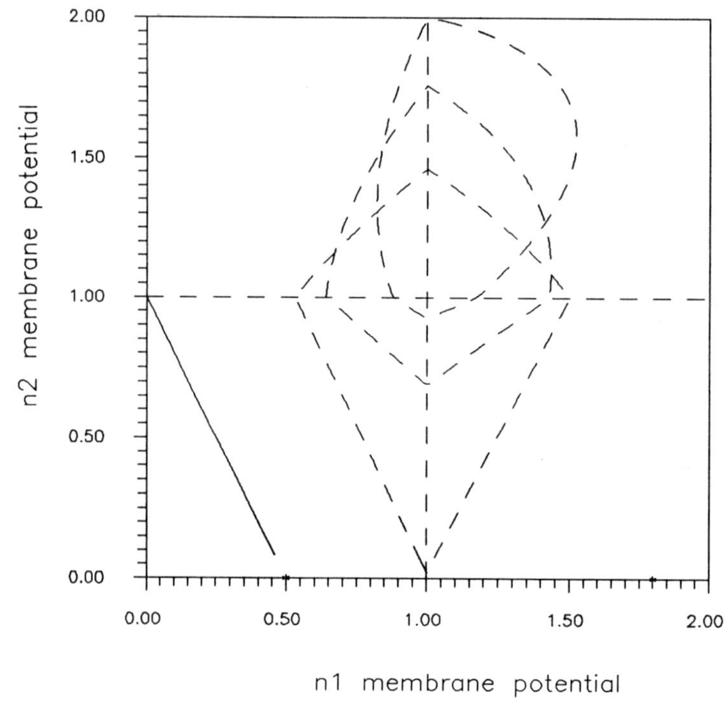

Figure 5. SNM Phase Plane Portrait. The closed trajectories represent oscillatory behaviors in the SNM, for different levels of activity in the element **T** when (**S-T** $\geq \theta_1$)(see Figure 4). The straight trajectory shows the condition when the equilibrium point associated to region R_1 becomes **real**, that is, (**S-T** $< \theta_1$).

B) Region $R_2 = \{(m_1, m_2) \mid m_1 \geq \theta_1, m_2 < \theta_2\}$, whose dynamic behavior is described by the linear system ($\mathbf{L_2}$),

$$\dot{m_1}(t) = \sigma_1 [- m_1(t) + S - T + w_{11}]$$

$$\dot{m_2}(t) = \sigma_2 [- m_2(t) + w_{12}]$$

C) Region $R_3 = \{(m_1, m_2) \mid m_1 \geq \theta_1, m_2 \geq \theta_2\}$, whose dynamic behavior is described by the linear system ($\mathbf{L_3}$),

$\dot{m}_1(t) = \sigma_1 [- m_1(t) + S - T + w_{11} - kw_{21}m_2(t)]$

$\dot{m}_2(t) = \sigma_2 [- m_2(t) + w_{12}]$

D) Region $R_4 = \{(m_1, m_2) \mid m_1 < \theta_1, m_2 \geq \theta_2\}$, whose dynamic behavior is described by the linear system (L_4),

$\dot{m}_1(t) = \sigma_1 [- m_1(t) + S - T - kw_{21}m_2(t)]$

$\dot{m}_2(t) = - \sigma_2 m_2(t)$

8 Analysis of SNM Dynamic Behaviors

Our approach offers several advantages. Within each one of the Phase Plane regions, we can get analytical solutions for the linear systems L_i (for i=1,2,3,4). With $m_1(0) = m_{10}$, and $m_2(0) = m_{20}$ as initial conditions, SNM solutions are:

A) for region R_1, $m_1(t) = S - T + (m_{10} - S + T)\, e^{-t\sigma}$
$m_2(t) = m_{20}\, e^{-t\sigma}$

B) for region R_2, $m_1(t) = S - T + w_{11} + (m_{10} - S + T - w_{11})\, e^{-t\sigma}$
$m_2(t) = w_{12} + (m_{20} - w_{12})\, e^{-t\sigma}$

C) for region R_3, $m_1(t) = S - T + w_{11} - kw_{12}w_{21} +$
$[m_{10} - S + T - w_{11} + kw_{12}w_{21} + kw_{21}(w_{12} - m_{20})\, \sigma t]\, e^{-t\sigma}$
$m_2(t) = w_{12} + (m_{20} - w_{12})\, e^{-t\sigma}$

D) for region R_4, $m_1(t) = S - T + (m_{10} - S + T - m_{20}kw_{21}\sigma t)\, e^{-t\sigma}$
$m_2(t) = m_{20}\, e^{-t\sigma}$

To simplify, in these equations we consider equal membrane time constants for both neurons; that is, $\sigma = \sigma_1 = \sigma_2$. Analyzing these solutions when $t \to \infty$, we have that, within all four regions, they tend to an equilibrium point given by

$(m_1^{(eq)}, m_2^{(eq)})_{R_1} = (S - T, 0)$ $(m_1^{(eq)}, m_2^{(eq)})_{R_2} = (S - T, w_{12})$

$(m_1^{(eq)}, m_2^{(eq)})_{R_3} = (S - T + w_{11} - kw_{12}w_{21}, w_{12})$ $(m_1^{(eq)}, m_2^{(eq)})_{R_4} = (S - T, 0)$

In our analysis an equilibrium point $\alpha_i = (m_1^{(eq)}, m_2^{(eq)})_{R_i}$ (for i=1,2,3,4), is relevant for the overall behavior of the global system when it is located within its corresponding region (R_i), since it becomes a *real* equilibrium point and the global activity of the network can converge to it. On the contrary, if the equilibrium point α_i falls outside its corresponding region (R_i), then it is defined as a *virtual* equilibrium point and none of the trajectories of the global system may converge to it (Ogata, 1970; Shinners, 1978). The SNM circuit presents an oscillatory behavior when all four regions have equilibrium points that are *virtual* (see Cervantes-Pérez,

1985). A set of relationships among parameters of the network under which the equilibrium states are virtual is

$(S - T) > \theta_1$; $w_{12} > \theta_2$ and $(S - T + w_{11} - kw_{12}w_{21}) < \theta_1$

That is, first, the combined effect of the activity (S) produced by the stimulus and the level of inhibition (T) coming from the modulatory path has to be strong enough to take the membrane potential of neuron **N1** above threshold; second, **N1 excitatory effect** must drive **N2** membrane potential above threshold; and, third, the combined inhibitory effect of (T) and **N2** must counteract the excitation coming from (S) and from **N1** recurrent axons, taking, in this way, **N1** membrane potential to subthreshold values.

Figure 5 shows how the overall behavior of SNM varies when the level of inhibition (T) coming from the modulatory path varies. Here we assumed there is no change, at any time, in the input signal; that is, we always have the same level of excitation (S) due to the continuous presence of the stimulus. In addition, we do not consider synaptic changes in the network.

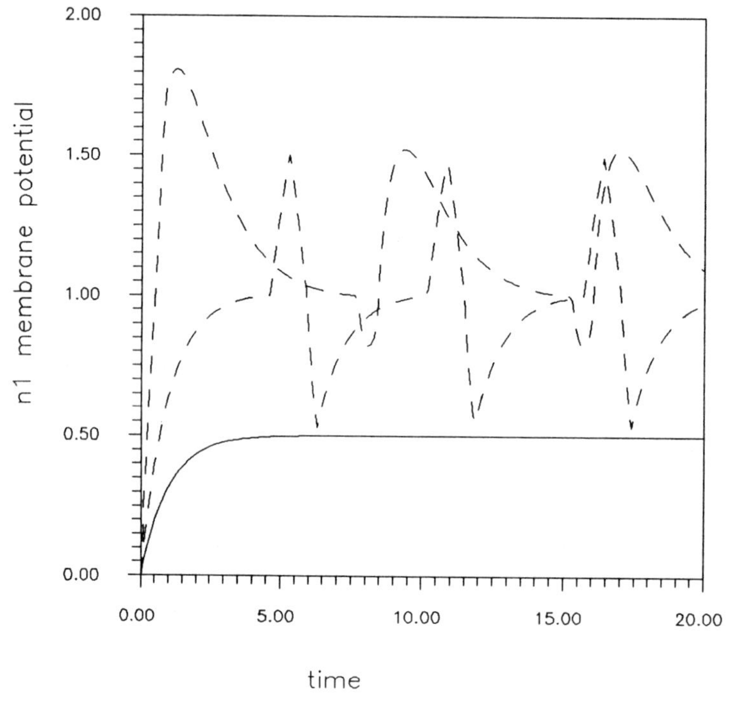

Figure 6. Temporal Dynamics of N1 Membrane Potential. The curves in dashed lines correspond to the closed trajectories of Figure 5. The temporal characteristics of the oscillatory behavior, in the net, can be controlled by the level of activity of T and the synaptic structure of the net.

If we analyze how the circuit activation dynamics depends on the level of activity in (T), we have that, as long as the value of (T) keeps all equilibrium points virtual, we get the same qualitative behavior in the network: an oscillation. The amplitude and time period of these oscillations change according to the value of (T); as (T) increases, the difference (S -T) decreases, and the location of the equilibrium point gets closer to the threshold value of **N1**, producing an increase in the amplitude and time period of the oscillation (see dashed lines in Figure 5). This quantitative change would account for the changes observed during the first part of the MRI learning process. Once the value of (S -T) gets smaller than θ_1, the equilibrium point associated with region R_1 becomes a *real* point, and a qualitative change takes place. Now, every trajectory would converge to a point where both neurons have subthreshold potentials, staying inactive after the equilibrium is reached. This qualitative change in the network dynamic behavior would explain the second stage of MRI, where the animals do not respond towards the prey-like stimulus.

Figure 6 shows the temporal behavior of **N1** membrane potential. It can be observed that not only do we get oscillations with different amplitudes and time periods, but we also get differences in the temporal characteristics of the membrane potential dynamics. The time period in which neurons are active, or inactive, can be controlled by the level of inhibition coming from **T**. Our analysis allows us to postulate that during the MRI learning process there are not structural (synaptic) changes in the neural network subserving prey-catching behavior but, rather, external inputs coming from the modulatory path yield a change in the network activation dynamics. In the MRI process, this may happen because the stimulus remains in the visual field.

9 Conclusions

We have presented a **Top-Down** study that interplays experimental data — ethological, anatomical and physiological — in toads with the analysis of conceptual and neural net models. In this analysis we have shown that in order to explain learning processes which are stimulus-specific, we must consider the interactions among the mechanisms of short-term and long-term memory.

Our study clearly shows that the retino-tectal-pretectal interactions have the computational properties to underlie qualitative and quantitative changes, which are analogous to those behavioral changes observed during the **MRI** learning process in toads. When the stimulus remains in the visual field despite the animal's efforts to catch it, processing of information within the *modulatory path* results in an increase in the level of pretectal inhibition upon tectum. A decrease in tectal activity reduces the probability of the current stimulus fitting the prey category.

Finally, based on the analysis of the SNM, we postulate that the activity within the *modulatory path* causes a change in the *activation dynamics* of the neural network subserving sensorial stimuli recognition, rather than changes in the *weight dynamics* related to its structural organization.

Acknowledgements. We thank Nydia Lara for valuable comments on the manuscript. This study was partially supported by the Dirección General de Asuntos del Personal Académico (DGAPA) de la UNAM, under grant IN-026189, and by PADEP under grant DING-9003.

10 References

Cervantes-Pérez, F. (1985) *Modelling and analysis of neural networks in the visuomotor system of anuran amphibian.* Ph.D. Dissertation, University of Massachusetts, Amherst, Massachusetts.

Cervantes-Pérez, F., R. Lara and Arbib, M.A. (1985) A neural model of interactions subserving prey-predator discrimination and size preference in anuran amphibia. *J. Theor. Biol.* 113:117-152.

Cervantes-Pérez, F. and Arbib, M.A. (1990) Stability and parameter dependency analyses of a Facilitation Tectal Column (FTC) model. *J. Math. Biol.* (in press).

Cervantes-Pérez, F. and Guevara-Pozas, D. (1990) Learning during prey-catching behavior in toads *Bufo marinus* horribilis. (Submitted).

Ewert, J.-P. (1984) Tectal mechanisms that underlie prey catching and avoidance behaviors in toads, In: *Comparative Neurology of the Optic Tectum* (H. Vanegas, ed.) Plenum Press, New York pp. 248-416.

Ewert, J.-P. and Ingle, D. (1971) Excitatory effects following habituation of prey-catching activity in frogs and toads. *J. Comp. Physiol.*, 77:3, pp. 369-374.

Ewert, J.-P. and Seelen, W.v. (1974). Neurobiologie und system theorie eines visuellen muster-erkennungsmechanisms bei Kröten. *Kybernetik* 14:167-183.

Ewert, J.-P. and Kehl, W. (1978) Configural prey-selection by individual experience in toad *Bufo bufo*. *J. Comp. Physiol.* 126:105-114.

Ewert, J.-P. and Siefert, G. (1974) Seasonal change of contrast detection in the toad's (*Bufo bufo* L.) visual system. *J. Comp. Physiol.* 94:177-186.

Ewert, J.P. (1980) *Neuroethology*: An introduction to the neurophysiological fundamentals of behavior. Springer-Verlag, Berlin, Heidelberg, New York.

Finkenstädt, T. (1989a) Stimulus-specific habituation in toads: 2DG studies and lesion experiments. In: Ewert, J.-P. and M.A. Arbib (eds)*Visuomotor Coordination Amphibians, Comparisons, Model and Robots*. Plenum, New York, pp.767-798.

Finkenstädt, T. (1989b) Visual associative learning: searching for behavioral relevant brain structures in toads. In: Ewert, J.-P. and M.A. Arbib (eds)*Visuomotor Coordination Amphibians, Comparisons, Model and Robots*. Plenum Press, New York, pp.799-827.

Grüsser, O.J. and U. Grüsser-Cornehls (1976) Neurophysiology of the anuran visual system, In: *Frog Neurobiology* (R. Llinás and W. Precht, eds.) Springer-Verlag, Berlin, Heidelberg, New York. pp.298-385.

Harris, J.D. (1943) Habituatory response decrement in the intact organism. *Psychol. Bull.* 40:385-422.

Ingle, D. (1982) Organization of visuomotor behaviors in vertebrates. In: *Analysis of visual behavior*, Ingle, D. M.A. Goodale and R. J. Mansfield (eds), MIT Press, Cambridge, Massachusetts pp.67-109.

Ingle, D. (1983) Prey selection by frogs and toads: a neuroethological model. In: *Handbook of Behavioral Neurobiology*, Vol. 6: Motivation, Satinoff, E. and P. Teitelbaum (eds). Plenum.

Ingle, D. J. (1977) Detection of stationary objects by frogs (*Rana pipiens*) after ablation of optic tectum. *J. Comp. Physiol. Psychol.* 91:1359-1364.

Ingle, D.J. (1973) Two visual systems in the frog. *Science* 181:1053-1055.

Kandel, E.R., Brunelli, M., Byrne, J., Castelluci, V. (1976) A common presynaptic locus for the synaptic changes underlying short-term habituation and sensitization of gill withdrawal reflex in aplysia. *Cold Sp. Harb. Symp.* 40: 465-481.

Lara, R. and M. Arbib (1985) A model of neural Mechanisms responsible for pattern recognition and stimulus specific habituation in toads. *Biol. Cybern.* 51:223-237.

Lara, R., M. Arbib, and A. Cromarty (1982) The role of the tectal column in facilitation of amphibian prey-catching behavior: A neural model. *J. of Neuroscience* 4(2): 521-530.

Lázár, Gy. (1984) Structure and connections of the frog optic tectum, In: *Comparative Neurology of the Optic Tectum*. Vanegas ed. New York pp. 185-210.

Leaton, R.N. and Tighe, T.J. (1976) Comparisons between habituation research at the developmental and animal-neurophysiological level. In: *Habituation, Perspectives from child development, animal behavior, and neurophysiology*. T.J. Tighe and R.N. Leaton (eds.). LEA Press, pp. 321-338.

Matsumoto, N. (1989) Morphological and physiological studies of tectal and pretectal neurons in the frog. In:*Visuomotor Coordination Amphibians, Comparisons, Model and Robots*. Ewert, J.-P. and M.A. Arbib (eds) Plenum, pp 201-222.

Ogata, K. (1970) *Modern control engineering*. Englewood Cliffs, N.J., Prentice Hall.

Schürg-Pfeiffer, E. (1989) Behavior-Correlated properties of tectal neurons in freely moving toads. In:*Visuomotor Coordination Amphibians, Comparisons, Model and Robots*. Ewert, J.-P. and M.A. Arbib (eds) Plenum, New York, pp 451-480.

Shinn, E.A. and J.W. Dole (1979) Evidence for a role for olfactory cues in the feeding response of western toads, *Bufo boreas*. *Copeia* 1979(1):163-165.

Shinners, S. (1978) *Modern control system theory and applications*. Massachusetts, Addison-Wesley.

Sokolov, E.N. (1976) The neuronal mechanisms of the orienting reflex. In: *Neuronal mechanisms of the orienting reflex*. E.N. Sokolov and O.S. Vinogradova (eds) Erlbaum, NJ, 217-235.

Thompson, R. and Spencer, W.A. (1966) Habituation. A model phenomenon for the study of neuronal substrates of behavior. *Psychol. Review*, 73:16-43.

Learning-Related Modulation of Toad's Responses to Prey by Neural Loops Involving the Forebrain

C. Merkel-Harff and J.-P. Ewert

Neurobiology, FB 19, University of Kassel
D-3500 Kassel, FR of Germany

Abstract. *The species-universal recognition of visual prey signals in toads takes advantage of the discrimination of moving configural objects. This is determined by a stimulus response-mediating circuit involving retinal, pretectal, tectal, and post-tectal integrative/interactive steps of visual information processing. In an individual toad, prey-catching can be influenced by various modes of learning (associative; non-associative). Our neuroethological studies show that learning proceeds in connection with information processing within extrinsic response-modulating circuits (neural loops) involving structures of the telencephalon that communicate with the tectum via diencephalic nuclei. If communication with certain telencephalic nuclei is interrupted in an experienced toad, this animal regains - and relies on - its species-universal naive knowledge. The question is open how learning related loops depend on the learning paradigm. Applying the [^{14}C]-2-deoxyglucose method, in the present study we map brain activity in a paradigm in which visual (prey) and olfactory (odor) cues are associated, whereby the visual prey-schema is extended ("generalized") in response to any moving object only in the presence of the prey-associated odor. These data are compared with previous ones obtained after prey/predator association, whereby the prey schema is "generalized" in response to any moving visual objects. For conditions of prey generalization, in both paradigms the trained toads vs. naive ones display a strong glucose utilization in the posterior ventromedial pallium, vMP, the "primordium hippocampi" of Herrick. However, whereas the data in prey/predator conditioned toads suggest that prey-selective properties in the tectum are reduced by vMP-mediated inhibition of pretecto-tectal inhibitory influences, in prey/odor conditioned toads, vMP may affect tectal properties via other thalamic and hypothalamic connections.*

1. Neural Loop Operations

1.1 Functional Brain Architecture

Visual information processing in the toad's brain can be interpreted from two points of view which are not mutually exclusive but rather helpful to understand brain organization and function. One view concerns the *parallel-distributed* mode of processing of visual input under various aspects involving divergence and a variety of stages of integration and interaction. This concept receives support from anatomical tracing the retino-central projections (Fig. 1a) [retino-tectal, retino-pretectal, retino-anteriorthalamic, retino-belloncinuclear, retino-basalopticnuclear] of which most display a retinotopic order (e.g., Lázár 1971; Ewert et al. 1974; Fite & Scalia 1976).

The other view focusses on the retino-pretectal/tecto-motor pathway. This primary stimulus-response pathway, in connection with different forebrain structures, is subject to *modulation* with respect to (i) arousing of prey-catching [involving tecto-lateroanteriorthalamo-striato-pretecto-tectal influences], (ii) stimulus-specific habituation of prey capture [involving tecto-anteriorthalamo-ventromedialpallio-dorsalhypothalamo-tectal influences], (iii) prey/predator association [involving tecto-thalamo-ventromedialpallio-pretecto-tectal influences] (see also contributions by Ewert; Wang, Arbib & Ewert; Ewert, Matsumoto, Schwippert & Beneke, this volume). The concept of stimulus-response mediating circuitry influenced by various loops (Fig. 1b) receives its support from anatomical connectional studies in terms of afferences and efferences of the central visual structures mentioned above (e.g., Kicliter & Ebbesson 1976; Northcutt & Kicliter 1980; Neary & Northcutt 1983; Wilczynski & Northcutt 1977, 1983 a, b; Tóth et al. 1985). To study the *pattern of activity* across these brain structures during a certain period of stimulation, the [^{14}C]-2-deoxyglucose(DG) technique is an appropriate tool.

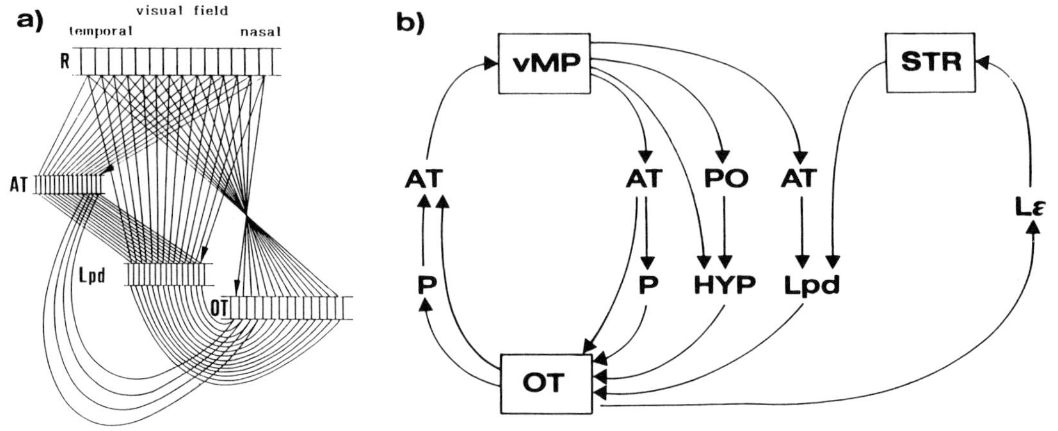

Fig. 1. Concepts of brain organization in anurans. a) Parallel/distributed processing of retinal input; examples. b) Loops by which forebrain structures may influence tectal properties; examples. AT, anterior thalamus including the anterior dorsal thalamic nucleus and the adjacent nucleus of Bellonci; HYP, hypothalamus; LA lateral anterior thalamic nucleus; Lpd, lateral posterodorsal pretectal thalamic nucleus; vMP, posterior ventral medial pallium; OT, optic tectum; P, postero-central pretectal thalamic nucleus; PO, preoptic area; R, retina; STR, ventral striatum.

1.2 Visual Learning

The notion of loop operations (Fig. 1b) has been put forward by Ewert (1987) and tested by means of [^{14}C]-2DG and lesion methods for associative and non-associative learning (Finkenstädt & Ewert 1988a,b; Finkenstädt 1989a,b). In the suggested functions, the telencephalic posterior ventromedial pallium (vMP) plays a significant role. One loop, concerning visual prey/predator association by hand-feeding (Brzoska & Schneider 1978; Ewert et al. 1983) involves pathway (iii) and leads to a prey *generalization* in such a manner that every moving visual object elicits prey-catching behavior, i.e., the species-universal prey-schema is extended. More specifically, the data suggest that this failure in object discrimination is due to associative vMP-induced inhibition of certain pretectal neurons whose inhibitory actions on tectal cells play a vital role in the visual feature analyzing neuroarchitecture (Ewert, this volume). The other loop, concerning stimulus-specific habituation (Ewert & Kehl 1978) involves pathway (ii) and leads to a prey *specification* (Wang, Arbib & Ewert, this volume). In the former case, the toad's species-universal prey-related object discrimination is permanently impaired due to associative storage of information. In the latter case, the non-associatively stored information only shows its influence when the toad is faced with the stimulus features according to which prey-catching was habituated; otherwise the species-universal object discrimination is intact. After lesioning vMP, in both cases of learning toads rely on their species-universal prey recognition.

2. Olfactory Conditioned Prey Motivation

2.1 A New Case Study

In the present study, we investigate a learning paradigm in which visual (prey) and olfactory (prey odor) cues are associated (Ewert 1965, 1968). This type of associative learning is interesting for three reasons: (1) in the absence of the prey-associated olfactory stimulus, toads display species-universal object discrimination during prey-catching; (2) only in the presence of the prey-associated odor is the activity of object discrimination strongly reduced, i.e. prey generalized; (3) in the absence of prey, this olfactory stimulus does not elicit prey-catching or any directed motor activity except of arbitrary intention head movements which ethologically are interpreted as a hint of the animal's readiness to perform a specific behavior pattern. We speak in this context of "conditioned prey motivation". Applying the [^{14}C]-2DG method, the aim of this pilot study was to compare learning-involved brain structures with those previously investigated in prey/predator association after hand-feeding (Finkenstädt & Ewert 1988b; Finkenstädt 1989b).

2.2 The Learning Paradigm

The experimental animals *Bufo bufo spinosus* were naive to the kind of odor to be associated with prey. During feeding with mealworms (not present in the toad's biotope), toads associated the odor from the mealworm excrements with prey. This learning was established after a few to several feeding sessions. It can be quantitatively evaluated by measuring (Fig. 2) the prey-catching activity to different moving visual objects in the presence of mealworm odor pre and post training, i.e., before and after feeding toads with mealworms. Comparable experiments were conducted with excrement-scentless mealworms scented either with farnesole (a chemical substance of mealworm excrements) or with cineole. These odorous substances, too, were effective in the association with prey. More specifically, in the training session the mealworm was presented to toads contingently (temporally and spatially paired) with the olfactory stimulus. For comparison, naive toads were treated in the same manner, but here the odor was presented temporally and spatially unpaired, whereby no association was built.

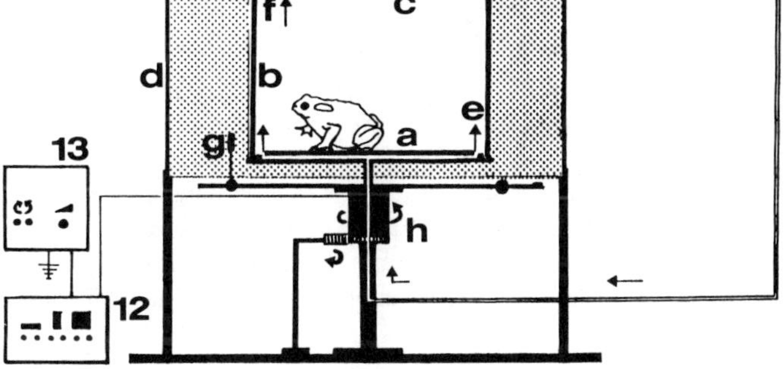

Fig. 2. Procedure for the test of prey-catching toward a moving visual object in presence or absence of odor: 1, unpurified air from the compressor; 2, pressure of controller; 3-7, purification of the carrier airstream [3, a washing flask with dest. water; 4, silica gel for drying; 5, charcoal activated; 6, molecular sieve; 7, paper filter]; 8, odor distributing system with four chambers controlled by valves; 9, capillary manometer; 10, chamber for behavioral investigation [a, arena base; b, cylindric glass vessel; c, top; d, cylindric box with a white background; e, odor inflow; f, odor outflow; g, moving black visual object; h, motor for moving the visual stimulus object; 11, variable odor/air-stream exhauster; 12, apparatus for changing the stimulus object; 13, control of direction and velocity of the visual object.

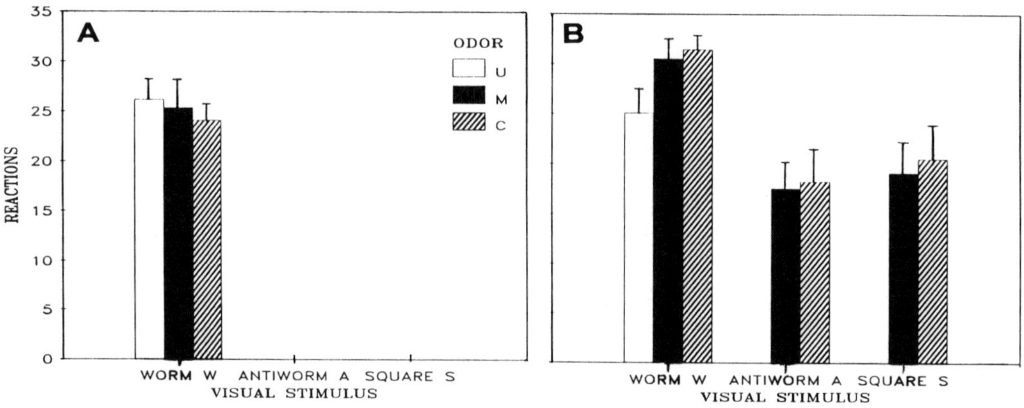

Fig. 3. Influence of prey-conditioned odor from mealworms (M), or cineole (C), on the toad's response to visual objects. Prey-catching orienting activity (reactions per 47 sec) toward different configural moving visual black objects (2.5 x 20 mm^2 stripe in worm-configuration *W*, same stripe in antiworm-configuration *A*, 80 x 80 mm^2 square *S*) before and after visual/olfactory conditioning, i.e., in the presence of odor (A) not associated with prey [n = 15 animals], or (B) associated with prey [n = 15 animals], or in M/C-scentless environment, *U*.

Figure 3 shows the quantitative behavioral data. Naive toads respond to (i) a black stripe of 2.5 x 20 mm^2 whose longer axis is oriented parallel to the direction of its movement (worm-configuration) with prey-catching behavior, to (ii) the same stripe if its longer axis is oriented perpendicular to the direction of movement (antiworm-configuration) without motor activity, and (iii) to a 80 x 80 mm^2 large black moving square with avoidance behavior, either in the presence or absence of an unconditioned odor (Fig. 3A). Trained toads (Fig. 3B), however, orient and snap also to (ii) and (iii), if the odor which they had previously associated with prey is present ($P < 0.01$, t-test); furthermore, prey-catching towards (i) is statistical significantly higher ($P < 0.01$, t-test) in the presence of a prey-associated olfactory stimulus.

2.3 Mapping of Brain Activity

In the ^{14}C-labeling experiment, [^{14}C]-2-deoxyglucose (15 uCi/ 100 g body weight) was injected into the toad's caudal lymphatic sacs. During the stimulation experiments, the animal sat in an arena shown in Fig. 2: the large black square was moved at an angular velocity of 7.6°/sec; the odor could be applied by a purified constant carrier airstream and exhausted after stimulation. Each toad received 30 stimulations, interrupted by pauses to prevent habituation, during a period of 90 min.

For mapping energy metabolism in the brain, autoradiographic x-ray film images of brain transverse sections were developed. The optical density was measured by means of a computer-assisted image analyzing system (Quantimet 920). For trained (T) and naive (N) toads, the concentration values [^{14}C](T) and [^{14}C](N) of single measurements of a brain structure - taken via ^{14}C-calibrated methylmethacrylate standards from optical density measurements - were expressed in relation to the concentration value [^{14}C]wh of a "white matter" (wh) reference brain structure, yielding X(N) = [^{14}C](N)/[^{14}C]wh and X(T) = [^{14}C](T)/[^{14}C]wh, respectively. This reference structure, an area below the nucleus ventralis nervi octavi, remained unaffected with regard to our experimental parameters, shown by means of the one-way analysis of variance *ANOVA*-test (for details see Finkenstädt & Ewert 1988a). The X(N,T)-values were calculated for a number *i* of histological transverse sections on which the brain structure appeared along its sagittal extension. For a brain structure under investigation, the bilateral *i* X(N)-values taken from a number *j* of "naive animals" (altogether $n_N = ij$ samples) were averaged, yielding X'(N), and in a comparable manner the X(T)-values from the trained (n_T samples) were averaged to X'(T). The comparison between the respective X'(N)- and X'(T)-values for a given brain structure was performed by means of the Kolmogorov-Smirnov-Omnibus *(KSO)*-test. This non-parametric statistical test is suitable for comparison between two independent samples of equal or unequal sizes. The test of significance was two-tailed.

2.4 Distribution of [^{14}C]-2DG Uptake

Learning-related changes in [^{14}C]-2DG uptake were investigated in the brains of thirty *trained* and thirty *naive* toads. Two experimental groups were formed. Group (1): ten trained and ten naive toads were binocularly blinded and then olfactorily stimulated with cineole, C, or dust from mealworm excrements, M. Group (2): twenty trained and twenty naive unblinded toads were stimulated with odor and large black square, S, simultaneously yielding the combinations {M + S} or {C + S}. Table 1 gives a qualitative survey about the brain-to-brain-comparison of glucose utilization "trained *vs.* naive" for each experimental group; quantitative data for Group (2) are shown in Fig. 4.

		Group (1) blinded		Group (2) unblinded	
Brain structures		M	C	{M+S}	{C+S}
	MOB	0	0	0	0
	AOB	+	+	+	+
	vSTR	0	0	0	0
	dSTR	0	0	0	0
TC	MS	0	0	0	0
	vMP	+	+	+	+
	dMP	0	0	0	0
	AL	0	0	0	0
	La	0	0	+	+
	P	0	0	+	+
	Lpv	0	0	+	+
DC	Lpd	0	0	0	0
	vHYP	-	-	-	-
	dHYP	0	0	0	0
	sOT	-	-	+	+
	vOT	-	-	+	+
MC	mOT	-	-	+	+
	dOT	-	-	0	0

Table 1. Brain-to-brain comparison "trained *vs.* naive" of [^{14}C]-2DG uptake during presentation of mealworm (M) or cineole (C) odor in post-training blinded toads (Group 1), or during presentation of the large moving visual object (S) in combination with M or C in unblinded toads (Group 2). Statistically significant differences ($P < 0.05$, *KSO*-test) of glucose utilization in cerebral structures are indicated by "+" (increase of metabolic activity), "-" (decrease), and "0" (no significant change). Abbreviations: TC, telencephalon; DC, diencephalon; MC, mesencephalon; MOB, main olfactory bulb; AOB, accessory olfactory bulb; vSTR, ventral striatum; dSTR, dorsal striatum; MS, medial septum; vMP, ventral part of the posterior medial pallium; dMP, dorsal part of the posterior medial pallium; AL, lateral amygdala; La, lateral anterior thalamic nucleus; P, postero-central pretectal thalamic nucleus; Lpv, lateral posteroventral pretectal thalamic nucleus; Lpd, lateral posterodorsal pretectal thalamic nucleus; vHYP, ventral hypothalamus; dHYP, dorsal hypothalamus; sOT, snapping area of the optic tectum; vOT, ventral OT (layers 1-5); mOT, medial OT (layers 6-7); dOT, dorsal OT (layers 8-9). Connections of brain structures are shown in Fig. 6.

Comparing [^{14}C]-2DG uptake (Table 1, Fig. 4) in both experimental Groups (1) and (2), four points are important to note: First, in both groups the telencephalic posterior ventromedial pallium (vMP) shows an increase of [^{14}C]-2DG uptake (cf. also Fig. 5b), but no significant change was found in the lateral amygdala (AL) and in other examined not primarily olfactory telencephalic structures. Second, the ventral hypothalamus (vHYP) displays in both groups a decrease in glucose utilization. Third, the optic tectum (OT) shows an opposite pattern of glucose utilization in both groups. Fourth, the tectum and visually sensitive diencephalic structures (La, Lpv, P) - except Lpd - display increase in glucose utilization in the unblinded animals, but in the accessory olfactory bulb (AOB) in both groups, whereas the main olfactory bulb shows no significant changes in both groups.

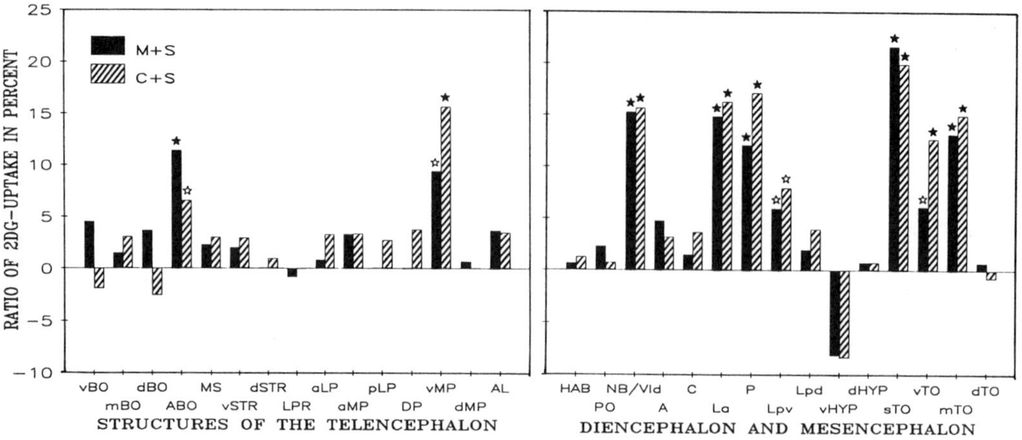

Fig. 4. Quantitative data of brain-to-brain comparison "trained *vs.* naive" according to Group (2) of Table 1: average percentage of the difference in [^{14}C]-2DG uptake across various brain structures. Toads were exposed to the large moving square (S) in combination with prey-associated odor; black bars refer to mealworm odor (M), hatched bars to cineole (C); white asterisks indicate a statistical significance of $P < 0.05$ and black ones $P < 0.01$ (*KSO*-test).

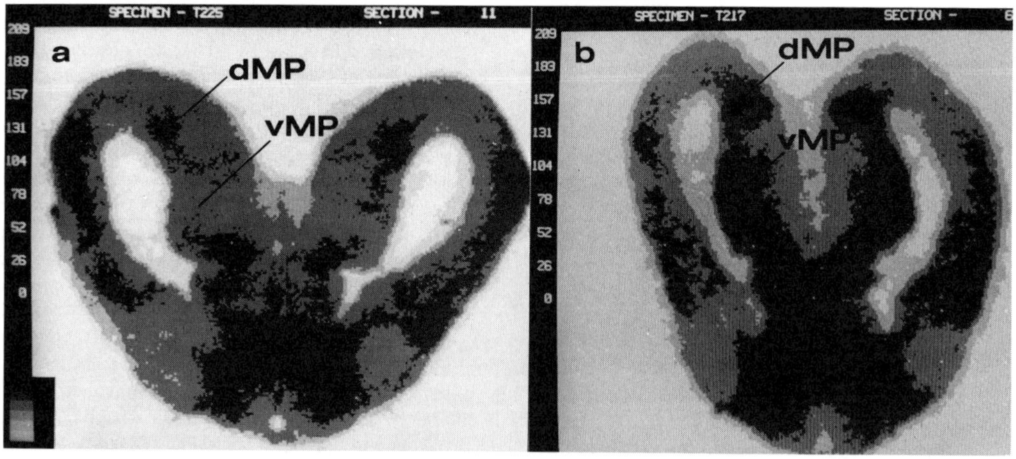

Fig. 5. Computer-processed autoradiographic images of transverse sections through the toad's brain at a posterior telencephalic level. (a) "Naive" cineole stimulated blinded toad. (b) Prey/cineole conditioned toad, subsequently blinded and thereafter stimulated with cineole. Note the difference in [^{14}C]-2DG uptake in the ventral medial pallium (vMP), while the dorsal medial pallium (dMP) displayed strong glucose utilization in both experimental groups.

3. Discussion

The olfactory sense in common toads is involved in motivational processes of feeding behavior (Heusser 1958). Odors, so far tested here, are associated with visual prey rapidly. After a few learning sessions the odor is recognized. Perceiving prey-associated odor, toads hunt moving visual objects that normally do not release prey-catching behavior (Ewert 1965, 1968; see also Sternthal 1974; Shinn & Dole

1978, 1979; Dole et al. 1981). The mapping of brain activity with [^{14}C]-2DG in prey-odor conditioned toads compared to odor-naive animals, raises the following discussion points:

(i) With respect to the concept of "conditioned prey motivation", we note that the learned odor presented alone does not elicit a conditioned motor response, but leads to an increased glucose utilization in a *not primarily sensory* telencephalic brain structure, the posterior ventral medial pallium vMP (Table 1, Group (2) blinded toads; Fig. 5b). This is most interesting, since previous [^{14}C]-2DG and lesion studies have shown that vMP - the "primordium hippocampi" of Herrick (1933) - is significantly involved also in other learning processes in toads (Finkenstädt & Ewert 1988a,b; Finkenstädt 1989a,b). A role of the mammalian hippocampus for the control of information storage was recently discussed by Schmajuk (1990). The data of combining prey associated odor with a large moving visual object are consistent in this context (cf. Table 1, Group (2) unblinded toads; see also Fig. 4). The ventral hypothalamus (vHYP), a linking structure of "limbic" forebrain and "limbic" midbrain areas (Nauta 1963), presumably exerts a modulatory function on learning-related processes. A suppression of glucose utilization in vHYP (Table 1, Fig. 4) has not yet been obtained in other learning paradigms in toads.

(ii) Learning coupled increases of brain activity are detected in *primary sensory* structures concerned with olfaction (AOB) and vision (OT, La, P, Lpv). The increases of glucose utilization in visual structures (Table 1, Group (2) unblinded toads; Fig. 4) can be interpreted in various ways: (a) coupling of excitatory influences between learning involved telencephalic structures (e.g., vMP) via diencephalic relays with tectal/tecto-motor systems; (b) increased visuomotor activity in response to the large object compared to naive toads; (c) intrinsic sensory processes concerned with learning (e.g., associative synapses). The increase of [^{14}C]-2DG uptake in the accessory olfactory bulb in response to prey-associated odor in blinded animals (Table 1) accounts for (c), i.e. an involvement of primary sensory structures. Comparable findings of metabolic changes in sensory structures related to olfactory associative learning (Sullivan et al. 1989) or to auditory habituation (Gonzales-Lima, Finkenstädt & Ewert 1989) are reported for rats.

(iii) The prey odor induced decrease of glucose utilization in the optic tectum of toads blinded after training (Table 1) leaves open a variety of explanations. One could be that learning-related activity of vMP via thalamic or hypothalamic nuclei inhibits tectal class T5.3 interneurons that are suggested to sharpen the prey-selectivity by inhibitory inputs to T5.2 tecto-bulbar/spinal projection neurons (cf. Ewert, this volume); in the unblinded toad, T5.2 neurons would then be "disinhibited" in response to the combination of visual and prey-associated olfactory stimuli, which may explain the average increase of glucose utilization in central tectal layers (Table 1, Fig. 4).

(iv) Table 2 gives a survey for comparison of glucose utilization in prey/predator and prey/odor conditioned toads, the former showing visual prey generalization permanently and the latter only in the presence of prey-associated odor. In common is mainly the strong [^{14}C]-2DG uptake in vMP in the presence of the conditioned stimulus. Different are the decreases in Lpd in prey/predator conditioned animals and in vHYP and OT in prey/odor conditioned ones if only the prey associated odor is present. Regarding the hypothalamus, there are also differences between both learning paradigms in the [^{14}C]-2DG uptake in vHYP *vs.* PO/dHYP. In common is the strong increase of glucose utilization in OT in response to the large visual object for conditions of prey generalization. We hypothesize that the prey-selectivity of tecto-bulbar/spinal T5.2 neurons in the two learning paradigms are reduced by different loop operations that both involve the vMP: due to vMP influences, one loop (prey/predator conditioning) facilitates the response of T5.2 neurons and impairs their selectivity to visual moving objects by reduction of pretecto(Lpd)-tectal inhibitory activity; the other loop (prey/odor conditioning) facilitates T5.2 and impairs their selectivity by inhibition of inhibitory tectal interneurons. Furthermore, in the latter paradigm we have to consider the decrease of glucose utilization in vHYP which might be explained by vMP-induced inhibition of certain hypothalamo-tectal connections which normally have a suppressive influence on the tectal prey-catching system.

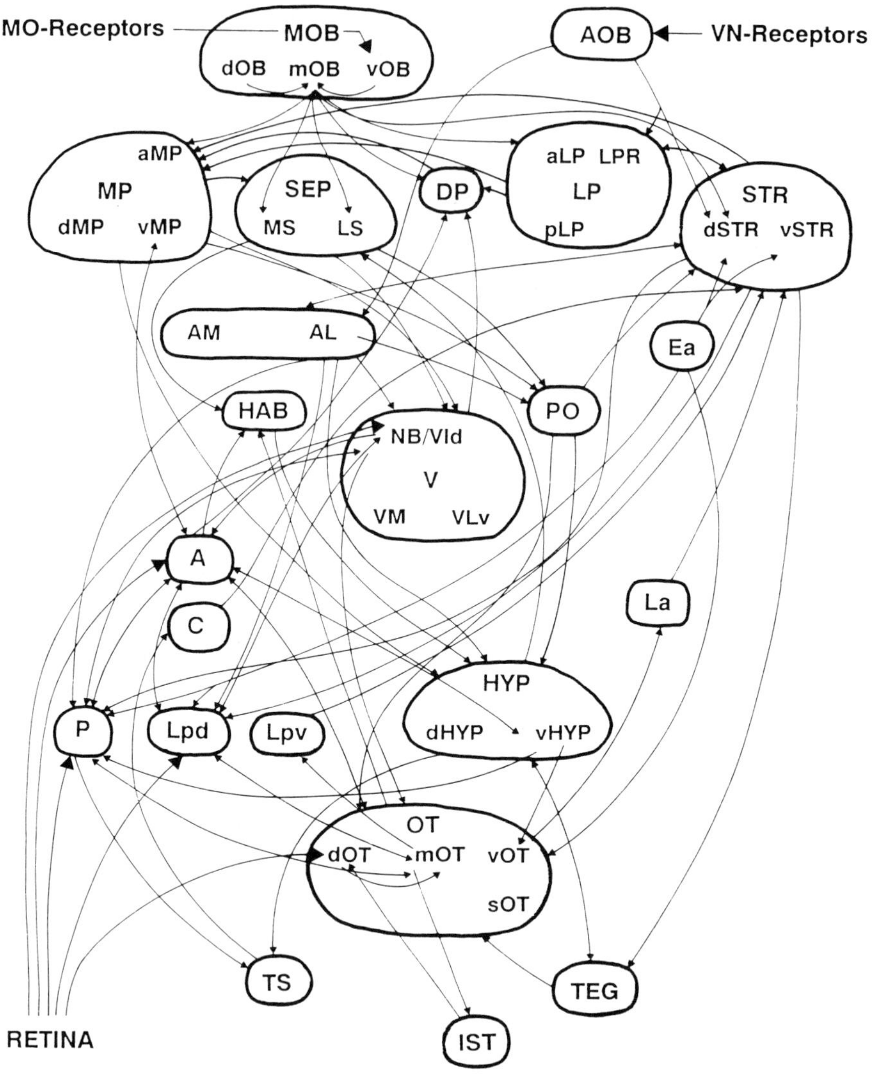

Fig. 6. Some connections between brain structures involved in visual and olfactory information processing. Abbreviations of brain structures: A, anterior dorsal thalamic nucleus; AL, nucleus amygdalae, lateral portion;; AM, nucleus amygdalae, medial portion; AOB, accessory olfactory bulb; C, central thalamic nucleus; DP, dorsal pallium; Ea, anterior entopeduncular nucleus; HAB, nucleus habenulae; HYP, hypothalamus; dHYP, dorsal hypothalamic area; vHYP, ventral hypothalamic area; IST, nucleus isthmi; La, lateral anterior thalamic nucleus; LP, lateral pallium; aLP, anterior lateral pallium; pLP, posterior lateral pallium; LPR, lateral prominence; LS, lateral septum; Lpd, lateral posterodorsal pretectal, thalamic nucleus; Lpv, lateral posteroventral pretectal, thalamic nucleus; MOB, main olfactory bulb; MO-Rec, receptors of the main olfactory epithelium; MP, medial pallium; aMP, anterior medial pallium; dMP, dorsal one-third of the posterior medial pallium; vMP, ventral two-thirds of the posterior medial pallium; MS, medial septum; NB/Vld, "neuropil" (nucleus) of Bellonci (NB) and ventrolateral dorsal thalamic nucleus (Vld); dOB, dorsal (granule cell layer) portion of the main olfactory bulb; mOB, middle (mitral cell layer) portion of the main olfactory bulb; vOB, ventral (glomerular layer) portion of the main olfactory bulb; OT, optic tectum; dOT, dorsal layers (layers 8/9) of the optic tectum; mOT, middle layers (layers 6/7) of the optic tectum; sOT, snapping evoking area of the optic tectum; vOT, medio-ventral layers (layers 1 - 5) of the optic tectum; P, posterior pretectal thalamic nucleus; PO, preoptic area; SEP, septum; STR, striatum; dSTR, dorsal striatum; vSTR, ventral striatum; TEG, tegmentum; TS, torus semicircularis; V, ventral thalamus; VLv, ventrolateral thalamic nucleus, ventral part; VM, ventromedial thalamic nucleus; VN-Rec, receptors of the vomeronasal epithelium.

Table 2. Comparison of [^{14}C]-2DG uptake in various brain structures (abbreviations see Fig. 6) for different learning paradigms; increase "+", decrease "-", and no change "0" in conditioned toads refers to brain-to-brain comparison "trained *vs.* naive", in naive toads to "S-stimulated" *vs.* "unstimulated". S, moving large black square object. (For quantitative data see Fig. 4; Finkenstädt & Ewert 1988b; Finkenstädt 1989b).

		[^{14}C]-2DG uptake in:						
Conditioning	Motor behavior	vMP	PO/dHYP	vHYP	P	Lpd	Lpv	v/mOT
"prey/predator" in presence of S	prey-catching	++	++	0	++	-	++	++
"prey/odor" in presence of odor	no response	+	0	-	0	0	0	-
in presence of odor and S	prey-catching	++	0	-	++	0	+	++
"naive" in presence of S	avoidance	-	(+)	0	(+)	++	(+)	+

We are aware that our present data hardly give *evidence* of functional connections. However, they provide us with ideas to test hypotheses by means of further investigation techniques. Recalling the concepts of parallel/distributed information processing and integrative/interactive loop operations (Fig. 1a,b), the schema in Fig. 6 illustrates the enormous complexity of possible (and probably not all considered) connections for bimodal visual/olfactory information processing, which is consistent with the notion that (almost) the whole brain - with different emphasis in the participation of its structures - is involved in learning.

ACKNOWLEDGEMENTS. We thank Mrs. Gerda Kruk and Mrs. Gisela Kaschlaw for technical assistance and animal care, and Mrs. Ursula Reichert and Mrs. Gudrun Frühauf for text processing. The work was supported by the Deutsche Forschungsgemeinschaft.

References

Brzoska J, Schneider H (1978) Modification of prey-catching behavior by learning in the common toad (*Bufo b bufo* L, Anura, Amphibia): changes in response to visual objects and effects of auditory stimuli. Behav Processes 3: 125-136

Dole JW, Rose BB, Tachiki KH (1981) Western toads (*Bufo boreas*) learn odor of prey insects. Herpetologia 37: 63-68

Ewert J-P (1965) Der Einfluß peripherer Sinnesorgane und des Zentralnervensystems auf die Antwortbereitschaft bei der Richtbewegung der Erdkröte (*Bufo bufo* L). Univ Göttingen Druck, Göttingen

Ewert J-P (1968) Der Einfluß von Zwischenhirndefekten auf die Visuomotorik im Beute- und Fluchtverhalten der Erdkröte (*Bufo bufo* L). Z Vergl Physiol 61: 41-70

Ewert J-P (1987) Neuroethology of releasing mechanisms: prey-catching in toads. Behav Brain Sci 10: 337-405

Ewert J-P, Kehl W (1978) Configurational prey selection by individual experience in the toad *Bufo bufo*. J Comp Physiol 126: 105-114

Ewert J-P, Hock FJ, Wietersheim A v (1974) Thalamus/Praetectum/Tectum: retinale Topographie und Physiologische Interaktionen bei der Kröte (*Bufo bufo* L). J Comp Physiol 92: 343-356

Ewert J-P, Burghagen H, Schürg-Pfeiffer E (1983) Neuroethological analysis of the innate releasing mechanism for prey-catching behavior in toads. In: Ewert J-P, Capranica RR, Ingle DJ (eds) Advances in vertebrate neuroethology. Plenum Press, New York London, pp 413-475

Finkenstädt T (1989a) Stimulus-specific habituation in toads: 2DG studies and lesion experiments. In: Ewert J-P, Arbib MA (eds) Visuomotor coordination: amphibians, comparisons, models, and robots. Plenum Press, New York London, pp 767-797

Finkenstädt T (1989b) Visual associative learning: searching for behaviorally relevant brain structures in toads. In: Ewert J-P, Arbib MA (eds) Visuomotor coordination: amphibians, comparisons, models, and robots. Plenum Press, New York London, pp 799-832

Finkenstädt T, Ewert J-P (1988a) Effects of visual associative conditioning on behavior and cerebral metabolic activity in toads. Naturwissenschaften 75: 95-97

Finkenstädt T, Ewert J-P (1988b) Stimulus-specific long-term habituation of visually guided orienting behavior toward prey in toads: a ^{14}C-2DG study. J Comp Physiol 163: 1-11

Fite KV, Scalia F (1976) Central visual pathways in the frog. In: Fite KV (ed) The amphibian visual system: a multidisciplinary approach. Academic Press, New York San Francisco London, pp 87-118

Gonzalez-Lima F, Finkenstädt T, Ewert J-P (1989) Learning-related activation in the auditory system of the rat produced by long-term habituation: a 2-deoxyglucose study. Brain Res 489: 67-79

Herrick CJ (1933) The amphibian forebrain. VIII: Cerebral hemispheres and pallial primordia. J Comp Neurol 58: 737-759

Heusser H (1958) Zum geruchlichen Beutefinden und Gähnen der Kreuzkröte (*Bufo calamita* Laur). Z Tierpsychol 15: 94-98

Kicliter E, Ebbesson SOE (1976) Organization of the "nonolfactory" telencephalon. In: Llinás L, Precht W (eds) Frog neurobiology. Springer, Berlin Heidelberg New York, pp 946-972

Lázár G (1971) The projection of the retinal quadrants on the optic centers in the frog: a terminal degeneration study. Acta Morph Acad Sci Hung 19: 325-334

Nauta WJH (1963) Central nervous organization and the endocrine motor system. In: Nalbandov AV (ed) Advances in neuroendocrinology. Univ Illinois Press, Urbano, pp 5-21

Neary T, Northcutt RG (1983) Nuclear organization of the bullfrog diencephalon. J Comp Neurol 213: 262-278

Northcutt RG, Kicliter E (1980) Organization of the amphibian telencephalon. In: Ebbesson SOE (ed) Comparative neurology of the telencenphalon. Plenum Press, New York London, pp 203-255

Schmajuk, NA (1990) The hippocampus and the control of information storage in the brain. In: Arbib MA, Amari S (eds) Dynamic interactions in neural networks: models and data. Springer,New York Berlin Heidelberg London Paris Tokyo, pp 53-72

Shinn EA, Dole JW (1978) Evidence for a role for olfactory cues in the feeding response of leopard frogs, *Rana pipiens*. Herpetologia 34: 167-172

Shinn EA, Dole JW (1979) Lipid components of prey odors elicit feeding responses in western toads *Bufo boreas*. Copeia 1979(2): 275-278

Sternthal DE (1974) Olfactory and visual cues in the feeding behavior of the leopard frog (*Rana pipiens*). Z Tierpsychol 34: 240-246

Sullivan RM, Wilson DA, Leon M (1989) Associative processes in early olfactory preference acquisition: neural and behavioral consequences. Psychobiology 17(1): 29-33

Tóth P, Csank G, Lázár G (1985) Morphology of the cells of origin of descending pathways to the spinal cord in *Rana esculenta*. A tracing study using cobalt-lysine complex. J Hirnforsch 26: 365-383

Wilczynski W, Northcutt RG (1977) Afferents to the optic tectum of the leopard frog: an HRP study. J Comp Neurol 173: 219-229

Wilczynski W, Northcutt RG (1983a) Connections of the bullfrog striatum: afferent organization. J Comp Neurol 214: 321-332

Wilczynski W, Northcutt RG (1983b) Connections of the bullfrog striatum: efferent projections. J Comp Neurol 214: 333-343

Dishabituation Hierarchies for Visual Pattern Discrimination in Toads: A Dialog between Modeling and Experimentation[1]

DeLiang Wang, Michael A. Arbib*, and Jörg-Peter Ewert***

**Center for Neural Engineering*
University of Southern California
Los Angeles, CA 90089-2520, U.S.A.
***Abteilung Neurobiologie, FB 19, Universität Kassel*
D-3500 Kassel, F.R.G.

Abstract. Toads exhibit stimulus- and locus-specific habituation. Instead of mutual dishabituation for different worm-like stimuli, toads exhibit a dishabituation hierarchy. In modeling the dishabituation hierarchy, we argue that the toad's visual discrimination is reflected in different average neuronal firing rates in a particular visual center, hypothetically anterior thalamus (AT). This theory is developed through a large-scale neural simulation, and the neural model predicts that the retinal R2 cells play a primary role in the discrimination via tectal small pear cells (SP) while R3 projections to AT refine the feature analysis by inhibition.

This theory predicts new dishabituation hierarchies based on reversing stimulus-background contrast and shrinking stimulus size. After the predictions were made, we designed a group of behavioral experiments to test them. In particular, we selected a pair of stimuli whose ordering in the dishabituation hierarchy we predict to be changed by contrast reversal, and the experimental result is as predicted. A size shrinking prediction was also tested and failed to be validated. Our further experiments concerning size effects suggest that visual pattern discrimination in toads is not affected by stimulus size. Finally, we discuss new insights into visual pattern discrimination of amphibians and the interplay between modeling and experimentation.

1. Behavioral Background

As in almost all other animals, toads and frogs show habituation of prey-catching behavior. That is, when presented repeatedly with the same prey dummy, toads and frogs orient less and less frequently to the stimulus. This visual habituation exhibits

---

[1] The research described in this paper was supported in part by grant no. 1RO1 NS 24926 from the National Institutes of Health.

locus specificity. After habituation to a worm-like stimulus at a certain visual location, the same stimulus can elicit a new prey-catching response when presented elsewhere (Eikmanns 1955; Ewert and Ingle 1971). Furthermore, Ewert and Kehl (1978) revealed that visual habituation in toads exhibits *hierarchical stimulus specificity.* After an extensive study, they found the dishabituation hierarchy shown in Fig.1, where a worm pattern higher in the hierarchy can dishabituate (i.e. release behavior despite habituation to) another worm pattern lower in the hierarchy. A stimulus lower in the hierarchy, however, cannot dishabituate habituation of a stimulus higher in the hierarchy. At the same level in the hierarchy, the left one can slightly dishabituate the right one, but not vice versa.

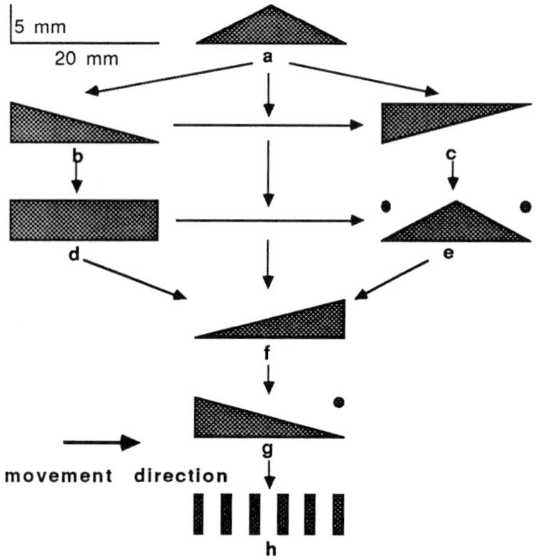

Figure 1. Dishabituation hierarchy for the worm patterns used in stimulus-specific habituation in toads. Between any pair of stimuli in the hierarchy, one stimulus is presented first, and after habituation to the stimulus, another stimulus is presented until it elicits no response. One stimulus can dishabituate all the stimuli below or to right of it, but not vice versa. In the experiment, a black worm dummy moves against a white background. Before any habituation, all the stimuli are about equally strong in releasing orienting responses (redrawn from Ewert and Kehl 1978).

The dishabituation hierarchy exhibited in toads is different from both invertebrates, like *Aplysia,* and mammals. In *Aplysia,* habituation seems to be independent of the specific patterning of stimuli (Kandel 1976), whereas in mammals, habituation is stimulus specific such that dishabituation is mutual: If stimulus *A* can dishabituate stimulus *B*, then stimulus *B* can dishabituate stimulus *A* as well (Thompson and Spencer 1966; Sokolov 1975). This characteristic of dishabituation in toads might represent an intermediate evolutionary step in the development of stimulus specific learning in mammals.

The biological significance of stimulus specific habituation may be to keep the prey-catching mechanism alert to "new" prey (Schleidt 1962). The hierarchical stimulus specificity exhibited in toads suggests that it is the *configurational cues* of the stimuli, not the pure "newness", which decide the animal's prey catching behavior (Ewert and Kehl 1978). The dishabituation hierarchy clearly demonstrates that the toad visual system is capable of discriminating fairly fine differences in objects.

2. Modeling Visual Pattern Discrimination in Toads

What are the neural mechanisms that underlie the dishabituation hierarchy? This is the question we try to answer through modeling, based on the known electrophysiology and anatomy of the toad brain. In the experiment of Ewert and Kehl (1978), all moving objects are 20 mm long and 5 mm high (see Fig.1), corresponding to 16° and 4° visual angles, respectively, from the viewing distance of 70 mm. The dots which are added to the triangular objects (see Fig.1) are 1 mm in diameter which is about 1°. Because all the moving objects have the same length and height, the critical cues are (1) leading edge (or the angle subtended by a leading edge); (2) trailing edge; (3) isolated dot; and (4) striped pattern. In what follows, we will give a summary of the model — the full exposition is provided in Wang and Arbib (1990).

The basic hypothesis of the model is that, in a certain neural structure of the toad visual pathway, a stimulus higher in the dishabituation hierarchy elicits a bigger firing response than one lower in the hierarchy (Lara and Arbib 1985). This hypothesis is mainly based on the observation that the dishabituation is unidirectional. We suggest that this structure is in the anterior thalamus, and offer a detailed neural model, summarized in Fig.2, which incorporates retina, tectum and anterior thalamus. In the figure, conical projections represent on-center off-surround convergence, while the cylindrical projection from the R2 layer to the small pear cell (SP) layer represents a 1 to 1 mapping. The connections from the receptor layer to both the depolarizing bipolar cell (BD) layer and the hyperpolarizing bipolar cell (BH) layer also constitute a small many-to-one convergence. The receptor layer contains 140x140 cells which correspond to a 70°x70° visual field. Bipolar and amacrine cell layers (ATD: on-channel, ATH: off-channel) consist of 140x140 cells respectively, in correspondence with the receptor layer. Three types of ganglion cells, R2, R3 and R4, have been modeled, each consisting of 25x25 cells which correspond to a 70°x70° visual field since the ganglion cells have 20° receptive fields (RF) and lie 2° apart. The R2 layer projects to the SP layer in the tectum, and the SP layer and the R3 layer together converge on the AT layer in the anterior thalamus, where, in our model, the discrimination of worm-like patterns is finally achieved. The entire simulation contains about 100,000 cells.

Our detailed retina model is based on the Teeters model (Teeters 1989) and more closely approximates the electrophysiological data. In particular, our model simulates both the on-channel and off-channel response of R2 cells and accepts bitmap stimuli directly. In quantitative terms, a moving stimulus is directly mapped onto the receptor layer, and the response of a neuron in a later layer is formed by a matrix of cells from the previous layer according to the *leaky integrator* model (following the style of modeling in Lara et al. 1982). The model responses of the retina model to various worm, antiworm and square stimuli are presented in Fig.3 for the R2 and R3 ganglion cells, together with the electrophysiological data (Ewert and Hock 1972; Ewert 1976). The average firing rate of a cell is computed by the temporal integration of its instantaneous firing rate divided by the time period during which a non-zero firing rate occurs consecutively. From the figure we can see a close match between the model responses and the experimental data.

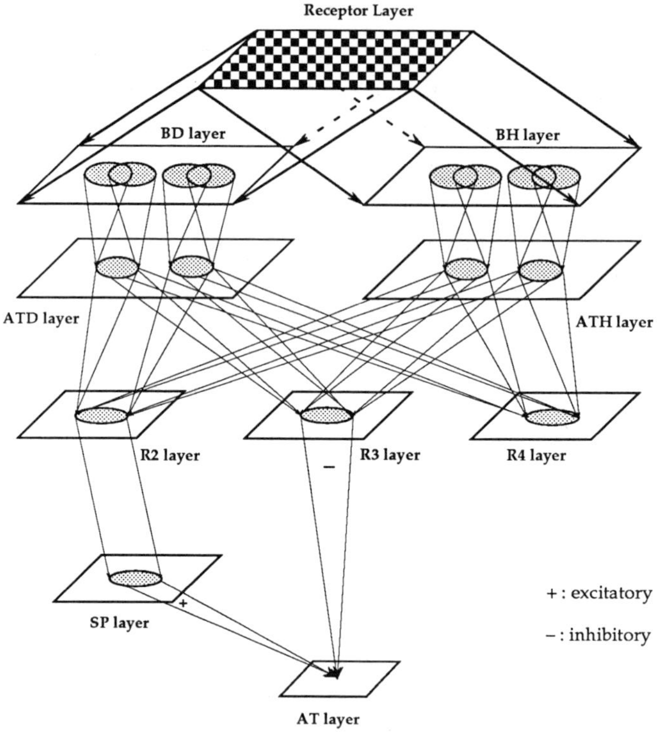

Figure 2. Anatomy of the entire model which incorporates retina, tectum, and anterior thalamus.

The pretectal/tectal network discriminates prey from nonprey like predators (Cervantes-Perez et al. 1985). However, in the present model we assume that this network does not play an important role in the discrimination of "subworms" (different worm patterns in Fig.1), but rather relays the R2 response via the small pear cells (SP, see Fig.2). SP cells are suggested to receive inputs from the R2 cells and project ascendingly to the anterior thalamus (Lázár et al. 1983).

The anterior thalamus receives R3 and R4 retinal projections (Scalia and Gregory 1970; Grüsser and Grüsser-Cornehls 1976) and SP tectal projections (Lázár et al. 1983). Among other ascending projections to the telencephalon, AT has a direct projection to the medial pallium (Scalia and Colman 1975; Neary and Northcutt 1983) which seems critical for habituation (Finkenstädt and Ewert 1988; Finkenstädt 1989). Based on the particular position of AT in the learning loops of the toad brain (Ewert 1987a, see also this volume), we hypothesize that it is the anterior thalamus where the subworm pattern discrimination is finally exhibited. Due to the lack of more detailed data, we presently use an array of neurons in AT, called ATH, for modeling the computational function of the anterior thalamus. ATH neurons receive excitatory-center inhibitory-surround inputs from tectal SP cells, and direct inhibitory inputs from R3 cells, as shown in Fig.2.

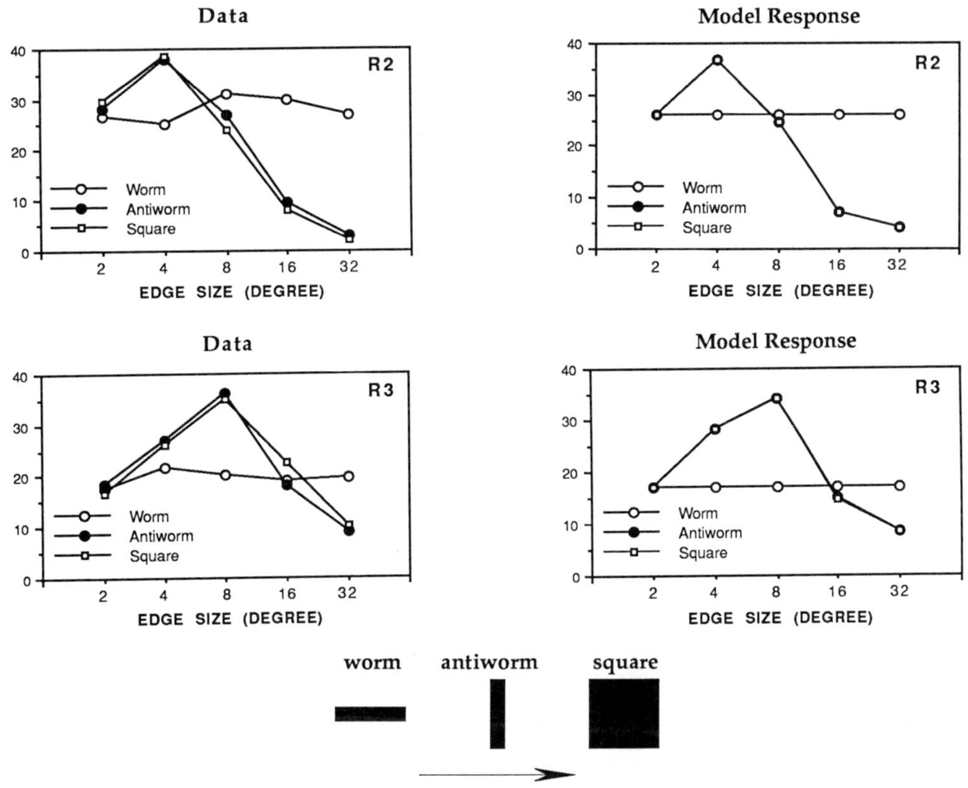

Figure 3. Left: The experimental data of R2 and R3 responses to worm, antiworm and square (redrawn from Ewert 1976); **Right**: The corresponding model response. A worm stimulus here is a rectangle with its elongated edge parallel to the direction of movement; an antiworm stimulus is a rectangle with its elongated edge perpendicular to the direction of movement. Each point in the figure represents a temporal average firing rate.

Fig.4 shows the average firing rates of a single ATH neuron to the 8 worm-like stimuli in Fig.1. The open symbols represent the response of the full model, while for a comparison the filled symbols provide the response without R3 inhibition. The result clearly matches the ordered dishabituation hierarchy in Fig.1. Not only do stimuli higher in the hierarchy generate larger ATH responses, but the stimulus pairs **b–c** and **d–e** which are on the same level in the hierarchy generate nearly equal responses. Note that the number of arbitrary hierarchies out of the 8 stimuli is 8! (40,320), and only one is supported by the data. The production of Fig.4 required considerable computer experimentation to find parameters for connections from retina and to ATH which matched the observed hierarchy, but no tuning of parameters of the retinal model was permitted. In the present paper, we will not model the habituation/dishabituation process (presumably in the medial pallium), and so the following discussions will be directly in terms of average firing rates produced by ATH.

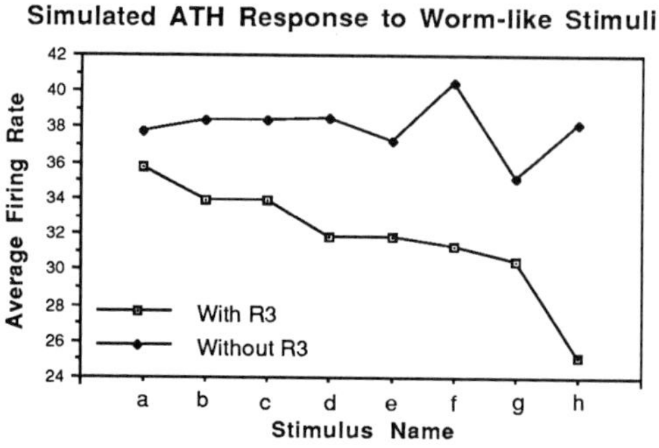

Figure 4. ATH response to the 8 worm-like stimuli shown in Fig.1. The 8 average firing rates are ordered, which corresponds to the ordered hierarchy in Fig.1. The figure also shows ATH response to the stimuli without inhibitory projection from R3 cells.

This model represents an integration of behavioral, physiological, anatomical, and theoretical studies on the brain in order to understand the neural mechanisms for pattern discrimination in toads. Based on the modeling, our explanation of the dishabituation hierarchy is:

(1) Both the leading edge and the trailing edge of a worm stimulus have to be taken into consideration.

(2) The receptive field of ATH neurons (25° ERF and 50° IRF in our model) is big enough to "see" both the leading and the trailing edge. Stimulus **a** elicits the biggest response because both diagonal edges elicit quite strong responses in R2 cells and these responses can be best integrated in ATH cells due to the relatively small distance between the midpoints of **a**'s leading and trailing edges. For stimulus **d**, for example, the response induced by the two edges counteracts each other in the ATH receptive field, due to the size of the ATH's ERF and IRF.

(3) Stimuli **b** and **c** are preferred to stimulus **f** because the inhibition by R3 cells, which have off-channel preference, is bigger for **f** than for **b** and **c**.

(4) Stimuli with dots appear lower in the hierarchy because they elicit smaller R2 response due to IRF interaction.

(5) A striped pattern elicits the smallest response in ATH neurons because of R3 inhibition. This is particularly clear when we compare the two curves in Fig.4.

This level of modeling enables us to offer specific predictions which form strict tests for the theory. Since, in the model, both the on-channel and off-channel responses are considered important, it is interesting to know whether the dishabituation hierarchy will be changed if stimulus-background contrast direction is reversed. After the model was completed, we presented the same set of stimuli to the model as in Fig.1 but reversed the contrast direction. Fig.5 shows the simulation result, where — on the basis of the simulated ordering of ATH responsiveness — a dishabituation hierarchy different from Fig.1 is predicted. The response of R2 cells is about the same with respect to contrast reversal, but now R3 cells show a trailing edge preference.

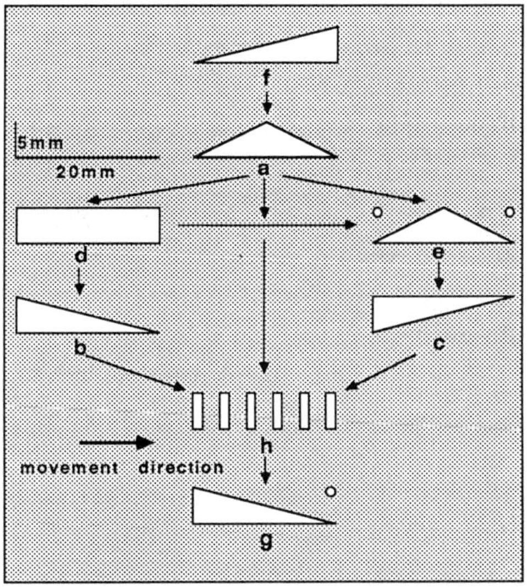

Figure 5. Dishabituation hierarchy predicted from this model by reversing the stimulus-background contrast direction. The same set of stimulus configurations is used as in Fig.1, but in contrast to Fig.1 white stimuli move against a black background.

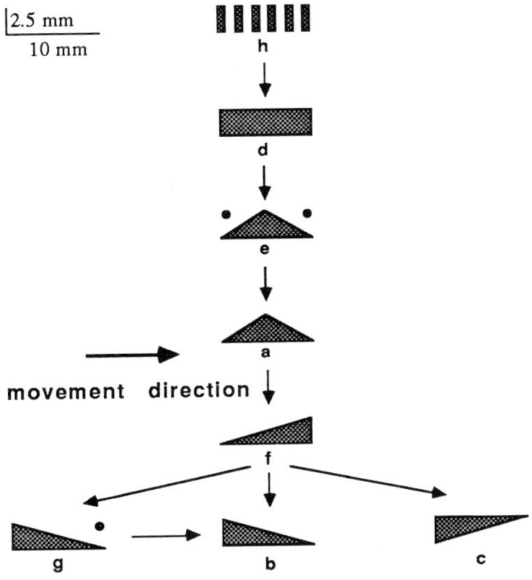

Figure 6. Dishabituation hierarchy predicted from this model by shrinking the stimulus size. The same set of stimulus configurations is used as in Fig.1, but all the stimuli are 10 mm long and 2.5 mm high, half of the size used in Fig.1.

Is the dishabituation hierarchy sensitive to stimulus size? The current model predicts that different hierarchies are produced based on different sizes of worm-like stimuli. As an example, we shrank the size of all the stimuli to 10 mm long and 2.5 mm high (this corresponds to 8° by 2° given the same distance to the animal) and tested these stimuli. The worm dummy that is 8° long and 2° high forms an optimal stimulus to T5.2 cells in the tectum which correlate well with prey-catching behavior (Ewert 1984). Fig.6 presents the dishabituation hierarchy predicted by this model. Our explanation of the remarkable difference is that since the stimulus size is halved compared to Fig.1, the previous IRF interaction in the R2 receptive field is converted into ERF interactions which strengthen overall response. This ERF interaction is particularly manifested by the striped pattern h. Note that the R3 inhibition in ATH neurons is relatively smaller than the excitation from SP cells, and thus cannot prevent stimulus **h** from inducing a strong ATH response in the final integration.

3. Back to Animal Behavior

After the modeling work was completed (Wang and Arbib 1990), behavioral experiments were designed to test part of the predictions of the model. About 200 Common Toads *Bufo bufo* (L.) were used in this behavioral investigation. The toads were kept in a constant temperature (20°C) room and fed regularly with mealworms (*Tenebrio molitor*). In the experiments, an electrically driven stimulus (prey dummy) was rotated at 20°/s from left to right (relative to the toad) on a horizontal plane in front of the background. The prey-catching orienting response was measured by counting the number of orienting turns within successive 1-minute intervals. The experimental apparatus and procedures were the same as in Ewert and Kehl (1978).

Only animals who exhibited more than 20 orienting turns in the first minute were were counted into the quantitative results. To determine whether the animal was able to discriminate two visual dummies *A* and *B*, the prey-catching orienting response of the toad was first habituated to stimulus *A*; then the response to *B* was tested and also habituated. Experiments were also repeated in the reverse order: first habituated to *B* and then *A* was tested and habituated. All experiments were repeated with 10 different toads to reach stable statistical results.

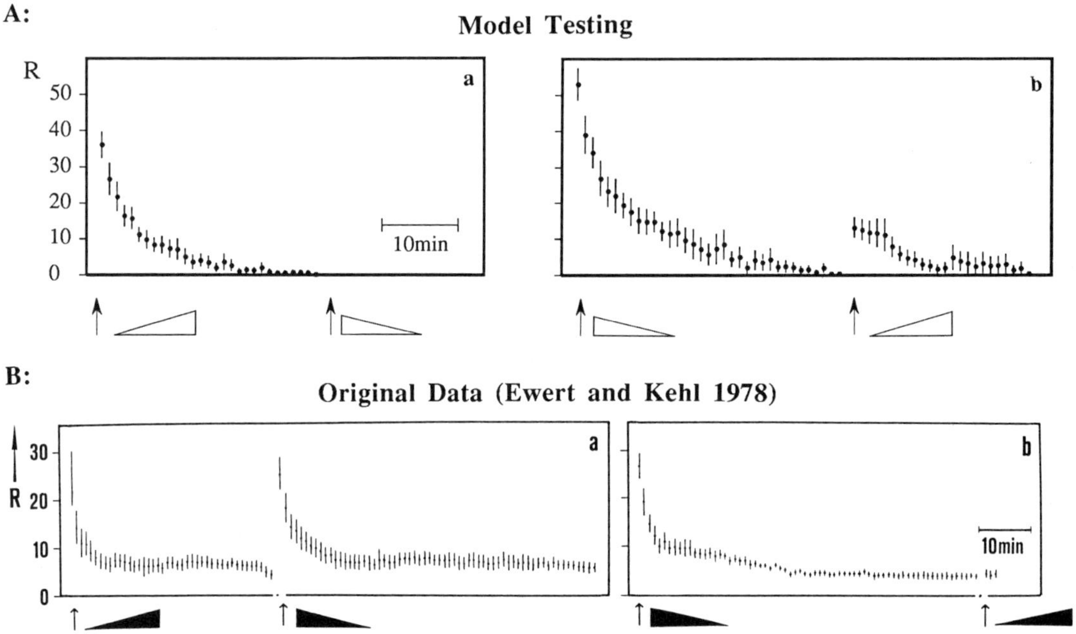

Figure 7 A. Dishabituation test between a white right-pointing triangle of 20x5 mm^2 area and its mirror image. The white stimuli were moved around the toad (sitting in a cylindrical glass vessel) at the constant distance of 70 mm and speed of 20°/s against a black background, from left to right relative to the animal. **Aa**. Habituation of the toad prey-catching orienting response first to the right-pointing triangle, and then immediately afterwards a test of the response (see vertical arrow) to the left-pointing triangle. **Ab**. Reverse order of presentation: Habituation first to the left-pointing triangle, and then a test of the response to the right-pointing triangle. Abscissa: Habituation time [min]; Ordinate: Successive number of orienting turns per minute (orienting activity, R). Each curve point represents an average value of 10 individuals, and the vertical bar represents a standard deviation. **B**. The corresponding dishabituation test with black stimuli on a white background (from Ewert and Kehl 1978).

3.1 Contrast Reversal

In order to test the prediction of the stimulus-background contrast effect on the dishabituation hierarchy, we selected a pair of stimuli **b/f**, which showed a strong difference between the black experimental hierarchy in Fig.1 and the predicted white hierarchy in Fig.5. In this group of experiments, a homogeneous black background was used, and white **b** and white **f** were 5 mm high and 20 mm long, as in Fig.1. The experimental results are presented in Fig.7A. For comparison, the corresponding experimental results with black **b/f** (Ewert and Kehl 1978) is shown in Fig.7B. In Fig.7Aa, white **b**, which is the right-pointing triangle, was presented first. Immediately after full habituation to white **b**, the left-pointing triangle, white **f**, was tested. A statistically significant increase in the responses was exhibited in the toads with the presentation of white **f**. In Fig.7Ab, the presentation order was reversed: white **f** was presented first, and immediately following full habituation to white **f**, white **b** was presented. All the experimented toads failed to release a new orienting response to the presentation of white **b**. In summary, the toad is able to distinguish between white **b** and white **f** of the same length and height, and it prefers white **f** to white **b** in dishabituation, the opposite of the effect shown with the corresponding black stimuli. With this selected pair, the experimental result is as predicted by the theory.

3.2 Size Effect

To test the prediction of size reduction shown in Fig.6, we again selected a pair of stimuli **b/d** from the figure (called small **b** and small **d** hereafter) which yields an opposite preference of dishabituation in Fig.6 to that in the original data of Fig.1. Both small **b** and **d** are 2.5 mm high and 10 mm long. The model testing results are presented in Fig.8A. In Fig.8Aa, small **b** was presented first, and immediately after full habituation to small **b**, small **d** was presented. Slight dishabituation was observed. However, if the order of presentation was reversed as in Fig.8Ab where small **d** was presented first and small **b** was tested next, remarkable dishabituation was exhibited in the toads. Comparing Fig.8Aa and Fig.8Ab, it can be concluded that the toad is able to distinguish small **b** and **d**, and that small **b** is still preferred by toads in dishabituation. In Fig.8B, the situation with the original size is shown for comparison. Although the stimulus size was halved, the same preference is established by toads, at least for this particular pair of stimulus configurations. The model failed to conform to this situation.

What went wrong with the model? Basically, the model predicts that the dishabituation hierarchy changes with stimulus size. One reasonable explanation for the failure of the model would be that toads may exhibit the same dishabituation hierarchy within a certain range of stimulus size. To test this conjecture of *size invariance*, the first set of experiments we did was to compare two different size stimuli of the same configuration. We choose configuration **b** of both 10x2.5 mm^2 and 20x5 mm^2, and the results are presented in Fig.9. In Fig.9a, small **b** was first presented and habituated and immediately afterwards big **b** was tested. A remarkable increase was exhibited in response to the presentation of big **b**. However, if big **b** was first presented and habituated, small **b** elicited almost no prey-catching response. These experiments demonstrate that by means of dishabituation toads are able to exhibit recognition of different sizes of a stimulus, and in particular big **b** has a preference to small **b**. No mutual dishabituation was found.

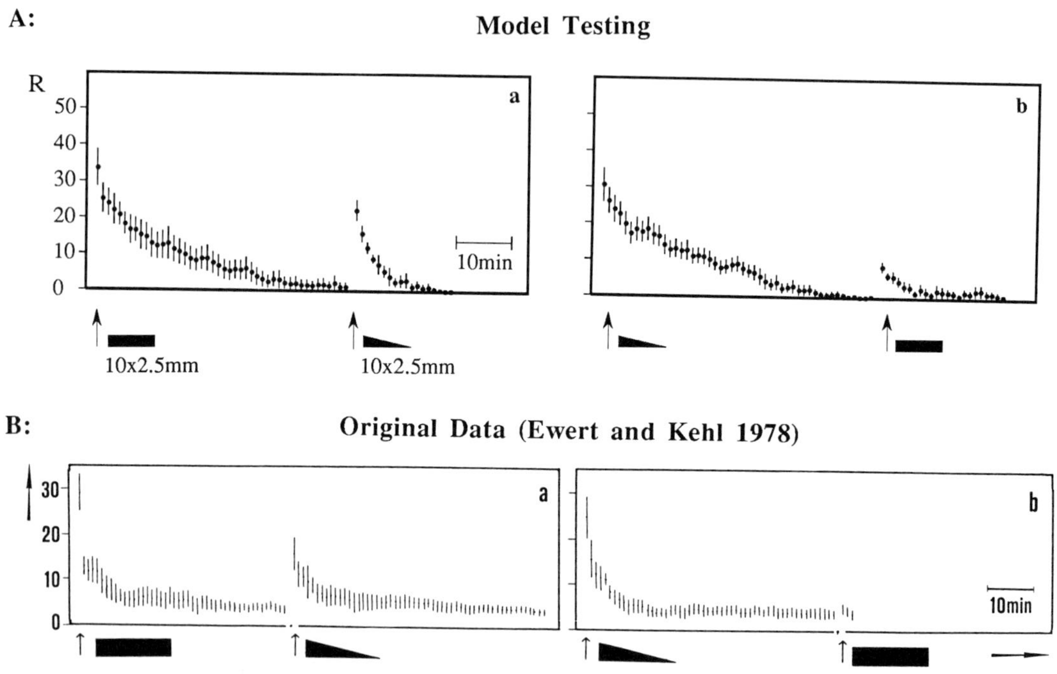

Figure 8 A. Dishabituation test between a black right-pointing triangle of 10x2.5 mm^2 area and a black rectangle of 10x2.5 mm^2 area (illustrated in the figure). **Aa**. Habituation of the toad prey-catching orienting response first to the triangle, and immediately afterwards a test of the response to the rectangle. **Ab**. Reverse order of presentation: Habituation first to the rectangle, and then a test of the response to the triangle. For other explanations see the legend of Fig.7. **B**. Test of the corresponding configurations with the size of 20x5 mm^2 (from Ewert and Kehl 1978).

The above experiments demonstrate that for toads the area of 20x5 mm^2 is preferable to that of 10x2.5 mm^2 in dishabituation (Note that no obvious difference in strength was found between big **b** and small **b** in releasing prey-catching behavior). The following experiments were to test dishabituation between small **b** (10x2.5 mm^2) and big **f** (20x5 mm^2). Two effects take place in this particular pair. Based on the size effect shown in Fig.9, one would expect that big **f** has a preference to small **b**, after a straightforward reasoning from Fig.9. From the perspective of configurational cues, however, shape **b** has preference to shape **f** as demonstrated in Fig.1, so one might expect to see the preference of small **b**. What actually happened is shown in Fig.10. In Fig.10a, small **b** was habituated first, and immediately afterwards big **f** was tested. No significant increase occurred in response to the later presentation of big **f**. However, as shown in Fig.10b, after habituation to big **f**, small **b** elicited strong prey-catching behavior.

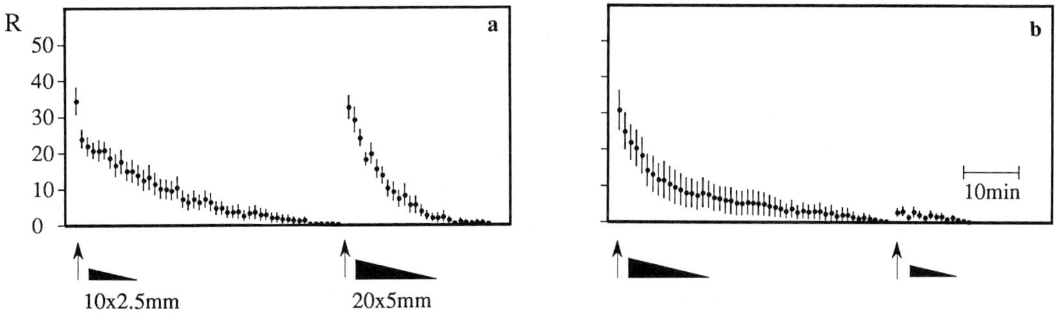

Figure 9. Test of size effect in dishabituation. Two different size stimuli of the same configuration were used. **a**. Habituation of the toad prey-catching orienting response first to the right-pointing triangle of 10x2.5 mm^2, and immediately afterwards a test of the response to the same shaped triangle of 20x5 mm^2. **b**. Reverse order of presentation: Habituation first to the big triangle, and then a test of the response to the small triangle.

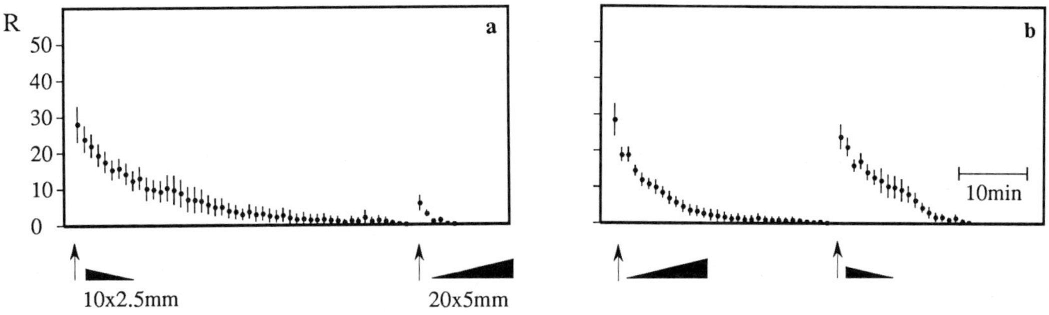

Figure 10. Dishabituation test between two triangles of different size and different configuration. **a**. Habituation of the toad prey-catching orienting response first to the right-pointing triangle of 10x2.5 mm^2, and immediately afterwards a test of the response to the left-point triangle of 20x5 mm^2. **b**. Reverse order of presentation: Habituation first to the big left-pointing triangle, and then a test of the response to the small right-pointing triangle.

The results in Fig.10 clearly demonstrate that toads prefer small **b** to big **f**, as exhibited by the ordering of dishabituation, and configuration plays the predominant role in this situation. The only explanation we can offer based on the results presented from

Fig.8 to Fig.10 is that visual object discrimination in toads is to a certain extent not affected by object size.

4. Discussion

In the modeling part we have suggested a pattern recognition paradigm for toads and frogs which uses intensity coding for representing different visual objects. The hierarchical stimulus specificity manifested in this paradigm represents an intermediate step between stimulus non-specificity found in several invertebrates (Kandel 1976) and full stimulus specificity demonstrated in mammals. Our analysis of the present model suggests that pattern recognition in anurans is based on visual cues like leading edge, trailing edge, dots, or striped patterns. This may be related to the fact that, in anurans, retinal ganglion cells respond to quite complex features of the stimulus.

It has been found behaviorally and physiologically that the edge preference (the leading edge or the trailing edge) of toads in prey catching behavior switches with reversal of contrast direction (Burghagen and Ewert 1982; Tsai and Ewert 1987; Ewert, this volume). The experimental results concerning contrast reversal (see Fig.7) demonstrate that the dishabituation hierarchy also changes with respect to stimulus-background contrast. In terms of sensitivity to contrast direction, our result is consistent with theirs. Furthermore, our hierarchy with contrast reversal predicts the precise preference between the eight stimuli. Our theory of toad pattern discrimination mechanisms is validated with respect to the particular pair of objects, **b** and **f** (see Fig.1) – two mirror images of a right triangle. The full test of the entire predicted hierarchy needs more extensive behavioral investigations.

The size variance prediction of our model (see Fig. 6) was not validated. Toads are able to recognize stimulus shapes by the dishabituation method, and their recognition does not seem to be affected by stimulus size, at least to some extent. Of course, toads will not elicit prey-catching behavior if a stimulus is too big, and toads will ignore a stimulus if it is too small. The new results from our experiments challenge further modeling, and our model of toad visual pattern discrimination will have to be adjusted to explain size invariance. Do toads form concepts in recognizing visual objects? Are toads the phylogenetically lowest animals that have developed size invariance? Why do toads form this specific shape preference of Fig. 1 in recognizing visual objects by dishabituation? These interesting questions need to be further studied.

The reason for the failure to exhibit size invariance in our model might be because the model relies on the specific properties of retinal cells, particularly the specific sizes of the receptive fields of R2 and R3 cells. Subworm pattern discrimination must come from a certain level of integration of retinal inputs. Perhaps direct reliance on R2 and R3 cells is too crude to exhibit size invariance in visual information processing. There might be intermediate networks between retina and anterior thalamus which generalize over certain retinal properties before retinal inputs arrive at AT. In order to solve this problem, we need to record how size change affects responses of visual neurons at various levels experimentally, and to figure out how to integrate new data and some size invariance into a future model. Many intriguing theoretical and experimental questions are triggered by this dialog, and answers to them are being pursued.

Our unpublished data show that toads are not only able to discriminate different visual objects, but also to store them in different neuronal substrates to avoid confusion. Toads can recognize different visual objects by dishabituation or conditioning (Cott 1936; Eibel-Eibesfeldt 1952; Brower and Brower 1962), and our experiments concerning size effects further suggest that toads recognize stimulus configurations regardless of stimulus size. At the same time, toads can identify stimulus size (see Fig.9). It can therefore be safely concluded that amphibians in evolution have developed fairly advanced learning capabilities. Due to their relatively simple visual system compared to mammals and fairly large amount of data available for various visual structures (Ewert 1984, 1987b), toads form an ideal model animal for investigating visual perception and pattern recognition.

Whether one is interested in understanding how the brain works or in developing more intelligent systems, understanding brain functions becomes very critical for neural network research (Arbib 1989). Since we know so little about the brain, it is not surprising to see so many neural network models which are claimed to be brain models. A brain model must be able to explain experimental data. But this is not sufficient. A good model, whether it be physical, chemical, or biological, has to be predictive. Predictions are crucial for setting up dialog between theoreticians and experimentalists. We hope this paper can serve as an example of establishing such a dialog. For the dialog to be possible, theoreticians have to constrain their neural models by experimental data and, more importantly, to make their models predictive, while biologists, on the other hand, should relate their data to modeling and face challenges from modelers by testing their predictions. We believe that this kind of dialog is both crucial and fruitful for understanding brain functions.

References

Arbib, M.A., 1989, *The metaphorical brain 2: Neural networks and beyond.* Wiley Interscience, New York.

Brower, J.V.Z., and Brower, L.P., 1962, Experimental studies of mimicry. 6. The reaction of toads (*Bufo terrestris*) to honeybees (*Apis mellifera*) and their dronefly mimics (*Eristalis vinetorum*). *Amer Natural* 96: 297-307.

Burghagen, H., and Ewert, J.-P., 1982, Question of 'head preference' in response to worm-like dummies during prey-capture of toads *Bufo bufo*. *Behav Proc* 7: 295-306.

Cervantes-Perez, F., Lara, R., and Arbib, M.A., 1985, A neural model of interactions subserving prey-predator discrimination and size preference in anuran amphibia, *J. Theor. Biol.*, 113:117-152.

Cott, H.B., 1936, The effectiveness of protective adaptations in the hive-bee illustrated by experiments on the feeding reactions, habit formation and memory of the common toad (*Bufo bufo bufo*). *Proc Zool Soc*, pp. 111-133.

Eibel-Eibesfeldt, I., 1952, Nahrungserwerb und Beuteschema der Erdkröte (*Bufo bufo L.*). *Behav* 4: 1-35.

Eikmanns, K.H., 1955, Verhaltensphysiologische Untersuchungen über den Beutefang und das Bewegungssehen der Erdkröte (*Bufo bufo* L.). *Z Tierpsychol* 12: 229-253.

Ewert, J.-P., 1976, The visual system of the toad: behavioral and physiological studies on a pattern recognition system. In: Fite, K. (ed), *The amphibian visual system: a multidisciplinary approach*. Academic Press, New York, pp. 141-202.

Ewert, J.-P., 1984, Tectal mechanisms that underlie prey-catching and avoidance behaviors in toads. In: Vanegas, H. (ed), *Comparative neurology of the optic tectum*. Plenum, New York London, pp. 246-416.

Ewert, J.-P., 1987a, Neuroethology: toward a functional analysis of stimulus-response mediating and modulating neural circuitries. In: Ellen, P., Thinus-Blanc, C. (eds.), *Cognitive processes and spatial orientation in animal and man*, part 1. Dordrecht, Martinus Nijhoff, pp. 177-200.

Ewert, J.-P., 1987b, Neuroethology of releasing mechanisms: prey-catching in toads. *Behav Brain Sci* 10: 337-405.

Ewert, J.-P., and Ingle, D., 1971, Excitatory effects following habituation of prey-catching activity in frogs and toads. *J Comp Physiol Psychol* 77: 369-374.

Ewert, J.-P., and Hock, F.J., 1972, Movement sensitive neurons in the toad's retina. *Exp Brain Res* 16: 41-59.

Ewert, J.-P., and Kehl, W., 1978, Configurational prey-selection by individual experience in the toad *Bufo bufo*. *J Comp Physiol* 126: 105-114.

Finkenstädt, T., 1989, Stimulus-specific habituation in toads: 2DG studies and lesion experiments." In: Ewert, J.-P., and Arbib, M.A. (eds), *Visuomotor coordination: amphibians, comparisons, models, and robots*. Plenum, New York, pp 769-797.

Finkenstädt, T., and Ewert, J.-P., 1988, Stimulus-specific long-term habituation of visually guided orienting behavior toward prey in toads: a ^{14}C-2DG study. *J Comp Physiol* A 163: 1-11.

Grüsser, O.J., and Grüsser-Cornehls, U., 1976, Neurophysiology of the anuran visual system. In: Llinás, R., and Precht, W. (eds.), *Frog neurobiology*. Springer, Berlin Heidelberg New York, pp. 297-385.

Kandel, E., 1976, *Cellular basis of behavior: an introduction to behavioral neurobiology*. Freeman, New York.

Lara, R., and Arbib, M.A., 1985, A Model of the Neural Mechanisms Responsible for Pattern Recognition and Stimulus Specific Habituation in Toads. *Biol. Cybern.*, 51:223-237.

Lara, R., Arbib, M.A., and Cromarty, A.S., 1982, The role of the tectal column in facilitation of amphibian prey-catching behavior: a neural model. *J Neurosci* 2: 521-530.

Lázár, Gy., Toth, P., Csink, Gy., and Kicliter, E., 1983, Morphology and location of tectal projection neuron in frogs: a study with HRP and Cobalt-filling. *J Comp Neurol* 215: 108-120.

Neary, T.J., and Northcutt, R., 1983, Nuclear organization of the bullfrog diencephalon. *J Comp Neurol* 213: 262-278.

Scalia, F., Colman, D.R., 1975, Identification of telencephalic efferent thalamic nuclei associated with the visual system of the frog. *Neurosci. Abstr* 1: 65.

Scalia, F., and Gregory, K., 1970, Retinofugal projections in the frog: location of the postsynaptic neurons. *Brain Behav Evol* 3: 16-29.

Schleidt, W., 1962, Die historische Entwichlung der Begriffe "angeborenes auslösendes Schema" und "angeborener Auslösemechanismus" in der Ethologie. *Z Tierpsychol* 19: 697-722.

Sokolov, E., 1975, Neuronal mechanisms of the orienting reflex. In: Sokolov, E., and Vinogradova, O. (eds), *Neuronal mechanisms of the orienting reflex*. Lawrence Erlbaum, Hillsdale, New Jersey, pp. 217-235.

Teeters, J.L., 1989, A simulation system for neural networks and a model for the anuran retina. Technical Report 89-01, Center for Neural Engineering, University of Southern California, Los Angeles.

Tsai, H.J., and Ewert, J.-P., 1987, Edge preference of retinal and tectal neurons in common toads (*Bufo bufo*) in response to worm-like moving stripes: the question of behaviorally relevant "position indicators". *J Comp Physiol* A 161: 295-304.

Thompson, R.F., and Spencer, W.A., 1966, Habituation: A model phenomenon for the study of neuronal substrates of behavior. *Psychol Rev* 73: 16-43.

Wang, D.L., and Arbib, M.A., 1990, How does the toad's visual system discriminate different worm-like stimuli? *Biol Cybern* (in press).